EXPERIMENTAL TOXICOLOGY
The Basic Issues

Second Edition

EXPERIMENTAL TOXICOLOGY
The Basic Issues
Second Edition

Edited by

Diana Anderson
The British Industrial Biological Research Association, Carshalton, Surrey

D. M. Conning
The British Nutrition Foundation, London

ROYAL
SOCIETY OF
CHEMISTRY

A catalogue record for this book is available from the British Library.

ISBN 0-85186-451-1 (Hardback)
0-85186-461-9 (Softcover)

First published in hardback 1988
First published in softcover 1990
Reprinted 1992
Second edition published 1993

Published by The Royal Society of Chemistry,
Thomas Graham House, The Science Park, Cambridge CB4 4WF

Printed and bound in the UK by Hartnolls Ltd., Bodmin

Contributors

W. N. Aldridge   *formerly The Robens Institute, University of Surrey*

H. E. Amos   *Pharmaco UK, Chelmsford*

Diana Anderson   *BIBRA Toxicology International, Carshalton*

W. H. Butler   *BIBRA Toxicology International, Carshalton*

D. M. Conning   *British Nutrition Foundation, London*

B. Copeland   *Barry Copeland Architects, London*

J. G. Evans   *Fisons plc, Loughborough*

S. D. Gangolli   *MRC Laboratories, Carshalton*

P. Grasso   *formerly The Robens Institute, University of Surrey*

R. I. Hawkins   *Coca Cola Northwest Europe, London*

A. B. G. Lansdown   *The Charing Cross and Westminster Medical School, London*

P. N. Lee   *P. N. Lee Statistics and Computing Ltd., Sutton*

D. P. Lovell   *BIBRA Toxicology International, Carshalton*

D. B. McGregor   *IARC, Lyon, France*

G. M. Paddle   *Department of Occupational Health, University of Birmingham*

J. C. Phillips   *BIBRA Toxicology International, Carshalton*

P. Rumsby   *BIBRA Toxicology International, Carshalton*

R. A. Riley   *formerly BIBRA Toxicology International, Carshalton*

F. J. Roe   *Consultant, 19 Marryat Road, Wimbledon Common, London*

A. B. Wilson   *Inveresk Research International, Midlothian*

Contents

Chapter 18   Statistics
By P. N. Lee   405

Chapter 19   Risk Assessment of Chemicals
By D. P. Lovell   442

CHAPTER 1

Introduction to Experimental Toxicology

D. M. CONNING

Toxicology is defined, classically, as the study of the adverse effects of chemicals on living systems. Originally this meant the study of poisons and that meaning remains the most satisfactory of definitions, essentially because it embodies the concept of the effect being proportional to the administered dose. Although the popular concept of poison concerns poisoning to death, the true study of poisons embraces the induction of morbid changes that are not necessarily fatal.

In this sense, the study of toxicology came to be regarded as an extension of the study of pharmacology and it is still so regarded by those who take a pedantic view of the topic. Of necessity, a pharmacological assay must explore the optimum dose in relation to the maximum therapeutic effect, and thus the dosage at which the effect is counter-productive of the desired result. In many ways the association of pharmacology and toxicology has been beneficial to the latter as a burgeoning science because it instilled two basic constraints which reinforced the scientific nature of the study. First was the recognition that pharmacology involved the perturbation of physiological function. That is, it was realised that a variety of detectable changes were compatible with normal living, and pharmacology sought to enhance those aspects which would be beneficial in the presence of disease or inhibit others for the same purpose. Second, the pharmacological study was a study of a defined function such as heart function, nerve transmission, or renal reabsorption, and never involved a less specific or more abstruse objective such as that embodied in the question 'Do any effects occur?' In other words, the pharmacological approach demanded an investigation of the way a particular physiological function could be chemically modified.

In recent decades, the definition of toxicology has been expanded in a way which has taken it firmly and, it seems, irrevocably out of the field of pharmacology. The first and most fundamental change was to add another purpose to the study. Thus toxicology came to be defined as the

1

study of the adverse effects of chemicals on living systems in order to predict chemical hazard to man. This had the effect of classifying toxicology as an ancillary to public or community health, and by extension to preventive medicine, always the poor brother of therapeutic medicine; and at the same time imposed impossible conditions on its practice as a science. Not only did the study of toxicology become a study of the effects of a chemical on any conceivable physiological function, defined or not, but under any conceivable circumstances because of the almost infinite variety of human activity and behaviour. Toxicology was expected thereby to predict the effect of a chemical in systems which themselves were not capable of being defined.

The problem was compounded by a further expansion of the definition to include 'chemicals or other agents' and the inclusion of 'man or his environment.' Thus toxicology is the study of the adverse effects of chemicals or other agents on living systems in order to predict hazard to man and his environment.

A number of very unsatisfactory consequences have resulted from these developments. The first was the birth of the concept of the 'no-effect level' and its embodiment in safety regulations. It is simply not possible at the present stage of development of toxicological knowledge to define with any precision the normal values for many biological activities and thus to define when abnormal values are detected. The best we can do is to define where the values in treated systems (*e.g.* the experimental rat) differ from those in untreated systems maintained in similar circumstances. We know only rarely whether any observed differences represent a toxic effect or an adaptive response. Sometimes we have great difficulty in determining if there are any differences at all, a problem which has given rise to a massive development in biometrics.

Our adherence to the 'no-effect level' has undermined our faith in epidemiology. Although the lack of epidemiological data and the relative insensitivity of epidemiological methods have themselves contributed to this outcome, it has seemed easier to put our faith in animal results which can be determined with some precision and therefore appear to be more easily judged. The result is that attempts to extrapolate the findings in animals, for example, to predictable effects in man have no basis in human experience and tend to assume that man's response will be the same as that of the animal.

Another consequence has been distortion of the economics of toxicological practice in that those who are involved with toxicological experiment spend so much of their time and resources generating data, there is very little available for scientific interpretation and further experiment. The construction and testing of hypotheses do not have a prominent role in toxicology.

All of this has come about for the best possible motives. The perception of the possible dangers consequent upon our chemical inventiveness has resulted in the appearance of potent forces to protect human communities against such consequences. The diligent pursuit of detail has extended very considerably the requisite observations before a 'no effect' dosage can be determined. It is a matter of profound regret that much of this invaluable data is not used to further our knowledge of biological function.

In this book we hope to lay the foundations on which future toxicologists can build the scientific practice of toxicology. Although we have acknowledged the demands of modern society and provided the basis for the career development of the toxicologist charged with the provision of data on which the acceptability of new chemical or physical agent can be defined, we hope this has been done in a way which does not stifle the needs of the enquiring mind. Despite the problems, toxicology still remains a science which promises real opportunities to unravel some of the fascinating problems of biology by identifying chemical and physical tools with which to probe living processes. In the end, it is this aspect of toxicology that will contribute most to our understanding of those features which determine the likelihood of human disease.

CHAPTER 2

Effects of Physical Form, Route, and Species

A. B. WILSON

1 Introduction

When investigating the literature in reference texts such as Sax (1979), which summarise toxicological information on a wide range of compounds, there is the impression of simplicity and unequivocality. More detailed texts (Clayton and Clayton, 1981) review the toxicology of classes of compound, indicate gaps in knowledge, and variable, even contradictory results. Hence there is well founded dispute about the effects of low doses of heavy metals, whether or not certain chemicals are carcinogenic, and the mechanisms of action of many compounds. This is further explored in more general toxicology textbooks (Klaassen, Amdur, and Doull, 1986; Hayes, 1983; Lu, 1985) and in toxicology journals such as: *Toxicology and Applied Pharmacology* (Academic Press, Duluth), *Fundamental and Applied Toxicology* (Academic Press, Duluth), *Food and Chemical Toxicology* (Pergamon Journals Ltd., Exeter), *Comments on Toxicology* (Gordon and Breach Science Publishers, London), *Archives of Toxicology* (Springer-Verlag, Germany), *Toxicology* (Elsevier, Ireland), *Regulatory Toxicology and Pharmacology* (Academic Press, Duluth), and *Human and Experimental Toxicology* (Macmillan Press, Basingstoke).

This chapter discusses the major factors that are causes of variability in results—physical form, vehicles, routes of exposure, and animal species as encountered in studies designed to screen compounds for general toxic potential. Few papers are available on the variations that actually occur in laboratory animal toxicology but information is published by Gaines and Linder (1986) with regard to acute toxicology of pesticides, Haseman (1983) with regard to carcinogenicity investigations, Lu (1985) with regard to a selection of factors, and by Rao (1986), Gartner (1990), Wollink (1989), and Vogel (1993) with regard to factors affecting laboratory animals.

4

2 Nature of Toxicant

A Physical Form

It can be expected that the physical nature of a material and the route of exposure will alter toxicity. Experiments in animals have to take account of this and batches of compound representative of normal production are usually employed so that purity, particle size, and other factors will match the real situation.

In general, a reduction in particle size improves solubility, increases absorption, and is likely, therefore, to increase toxicity. The effects of particle size are probably greatest on inhalation toxicity and that is covered separately. Solids will, of course, encounter body fluids in the gastro-intestinal tract, lungs, eyes, and even the skin and the therefore the opportunity exists for the material to be dissolved and for absorption to be improved. Insoluble materials are often regarded as inert and non-toxic. There are exceptions: asbestos and coal dust induce toxic effects by virtue of their insolubility and non-absorption which lead to residence in tissues such as the lung.

The pharmacist, pharmacologist, and experimental toxicologist are more likely to be concerned with capsules (perhaps enteric coated to carry them through the stomach intact), modification or buffering of pH to alter dissociation, and selection of soluble salts (*e.g.* soluble aspirin) to reduce local effects. All these measures and many more can have potential profound effects upon the toxicity of a compound.

Liquids do not have variations in terms of particle size in quite the same manner as do solids. However, liquids can become droplets (aerosols) with varying dimensions, which may be of some relevance if these are immiscible and of substantial significance if exposure is by inhalation. Liquids may have significant vapour pressure so they can sometimes be considered in much the same way as gases.

Gases and vapours present their hazards by being readily inhaled or by penetrating the skin. They too can be dissolved in liquids or adsorbed onto solids; indeed some show a remarkable affinity for solid surfaces, a property exploited in the manufacture of plastic strips containing pesticides such as dichlorvos.

The physical form of a compound will alter the potential for the material to enter the body, the extent of likely absorption, and therefore the eventual toxicity and hazard.

B Vehicle and Concentration

It has already been mentioned that gases, liquids, or solids may be presented in a vehicle as mixtures, solutions, or suspensions. If the vehicle is well absorbed, there is a tendency for the solute to be more rapidly and completely absorbed and *vice versa*. Many complex factors are involved, including the permeability of the surface to vehicle and compound, the partition coefficients, the extent of ionisation, and the effect of the material on local blood flow.

The concentration of material in the vehicle can alter toxicity. Often the more dilute the solution, the more complete the absorption of the solute (presuming the vehicle itself is absorbed) though the rate of absorption may be slower if dilution is substantial, as happens with alcohol. A rapidly detoxified or excreted compound will show less toxicity if absorption is slow. If large volumes are administered, it is worth bearing in mind that the vehicle itself might contribute to the toxicity or variations between results, for example, the interaction of corn oil in gavage studies (Nutrition Foundation, 1983). More than one investigator has thought he was looking at a concentration effect when it was probably vehicle toxicity.

High concentrations are more likely to cause local reactions such as irritation, ulceration, altered blood flow, or necrosis, which will themselves affect absorption and systemic toxicity, most obviously in dermal studies but frequently by other routes. The local effects are, of course, expressions of toxicity in their own right.

The interactions between vehicle and compound are exploited in the art and science of formulating drugs. By skilful means the therapeutic action, pharmacokinetics, and toxicity can be balanced to favour the desirable over the adverse (Prescott and Nimmo, 1979; Prescott and Nimmo, 1985).

3   Routes of Exposure

A   Frequently Encountered Routes

The most frequently encountered exposures to industrial chemicals are by the dermal and inhalation routes but these predominate in the place of work where much can be done to limit and control the actual risks. In fact, most of the experimental toxicology on laboratory animals is carried out *via* the oral route because:

(a) The oral route is the normal model of exposure (*e.g.* for food additives and for many pharmaceuticals).

(b) The widest exposure to herbicides and pesticides is likely to be as residues in crops or meat, which will be ingested.

(c) The oral route is simple to use and expedient. If reasonable evidence exists to show that differences in toxicity related to route are quantitative rather than qualitative then the bulk of the investigations will employ oral administration.

The route by which a chemical encounters the body can alter very substantially its effects and, in particular, the quantity required to cause that effect.

The physical presentation of a chemical and the route of exposure interact to influence greatly whether a chemical will be absorbed, its rate of absorption, tissue distribution, removal, and excretion. The integrity of the exposed surface, the residence time of the potential toxicant, and the metabolic activity of the surface are amongst the relevant factors. It may be reasonable to state that the route is likely to alter primarily the rate of absorption and the quantity absorbed (hence the dose required to cause toxicity) rather than the nature of the toxic effect. The exceptions that occur can generally be attributed to effects at the point of application or to rapid metabolism in the organs first encountered after absorption, *e.g.* by the liver after absorption through the gut.

With regard to route of exposure, coal dust on the skin does not present much of a hazard, but inhaled (as mentioned above) it can result in a severe and debilitating disease. However, the skin is not a universal barrier and many compounds, particularly liquids and gases, will penetrate quite readily. It must also be noted that the surface presented to a compound is particularly vulnerable to local effects as a result of high concentrations over a potentially small area, such as are likely to occur in mouse skin carcinogenicity studies (Wilson and Holland, 1982). One must therefore consider toxicity to the surface as well as through the surface. The local effects of an irritant or caustic material on the skin or gut might be quite dramatic but the systemic effects might only be marginal. Alternatively, compounds such as DDT are likely to have relatively insignificant local effects upon the skin but can result in very severe systemic toxicity and death if administered topically.

For completeness, it should also be remembered that toxicity to the skin or lungs does not necessarily depend upon the chemical being administered by that route—thallium does not have to be given topically to cause baldness nor does paraquat have to be inhaled to result in lung lesions.

The route of administration governs which organ is first encountered by the compound. That organ will generally meet the highest concentrations of the material and might therefore be the most seriously affected. It will potentially metabolise and perhaps detoxify the material, hence reducing effects on other organs. If propranolol is given intraperitoneally, it is absorbed through the hepatic-portal circulation and rapidly metabolised by the liver so that its toxic (or pharmacological) effects are greatly diminished. A different route of administration for such compounds (*e.g.* intravenous injection) will allow distribution and hence toxicity to other organs and tissues before metabolism can take place.

An equivalent effect occurs with the insecticide, dichlorvos, which is very rapidly hydrolysed and detoxified in the stomach. Its oral toxicity is therefore low but when the vapour is inhaled, hydrolysis and metabolism are sufficiently delayed for toxicity to be quite significant (though not at the concentrations likely to be encountered by man).

The tissue or surface first encountered has its influence in other ways too. The high vasculature of some organs assists the degree and extent of absorption. Hence absorption *via* the lungs is extremely rapid and can be similar in rate and nature to that of intravenous injection. The poorer vascularity of the skin is a factor in reducing the extent of absorption and its speed.

The very rapid achievement of high blood concentrations has substantial effects upon the pharmacokinetics, with the distinct possibility that peak concentrations for a short time will enable a material to overload normal metabolic pathways and produce different metabolites or reach different organs. In recent years there has been a sensible and welcome increase in the numbers of animal toxicity studies that include measurement of blood (and tissue) levels of compound during their course.

The effects of route of exposure are sources of opportunity for those concerned with targeting potent drugs to specific organs while minimising less desirable effects (Prescott and Nimmo, 1979). To the toxicologist they tend to cause confusion (or at best a challenge and a lesson) when a material that is toxic by one route is safe by another.

B Oral

Toxicological investigations make extensive use of oral administration. The material is often administered as a liquid (solution, suspension, or undiluted) by gavage, *i.e.* a tube is passed down the oesophagus and the material is injected into the stomach *via* the tube. Gavage is a particularly simple procedure in rats and mice and the method used almost exclusively in the conduct of oral LD_{50} studies and other short term oral studies in rodents. Vehicles used frequently for gavage experiments include: water, vegetable oil (*e.g.* corn oil), or water and suspending agent (*e.g.* carboxymethyl cellulose).

Gavage dosing of rodents over long periods (*e.g.* daily over two years) is quite practicable (Nutrition Foundation, 1983). However, incorporation into the diet is less labour intensive when administration to rodents is intended to be over weeks or months. The compound is carefully mixed in the diet to a predetermined concentration and the animal is allowed to eat in the normal way. For non-rodents, capsules and tablets are obvious alternatives and are frequently used in studies with dogs and primates, whether acute or longer term.

The rodent is normally fed *ad libitum* and eats over a period of several hours, mainly at night. Therefore intake of a toxicant given in the diet will be more gradual than by gavage. A rapidly metabolised or short acting compound might thus have to be given in considerably higher doses than by gavage to achieve the same effect. For a longer acting, less quickly metabolised or excreted compound, there might be little difference in dose required (assuming the compound is equally available to the animal by both methods). Note that dogs are usually given a fixed amount of food per day, all of which is likely to be eaten in a short period of time.

The units of dosage by the oral route are normally expressed as: mg (or g) of compound per kg body weight of animal per day, *e.g.* $mg\,kg^{-1}\,day^{-1}$, or parts by weight of compound per million parts by weight of diet, *i.e.* parts per million (p.p.m.).

When dosing by the diet, the same dietary concentration (p.p.m.) is sometimes given over a period of months or years but rodents will eat substantially less per unit body weight as they grow older, effectively reducing the dose in terms of $mg\,kg^{-1}\,day^{-1}$. Feeding studies with herbicides or pesticides are frequently conducted using a constant concentration in the diet throughout the experiment. Alternatively, the concentration of the diet can be adjusted regularly according to food intake and body weight so that a reasonably constant dose (in terms of $mg\,kg^{-1}\,day^{-1}$) is achieved. Convention has it that doses of pharmaceutical products are adjusted in this way when given in the diet.

A formula for conversion between the two systems is:

$$\text{Dose Administered (mg cpd kg body weight}^{-1}\,day^{-1}) = \frac{(\text{kg diet eaten day}^{-1}) \times (\text{mg cpd kg diet}^{-1})}{\text{Animal's Body Weight (kg)}}$$

C Dermal

The skin can act very successfully as a barrier but its status in that regard varies according to its thickness, its vascularity, its degree of hydration (hydration tends to increase its permeability), whether or not it is intact (*i.e.* abraded, ulcerated or otherwise damaged), and, of course, the compound in question. A fuller account of such factors can be read in Klaassen *et al.* (1986).

The skin has a wide selection of glands and other components so compounds can enter through sebaceous glands, sweat glands, or hair follicles, as well as *via* the epidermis itself. The variability of these components in different parts of the body and from species to species presents difficulties in predicting accurately the rate of absorption and quantity of material likely to be absorbed.

Irritation, inflammation, and ulceration are frequent and important reactions of the skin to insult by foreign chemicals and these can be investigated using laboratory animals. The rat and the rabbit are the species most frequently used. An area of the back is shaved of its hair (generally with clippers) and the compound is applied. The animal has to be prevented from ingesting the material so a collar may be put on. Alternatively, the site can be wrapped in an occlusive (impermeable) dressing. The latter technique normally serves to increase the sensitivity of the test. In a skin irritation test, the compound is usually applied only once for a few hours before it is wiped off and the skin is examined. Any reddening, oedema, ulceration, or other reactions of the skin are carefully recorded over the next hours and days, to assess both delayed reactions and the ability of the skin to recover. A typical scoring system is described by Kay and Calandra (1962).

The most significant variables in this type of test are the concentration of the compound, whether it is applied as a solid or a liquid, the vehicle used, the duration of application, and whether or not the skin is occluded. Within sensible limitations, the dose of compound is less relevant to irritation than is its concentration.

Unfortunately *in vitro* studies are not yet at the stage in which they can satisfactorily substitute for this type of work in animals.

The skin is sometimes used to investigate carcinogenicity—topical or systemic. Conventionally it is the mouse that is employed, the hair of the back is shaved off regularly and the material is painted on repeatedly (*e.g.* daily or weekly) over a period of months (Toben and Kornhauser, 1982; Slaga and Nesnow, 1985). For some tests a single dose of initiator (*e.g.* DMBA) is applied first. The skin and internal organs are examined for tumours. This type of protocol is most often employed for testing of compounds of similar chemical structure and making comparisons between them.

Broadly similar methods are used to investigate toxicity as a result of absorption through the skin. The rat is the species of choice. While very frequently it is the toxicity of a single dose that is investigated, sometimes repeated applications are made over days, weeks, or months. Longer term dermal experiments pose particular practical problems as they require the animal to be clipped or shaved regularly (*e.g.* twice per week) and occasional skin damage can result from this. If an occlusive dressing is applied, its frequent application and removal can cause adverse effects. Other methods of restraint, while appropriate and reasonable for a few applications, might be difficult to justify over long periods. If no occlusion or restraint is applied then ingestion through grooming is a potential complicating factor.

It is sometimes relevant to abrade the skin for toxicity experiments, to simulate the effect that cuts or grazes might have in man. Damage can greatly increase permeability and therefore toxicity (Klaassen *et al.*, 1986).

The variables to be considered in skin toxicity are dose, concentration, total area of application, vehicle, duration of exposure, number of exposures, occlusion, and abrasions. Note, however, that a high concentration of a compound might prove to be irritant and this would be likely to improve absorption and hence increase toxicity.

D   Inhalation

General Principles of Atmosphere Generation

The subject of inhalation toxicity is exceedingly complex and has resulted in a sizeable library of chapters and books, addressing themselves to the relevant theory, practice, and experimental techniques (Willeke, 1980; Phalen, 1984). This latter publication could be considered to be the standard reference on inhalation toxicology. While it is relatively simple to prepare a compound for dosing orally or topically and to hope to approximate the human situation, it is much more difficult by the inhalation route.

The compound needs to be mixed homogeneously with air so that each animal is exposed to the same concentration of material. Chamber designs, cage designs (normally mesh), air flow patterns, and atmosphere generation systems have all to be considered with this in mind. Even then it is common to move animals from one position in an inhalation chamber to another on successive exposures, in order to balance the effects of position and minimise variations attributable to location.

There are further complications in that inhalation exposure is particulary relevant to the work place and all too often the hazard there is in the form of a mixture of materials with different volatility or dispersion characteristics. Experimental design becomes difficult and atmosphere generation often requires unusual techniques, for instance, diesel engines might need to be run to provide exhaust gases.

Ultimately, the atmosphere needs to be measured. On first consideration, one might think that weighing the material used and measuring the air flow would give the concentration for exposure. However, material tends to adhere to pipework and the walls of inhalation chambers considerably reducing the atmospheric concentration during exposure and sometimes forming a source of toxicant after the exposure has supposedly finished. For these reasons, analysis of atmospheres is particularly important. This permits definition of three concentrations:

(a) nominal (*i.e.* what is desired or nominated),
(b) calculated (*i.e.* calculated from quantity of compound used, air flow, *etc.*), and
(c) actual (*i.e.* what is present as determined by analysis of the atmosphere).

Frequently what are described above as 'nominal' and 'calculated' concentrations are both termed 'nominal'.

Generation of Gases, Vapours, and Aerosols

Unless a gas has peculiar properties, generation of the desired atmosphere can be quite straightforwardly achieved by feeding a supply of the gas (from a gas cylinder) into the airstream and analysing the attained concentration.

There are several methods of creating the desired atmosphere from liquids. Most frequently the air is bubbled through the liquid, droplets are filtered out, and the resultant concentrated mixture of air and compound is diluted as required by mixing with more air. Although it is possible to increase concentration by warming the vessel containing the liquid, if the saturated vapour pressure is exceeded, the compound will form droplets and some will condense on the pipework. It will not have been the first inhalation experiment in the literature to report unintentionally on the toxicity of vapours above their saturated vapour pressure.

Other methods of generation involve blowing air over the liquid which is adsorbed onto charcoal or filter paper. This avoids droplet formation and, if the air flow is adjusted correctly, creates an almost saturated atmosphere of test compound, the concentration of which remains very stable. This method requires more initial setting up than the former.

The term 'aerosol' embraces both droplets of liquid and particles of solids. The generation of aerosol atmospheres needs to control both concentration and particle size. It abounds with hardware, such as atomisers, spinning discs, dust feed mechanisms, cyclones, and fluidised beds. All too often the hardware has to be adapted to the particular compound, making accurate comparison of work between laboratories and test materials even more troublesome than usual.

Within the inhalation chamber different particle sizes follow different air currents and so homogeneity is difficult to obtain. Deposition upon surfaces is inevitable, so direct measurement rather than calculation of concentration is essential. At its simplest, liquids are drawn into a solvent for analysis, while solids are caught on filters, but again the behaviour of particles in different air currents means that misleading and selective sampling of different-sized particles can occur. Other methods are numerous and include measuring optical density and exploitation of aerodynamic differences between particles.

Particle size, too, should be measured and care taken that influence of the sampling method on the particle size recorded is taken into account. Some techniques depend upon visual microscopic measurement of samples, others depend upon aerodynamic properties (*e.g.* cascade impactors), still others are selective filters (*e.g.* Andersen), and so on. Particle size is expressed as 'mass median diameter', *i.e.* the particle size at which there is the greatest mass (not necessarily number) of particles, so a few large particles count as much as many more small particles. It is better, and more normal, to indicate the range of particle sizes and the proportion by weight of particles in each size range.

Particle Size as a Variable

Particle size matters because different sizes of aerosol deposit in different parts of the respiratory tract and can, therefore, result in different toxic responses, sometimes through local effects and sometimes through differential absorption.

Generally in man, particles above about 20–30 μm in diameter are not inhaled and those between 7 μm and 20 μm deposit at various levels in the upper respiratory tract to be absorbed or swept along by muco-ciliary motion, coughed up, and swallowed.

Unless particles are below 5–7 μm diameter, few or none will penetrate to the alveoli. Below that diameter an increasing proportion reaches the alveoli but once the size is in the region of 0.1 μm and less, a very high proportion is exhaled. The dimensions given refer to spherical particles of unit density. Differently shaped particles will, broadly speaking, behave similarly to a spherical particle with equivalent aerodynamic characteristics. Fibres, such as asbestos, can have special characteristics. Whilst the above comments on deposition of particles are true for man, the respiratory tract and rates of flow of air in the tract are different in laboratory animals. Thus, absorption and deposition of gases and specified particles varies between species (Morris *et al.*, 1986).

Inert solids, depositing above the alveoli will be removed by muco-ciliary action, with little or no significant toxicity. If they reach the alveoli, muco-ciliary clearance does not occur and it is left to the cells of the reticulo-endothelial system to remove the particles. This is much less effective and brings the compound into intimate contact with cells resulting in many forms of toxicity, attributable to the physical rather than the chemical nature of the compound. A selection is:

Asbestos—Asbestosis, lung cancer, mesothelioma
Coal Dust—Pulmonary fibrosis
Kaolin—Fibrosis
Talc—Fibrosis, pleural sclerosis

Research on asbestos is apparently showing that fibre length and diameter have profound effects upon its toxicology (Bernstein, 1982).

Droplets of soluble or chemically active materials may be toxic by inhalation even if their size is greater than 5–7 μm. They can have local effects and penetrate the nasal, tracheal, or bronchial mucous lining to cause systemic toxicity. The term 'respirable' is commonly used to distinguish particles that will penetrate to the alveoli, *i.e.* particles <5–7 μm diameter are classed as respirable to man. This division is highly relevant in assessing toxic potential of inert solids, but is of less relevance for other materials, though curiously, some regulatory bodies have required no inhalation toxicology if aerosol diameter is >7 μm, whatever the nature of the compound.

Method of Exposure

There are two main methods of exposure, 'whole body' and 'nose only' (plus occasionally 'head only'). The terms are reasonably self-explanatory. For whole body exposure the animals are placed in mesh cages in an inhalation chamber and a mixture of air and the material to be tested is passed through the chamber, so not only do the animals breathe the toxicant but the skin is exposed so dermal absorption, and ingestion by grooming are also possible. If continuous 24 hours a day exposure is required, then food and water must be provided and contamination of these can occur.

'Nose only' exposure is based upon restraining the animal so that only the nose protrudes into the test atmosphere. This reduces contamination of the skin but is, of course, practicable for only a few hours during each day. Although rats soon become accustomed to the restraint, the stress of the first minutes of an exposure could distort results, *e.g.* the toxicity of cholinesterase inhibitors is likely to be increased by muscular activity. For larger animals (*e.g.* dogs) then either whole body exposure or specially designed masks are used.

Most inhalation experiments are conducted in a 'dynamic' atmosphere, *i.e.* one in which there is a continuous flow of atmosphere into (and out of) the chamber, with little or no recirculation. It has the disadvantage of requiring large quantities of material and, of course, this has to be disposed of. Very occasionally a static atmosphere is used, in which case there is no flow through the chamber. This conserves toxicant (*e.g.* radiolabelled compound) but is only practicable for short exposures. There can be difficulties in maintaining a homogeneous mixture in the inhalation chamber.

Design of Inhalation Experiments

The variable most peculiar to inhalation toxicology is, therefore, the duration of each exposure period. Convention is that single exposure of 4–6 hours (either 'whole body' or 'nose only') is the accepted duration for investigating the acute toxicity and deriving the LC_{50} (lethal concentration to 50% of the animals, the inhalation equivalent of the LD_{50}). The Society of Toxicology (1992) reviewed the design of acute inhalation studies. Regular exposures of 7 to 8 hours a day are taken as being equivalent to the working day for man and are often selected when there is a need to simulate the period of an individual's exposure in the work place. For the same reason (and also for expediency) these studies are often run for 5 days out of 7 each week. When the hazard is likely to involve continuous exposure, as for an environmental contaminant then exposure for close to 24 hours per day, 7 days each week is necessary.

The units used in inhalation are rather different from oral or dermal studies:

$$\text{parts per million (p.p.m.)} = \frac{\text{no. of parts by volume of the compound}}{10^6 \text{ parts by volume of air}}$$

i.e. 10 p.p.m. means 10 ml of the vapour in 10^6 ml of air. It is a volume: volume measurement appropriate to gases and vapours but not to droplets or particles.

$mg\,m^{-3}$ and $mg\,l^{-1}$ are weight:volume measurements suitable for any materials *i.e.* $10 \, \text{mg m}^{-3}$ means $10 \, \text{mg}$ of the compound (vapour, liquid, or solid) in $1 \, \text{m}^3$ of air.

The two sets of units can be converted using the following formula:

$$\text{mg m}^{-3} = \frac{\text{p.p.m.} \times \text{molecular weight}}{24.5}$$

E Intravenous and Other Routes

The potential for modifying the activity of pharmaceutical products by altering the route of administration means that many and special methods need to be developed, most frequently with the object of imitating the intended method of administration to man. Thus studies are conducted by intravenous, intraperitoneal, intramuscular, and subcutaneous injection. Also by the use of implants and suppositories.

Intravenous techniques (bolus injection or continuous infusion) have come increasingly to the fore with the development of biotechnology products (Wilson, 1987). Many of these are potent but labile peptides and proteins which will be rapidly metabolised, therefore requiring special techniques for administration.

F Dose

The calculation of dosage for oral or parenteral routes of administration is normally straightforward—merely a measurement of the weight of compound administered, usually expressed in terms of g of compound per kg of animal. The extent of absorption is not taken into consideration. However, dosage calculation presents a very different problem in inhalation studies. It should be remembered that a 100 g rat inhales about 0.071 per min which compares with 7.51 per min for a 70 kg man. This means that it will inhale roughly 6 times as much volume (and therefore toxicant) per unit body weight as a person. Dosage is difficult to estimate for inhalation studies (Dahl, Schlesinger, Heck, Medinsky, and Lucier, 1991), so the extent of exposure is stated instead by giving the atmospheric concentration and the duration of exposure. With some degree of caution, when these variations are not great, these can be combined in the equation: concentration × time. It becomes less valid when the time (or concentration) figures differ significantly. Atherley (1985) discusses this in the context of human industrial exposure situation.

The results of toxicology studies are frequently used in the calculation of safety factors. The dose administered (or level of exposure) becomes relevant but it is not necessarily true that the quantity administered per unit body weight is a valid basis for extrapolation across species and to man (Davidson, Parker, and Beliles, 1986), even if factors such as extent of absorption, route of metabolism, and susceptibility are comparable.

4 Choice of Species

A General Principles

It is obvious that the preferred experimental animal in terms of imitating man is man. However, even if that were practicable, man would have many undesirable features as a model: expensive to maintain, needing large premises, requiring large quantities of test compound, slow growing, long gestation period, and somewhat variable in nature. The last mentioned point can be illustrated by racial differences in metabolism (Bailey, 1983), *e.g.* lactose metabolism and rate of acetylation.

This indicates that no single strain or species is likely to be ideal and that an understanding is required for the mechanisms of toxicity so that a prediction can be made of how different populations will react to compounds. The routine procedure is for two species, one rodent and one non-rodent, to be used for most of the pivotal subacute and chronic studies.

It is assumed that if the two species react in the same way at approximately the same dose level (and allowing for some skilled judgement), then man is likely to respond in the same manner. A reasonable safety factor is usually allowed. When species behave differently then much more has to be known about metabolism, kinetics, mechanisms, *etc.*, before there is any confidence in extrapolation to man. Zbinden (1993) debates this point.

An excellent general reference for information on laboratory animals is Poole (1987). Relevant journals include *Laboratory Animal Science* (Laboratory Animals, UK Ltd.) and *Laboratory Animals* (American Association for Laboratory Animal Science, Cordona).

B Rat

The rat is probably the most commonly used species for regulatory toxicology studies. Its reactions to chemicals are therefore more likely to be sensibly interpreted than those of most other species. It is a uniform, easily bred animal, weighing only a few hundred grams, requiring little space and an undemanding diet. Its size limits the amount of blood that can be taken which is a distinct disadvantage and is reflected in the design of studies, even although modern analytical techniques can be very small quantities indeed. Toxicokinetic studies on rats generally require animals to be killed at each time point so that sufficient blood can be obtained for analysis.

The rat is used extensively in most routine studies—acute and chronic, oral and parenteral. Its short life span (24–30 months) means that it can be used for lifespan studies and for investigation of carcinogenicity. However, some strains have high incidences of specific lesions which can confound interpretation (*e.g.* mammary tumours in Sprague–Dawley rats and testicular tumours in the Fischer 344 strain). The nature of the diet substantially affects longevity and can alter tumour incidence. European laboratories carrying out lifetime studies (such as carcinogenicity studies) use specifically formulated maintenance diets with low levels of protein (*e.g.* 14%) and a low calorific value.

There is a wide variety of strains. Albino rats of the Sprague-Dawley (such as the Charles River CD), Fischer 344 (F344), or Wistar strains are those most frequently encountered but there are many others, sometimes with specialised applications.

C Mouse

The small size of the mouse (20–30 g) is a major disadvantage and more than outweighs the virtues of cheapness, ease of housing, and ease of handling. Although useful quantities of blood can be taken from mice without causing death, it is probable that the animal's physiology will be sufficiently disturbed to be uncertain of the reliability of subsequent observations. Therefore, the mouse is normally killed if blood samples are needed.

It has a slightly shorter lifespan than the rat, 21–27 months. The combination of strengths and weaknesses means that the mouse is used for acute, LD_{50} type studies and as a second species in carcinogenicity investigations. While intermediate length studies are conducted, in most cases their prime aim is for dose selection for the long term carcinogenicity work. Its value in carcinogenicity studies has been in doubt since the work of Thorpe and Walker (1973) indicated its propensity to develop an increased incidence of liver tumours when dosed with test materials for which there was little or no other evidence of carcinogenicity. Debate and discussions continue but a direction was perhaps established at the November 1991 International Conference of Harmonisation in Brussels during which presenters indicated that the mouse should not be used routinely for carcinogenicity studies and that a single species (the rat) would generally suffice.

The mouse is the species of choice (by custom) for skin carcinogenicity investigations. Its position may be owed to the use of the mouse in the 1960s as a screen to detect carcinogenic hydrocarbons associated with certain oils. The suspect material is painted onto the skin regularly for many months, sometimes even the life time of the mouse, and dermal tumours are counted. A variation is to administer an 'initiator' before giving the test material and to expect tumours to develop much sooner. This relates to the 'initiation' and 'promotion' phases of carcinogenicity—a concept which was useful in its time but is now probably of less value. As with rats, a variety of strains are used but the Charles River

CD-1 is probably the most widely used for toxicology. Inbred and crossbred strains are also employed, *e.g.* B6C3F1 for carcinogenicity studies in the USA National Toxicology Programme (Rao, Birnbaum, Collins, Tennant, and Skow, 1988). An introduction to the relative merits and variability of inbred and random bred strains is given by Rice and O'Brien (1980).

D   Guinea Pig

The most frequent use of this species in toxicology is for the prediction of Delayed Hypersensitivity reactions, Type IV allergic reactions, by such test methods as described by Magnusson and Kligman (1969) and Buechler (1965). The mouse is also beginning to be used for this work (Botham, Basketter, Maurer, Mueller, Potokar, and Bontinck, 1991) and is gaining acceptability with regulators. It remains to be seen whether it will replace the guinea pig entirely.

The guinea pig's lifespan is too long for it to be employed in carcinogenicity work and its gestation period also long (about 9 weeks) which means it is hardly ever used in reproduction investigations.

E   Rabbit

Albino rabbits (usually New Zealand White) are used to investigate the irritant potential of compounds to the skin and to the eyes. The large, exposed, unpigmented eyeball and low blink rate mean that changes can be readily detected and an adequate safety factor incorporated in the experimental study.

New Zealand White or other strains of rabbits are frequently used in teratogenicity investigations. Rabbits show clearly the teratogenic effect of thalidomide, modelling its effect in man. It is, however, very difficult to display thalidomide teratogenicity in the rat. It might be that the fewer layers in the rabbit placenta have led experimenters to believe it will allow toxicants to pass more readily to the foetus. One way or another, it is firmly established as the second, (non-rodent) species for teratogenicity work.

F   Dog

Virtually all dogs used in toxicology studies are purpose bred Beagles. It is alleged this breed was chosen because sizeable groups were available, they accepted confinement very readily, they rarely fought, were easy to handle, they were short haired, and about the right size. They can be readily trained to wear face masks for inhalation studies or to stand in slings without apparent distress. Even if starting afresh, it would be difficult to think of a more appropriate breed and certainly the extensive background knowledge now available means it is unlikely to be replaced.

The Beagle fills the role of a second species as a non-rodent in toxicity studies lasting from 1 month to 2 years. It is very rarely employed in acute work (except for selection of doses for longer term studies) and since its lifespan is 11–15 years it is impracticable to use it in crcinogenicity work. The latter has been done though more often than not the studies have turned out to be unsatisfactory.

Repeated blood samples, several ml in volume, can easily be taken while causing minimal physiological disturbance or stress. This very valuable asset means the dog is used frequently in metabolism and toxicokinetic work. The dog does vomit particularly readily and therefore presents practical difficulties of certain compounds particularly by the oral route.

G   Primates

Whilst virtually all other species used in toxicology are bred specially for the purpose, it is regrettable that the primate was so frequently caught in the wild and imported for the purpose. This situation is changing rapidly. Breeding colonies of Macaques have been established in China, the Philippines, Mauritius, and USA. Supplies of purpose bred animals are now becoming available, so that studies can be conducted using clean animals of known parentage and history.

Marmosets have been used but are rather smaller than is ideal, being only a few hundred grams in weight. This limits the volume of blood available and considerably restricts their use in kinetic investigations. Supplies available to toxicology were very restricted during the 1980s so only limited background data are available.

The Cynomolgus monkey is the species that has been used frequently. The role of the primate in toxicology is similar to that of the dog (*i.e.* as a second species for subacute and chronic work) though, in addition, they are occasionally used in teratology and some other reproductive studies. It is curious that primates are almost never chosen for toxicology studies on herbicides, pesticides, or industrial chemicals, but are chosen on a very substantial number of occasions as the non-rodent species in subacute or chronic studies on pharmaceutical products.

Primates can be difficult to handle, used to be of unknown parentages or history, and were often diseased, all of which tended to militate against their use. Furthermore, the diseases include very significant zoonoses, tuberculosis, salmonellosis, virus B, and rabies, which complicates matters even more. The primate is a higher species and there is concern that its use in toxicology is carefully justified on each occasion. It is assumed, and possibly correctly, that primates are more similar to man in their physiology and metabolism than are other species. However, if studies on the rat and dog display substantially different toxicology in each, it would be a brave man to assume that the results in primates would provide the definitive answer—at least not without knowing much more about the mechanisms of action of the compound.

H　Other Species

While species other than the above are occasionally used, it is usually for a specific reason related to the actions or metabolism of a particular compound and because some laboratories are a little more adventurous in their choice. One can find the ferret used frequently in the occasional laboratory as an alternative to dogs and primates (it is a non-rodent), Syrian hamsters used for carcinogenicity work instead of mice, Chinese hamsters employed in cytogenetic studies (they have only 11 pairs of chromosomes), and hens used to investigate delayed neurotoxicity associated with organophosphorus compounds. The mini-pig and micro-pig have been investigated over many years and are being used more frequently (but not commonly) as non-rodent species.

Background data are very valuable in aiding the interpretation of ambiguous and unusual results so there is a built-in inertia which discourages the use of more unusual species.

Last, but by no means least, it is vital to be aware of the significant advances being made in the use of bacteria, yeasts, cell cultures, and other methods by which the use of animals can be appreciably reduced. They are widely exploited in the prediction of genotoxic effects and much research is on-going to develop suitable techniques for skin and eye irritation and for teratology (Balls, Riddell, and Worden, 1983). Each test tends to work satisfactorily for a particular group of materials and may be sensitive to a specific mode of action (Green, Chambers, Gupta, Hill, Hurley, Lambert, Lee, Lee, Liu, Lawther, Roberts, Seabaugh, Springer, and Wilcox, 1993). They currently show most potential for classifying closely related chemicals and for indicating potential activities.

5　Conclusions

Thus, it can be appreciated that seemingly very different toxicological results can be obtained for what is apparently the same material, depending upon its nature when presented to the animal, the route by which it is presented, and the species used.

Haseman (1983) summarises a selection of the results from long term studies conducted for the USA National Cancer Institute and produces data on the variations between laboratories and between animal suppliers. This illustrates how much there is to be done to reduce variability in animal toxicology studies. Biologists tend to appreciate and accept this, perhaps too apathetically. The chemist and the statistician find it puzzling and frustrating, as illustrated by Horowitz (1984). Toxicological studies will be extrapolated to man in whom there will be yet another source of variability (Bailey, 1983).

References

Atherley, G. (1985). A critical review of time-weighted average as an index of exposure and dose, and of its key elements. *Am. Ind. Hyg. Assoc. J.*, **46**, 481–487.

Bailey, K. (1983). Physiological factors affecting drug toxicity. *Regulat. Toxicol. Pharmacol, 3,* 389–398.

Balls, M. J., Riddell, R. J., and Worden, A. N. (1983). (*Eds*). 'Animals and Alternatives in Toxicity Testing'. Academic Press, London.

Bernstein, D. M., Drew, R. T., and Kuschner, M. (1982). The pulmonary response to sized glass fibres in rats. *Toxicologist, 2,* 59.

Botham, P. A., Basketter, D. A., Maurer, T., Mueller, D., Potokar, M., and Bontinck, W. J. (1991). Skin sensitisation—a critical review of predictive test methods in animals and man. *Food Chem. Toxicol., 29,* 275–286.

Buehler, E. V. (1965). Delayed contact sensitivity in the guinea pig. *Arch. Dermatol, 91,* 171–175.

Clayton, G. D. and Clayton, F. E. (1990). 'Patty's Industrial Hygiene and Toxicology', 4th Edn., Wiley (Interscience), New York.

Dahl, A. R., Schlesinger, R. B., Heck, H. d'A., Medinsky, M. A., and Lucier, G. W. (1991). Comparative dosimetry of inhaled materials: differences among animal species and extrapolation to man. *Fundam. Appl. Toxicol., 16,* 1–13.

Davidson, I. W. F., Parker, J. C., and Beliles, R. P. (1986). Biological basis for extrapolation across mammalian species. *Regulat. Toxicol. Pharmacol., 6,* 211–237.

Gaines, T. B. and Linder, R. E. (1986). Acute toxicity of pesticides in adult and weanling rats. *Fundam. Appl. Toxicol., 7,* 299–308.

Gartner, K. (1990). A third component causing random variability beside environment and genotype. *Lab. Anim., 24,* 71–77.

Goodman, J. I., Ward, J. M., Popp, J. A., Klaunig, J. E., and Fox, T. R. (1991). Symposium overview. Mouse liver carcinogenesis: mechanisms and relevance. *Fundam. Appl. Toxicol., 17,* 651–665.

Green, S., Chambers, W. A., Gupta, K. C., Hill, R. N., Hurley, P. M., Lambert, L. A., Lee, C. C., Lee, J. K., Liu, P. T., Lowther, D. K., Roberts, C. D., Seabaugh, I. M., Springer, J. A., and Wilcox, N. L. (1993). Criteria for *in vitro* alternatives for the eye irritation test. *Food Chem. Toxicol., 31,* 81–85.

Haseman, J. K. (1983). Pattern of tumour incidence in two-year cancer bioassay feeding studies in Fischer 344 rats. *Fundam. Appl. Toxicol., 3,* 1–9.

Hayes, A. W. (1983). (*Ed.*). 'Principles and Methods of Toxicology.' Raven Press, New York.

Horowitz, W. (1984). Decision making in analytical chemistry. *Fundam. Appl. Toxicol., 4,* S309–S317.

Kay, J. H. and Calandra, J. C. (1962). Interpretation of eye irritation tests. *J. Soc. Cosmet. Chem., 13,* 281–284.

Klaassen, C. D., Amour, M. D., and Doull, J. (1986). 'Cassarett and Doull's Toxicology'. 3rd Edn. Macmillan, New York.

Kimber, L. and Botham, P. A. (1993). Results with OECD recommended positive control sensitizers in the maximisation, Buehler and local lymph node assays. *Food Chem. Toxicol., 31,* 63–67.

Lu, F. C. (1991). 'Basic Toxicology'. Hemisphere Publishing Corporation, New York.

Magnusson, B. and Kligman, A. M. (1969). The identification of contact allergens by animal assays: the guinea pig maximisation test. *J. Invest. Dermatol., 52,* 268–276.

Nutrition Foundation Report of the *Ad Hoc* Working Group on Oil/Gavage in Toxicology. The Nutrition Foundation, Washington (1983).

Morris, J. B., Clay, R. J., and Cavanagh, D. G. (1986). Species differences in upper respiratory tract deposition of acetone and ethanol vapours. *Fundam. Appl. Toxicol.*, **7**, 671–680.

Phalen, R. F. (1984). 'Inhalation Studies, Foundations and Techniques.' CRC Press.

Poole, T. (1987). 'The UFAW Handbook on the Care and Management of Laboratory Animals', 6th Edn. Longman Scientific and Technical, Harlow.

Prescott, L. T. and Nimmo, W. S. (1979). 'Drug Absorption Proceedings of the Edinburgh International Conference'. ADIS Press, New South Wales.

Prescott, L. T. and Nimmo, W. S. (1985). 'Rate Control in Drug Therapy'. Churchill Livingstone, Edinburgh.

Rao, G. N., Birnbaum, L. S., Collins, J. J., Tennant, R. W., and Skow, L. C. (1988). Mouse strains for chemical carcinogenicity studies: overview of a workshop. *Fundam. Appl. Toxicol.*, **10**, 385–394.

Rice, M. C. and O'Brien, S. J. (1980). Genetic variance of laboratory outlined Swiss mice. *Nature (London)*, **283**, 157–161.

Sax, I. N. (1979). 'Dangerous Properties of Industrial Materials', Van Nostrand Reinhold Co., New York.

Slaga, T. J. and Nesnow, S. (1985). *In*: 'Handbook of Carcinogen Testing'. Noyes Publications, New Jersey.

Society of Toxicology, Technical Committee of the Inhalation Specialty Section. (1992). Recommendations for the conduct of Acute Inhalation Limit Tests. *Fundam. Appl. Toxicol.*, **18**, 321–327.

Tobin, P. S., Kornhausser, A., and Scheuplein, R. J. (1982). An evaluation of skin painting studies as determinants of tumorigenesis potential following skin contact with carcinogens. *Regulat. Toxicol. Pharmacol.*, **2**, 33–37.

Thorpe, E. and Walker, A. I. T. (1973). The toxicology of HEOD, II. The comparative long term oral toxicity studies in mice with dieldrin, DDT, phenobarbitone, beta-BHK, and alpha-BMC. *Food Cosmet. Toxicol.*, **11**, 433–442.

Vogel, W. H. (1993). The effect of stress on toxicological investigations. *Hum. Exp. Toxicol.*, **12**, 265–271.

Willeke, K. (1980). 'Generation of Aerosols and Facilities for Exposure Experiments.' Ann Arbor Science, Ann Arbor.

Wilson, A. B. (1987). The toxicology of the end products from biotechnology processes. *Arch. Toxicol.*, Suppl. 11, 194–199.

Wilson, J. S. and Holland, L. M. (1982). The effect of application frequency on epidermal carcinogenesis assays. *Toxicology*, **24**, 45–53.

Wollink, F. (1989). Physiology and regulation of biological rhythms in laboratory animals: an overview. *Lab. Anim*, **23**, 107–125.

Zbinden, G. (1993). The concept of multispecies testing in Industrial Toxicology. *Regulat. Toxicol. Pharmacol.*, **17**, 85–94.

CHAPTER 3

Influence of Animal Species, Strain, Age, Hormonal, and Nutritional Status

F. J. C. ROE

1 Introduction

The use of laboratory animals in toxicology and safety evaluation is based on the assumption that they are models for man. For many potent toxins this assumption is well founded in respect of qualitative findings. For weak toxins, for the purposes of safety evaluation and for quantitative prediction of toxic risk, the value of tests in laboratory animals is much more questionable. This is because the animals used are intrinsically poor models and because tests are undertaken under variable conditions which influence the results. The present chapter addresses these problems.

There is, throughout evolution, a continuity in the mechanisms that underlie life processes, and the extent of similarity in body structure between different mammals is striking. Nevertheless, there exist huge differences between species, particularly in the spectrum of foods which they eat and their lifespan. For example, a three year old rat or mouse is equivalent to a human centenarian and yet, within their relatively short lifespans, those species manage to manifest many of the diseases to which mankind is prone, including some degenerative conditions and a wide array of cancers. Higher metabolic rate and faster turnover of cells have been evoked to explain differences in longevity between species. However, these are unlikely to be complete explanations since cells derived from laboratory rodents, for example, behave similarly in *in vitro* cell cultures, and have similar growth requirements, to human cells.

2 Genetic *Versus* Environmental Influences

There has been much investigation of the ageing process but no satisfactory analysis or explanation of the process yet exists. In man, it is a matter of common observation that age-associated changes occur earlier in some individuals than others. Moreover, longevity and early and late ageing seem to be familial. Nevertheless, it is not clear whether genetic or environmental factors are primarily responsible for these differences. Ageing is not a single entity. It is, rather, a progressive accumulation of structural faults and defects in bodily functions. Although either genetic or environmental factors may determine the time of onset of these faults and defects, it is more probable that genetic and environmental factors *interact* as determinants of the time of onset of age-associated diseases. Undoubtedly, genetic constitution often determines susceptibility to particular kinds of ageing defect. But environmental influences serve to reveal this susceptibility. Thus, α_1-antitrypsin deficiency renders humans susceptible to emphysema, but if deficient individuals totally avoided exposure to lung-damaging agents they may never develop the disease.

In the past, premature ageing and a wide variety of animal diseases have been attributed to defective genetic constitution but it is now clear that these may have been false attributions. By changing the quality of the laboratory environment, for example, many of these diseases can be made to disappear. The first major step in this direction came with the development of specified pathogen-free conditions of husbandry. Diseases such as ectromelia in mice and debilitating chronic respiratory disease, complicated by bronchiectasis, in rats markedly reduced in incidence or disappeared. (Bronchiectasis was once regarded as a strain-characteristic and as a 'normal' age-associated change by some experimentalists). The second major development has been a growing recognition that many spontaneous and age-associated diseases of laboratory animals may be due to gross over-feeding. This is, for instance, true of the common chronic nephropathy of rats and of many types of tumour of both rats and mice. Unfortunately, this growing recognition has not yet led to any major changes in diet formulation or feeding schedules in experimental laboratories. But it is certain that such changes must be made. The distinction between genetic and environmental causes of diseases in laboratory animals is difficult and unreliable. The influence of environment on the risk of developing various diseases continues to be underestimated.

3 Inbred *Versus* Outbred Strains

Sequential inbreeding serves to reveal genetic faults by progressively increasing the risk that offspring inherit defective genes from each parent. Similarly, first generation hybrids of two pure inbred strains are often more vigorous and longer-lived than their parents because hybridisation reduces the chances of inheriting pairs of defective genes.

The experimentalist seeking a model has to choose between two needs. On the one hand, he will desire to improve the experimental design by using animals which are genetically identical, since the probability that effects are due to test agents is improved if 'like' is being compared with 'like', except for exposure to the test agent itself. On the other hand, the experimentalist knows that man is not inbred and that if he chooses a particular inbred strain it may, because of its peculiar genetic make up, give quite misleading information concerning the effects of a particular test agent.

For the purposes of evaluating new chemicals for safety for man, it has been argued that it is better to use outbred strains in the hope that at least a few of the animals included in the study will respond in a way that is predictive for man. This is a fallacious argument since there is no way of predicting which observation is pertinent to man. Since different experiments have different objectives, there is no single answer to the problem.

4 Choice of Species

Absolute qualitative differences in the way species respond to chemicals are uncommon but quantitatively very big differences are well documented. For the most part such differences are due to differences in absorption and metabolism, dependent on the presence of particular enzymes and on other conditions which pertain in the gut, the liver, the kidneys, and other organs. Where there is adequate information from studies in several species, which is rarely the case, it is common to find that different species share a spectrum of metabolic pathways but that the predominant pathway in one species is different from that in another. The reasons for such differences are not necessarily genetic. Lead acetate can be fed to rats throughout their lives at dose rates well above ones that would be lethal for man (Van Esch, Van Genderen, and Vink, 1962). This is because the proportion of lead absorbed from the gut is very low in rats fed a standard chow. Changing to a high milk diet multiplies the extent of lead absorption 40-fold and renders lead almost as toxic for the rat as for man (Kostial and Kello, 1979).

Ideally, for safety evaluation purposes, one should choose the species that most closely mimics man in the way that it handles a test material metabolically. In practice, this ideal may be quite unattainable. Firstly, metabolism studies can be difficult, time-consuming, and costly, even if interest is confined to a short list of regularly used laboratory species. Secondly, the studies undertaken may show that man is seemingly unique and that there is *no* good laboratory animal model. Thirdly, it may simply not be known whether other strains behave similarly or how much the available observations are dependent on factors peculiar to the set of conditions of the experiment. Regrettably, much of the literature on comparative metabolism is inadequate and, in addition, the information available for man is usually scanty and not comparable with that for the laboratory species studied. Thus, for ethical reasons the only human studies performed are often for single, very small doses, whereas there are data for animals which have been exposed to larger and repeated dosage.

Furthermore, in tests for chronic toxicity and carcinogenicity, the choice of species is limited if there is to be a reasonable chance of detection of more than the strongest effects. To achieve sensitivity, relatively large numbers of animals (*e.g.*, groups of 50 males and 50 females) are required, particularly if effects in animals exposed to at least 2, and preferably 3, different dose levels are to be compared. This results in huge experiments that cannot even be contemplated except for small animals such as rats, mice, and hamsters. Since weak toxic and carcinogenic effects may not become manifest until animals are well through their available lifespan, it is usually necessary to choose short-lived species. In practice this tends, again, to limit the choice for chronic toxicity testing to rats, mice, and hamsters.

Finally, there is yet one further important consideration. Meaningful interpretation of tests on laboratory animals depends on the existence of adequate information on the array of 'spontaneous' lesions to which the test species is prone. The pathologist trying to evaluate what he sees at necropsy and later during the histological examination of tissues, may not know whether particular appearances are a result of exposure to the test material or manifestations of a spontaneous disease too uncommon to have appeared in the small number of controls.

Thus, the choice of species for the ideal experiment can be very difficult. Undoubtedly, man is the best model for man, but even here, intraspecies variation in response plus interference from environmental factors prevent really confident extrapolation from one man to another. These sources of uncertainty are magnified by species difference but somewhat reduced by the use of large numbers of animals which is possible in the case of small rodents. Comparative metabolism studies are of little practical value except to identify an animal species as *inappropriate* as a model for man.

5   Numbers of Animals: Safety Evaluation as Distinct from Mechanism Studies

Most toxicologists would regard it as more important and worthwhile (and certainly more interesting) to study mechanisms of toxicity than to screen new chemicals for possible toxicity. Regulatory authorities on the other hand (under pressure from an inadequately informed public) require extensive safety testing which utilises most of the available resources. To make matters even worse, it can be very difficult to persuade a regulatory authority that a substance which is in some way toxic for laboratory animals is safe for man, even if some studies of mechanisms have been completed.

The experimental requirements for mechanism studies and safety evaluation differ substantially. The former may require no more than a few animals in which specific measurements are made using sensitive methods. Arguably, studies of mechanisms are by far the most reliable way to characterise underlying changes that result in toxicity and to establish the basis of reliable prediction of human effects. The field is one that is ripe for fruitful expansion. As an example, there is in progress rapid expansion in our knowledge of regulatory peptides (Bloom, Polak, and Lindenlaub, 1982) and the mechanisms involved in homeostasis. This new knowledge is being turned to good effect by the pharmaceutical industry but has yet to be assimilated by toxicologists. Most of the observations routinely made by toxicologists (*e.g.*, haematological, urinalysis, the examination of haematoxylin and eosin-stained sections of tissues, *etc.*) are no more than fairly crude and insensitive screens for possible effects of xenobiotic agents.

In contrast, to establish the apparent safety of a substance for human use it is usually necessary to study large numbers of animals because of the insensitivity of the methods used, and yet the results are regarded as more pertinent than knowledge of toxic mechanisms. In the case of teratogenic and carcinogenic effects, for example, a positive response at any dose level may lead to the test substance being proscribed for human use. Thus the test is essentially designed to demonstrate the *absence* of activity, *i.e.*, to *prove a negative effect*—which is an impossibility. In practice, the negative results required are listed, along with details of the methods to be used, as 'Guidelines' by various authorities (*e.g.*, OECD Guidelines, 1981).

One consequence of the development of new and more sensitive tests, many of them applicable to living animals, will be to highlight the problem of how to distinguish between an 'effect' and a 'toxic effect'.

6 Animal Susceptibility

Rats and mice in the wild carry microbial and parasitic diseases and, like humans, are prey to epidemics of infectious diseases. Before the development of barrier conditions within which disease-free laboratory animals could be given pathogen-free food, endemic and epidemic disease rendered long-term experimentation frustratingly difficult. The cleaning up of animal laboratories has not, however, completely solved the problem of there being an unacceptable level of background disease, although the spectrum of diseases that most commonly occur has changed.

The inability to eradicate natural disease processes raises the question as to how far such conditions render the experimental animal more (or less) susceptible to the adverse effects of chemical treatments and how this affects the process of extrapolation to man.

It is possible that a high incidence of a particular disease in untreated animals is indicative of susceptibility to the induction of that disease by exogenous agents. This is of particular relevance to specified types of tumour. For example, it is seemingly easier to induce adenomatous tumours of the lung in strains of mice that exhibit a high incidence of such tumours 'spontaneously' than in low 'spontaneous' incidence strains. At one time breeding programmes, aimed at developing high tumour incidence strains of mice (*e.g.,* the B6C3F$_1$ hybrid) in the hope that such strains would prove to be sensitive tools for detecting carcinogenicity, were initiated. There is no good evidence that such efforts improve safety prediction for man.

Ideally an animal model should exhibit the spectrum of morbid and fatal diseases of man (*e.g.,* high incidences of cardiovascular disease and cancers of epithelial cell origin, *etc.*). At present no such animal model is available. Many strains of the laboratory rat show very high incidences of endocrine tumours (*e.g.,* 100% incidence of mammary tumours in females, 100% incidence of Leydig cell tumours in males, up to 80% incidence of pituitary tumours in both sexes, high incidences of adrenal, thyroid, and parathyroid tumours, *etc.*) (Roe, 1981); and many strains of laboratory mouse, including the B6C3F$_1$ hybrid referred to above, exhibit high incidences of malignant lymphoma, liver, and lung tumours. In addition, rats aged two years or more often develop severe progressive nephropathy which affects both glomeruli and tubules to varying degrees, and which is associated with increased low-molecular weight proteinuria. Nor does the hamster offer a way of avoiding these deficiencies because under laboratory conditions, these animals tend to develop very high incidences of amyloid degeneration of the kidney and of septic atrial thrombosis.

Overall, the position is unsatisfactory in that animals in control groups commonly develop high incidences of diseases which are uncommon or even rare in man and very few animals die spontaneously of any of the diseases which most commonly affect humans. There are two serious corollaries of this. First, if the aim is to reduce man's burden of disease, then the tests should aim at detecting environmental factors or agents which induce or exacerbate the common diseases of man (*e.g.* arthritis, heart disease, stroke). Secondly, if a high spontaneous incidence of a disease is indicative of easy inducibility, the animal models we presently use are of little value for this purpose and others should be sought. Moreover, the high incidence of irrelevant spontaneous diseases render animal models unsuitable for detecting subtle manifestations of toxicity of certain kinds. It would not, for instance, be possible to detect a weak chronic toxic effect on the kidney under experimental conditions in which all the untreated control rats develop serious spontaneous renal disease.

7   The Effects of Overfeeding

When the spectra of neoplastic and non-neoplastic diseases that have been encountered in untreated animals in conventional chronic

Table 1 *A few of the many kinds of background pathology encountered in a group of 60 untreated male Sprague-Dawley rats that constituted the controls in a carcinogenicity study of 2 years duration*
(Unpublished data from personal files)

	%
Moderate to severe	
chronic progressive nephropathy	67
Parathryoid hyperplasia	67
Calcification of aorta	34
Adrenal medullary	
– hyperplasia/neoplasia	32
– neoplasia	20
Chronic fibrosing myocarditis	83

toxicity/carcinogenicity tests (Tables 1–3) are compared with the effects of controlling dietary intakes that have been observed in different studies (Tables 4–6), it is clear that much of the background disease which is presently such a prominent feature of long-term rodent studies is associated with overfeeding and is probably avoidable. At present, the

Table 2 *Incidence of certain non-neoplastic and neoplastic diseases in groups of 86 male and female untreated Sprague-Dawley rats*
(Reproduced by permission from Kociba *et al.*, 1979, *Food Cosmet. Toxicol.*, **17**, 205–221)

% of rats with:	Males	Females
Mineral deposition kidney	—	29
Moderate or severe renal disease	65	7
Mineral deposition in gastric		
mucosa and muscularis		
secondary to kidney disease	29	—
Multiple foci of hepatocellular		
alteration	—	17
Focal hepatocellular cytoplasmic		
vacuolation	40	—
Periarteritis	23	—
Neoplasms of:		
Pancreas – exocrine	33	—
Pancreas – endocrine	16	9
Pituitary – pars distalis	31	62
Adrenal – medulla	51	8
Thyroid – C-cell	8	8
Mammary gland – fibroadenoma	1	76
– adenocarcinoma	3	8
Any site	88	97

Table 3 *Percentages of rats with certain endocrine tumours in the control groups in 3 separate 2-year carcinogenicity studies on a prolactin-releasing drug. (The arrows in brackets indicate the effect of the drug on tumour incidence, if any*)*
(From confidential data in personal files)

Study No.	1		2		3	
Strains of rat	Wistar		Wistar		Sprague-Dawley	
Sex	♂	♀	♂	♀	♂	♀
Pituitary	22(↑)	62	17	53	41	46
Benign mammary	0(↑)	86(↑)	2	11	8(↑)	77
Malignant mammary	0	10	0	0	2	0
Phaeochromocytoma	18(↑)	22	0	0	2	0
Adrenal cortex	10	10	0	0	0	3
Thymoma (endocrine type)	4(↑)	10(↑)	0	0	0	0
Thyroid – follicular	26(↓)	18	0	0	0	0
– C-cell	0	6	9	5	1	0
Pancreas – islet cell	4(↑)	0(↑)	4(↑)	3(↑)	4	5
Parathyroid hyperplasia	25	7	0	0	0	0

* If only study No. 3 had been carried out, the drug would have run into fewer regulatory problems than it did!

simplest way of avoiding overfeeding is to limit the time during which food is freely available. Whether the same effect can be achieved by reducing the nutritive value of the diet provided throughout the 24 hours of the day, is not yet certain. Of special interest is the observation (Ross, Lustbader, and Bras, 1982) that the amount and kind of food an animal eats shortly after weaning has a seemingly indelible effect on its subsequent body growth and the incidence of tumours. Clearly the effects

Table 4 *Effect of restricting the access of Wistar rats to a standard laboratory chow (PRD formula) from 24 hours per day to $6\frac{1}{2}$ hours per day on incidence of chronic progressive nephropathy at 2 years*
(Reproduced from Salmon *et al.*, 1990)

Sex:	Male		Female	
Hours of access to food per day:	24	$6\frac{1}{2}$	24	$6\frac{1}{2}$
% of rats with moderate or severe nephropathy:	65	5	60	0

Table 5 *Effect of 20% diet restriction by weight on tumour incidence before 2 years in groups of 50 male and female ICI Wistar rats*
(Reproduced by permission from Tucker, 1979, *Int. J. Cancer*, **23**, 803–807)

	Male		Female	
	Ad lib.	20% Restricted	*Ad lib.*	20 Restricted
% survival to 2 years	72	88	68	88
% of rats which developed tumours at any site	66	24***	82	57**
% of rats which developed tumours at more than one site	22	2**	26	10*
Total number of all sites of tumours in group of 50 rats	47	14***	59	37**
% of rats that developed mammary tumours	0	0	34	6***
% of rats that developed malignant mammary tumours	0	0	12	2
% of rats that developed pituitary tumours	32	0***	66	38**

* $p < 0.05$
** $p < 0.01$
*** $p < 0.001$

of diet composition, food-scheduling, and eating patterns on spontaneous disease in laboratory animals requires more intensive investigation.

The data presented illustrate the extent to which environmental factors influence the incidence of diseases which, in the past, many experimentalists considered to be primarily determined by genetic constitution.

Table 6 *Effect of diet restriction on longevity and tumour incidence in mice*
(Reproduced by permission from Conybeare, 1980, *Food Cosmet. Toxicol.*, **18**, 65–75)

	Ad lib.	Restricted (75% of ad lib.)	*Ad lib.*	Restricted (75% of ad lib.)
Number of mice	160	160	160	160
% survival to 83 weeks	58	66	62	77*
% Lung tumours	19	12*	15	5**
% Liver tumours	29	8***	4	0.6*
% Any tumour at any site	44	23***	31	11*
% Any malignant tumour at any site	11	4	14	4**

* $p < 0.05$
** $p < 0.01$
*** $p < 0.001$

However, it is clear that the problem of high incidences of background disease cannot be fully resolved by attention to diet. Although diet restriction has been shown to have major beneficial effects on the incidence of kidney disease, endocrine changes, and neoplasia of various kinds, *etc.*, there remains in the restricted-fed animals an unacceptably high incidence of such lesions (particularly hyperplasia and neoplasia of the anterior pituitary gland in rats, and malignant lymphoma in mice). This suggests that the contributions of other aspects of the laboratory environment to the causation of disease in untreated animals should be explored. The suggestion that enforced celibacy may be detrimental (Roe, 1981) has not been supported by a recent small-scale study (Salmon *et al.*, 1990). It is probable that lack of exercise or general boredom will be found to be important. A study of the effects of providing an exercise wheel on the development of ageing and other diseases in mice was reported by Conybeare (1988). Behavioural scientists are probably best placed to devise protocols for testing the effect of general boredom!

8   Future Development in Long-term Testing

At present, the laboratories that carry out tests and the regulatory authorities who judge the results are understandably reluctant to use, or recommend the use of, diet restriction in the design of chronic toxicity/carcinogenicity tests. A fundamental change of this kind might devalue the banks of accumulated data derived from earlier conventionally-designed tests, and there is the fear that by reducing background disease, the model will be made less sensitive to the induction of those diseases relieved by dietary control. It is of critical importance that these objections are examined by properly conducted scientific experiment. It may then be possible to devise conditions which maintain laboratory animals in good health, free of infection, obesity, endocrine disorder, and many forms of neoplasia, and which are all properly monitored and controlled. It will then only be a short step for regulatory authorities to *require* that all tests for toxicity and carcinogenicity shall be undertaken in *hormonally normal animals* rather than in animals that display a mixed medley of laboratory artefacts.

9   Summary

This chapter has addressed a number of inter-related aspects of basic toxicology.

(a) There is no clear understanding of the 'ageing' process because of widespread confusion between diseases that are simply more common in old age (*i.e.*, intrinsically due to an ageing process) and diseases which are caused by avoidable environmental exposures.

(b) Genetic constitution influences longevity where inbreeding results in offspring receiving defective genes from both parents. Hybrid vigour illustrates an escape from this handicap.

(c) Many so-called 'strain characteristics' are wholly or partly environmentally determined.

(d) The advantages and disadvantages of using inbred as distinct from outbred strains must be considered in relation to the type and purpose of the experiment.

(e) The use of a species and/or strain of animals that handles the chemical to be tested in the same way as man is usually impracticable, either because the comparative metabolic data do not exist, or because no species which it is feasible to use is sufficiently akin to man.

(f) The principles underlying the investigation of mechanisms of toxicity are quite different from those underlying the design of safety evaluation tests. More studies of mechanisms are necessary to improve interspecies extrapolation.

(g) The spectrum of diseases which afflict man is quite different from the spectra which afflict rodent species. The relevance of rodents as models for the more common diseases of man merits further study.

(h) The high incidence of obesity and of many kinds of background disease in the laboratory rodents used in chronic and carcinogenicity testing today and, in particular, the high incidence of various endocrine disturbance and neoplasia of endocrine glands are unacceptable. Over-feeding appears to be a major factor in the causation of chronic progressive nephropathy, periarteritis, cardiomyopathy, and various endocrine tumours in rats and of malignant lymphoma, liver, and lung tumours in mice.

(i) Further research into ways of maintaining laboratory animals in normal endocrine status throughout their natural lives is urgently needed.

10   1993 Update

Since the above chapter was written for the first edition of this book two major developments have occurred in relation to the role of non-genotoxic mechanisms in carcinogenesis.

In relation to the first of these developments, the reader is referred to two books: Weindruch and Walford (1988) and Fishbein (1991). In the latter, there is a report of the results of an experiment involving 1200 rats exposed to various dietary regimes (Roe, 1991) and some of the findings in this experiment have been summarised in Roe (1993). Particularly noteworthy are the highly significant correlations between body weight at the age of 29 weeks and risk of premature death and/or malignant tumour development before the age of 133 weeks.

The second development concerns the growing recognition of the role of endogenous mutagens and the importance of increased cell turnover rates as determinants of premature ageing and increased cancer risk. This development is also discussed briefly in Roe (1993).

References

Bloom, S. R., Polak, J. M., and Lindenlaub, E. (1982). 'Systemic Role of Regulatory Peptides.' pp. 1–547. Schattauer Verlag, Stuttgart/New York.

Conybeare, G. (1980). Effect of quality and quantity of diet on survival and tumour incidence in outbred Swiss mice. *Food Cosmet. Toxicol.*, **18,** 65–75.

Conybeare, G. (1988). Modulating factors: challenges to experimental design. *In*: 'Carcinogenicity: the design, analysis and interpretation of long-term animal studies'. *Eds* H. C. Grice and J. L. Ciminera. ILSI Monograph. Springer-Verlag, New York, pp. 149–172.

Fischbein, L. (1991). 'Biological effects of dietary restriction'. Springer-Verlag, Berlin, 349pp.

Kociba, R. J., Keyes, D. G., Lisowe, R. W., Kalnins, R. P., Dittenber, D. D., Wade, C. E., Gorzinski, S. J., Mahle, N. H., and Schwetz, B. A. (1979). Results of a two-year chronic toxicity and oncogenic study of rats ingesting diets containing 2,4,5-trichlorophenoxyacetic acid (2,4,5-T). *Food Cosmet. Toxicol.*, **17,** 205–221.

Kostial, K. and Kello, D. (1979). Bioavailability of lead in rats fed 'human' diets. *Bull. Environ. Contam. Toxicol.*, **21,** 312–315.

OECD Guidelines for Testing of Chemicals (1981).

Roe, F. J. C. (1981). Are nutritionists worried about the epidemic of tumours in laboratory animals? *Proc. Nutr. Soc.*, **40,** 57–65.

Roe, F. J. C. (1991). 1200-Rat Biosure Study: Design and overview of results. Chapter 26 *in*: 'Biological effects of dietary restriction'. *Ed.* L. Fishbein. Springer-Verlag, Berlin, pp. 287–304.

Roe F. J. C. (1993). Recent advances in toxicology relevant to carcinogenesis: seven cameos. *Food Chem. Toxicol.*, **31,** in press.

Ross, M. H., Lustbader, E. D., and Bras, G. (1982). Dietary practices in early life and spontaneous tumours of the rat. *Nutr. Cancer*, **3,** 150–167.

Salmon, G. K., Leslie, G., Roe, F. J. C., and Lee, P. N. (1990). Influence of food intake and sexual segregation on longevity, organ weights and neoplastic diseases in rats. *Food Chem. Toxicol.*, **28,** 39–48.

Tucker, M. J. (1979). The effect of long-term food restriction on tumours in rodents. *Int. J. Cancer*, **23,** 803–807.

Van Esch, G. J., Van Genderen, H., and Vink, H. H. (1962). The induction of renal tumours by feeding of basic lead acetate to rats. *Br. J. Cancer*, **16,** 289–297.

Weindruch, R. and Walford, R. L. (1988). Retardation of ageing and disease by dietary restriction. C. C. Thomas, Springfield, Illinois, 436pp.

CHAPTER 4

Experimental Design

A. B. WILSON

1 Regulatory and Experimental Toxicology

Regulatory toxicology includes a wide range of procedures (Tables 1 and 2), conducted with the main objective of screening compounds and meeting the routine demands of regulatory bodies in a variety of countries. Chapter 22 will highlight the nature of these requirements for food but they result in a spectrum of studies of fairly standard design. In practice, most government authorities provide considerable scope for variations to provide for the inevitable differences between compounds, their uses and effects. This chapter will outline some of the standard designs used and discuss a selection of design problems associated with these investigations.

Investigational toxicology arises from a desire to study toxic effects more deeply and specifically than routine studies permit. Whilst regulatory toxicology is mainly a screen and a first stage, investigative work often starts from there. The screen identifies an effect which is of such a nature or exposure level that further investigations are warranted. Sometimes these will be adequately covered by other parts of the screen, *e.g.* use of a second species, studies on the metabolism of the compound in man and animals, and pharmacokinetics. On other occasions, the designs of screening studies will be adjusted to provide additional information on particular effects, *e.g.* electron microscopy of the liver after hepatoxicity or behavioural tests after observing CNS signs.

The potential range of disciplines involved in investigative toxicology is extremely wide, and can involve tissue culture and tissue homogenates, electron microscopy and histochemistry, autoradiography and pharmacokinetics, genetics and molecular biology, epidemiology and evaluation of information from accidental exposures, agonists and antagonists, and many other approaches. One of the skills is to select which of these are really likely to provide useful information, but further investigations generally depend upon the development of a pertinent model of the lesion or effect that is to be explored.

In conducting these studies one can only recommend sound scientific practice as required by Good Laboratory Practice (GLP) Regulations: evidence of thought, reflection, justification, and planning of a study; adequate resources (people, equipment, premises) both in quantity and in quality; and an ability to reconstruct the experiment and the process leading from the original data to the conclusions.

Chapter 24 will explore how these are achieved in the context of toxicology but the fundamentals, as stated above, are relevant to most branches of science and certainly to both regulatory and investigational toxicology. The intelligent application of GLP has transformed toxicology into a branch of science with an exceptionally high standard of experimental conduct.

2   Object of Study

A   Reflection

Clark (1977) in his chapter on inhalation toxicology describes reflection as the most necessary but most neglected first step in designing a study. This corresponds with the emphasis GLP places upon planning and justification for actions. It has been said that organisation is the enemy of research and, while it is certainly true that organisation blunts spontaneity, there are a great many spontaneous actions which are later regretted. Certainly, in regulatory toxicology, an orderly and disciplined approach is of more value than bursts of occasional brilliance.

The first step, before becoming more involved with experimental design, is to assess the knowledge already available on the compound in question, *i.e.* the problem being addressed and the techniques likely to be pertinent. The wide range of computer-based literature searching facilities available (Wexler, 1987) provides a good basis from which to start so that, on a novel compound, structure–activity relationships, properties of similar compounds, and strengths and weaknesses of the test methods can receive a preliminary assessment.

B   Definition of Object

The next step, or perhaps in parallel with the first, is to define the object of the investigation. This is not as easy as it first sounds and the omission, ambiguity, or superficiality of many of the 'Object' paragraphs of regulatory toxicology protocols and reports too often reflect inadequate considerations of the matter and can result in a poor or wasted experiment. A few examples might be relevant. It is common practice to conduct animal studies involving dosing for 90 days, embracing a range of clinical, haematological, clinical chemical, and pathological examinations, the object being described as 'to investigate the effects of dosing compound X for 90 days). This may be only half the story. The object may also be to satisfy a perceived regulatory requirement, which would explain a few of the observations required which may otherwise be viewed with curiosity.

Sometimes special investigations (*e.g.* for effects upon the kidney) are coupled with such studies. This is when complications arise, since there would then be 3 probable objects:

(a) to investigate the toxicity of compound X over 90 days,

(b) to satisfy regulatory bodies, and

(c) to gather further data on the renal toxicity of compound X.

The last mentioned object is very loosely phrased but is often what occurs and might indeed be an adequate description. On many occasions, certain specific techniques are considered which might compromise other declared objects. The combining of several objects can result in a potentially unhappy compromise through which none is particularly well satisfied.

The type of conflict that occurs in the above example can arise from interference with an animal for certain observations, and hence cause a change. Anaesthesia, restraint, withdrawal of food, and frequent blood sampling (Cardy and Warner, 1979), can confound routine observations fundamental to the other objects. Tests for behavioural changes (Alder and Zbinden, 1983) are particularly difficult to incorporate for these reasons. The more objects there are, the more frequently such compromises will occur.

Many toxicology experiments are organisationally somewhat complicated, involving the co-operation and integration of a wide variety of disciplines over months and even years, resulting in tens and hundreds of thousands of items of data recording. Any attempt to further complicate the issue is to court disaster since the organisational system might not cope.

It is essential to keep study designs sufficiently simple and robust for successful accomplishment of the original objects to be assured—scientists are not always the best managers or organisers and *vice versa*. There are, thus, two reasons for keeping study objects simple:

(1) to avoid unsatisfactory compromises in scientific approach, and

(2) to allow the organisation to cope.

It is, therefore, far better to divide a complicated study into different portions, each simple, thorough, and relevant, than to try to create a complicated experiment. As an experienced researcher remarked, "There is always time to repeat an experiment but never time to do it once properly."

For example, in the field of reproductive toxicology there are two conventional approaches. One, favoured by regulatory bodies involved with agricultural chemicals, is to breed two or three successive generations of rodents, looking for effects on fertility, foetal development, post-natal development, and sometimes incorporating 90 day toxicology. Reviews of such studies show, not surprisingly, that very few have been pivotal to the assessment of a compound (Food and Drug Administration, 1970). The alternative approach, generally used for testing of pharmaceutical products, is to treat each portion separately—a fertility, a teratology, and a peri-post-natal study. This latter approach uses no more animals, takes less time (studies can be overlapped), results from one study can be used to modify the design of the succeeding study, and if there are practical problems with one component it does not mean repetition of the full sequence of studies to rectify matters.

C Confirmation of Design

Having attained a provisional object and study design, it is as well to check what action will be taken if various combinations of results occur. Thus, special investigations of renal function within a 90 day screening study may result in the following: tests negative, perhaps not sufficiently sensitive (design compromise) so a specifically designed repeat study is required, or tests positive, therefore specifically designed repeat studies are justified, *i.e.* the same endpoint may be reached whatever the result, making one question the need for the original experiment.

Three questions can be asked when the results of a study appear.
(a) Do I believe this effect would recur if I repeated the experiment in the same way, *i.e.* is it reproducible?
(b) Do I believe the effect was attributable to the compound?
(c) Do I believe the effect is biologically significant?

It is sensible to look at the experimental design in anticipation of these questions and to try out various combinations of results to check what conclusions or further actions will follow. In practice most major studies follow well established patterns. Study design difficulties occur more frequently with shorter studies, preliminary studies, and special investigations. Though smaller in terms of numbers of animals used and cheaper in terms of man hours, they are more demanding in terms of forethought.

D Historical Data

Data from previous studies are frequently and rightly used to aid in the interpretation of study results. There needs to be a degree of caution since work carried out, even under apparently similar circumstances, can produce different results. However, they can indicate whether (for example) performance of the control group on a study was in any way exceptional—perhaps having a lower incidence of a particular tumour and giving the spurious impression of an increased incidence caused by the compound in question.

Background data form the basis of 'normal ranges' and these are usually taken to mean the range of values generally encountered in equivalent control animals maintained under similar conditions. If a statistically significant change occurs in a study but the control and test values both lie within 'normal limits', it can be suggested that the change is, of itself, of doubtful toxicological significance, so long as it does not point to another more profound effect. For example, a slightly lower haemoglobin content might lie within normal limits for that laboratory and so it could be suggested that those animals could live and function entirely satisfactorily with that slight change. However, the change might be the result of hepatotoxicity, bone marrow disfunction, or another serious effect of the chemical under investigation. The mere fact that the change lies within normal limits does not mean it can be dismissed as irrelevant.

3   Customary Study Designs

A   Principles

Most routine studies, orientated towards general, non-specific investigations and satisfying regulatory bodies, have a similar pattern: they attempt to imitate the routes of exposure likely to be experienced by man; they tend to exaggerate the likely exposure, both in terms of duration and dosage; and they have a control group and three to five dose groups with increasing dose levels. The question of route of exposure is covered more fully in Chapter 2.

Exaggeration of exposure works on the simple principle that a high dose is more likely to show up the potential effects, although effects never likely to occur in man could also be encountered. While it is probably a valid approach, the process of exaggeration is similar to using a magnifying glass—it often distorts the image and it focuses upon detail. It takes an expert judgement to decide whether the detail is relevant and to allow for the distortion.

The highest dose used in longer term and carcinogenic studies is generally the 'Maximum Tolerated Dose' (MTD). The definition varies inevitably with the circumstances of each study (Haseman, 1986; Feron and Kroes, 1986). It reflects an attempt to administer as high a dose as possible without the toxicity of the compound distorting the animal model excessively. However, the effects contributing to a judgement of MTD are often clinical (perhaps bodyweight or laboratory tests or pathology). Therefore, some toxic effects in longer term studies may arise from overload of the normal routes of metabolism of a compound with resultant unrealistically high concentrations of metabolites (or the parent compound). The International Conference on Harmonisation (1991) recommended greater attention to toxicokinetics in the selection of dose levels for carcinogenicity studies. Coupled with a series of pertinent publications (Carr and Kolbye, 1991; Goodman and Wilson, 1991; Monro, 1992) it seems possible that there will be a shift in thinking so far as carcinogenicity studies are concerned.

Again, it is important to address the original object of the study, whether it was to show the toxic potential of a compound or to help in assessing safety. The differences between these can be critical to the experimental design, particularly with compounds that are relatively non-toxic at high dose levels. Under such circumstances, it is reasonable to expect a safety factor of perhaps 100 to 1000 fold to be adequate, and dosing beyond that to be irrelevant to the assessment of safety, though not irrelevant to the investigation of potential toxicity.

This is illustrated in the different approaches of regulatory bodies. Some tend to request that sufficient compound be administered in a rodent carcinogenicity study to provoke a clear effect, even if that dose is several hundred times the human exposure. The regulatory body is, therefore, investigating the carcinogenic potential of the compound, but may presumably choose to interpret carefully a toxic result at a high dose level. Other bodies, in addressing themselves to the problem of the safety of the compound, do not believe that heroic doses can normally be justified in such an experimental design.

B Control Groups

In many conventionally designed toxicology studies, the compound is given to animals in various treatment groups and results are compared with the results of a control group. A control group is not always required where its response can be predicted with adequate certainty. For instance, in oral rat LD_{50} studies, where clear clinical signs and deaths are expected, most laboratories will have sufficient experience to know that rats and most other species rarely die or show clinical signs under those conditions unless the compound is responsible. If, however, an unusual vehicle is being used or more subtle observations made, comparison with background data may be inadequate and control animals would then be needed.

It is apparent that a control group can take more than one form. Normally the control group is treated in precisely the same manner as treatment groups, except for administration of the test compound, *i.e.* animals are handled, weighed, observed, fed, and sham dosed (generally with vehicle) in an identical manner to dosed animals. Thus it is possible to identify effects attributable to the compound. On occasions, the manipulation of control animals is likely to have a substantial effect upon them, as can occur in restraint for nose-only exposure of rats (Chapter 2), the feeding of an artificially high protein diet to imitate the protein content of an enzyme (contributing protein to the diet) under investigation, or the dosing by gavage with substantial quantities of vegetable oil as the vehicle. In such instances a second control group, maintained under more normal undosed or unrestrained circumstances, can be considered. Its value is limited since its performance is compared (usually subjectively) to background control performance, with the object of assessing whether the routine factors of the study were being conducted satisfactorily and showing the extent to which the experimental technique altered the animal model. A placebo group is another type of control but appropriate only to human studies.

The control groups described above are sometimes termed 'negative controls' since they are not treated with compound. If the group is dosed with the vehicle used for suspending/dissolving the test compound, then it is frequently termed the 'vehicle control'. A 'positive control' refers to the administration of a compound to a further group of animals. The compound is selected as having the effect being investigated, for example, mutagens such as 2-amino anthracene or sodium azide are used in bacterial mutagenicity studies. Alternatively, the control compound might be chosen because it has either similar biological activity to the test compound or a similar structure, in both instances its toxicology would normally be known. In such circumstances it might be more correctly termed a 'reference compound'. Positive controls give an indication as to whether or not the test procedure has been succesfully carried out and are more appropriate to studies for specific effects (*e.g.* mutagenicity) than for general screening investigations. There are other methods of achieving the same end so positive controls are not used all that frequently, except in short term tests for identifying mutagenic or genotoxic compounds (generally '*in vitro*' tests). The use of higher dose levels in screening studies may be regarded as equivalent to a positive control in many respects.

The value of a positive control group is sometimes misjudged. For instance, the inclusion of a group of animals dosed with CCl_4 (a hepatotoxin) will give some added confidence that liver toxicity would have been detected if present. It does not, however, prove the susceptibility of that strain of animal or suitability of that test protocol for all potential hepatotoxins.

The control group, particularly the 'negative control', is crucial since each dose group is compared to it statistically as well as biologically. If the control group 'misbehaves' in some way it could give rise to misleading conclusions. A subjective assessment therefore indicates that the control group should be larger than any one of the treatment groups (see Chapter 18).

Despite the statistical arguments and basic logic, it is the exception rather than the rule for negative control groups in toxicology studies to be larger in number than treatment groups. Rodent carcinogenicity studies do, sometimes, include a large (often double size) control group, and dog or primate experiments where group sizes are generally very small (3 to 6 per sex per group) sometimes have larger control groups.

C   Selection of Dose Levels

As previously described, the vast majority of toxicology studies comprise a control group and three or four dose groups, *e.g.*:

Group	Dose ($mg\,kg^{-1}$)
Control	0
Low Dose	1
Intermediate Dose	2
High Dose	4

The first stages of dose selection are covered under sections 2A and 2B of this chapter. For instance, there should be a review of the available information and, in many instances, preliminary studies will have been carried out with the express object of discovering at what dose levels effects are likely to occur or not to occur. These might well have commenced with a form of acute investigation and progressed through various study durations as described later.

High Dose

Potency, palatability, solubility, and practicalities of administering a larger dose generally restrict the maximum dose that is administered. It is normally essential that this dose is sufficiently high to cause a clear and indisputable toxic effect. The effect might be reduced body weight gain, unusual clinical signs, or more subtle effects observed by the pathologists, haematologists, or clinical chemists. Sometimes it is governed by metabolic capacity or kinetics.

Views differ as to the desired severity of the effect and, of course, the object of the experiment might be a determinant in that regard, *e.g.* an LD_{50} study requires death to be the effect. A key question is, 'If no effects are seen at the highest dose level will the experiment need to be repeated?'.

There are pressures to limit as the highest dose given, to the lowest dose level that will cause a significant and unambiguous effect. Keeping the dose low reduces the span of doses to be covered in the study.

Low Dose

It is normal to select a dose at which no effect is likely to be discerned and to wish to place this as high as possible, consistent with no effect being observed. Another frequent determinant is the minimum acceptable safety factor. If the human dose of a drug is to be 1 mg kg^{-1} it might be that a minimum safety factor of 5 is acceptable for certain parameters in which case there is little point in selecting a dose under 5 mg kg^{-1}. If this sort of judgement is made, the scientist in the example has to be sure that if a marginal effect occurs at 5 mg kg^{-1} he will not be obliged to repeat the experiment at a still lower dose level. The ratio between the toxic dose and the therapeutic dose is the 'therapeutic index'—the pharmaceutical equivalent of the 'safety factor'.

Intermediate Dose

In practice, it is usual to select both high and low dose before the intermediate dose. Doses may be spread arithmetically, so that the intermediate dose is the mean of the low and high dose (relatively uncommon). More frequently an attempt is made to spread the doses geometrically, so that the intermediate dose is the same multiple of the low dose as the high dose is of the intermediate dose. The statistician may plead for a logarithmic spread of dose levels; however, the scientist may not be interested in constructing a dose–response curve. He is often more interested in setting the highest dose he can at which no effect is discerned, in which case he may place the intermediate dose group rather lower than otherwise. The intermediate dose group is often an attempt to salvage something of value from the experiment when the high dose is too toxic or the low dose too low. A dose–response curve can be of considerable assistance in judging the 'safety factor'. A steep curve implies the effects will occur over a narrow range of dose levels so the 'no observed effect level' can be defined more precisely than a shallow dose–response curve would permit.

Experiments may be designed to display the carcinogenic or teratogenic potential of the compound. The top dose is normally selected to have a slight toxic effect on the animals (the MTD, *vide supra*), but a significant toxic effect could lead to teratogenic effects secondary to significant maternal toxicity or carcinogenic effects related to altered growth rate or maturity. Under such circumstances, the intermediate dose may be placed quite high, as a reserve high dose in case the high dose chosen proves to be so toxic that results are difficult to interpret. The relevance of the MTD and of safety factors in reproductive toxicology is well dicussed by Johnson (1988).

In summary, for most animal toxicology studies, the experimenter has to be very sure that the high dose will have some distinct and observable effect and the low dose will have none. It is probably true to say that many more errors are made by having doses too close together than by placing them too far apart.

It is curious that some regulations and guidelines request that lower dose levels be set as fixed fractions of the top dose level. Such dictates certainly curtail discussion of dose levels but they make no acknowledgement of steep or shallow dose–response curves, or of the arguments produced above, and place conformity before good sense (Mackay and Elliott, 1992).

D Selection of Species and Number of Animals

Chapter 2 provides some basic information on the merits, limitations, and conventions in the use of specific species of laboratory animals in toxicology screening studies. Familiarity and availability have a great deal to do with the initial selection of the species for testing. Following that, the selection should preferably be for the one most likely to be relevant to man, not necessarily the most susceptible species. There are substantial differences between species in their anatomy, physiology, and metabolism, any or all of which might render the species either anomalous or the species of choice. Homburger (1987) presents a series of interesting examples as well as some of the strengths and limitations of *in vitro* studies. However, if the species turns out to be a somewhat exotic animal, which is difficult to use and is poorly understood, then there are sound arguments for remaining with the more normal species and interpreting results more carefully. Frequently, the species selected for testing of pharmaceuticals is the species which shows similarity to man in its pharmacological reaction to the drug. Primates are often selected for studies on products of biotechnology (*e.g.* hormones and monoclonal antibodies) because of species specificity as well as reduced antigenicity. Similarities regarding pharmacological activity do not mean that species will necessarily react appropriately regarding other actions of the drug.

The choice of numbers of animals in a study provides some interesting contradictions. On first consideration one might select the number based upon the degree of change to be detected and its background variability (Plaa, 1982). Tables 1 and 2 indicate the numbers used in routine studies and they do not entirely tally with this logic. Dogs and primates have much more variability than rats or mice, yet substantially fewer are used in each dose group, 3 to 6 per sex, whereas for an equivalent study 10 to 25 rats are used. Expense, expediency and convention are the main reasons.

If very large numbers of animals are employed, precision should increase, but the mean values might be no different (Muller and Kley, 1982) and more changes will be identified which are statistically significant. However, each change of statistical significance has to be interpreted biologically. If the finding is slight (*e.g.* 2% alteration in body weight gain), this will probably be quite uninterpretable on its own. There is, therefore, little point in identifying such a change (except that it might be part of a larger picture or an indicator of more significant effects in other organs). It is, therefore, unusual in screening studies to take blood samples from more than about 10 or 12 rats per sex per group for routine haematology or clinical chemistry. Histopathology is usually adequate from 10 to 15 per sex per group, unless a low but relevant incidence is suspected, *e.g.* tumours. Organ weight data are not easy to interpret in any case, so it is difficult to justify weighing organs from above 10 per sex per group. If a compound is deemed to have caused the change, but the change is interpreted as being of no biological or toxicological relevance, then the merit of identifying it has much diminished. Chanter *et al.* (1987) used data from a wide range of routine toxicology studies to indicate the precision associated with many routine parameters and, in particular, the likelihood of false negative results.

Debate over the precision necessary in LD_{50} studies has highlighted these points. Muller and Kley (1982) and many others confirmed that the use of fewer animals does not generally alter the mean value obtained for the LD_{50}, so it is difficult to justify the use of more than 5 animals per sex per group under most circumstances. Even then, it is reasonable to obtain the value for one sex before investigating the other—females are, in

Table 1 *Rat and Mice: Typical Study Designs for Toxicity Studies*

Duration	Purpose	Number of groups inc. control	Animals/ sex/ group	Body weight	Food consumption	Haematology/ clinical chemistry	Gross autopsy	Histopathology	Months from autopsy to report
14 days or 28 days	To select doses for longer studies	4–6	5–10	Weekly	Weekly	At end	All animals	Major organs and target organs*	1–3 mth
28 days	To provide definitive information	4–5	10	Weekly	Weekly	At end	All animals	Wide selection of organs*	1–3 mth
90 days	To provide definitive information	4	15	Weekly	Weekly	Week 6 and end	All animals	Wide selection of organs†	3–6 mth
26 weeks or 52 weeks	To provide definitive information	4	15–20	Weekly for 13 weeks then every 2–4 weeks.	Weekly for 13 weeks then every 2–4 weeks.	Weeks 6, 12, 26 and end	All animals	Wide selection of organs†	4–9 mth
18–24 mth (mouse) 24–30 mth (rat)	Carcinogenicity	4	50 or more	Weekly for 13 weeks then every 2–4 weeks.	Weekly for 13 weeks then every 2–4 weeks.	—	All animals	Wide selection of organs†	7–15 mth

* Histopathology on control, top dose and selected animals/organs from other groups
† Histopathology on all animals
The above designs are intended only as examples.

Table 2 *Dog and Primates: Typical Study Designs for Toxicity Studies*

Duration	Purpose	Number of groups inc. control	Animals/ sex/ group	Body weight	Food consumption	Haematology/ clinical chemistry	Gross autopsy	Histopathology	Months from autopsy to report
28 days	To provide definitive information	4	3–4	Weekly	Weekly	Pretrial and at end	All animals	Wide selection of organs on all animals	1–3 mth
90 days	To provide definitive information	4	3–6	Weekly	Weekly	Pretrial, weeks, 6 and 12	All animals	Wide selection of organs on all animals	1–3 mth
26 week or 52 week	To provide definitive information	4	4–6	Weekly	Weekly	Pretrial, weeks, 6, 12, 26, and 52	All animals	Wide selection of organs on all animals	3–6 mth

general, a little more susceptible than males. The lack of precision makes it more difficult to compute a dose–response slope with confidence. However, a great many LD_{50} tests are carried out either for labelling purposes (just for classification of a chemical) or for selection of dose levels for longer studies (increasingly flexible approaches are now being used and LD_{50} studies as dose ranging investigations are relatively rare). In the former instance, dose response is not required and, in the latter instance, it would often be more useful to extend the range of doses selected at the next phase of testing than to increase the precision of the acute study by doubling animal numbers. It is, as ever, important to be quite clear about the object of a study so that the design can produce the most pertinent results from the minimum numbers of animals: Zbinden and Flury-Roversi (1981) discuss this and other topics related to the conduct and interpretation of LD_{50} studies most thoroughly. Alternatives to the LD_{50} are gaining credibility (van den Heuvel, Clark, Fielder, Koundakjian, Oliver, Pelling, Tomlinson, and Walker, 1990; Yam, Reer, and Bruce, 1991).

E   Study Duration

Ideally, a study should be conducted for as long as is required to identify the toxicological events of significance to man. Short term studies of high dosage give valuable indications of the effects of acute poisoning, which may be of value in the treatment of poisoning. Often the exposure of man is over a much longer period and to smaller dosages. The response may then be very different and there is no fixed relationship between the dosage causing an acute effect and that causing a chronic effect (Plaa, 1982). It is therefore necessary to conduct both single and multiple dose studies in animals. The rat and mouse may be dosed for their full lifespan (1.5 to 2.5 years) on the assumption that this will equate to man. The evidence is tenuous but the conclusion convenient.

Two factors are generally relevant to the selection of study duration: (a) convention has dictated a series of arbitrary study durations from acute to chronic, and (b) the shorter studies are required for selection of dose levels before longer studies can be conducted.

Following a single dose (acute) study, sub-acute is the next duration frequently used, *i.e.* 14 or 28 days of daily dosing (or such intervals of dosing as are indicated by the likely use). The 14 day period is rarely mentioned in regulations and is generally used only for screening studies or as a preliminary to longer studies. However, it may be the appropriate duration of study for a material to which man is likely to be exposed very infrequently and for a very short duration, *e.g.* a vaccine or diagnostic aid.

Twenty-eight day studies (4 weeks, 1 month) are mentioned in regulations and are often the longest study carried out for many industrial chemicals with limited production volume and limited numbers of people exposed. While there is nothing to prevent a study being designed with a length between 1 month and 3 months, convention dictates that 90 days, (3 months or 13 weeks) is the next step, often called sub-chronic.

Six months (26 weeks) or 12 months studies, the duration again dictated by convention and regulations, are regarded as chronic toxicity studies. It is unusual for a toxicity study (as distinct from a carcinogenicity study) to be extended beyond 12 months—indeed there are very few compounds that show effects after 6 months or 12 months that do not show evidence of the effect at 3 months (Heywood, 1983; Walker, Schuetz, Schuppan, and Gelzer, 1984). Ocular toxicity, and cataracts in particular, represents one of the few changes requiring study durations to extend beyond 3 months. European countries rarely request toxicity studies longer than 6 months, though Japan and the USA prefer 12 month studies. This is a subject that was debated at the International Conference on Harmonisation (1991). Whilst there was much sympathy for reducing the duration of chronic studies to 6 months, there is, in practice a reluctance from regulators to put this into practice.

The above implies a distinction between carcinogenicity testing and toxicity testing. Without wishing to discuss the justification of that separation, it is true that the jargon of toxicology does draw such a distinction. Carcinogenicity studies are designed with the main object of assessing the carcinogenic potential of a compound and normally last most of an animal's lifespan—in mice that means 18 months to 24 months and in rats 24 months to 30 months. At 18–30 months, rodents are developing a great many lesions associated with old age and therefore it can be very difficult to distinguish between toxic changes and background lesions. This is one reason why toxicity studies of this length are of limited value. There is a 'Catch 22' for carcinogenicity studies: it is assumed that the animals must reach a certain maturity (*i.e.* be approaching death) before a judgement on lack of carcinogenicity can be made. This means that minimal durations of 18 months (mouse) and 24 months (rat) are normally specified. However, various regulatory bodies prefer a high survival at that time and some pathologists (*e.g.* Roe, 1983) complain of the background lesions present. The animal husbandry experts, therefore, improve conditions, lower dietary protein (Tannenbaum, 1940; Ford and Ward, 1983; Ross, 1959), select longer lived strains of animals, and adjust husbandry methods to achieve greater longevity. The regulators then say the survival at 18 months or 24 months is too high so that studies must be lengthened to ensure the animals are near their maximal lifespan, which results in a high incidence of background lesions, and so on. The solution to the dilemma lies in better characterisation of the end points to be identified and for pathologists and toxicologists to define what they want the ageing animal to look like or die of. This will involve an acceptance that not all biological events are of toxicological significance and studies should not be designed on the assumption that they are.

F Study Designs

Preceding paragraphs have alluded to various conventional study designs employed for screening purposes in animal toxicology. Tables 1 and 2 are brief outlines of examples of this approach. The examples are not meant to be exhaustive. Table 3 gives typical lists of haematology, clinical chemistry, urinalysis, and pathology observations and again there are many variations possible. Lumley and Walker (1985), Heywood (1981) and Grandjean, Sandoe, and Kimbrough (1991) provide information on the organs and observations that most frequently show evidence of a toxic response.

4 Randomisation and Replication

It is generally advised that the statistician be involved at the very outset of a study's design but this is often ignored because:
(a) the design may be standard or be dictated by other factors;
(b) the scientist has not defined his objects clearly enough to consult a statistician;
(c) the statistician talks a 'foreign language'; and
(d) the statistician is likely to explain that the study requires twice as many animals to meet its objectives.

There are, however, a few basic principles relating to the avoidance of bias that should be followed, replication and randomisation being amongst them. If animals are not formally randomised (as opposed to haphazardly allocated) to groups, then bias can occur—perhaps due to placing all the heavy animals in one group, all the fast moving animals in another, *etc.* If small numbers of animals are being used it is very possible that, despite formal use of random number tables, *etc.*, the various groups will be obviously different from each other in one or more key parameters.

Two ways round this are to re-allocate animals and/or substitute a few spares for animals already allocated, and thus balance the groups. These solutions are not entirely satisfactory since animals will not have been allocated randomly and the basic assumption behind the statistical tests applied is that random allocation has occurred. Another solution is to measure the critical parameter(s), say body weight, and rank the animals in body weight order or place them in a series of batches each containing a specific range of body weights. The animals may then be randomised by mixing one member of one batch with another member from each other batch such that each treatment group will have similar numbers of animals from each weight range. The means of each treatment group will thus be similar. This is the first stage of 'replication' whereby blocks of animals, or replicates, are identified.

The practicalities should not be forgotten and, if 600 rodents are to be allocated to several different treatment groups, they will probably have to

Table 3 *Typical Lists of Requirements for Laboratory Investigations and Pathology in Toxicology Studies*

HAEMATOLOGY

Erythrocyte count
Haemoglobin content
Packed cell volume
White blood cell count
Differential white cell count
Platelet count

Selected measurements
of clotting function

CLINICAL CHEMISTRY

Blood urea nitogen
Total protein
Albumin
Albumin/Globulin ratio
Aspartate aminotransferase (AST)
Alanine aminotransferase (ALT)
Alkaline phosphatase (AP)
Sodium
Chloride
Potassium
Blood glucose

URINALYSIS

Volume
Appearance
pH
Sediments
S.G.
Blood pigments
Protein
Bilirubin
Ketones
Glucose

PATHOLOGY

Adrenal*
Any abnormal lesions
Aorta
Bone with Marrow (Sternum)
Brain*
Caecum
Colon
Eye
Heart*
Intestine (Small and Large)
Kidney*
Liver*
Lung and Bronchus*
Lymph Nodes (Selected)
Muscle
Oesophagus
Pancreas
Pituitary*
Prostate/Uterus
Salivary Gland
Sciatic Nerve
Skin with Mammary Gland
Spinal Cord
Spleen*
Stomach
Testis/Ovary*
Thymus
Thyroid with Parathyroid
Trachea

*Organs frequently weighed.
These lists can be extended very considerably if justified by the nature of the compound and its likely effects.

be numbered provisionally, randomised, moved to different cages, given new numbers, allocated to groups, and labelled accordingly. One has to be sure that a refined and elaborate randomisation system will not cause confusion in its operation. If it does, a similar but less 'pure' method may be better. Some practical methods of randomisation are described in one of a series of volumes on Standard Operating Procedures in Toxicology (Paget and Thomson, 1979).

Having randomised animals to groups it is important to consider location of cages or pens. Different locations will have different temperatures, different light intensities, different times of dosing, different times of observation, different noise levels, and different airflows. Even in a well designed animal room, the top and bottom of a rack of rat cages can differ in temperature by 5 °C so the method sometimes encountered of putting each dose at a different level in a rack cannot normally be justified. Usually the randomisation process used to allocate animals to groups also serves to place them in specific locations; there is no significant advantage in employing two separate procedures.

For larger rodent studies, using several hundred animals, there is often considerable debate as to whether dose groups should be scattered randomly within each rack of cages or whether one rack should contain only one dose group. The former is better, so long as the random positioning does not cause confusion to the operator and result in administration of the wrong dose to a particular animal or contamination by dust in feeding studies. The latter randomisation method probably presents less chance of a dosing error, although it is arguable that if it occurs, all the animals in at least one group will be misdosed. A compromise is to employ one rack per dose group and to move the racks to different positions in the room at regular intervals. Haseman, Winbush, and O'Donnell (1986) reviewed data from a large number of rodent carcinogenicity studies with dual control groups in order to assess the incidence of false positive results. One of the two main recommendations was to distribute dose groups systematically or randomly.

Similarly, there is debate as to whether or not animals should be dosed in a random order. This is, of course, desirable, but if large numbers are involved, the chances of dosing an animal with the wrong material are much increased. Often a compromise is reached whereby each group is dosed in its entirety but the order of dosing is changed on different days or weeks, *i.e.* the potential operator errors inherent in the better system are deemed greater than the bias introduced by dosing all of one group at one time.

Rodents are often housed about 4 or 5 animals to a cage. If a study requires only 10 animals per group the numbers per cage should normally be reduced otherwise 'cage effects' rather than compound effects might be measured.

There are several other reasons for considering the use of replicates in a study. Replication ensures that fortuitous allocation of animals in an unbalanced manner should not occur, *e.g.* one could, by chance, have all of one dose group placed in adjacent cages when in fact one had hoped to distribute them fairly evenly. In large studies it is not always possible to carry out all of a day's procedures within a single 24 hour period, *e.g.* taking and examining blood samples from 150 animals or carrying out a detailed autopsy on 500 rats. If replicates are used then a suitable number of blocks can be observed, bled, or autopsied each day, whilst ensuring that a balanced number of each group is present and in a random order within each block. The second of the recommendations by Haseman *et al.*

(1986) addressed pathology and included the suggestion that histopathological evaluation should contain an element of evaluation in which the pathologist is unaware of the group from which the slide derived. This emphasises the importance of evaluation being a random order and 'blind'. That is, 'blind' in relation to the group, not necessarily in relation to the lesion in question. Aspects of experimental design such as replication, factorial arrangement of treatments, and strain variability are reviewed by Festing (1992).

Some experiments have particular problems. Inhalation studies often require each dose level to be in a separate inhalation chamber. If there are environmental differences between the chambers, these could confuse the results. For instance, efforts are often made to generate atmospheres of test material with very clean, oil free, dry air and diluting this concentrate with room air. In high dose groups, the dilution is likely to be much less and the relative humidity in that group's chamber will be lower. In extreme circumstances this can cause ringtail in rats. The scientist thus has to strive to take account of the potential differences and attempt to minimise them. In this respect, it is important to house or restrain control animals in the same manner as the test groups if observations much beyond gross clinical signs are to be measured.

Normally, the range of doses from low to high is within a span of about 10 fold and it is most exceptional for the range to be greater than 100 fold. If this range is very wide (*e.g.* 1000 fold) there is a real possibility of contamination (from vapour, dust, or compound in excreta) of control groups from high dose groups to a biologically significant extent. If the dose range is within the more usual limits then the contamination is likely to be well below levels of any consequence, unless the compound is in the diet and is exceptionally volatile or dusty. Under such circumstances each dose group can be held together and separated from the other groups, the errors likely to be induced by contamination being greater than the bias likely to arise from lack of positional randomisation.

Multigeneration reproduction studies and similar studies illustrate other difficulties. If there is a slight chance effect that alters one factor at an early stage, the knock-on effects can continue throughout the study. Thus a reduction in litter size could increase weight gain in the young, with possible effects on age of reaching sexual maturity and effects on haematology, clinical chemistry, and organ weights. An example of differences in F_1 body weights causing confusion in a carcinogenicity study is given in Olsen, Wurtzen, and Meyer (1983). The statistician will feel uncomfortable because F_1 and succeeding generations will not have been allocated randomly to their dose groups but will have been allocated according to their parents' dose group.

Teratology study results are handled in two ways: (a) using the litter as the unit, *i.e.* comparing the number of litters containing abnormal young, or (b) using the foetus as the unit, *i.e.* comparing the number of abnormal foetuses. Both have validity and, in general, both approaches are used in the evaluation of results in any one study.

5 Conclusion

It should be remembered that the toxicological investigation of a compound rarely rests upon one study on its own. Although each experiment has to stand alone, the assessment of toxicity will mean an overview of the full range of studies: acute and chronic, rodent and non-rodent, *in vitro* work and human experience, metabolism and literature searching, reproductive effects and pharmacokinetics.

Regulatory bodies have a dominant influence upon the design of toxicological investigations for new and existing compounds. It is pleasing to note that there is now a very high degree of accord between different countries in the studies and study designs they require. This has almost eliminated repetition of animal work for idiosyncratic rather than scientific reasons. However, achievement of improvement at any speed is likely to compromise this hard earned uniformity and interesting conflicts will thus arise. It is possible that the novel challenges presented by the toxicological evaluation of peptides and proteins resulting from biotechnology will be a catalyst for a constructive evaluation (Petricciani, 1983; Lasagna, 1986, and Wilson, 1987).

In the context of toxicology, the experimenter would do well to consult the papers of Brusick (1987), Oser (1987), and Zbinden (1991) before making judgement as to whether toxicological screening studies as currently conducted can match the objectives they set out to achieve and whether substantial improvement is necessary.

References

Alder, S. and Zbinden, G. (1983). Neurobehavioural tests in single and repeat dose toxicity studies in small rodents. *Arch. Toxicol.,* **54,** 1–29.

Brusick, D. (1987). Current approaches to toxicity testing—status and prospects. *Comments Toxicol.,* **1,** 257–264.

Cardy, R. H. and Warner, J. W. (1979). Effect of sequential bleeding on body weight gain in rats. *Lab. Anim. Sci.,* **29,** 179–181.

Chanter, D. O., Tuck, M. A., and Coombs, D. W. (1987). The chances of false negative results in conventional toxicology studies with rats. *Toxicology,* **43,** 65–74.

Clark, D. G. (1977). Long-term inhalation toxicology studies. *In:* 'Current Approaches in Toxicology', *Ed.* B. Ballantyne. John Wright and Sons, Bristol, pp. 105–114.

Feron, V. J. and Kroes, R. (1986). The long-term study in rodents for identifying carcinogens: some controversies and suggestions for improvements. *J. Appl. Toxicol.,* **6,** 307–311.

Festing, M. F. W. (1992). The scope for improving the design of laboratory animal experiments. *Lab. Anim.,* **26,** 256–267.

Food and Drug Administration Advisory Committee on Protocols for Safety Evaluation: Panel on Reproductive Report on Reproductive Studies in the Safety and Evaluation of Food Additives and Pesticide Residues (1970). *Toxicol. Appl. Pharmacol.,* **16,** 264–296.

Ford, D. J. and Ward, R. J. (1983). The effect on rats of practical diets containing different protein and energy levels. *Lab. Anim.*, **17**, 330–385.

Grandjean, P., Sandoe, S. H., and Kimbrough, R. D. (1991). Non-specificity of clinical signs and symptoms caused by environmental chemicals. *Hum. Exp. Toxicol.*, **10**, 167–173.

Haseman, J. K. (1986). Issues in carcinogenicity testing: dose selection. *Fundam. Appl. Toxicol.*, **5**, 66–78.

Haseman, J. K., Winbush, J. S., and O'Donnell, M. W. (1986). Use of dual control groups to estimate false positive rates in laboratory animals carcinogenicity studies. *Fundam. Appl. Toxicol.*, **7**, 573–584.

Heywood, R. (1981). Target organ toxicity. *Toxicol. Lett.*, **8**, 349–358.

Heywood, R. (1983). *In*: 'Animal and Alternatives in Toxicity Testing.' *Eds* M. Balls, R. J. Riddell, and A. N. Worden. Academic Press, London, pp. 79–93.

Homburger, F. (1987). The necessity of animal studies in routine toxicology. *Comments Toxicol.*, **1**, 245–255.

Johnson, E. M. (1988). Cross-species extrapolations and the biological basis for safety factor determinations in developmental toxicology. *Regulat. Toxicol. Pharmacol.*, **8**, 22–36.

Lasagna, L. (1986). Clinical testing of products prepared by biotechnology. *Regulat. Toxicol. Pharmacol.*, **6**, 385–390.

Lumley, C. E. and Walker, S. R. (1985). The value of chronic animal toxicology studies of pharmaceutical compounds: a retrospective analysis. *Fundam. Appl. Toxicol.*, **5**, 1007–1024.

Mackay, J. M. and Elliott, B. M. (1992). Dose ranging and dose setting for *in vivo* genetic toxicology studies. *Mutat. Res.*, **271**, 97–99.

Muller, H. and Kley, H. P. (1982). Retrospective study on the reliability of an 'Approximate LD_{50}' determined with a small number of animals. *Arch. Toxicol.*, **51**, 189–196.

Olsen, P., Wurtzen, G., and Meyer D. (1983). BHT, Long-term toxicity. *Toxicol. Forum*, Geneva, Oct 18–22, 1983.

Oser, B. L. (1987). Toxicology then and now. *Regulat. Toxicol. Pharmacol.*, **7**, 427–443.

Paget, G. E. and Thomson, R. (1979). 'Standard Operating Procedures in Toxicology', Vol I., MTP Press, Lancaster.

Petricciani, J. C. (1983). An overview of safety and regulatory aspects of the new biotechnology. *Regulat. Toxicol. Pharmacol.*, **3**, 428–433.

Plaa, G. L. (1982). Principles of toxicology. *In*: 'Survey of Contemporary Toxicology', Vol 2, *Ed.* A. T. Tu. John Wiley, New York, pp. 203–225.

Roe, F. J. C. (1983). Carcinogenicity testing. *In*: 'Animals and Alternatives in Toxicity Testing'. *Eds* M. Balls, R. J. Riddell, and A. N. Worden. Academic Press, London, pp. 127–130.

Ross, M. H. (1959). Protein, calories, and life expectancy. *Fed. Proc.*, **18**, 1190–1207.

Tannenbaum, A. (1940). The initiation of the growth of tumours. 1. Effects of underfeeding. *Am. J. Cancer*, **38**, 335–350.

Van den Heuvel, M. J., Clark, D. G., Fielder, R. J., Koundakjian, P. P., Oliver, G. J. A., Pelling, D., Tomlinson, N. J., and Walker, A. P. (1990). The international validation of a Fixed-dose Procedure as an alternative to the classical LD_{50} test. *Food Chem. Toxicol.*, **28**, 469–482.

Walker, S. G., Schuetz, E., Schuppan, D., and Gelzer, J. (1984). A comparative retrospective analysis of data from short- and long-term animal toxicity studies on 40 pharmaceutical compounds. *Arch. Toxicol.*, Suppl 7, 485–487.

Wexler, P. (1987). 'Information Resources in Toxicology', 2nd edn. Elsevier, New York.

Wilson, A. B. (1987). The toxicology of the end products from biotechnology Processes. *Arch. Toxicol.*, Suppl 11, 194–197.

Yam, J., Reer, P. J., and Bruce, R. D. (1991). Comparison of the up-and-down method and the Fixed-dose Procedure for acute oral toxicity testing. *Food Chem. Toxicol.*, **29**, 259–263.

Zbinden, G. and Flury-Roversi, M. (1981). Significance of the LD_{50} Test for the toxicological evaluation of chemical substances. *Arch. Toxicol.*, **47**, 77–99.

Zbinden, G. (1991). Predictive value of animal studies in toxicology. *Regulat. Toxicol. Pharmacol.*, **14**, 167–177.

CHAPTER 5

The Biochemical Principles of Toxicology

W. N. ALDRIDGE

1 Introduction

Toxicology is the science of the action of substances which when introduced into an organism cause change in its structure and/or function. In this chapter, the substances discussed are restricted to low molecular weight chemicals (proteins, *etc.*, are excluded) and to mammals or their constituent parts.

Toxicology for its scientific development requires the integration of information acquired by the use of techniques used in many different scientific disciplines (see Figure 3). To speak of the 'biochemical principles of toxicology' is thus in some way a contradiction. Toxicological problems can be tackled by applying biochemical techniques or biochemical hypotheses to many biological systems. Advances in toxicology result as the information is shown to be relevant to toxicity in whole animals.

This chapter, therefore, is concerned with those hypotheses (principles) which have evolved from research aiming to solve questions concerning the molecular basis for toxicity. It should be self-evident that the action of a chemical on a biological system ought to be explainable in chemical terms. That it is not always so is a reflection that many areas of normal biology are not yet understood in molecular terms (Aldridge, 1980, 1981).

2 Types of Chemical Interaction

The receptor theory in pharmacology is now supported by the isolation of receptor proteins. In toxicology there are primary reactions with macromolecules which initiate a toxic reaction; these are sometimes called targets and may be proteins, nucleic acids, components of membranes, enzymes, *etc.* The reactions between a toxic chemical and macromolecules are chemical reactions and can be examined by the normal methods of chemistry.

Reactions can be classified into two categories: reversible or covalent interactions:

$$M + AB \underset{k_{-1}}{\overset{k_{+1}}{\rightleftharpoons}} M \cdot AB \tag{1}$$

$$M + AB \xrightarrow{k_2} B + MA \dashrightarrow \tag{2}$$

$$M = macromolecule \qquad AB = toxic\ chemical$$

These two types of reaction govern dose–response relationships at the target. The kinetic behaviour expected depends theoretically on a knowledge of the concentration of both reactants. As will become clear, these concentration terms are influenced by many factors.

A   Reversible Reactions

The relationship between binding and concentration of reactant in reaction (1) is shown by:

$$K_{aff} = \frac{k_{+1}}{k_{-1}} = \frac{[M \cdot AB]}{[M][A]} \tag{3}$$

The affinity constant (K_{aff}) defines the concentration for 50% reaction to take place. The relationship between percentage binding and concentration when K_{aff} is 10^5 M^{-1} is shown in Figure 1A under the common condition when the initial concentration of the chemical ($[AB]_0$) is much greater than that of the macromolecule ($[M]_0$). From this relationship and for macromolecules where there may be more than one binding site several graphical methods have been derived for handling experimental data. The Scatchard (1949) method is perhaps the most common and is based on the following equation:

$$\frac{B}{F} = nK_{aff} - BK_{aff} \tag{4}$$

where
B = bound chemical expressed in various concentration terms such as mol mol^{-1}, mol mg^{-1} protein, *etc.*
F = free (unbound) chemical expressed as a concentration.
n = the maximum concentration of binding sites of affinity, *K,* and expressed as n mol^{-1}, nmol mg^{-1} protein, *etc.*

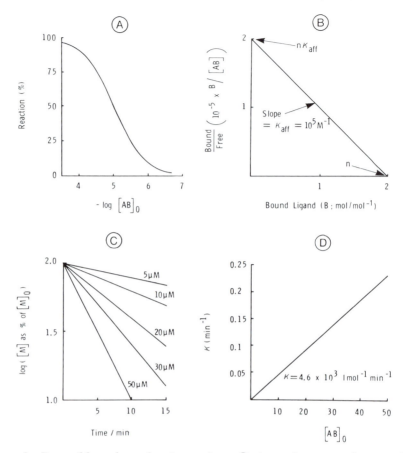

Figure 1 *Reversible and covalent interactions.* Ⓐ *shows the extent of a reversible reaction when* $[AB]_0 \gg [M]_0$ *(in text, Equation 3) and* K_{aff} *is* $10^5 \, M^{-1}$.

Ⓑ *is a Scatchard plot for a reversible reaction of* AB *with a protein containing two binding sites with a* K_{aff} *of* $10^5 \, M^{-1}$ *(in text, Equation 3).*

Ⓒ *and* Ⓓ *represent the rate of a covalent reaction when* $[AB]_0 \gg [M]_0$ *and the second order rate constant* k *is* $4.6 \times 10^3 \, l \, mol^{-1} min^{-1}$ *(in text, Equation 5).* Ⓒ *is the rate of reaction for increasing* [AB] *and* Ⓓ *is the relationship between the first order rate constants (k derived from* Ⓒ*) and* [AB].

A simple case where there is one class of binding site (2 sites with the same affinity for the ligand) on a macromolecule is shown in Figure 1B.

For many situations the degree of binding of a toxic chemical cannot be measured directly. Changes in biochemical and/or physiological parameters are often the only quantitative measures of binding to a target. In the worst case the concentration of chemical, [A] related to the binding to the target has to be assumed to be:

$$[A] = f \, [dose \, administered]$$

The exact mathematical relationship is always complicated since in whole animal studies, steady state conditions are rarely attained; only if input equals the output of the chemical is this so.

The most important feature of reversible interactions is that the substance AB is not changed chemically by the interaction; if the concentration of AB is reduced by excretion and/or metabolism the reaction is reversed and $M \cdot AB$ dissociates by the same route (k_{-1}) as it was formed (k_{+1}).

For a more extensive discussion of reversible interactions see Goldstein, Aronow, and Kalman (1974).

B Covalent Interactions

The covalent interaction shown in Equation (2) is progressive and will be defined by the second order kinetics of a biomolecular reaction.

It is often the case, as for reversible interactions, that $[AB_0] \gg [M_0]$ and under these conditions the second order equation simplifies to

$$k_2 = \frac{1}{[AB_0]t} \ln \frac{[M_0]}{[M]} \tag{5}$$

in which k_2 is the bimolecular rate constant (l mol^{-1}), $[AB_0]$ and $[M_0]$ are the initial concentrations of AB and M, and t is the time. The relationships between extent and rate of reaction to time and concentration of AB is shown in Figure 1C and D.

It is obvious that under this condition $([AB_0] \gg [M_0])$, and if the time is prolonged, then almost all of M will become MA (Equation 2). This condition may apply in *in vitro* systems but *in vivo* there are often processes which remove AB.

An important distinction between reaction (2) and the reversible reaction (1) is that when one molecule of M reacts with one molecule of AB to form MA, a molecule of AB is destroyed. Formation of M from MA is almost always slow and is not by a reversal of reaction (2) but by an entirely different reaction.

For a more extensive discussion of covalent interactions see Aldridge and Reiner (1971).

3 Types of Toxicity

Definitions of acute and chronic toxicity should be stated in a form which reflects what we know of molecular mechanisms and distinguished from operational criteria, such as the mode of administration.

A Acute Toxicity

Acute toxicity is that toxicity which arises soon after administration and, unless death occurs, recovery is complete. Examples are the toxicity of cyanides or organophosphorus compounds. The interaction of cyanide with cytochrome oxidase is dissociable and recovery is rapid as cyanide is detoxified or excreted. Organophosphorus compounds react covalently with acetylcholinesterase. Recovery may be complete but is slow, since in many cases it depends on resynthesis of new enzyme.

B   Chronic Toxicity

Chronic toxicity may be divided into two main types
(a) that which requires prolonged or repeated administration of the compound before the toxicity becomes apparent; and
(b) that which is long lasting but can be brought about by one or very few doses of the chemical.

Prolonged Administration of the Chemical

Chronic toxicity, *i.e.* that toxicity which may take a long time to appear, can be subdivided into several types. Repeated or continuous dosing of a substance may be required to increase its concentration or a metabolite of it at its site of action. Cadmium is a good example, where over several years its concentration may increase in the kidney cortex up to 300 μg g^{-1} and thus cause functional deficiency (see page 72). Cadmium will produce nephropathy after an acute dose but since it is highly cumulative in mammals, it can produce toxicity to the kidney after exposure to low concentrations for a long time.

In many instances of the above type of chronic toxicity, the tissue repairs and function may become, if not normal, then adequate when the exposure to the chemical ceases. However, if the tissue contains cells which do not undergo mitotic division then a chemical insult which causes death of these cells will lead to a permanent deficit. The consequences may not be severe. Death of striated muscle cells, for example, may not be very important, since, although they will not be replaced, those remaining will hypertrophy so as to maintain normal function. If the cells dying are neurons then a permanent deficit remains for the rest of the animal's life (see page 64).

Repeated administration of a chemical may cause necrosis of cells, *e.g.* carbon tetrachloride causes necrosis of hepatocytes. The remaining hepatocytes divide so as to replace those damaged but, in addition, the systems concerned with the structural elements in the liver are activated and lead to the laying down of an abnormal proportion of collagen (liver cirrhosis; Cameron and Karunaratne, 1936). The different molecular forms of collagen (types I, III, and β) are all increased in concentration (Rojkind and Martinez-Palomo, 1976; Biempica *et al.*, 1980; Seyer, 1980). In this way the functional capacity of the liver is permanently reduced. For example, dimethylnitrosamine which, besides being a potent carcinogen (Magee and Barnes, 1956) also causes centrilobular necrosis in the liver, was first suspected of being a toxic compound because of the diagnosis of liver cirrhosis in men working in a pilot plant for its manufacture (Barnes and Magee, 1954). Another example is the pulmonary fibrosis diagnosed in some coal miners after many years exposure to coal dust.

One or Few Doses of the Chemical

Chronic toxicity may be a long lasting biological effect produced by one or very few doses of unstable compounds. The chemical, although present for a very short time, initiates a biological 'cascade' leading to biological change sometime later. Dimethylnitrosamine given as a single oral dose to rats on a particular dietary regime produces tumours in the kidneys of all the rats many months later (Swan and McLean, 1968). Delayed neuropathy produced by organophosphorus compounds is another example (see page 70).

4 Physical and Chemical Properties of Toxic Chemicals

The essence of organised biological systems is the integrated structure in which various functions are exercised by specific organs; within organs chemical reactions occur in specific compartments. Their functions and biochemistry involved could not be smoothly organised without a multitude of compartments surrounded by membranes. While these membranes are essential for the orderly function of the organism, they also provide barriers to extraneous chemicals. The properties conducive for the penetration of membranes containing protein and lipid layers have been well worked out. Rapid penetration of chemicals is best achieved when the substance has both lipophilic and hydrophilic properties. Two substances serve to illustrate this point. Dimethylnitrosamine, a potent carcinogen, is soluble in lipid and water and on oral administration is rapidly absorbed from the gastrointestinal tract and mixes with body water (Magee, 1956). O,S,S-Trimethyl phosphorodithioate, which causes death due to lung damage, has the same physical properties, is very rapidly absorbed after oral dosing, and mixes so rapidly with the body water that the first passage through the liver where it is metabolised does not appreciably influence its final concentration in the blood and tissue. There is no doubt that both of these substances will be taken up rapidly by other routes, *e.g.* skin, lungs, mouth, *etc.* (Aldridge, Verschoyle, and Peal, 1984).

For substances either more lipophilic or more hydrophilic, constraints on the rate of absorption are found. For example, it is difficult, if not impossible, to administer by the oral route a toxic dose of the lipophilic DDT as a water suspension of finely divided compound; it is, however, absorbed rapidly if given as a solution in vegetable oil. DDT is soluble in oil but has a very low solubility in water.

Negatively charged substances are not absorbed well but may be absorbed from the acid conditions in the stomach where sufficient of the compound is in the uncharged form, *e.g.* the uncoupler, hexachlorophane (pK of hydroxyl groups 5.4 and 10.85; Mahler, 1954). Positively charged species are not well absorbed. Nevertheless, none of these rules are absolute and the fact that compounds such as paraquat are absorbed from the gut may well be due to specialised transport systems (Figure 2).

After substances have been absorbed, there are many membranous impediments to their penetration into different tissues. For example, substances which are positively charged do not easily penetrate the central nervous system (CNS), *e.g.* paraquat does not penetrate cells in the central nervous system and other tissues. It is toxic to the lung because Type 1 and Type 2 pneumocytes possess a transport system for putrescine and this transport system can be utilised by paraquat (Smith and Wyatt, 1981). From work on tissue slices it appears that paraquat is actively taken up by cells of the brain. However, in the intact animal these cells are protected because paraquat does not penetrate the blood brain barrier.

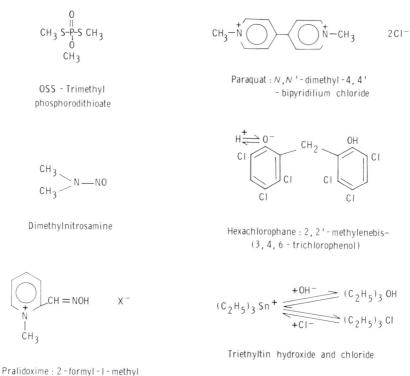

Figure 2 *Structural formulae of chemicals showing differing penetration of animals, organs, and cells*

The positively charged organophosphorus compounds produced by storage of metasystox in water (Heath and Vandekar, 1957) do not penetrate the CNS as shown by lack of inhibition of acetylcholinesterase. Death occurs due to inhibition of this enzyme in the peripheral nervous system. The therapeutic agent, pralidoxime, which is used for the treatment of poisoning by organophosphorus compounds and which reactivates the phosphorylated acetylcholinesterase, is ineffective for the enzyme in the brain because it does not penetrate the central nervous system.

Organotins, such as triethyltin sulphate, under mildly acid conditions, are ionised into the positively charged triethyltin ion. Nevertheless it is rapidly absorbed on oral administration because the hydroxide and chloride (K_{aff} for hydroxyl ion: $2 \times 10^7 \, M^{-1}$ and for chloride: approximately $10^1 \, m^{-1}$; Aldridge, Street, and Skilleter, 1977) are lipid and water soluble.

The way properties of related compounds influence their distribution is well shown by mercury and the organomercurials. Mercuric ions (Hg^{2+}) do not penetrate the nervous system but concentrate in and damage the kidney. Methyl- and phenylmercuric salts, however, are lipid soluble and do penetrate the central nervous system. Although methylmercuric salts are much more stable than phenylmercuric salts, and while methylmercury salts are toxic to the nervous system, phenylmercuric salts, by virtue of the mercuric ions released by its metabolism, are toxic to the kidney.

Metallic mercury is lipophilic. Its penetration of the nervous system is dependence on the supply of blood to the brain and its rate of oxidation in the blood to mercuric ion (Magos, 1981; 1982).

Thus, there are physico-chemical properties of chemicals which facilitate their penetration into an organism and its constituent cells. Selective toxicity, however, may also be due to special uptake mechanisms which are there to transport an essential substance into cells but which can also accommodate the foreign chemical.

For a more extensive discussion of the influence of physico-chemical properties on toxicity see Albert (1985).

5 Selective Toxicity and Selectivity

The term selective toxicity is often used to indicate a differing toxicity to different organisms or species (Albert, 1985). Thus, penicillin is selectively toxic to certain bacteria and has a low toxicity to mammals. Malathion is toxic to many insects and yet has very low toxicity to mammals.

Within an animal, the damage a chemical induces is often predominantly to one organ. Beryllium compounds, for example, produce necrosis of the liver, less damage in the spleen and kidney, and no changes in other organs. This is because, although beryllium is toxic to

other organs, it does not enter the cells. Similarly, paraquat damages the
lung because certain cell types concentrate it—paraquat is toxic to all
mammalian cells but is concentrated in very few. Trimethyltin is
selectively toxic to the brain of mammals, primarily in the hippocampus,
with less damage in the amygdaloid nucleus and pyriform cortex (Brown,
Aldridge, Street, and Verschoyle, 1979). In this case, the selectivity is

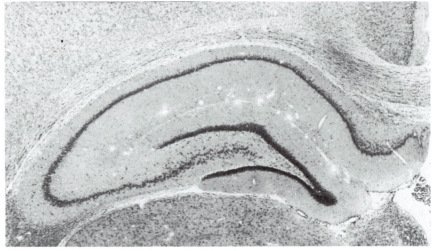

Figure 3A *Neuronal necrosis in the rat hippocampus produced by trimethyltin.
Control rat hippocampus Haematoxylin and eosin × 42*

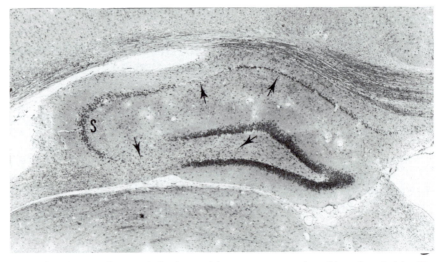

Figure 3 *Neuronal necrosis in the rat hippocampus produced by trimethyltin. Ten
weeks after a single dose of trimethyltin chloride (10 mg kg^{-1}). Shrunken hippocampus with marked neuronal loss in the pyramidal cell band (arrows), less
damage in the fascia dentata and portion of the Sommer sector (S). Haematoxylin
and eosin × 46*

even greater for it is only neurons in particular locations in the hippocampus which are affected (Figure 3). The evidence currently available suggests that this is not due to selective accumulation of trimethyltin in these areas of the brain.

Thus, selective toxicity is mediated by changes brought about in only one organ and often by certain cells in that organ. As explanations are sought for such selective action, an affinity of the chemical for certain subcellular organelles, enzyme system, or membranes may be found; once the target has been identified this selectivity at the molecular level may allow the structure–activity relationships for the target to be disentangled from that for the whole animal. For a discussion of some of the principles and definitions of selective toxicity, see Aldridge (1987).

6 Phases of Developing Toxicity

It is convenient to consider toxicity as four interconnecting phases (Figure 4). They are, in order: (1) all systems which affect the delivery of the chemical to its site(s) of action; (2) its primary reaction with macromolecular targets; (3) the cascade of biological (biochemical, physiological, morphological) responses resulting from reaction with the primary targets; and (4) the final clinical signs, symptoms, or syndrome in animals or in man. This scheme is highly simplified and many examples can be found where more elaboration of the diagram might be desirable.

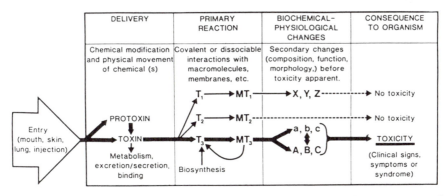

Figure 4 *Scheme illustrating phases of developing toxicity following entry of chemicals into an organism. Heavy lines and letters lead to toxicity whereas light lines are events which are unproductive.* T_1, T_2, T_3 *and* MT_1, MT_2, MT_3 *are targets before and after modification by the chemical, respectively, and* $A, B, C, X, Y, Z, a, b, c$ *represent an undefined and sometimes unconnected number of changes which may or may not be sequential. Acute and chronic toxicity may be differentiated by the time in one or several of the phases. Recovery from changes in function or morphology can occur by regeneration of* T_3 *and by disposal and repair of modified tissue. Prophylaxis or therapy is theoretically possible in several of the phases.*
(Reproduced by permission from Aldridge, 1981, *Trends Pharmacol. Sci.*, **2,** 228–231).

It does, however, provide a framework for thought and perhaps to suggest testable hypotheses for experimental studies.

A Delivery

The properties of chemicals which allow their ready absorption through the gastric mucosa, lungs, intestine, skin, mouth, *etc.* have been previously discussed (p. 61). Once a chemical enters the circulation its physical properties determine its penetration of cells. A major determinant of the toxicity of the compound is the rate at which it is metabolised.

A discussion of the action of Malathion in the rat will illustrate the principles involved. Malathion is a very effective organophosphorus insecticide and when pure has an oral LD_{50} to rats of 10 g kg^{-1} (Figure 5). The rats die with symptoms caused by inhibition of acetylcholinesterase (salivation, muscular fasciculation, red tears, respiratory distress). Malathion does not inhibit acetylcholinesterase *in vitro*. The thionophosphorus grouping (P=S) is oxidised to the oxon (P=O) predominantly by the liver. The product, Malaoxon, is a direct inhibitor of acetylcholinesterase and has an oral toxicity of approximately 160 mg kg^{-1}. The major reason for the low toxicity of Malathion is that in the liver there is a carboxylesterase which removes predominantly the α-ethyl group to produce a non-toxic ionised compound (Figure 5). This system is highly effective since when this carboxylesterase is inhibited by certain other organophosphorus compounds, thereby preventing the detoxification of malathion, its toxic dose will be as low as 20 mg kg^{-1}. The rat becomes as sensitive to malathion as insects which do not possess this detoxification pathway. These findings indicate that from an oral dose only about 1 molecule of malathion in 500 is converted to malaoxon to cause toxicity (for discussion and references, see Aldridge, Miles, Mount, and Vers-

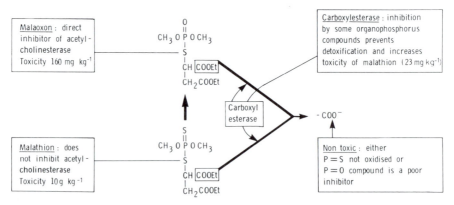

Figure 5 *Pathways of metabolism of malathion which influence its toxicity and biological activity*

choyle, 1979). This example illustrates the two general consequences—
that the toxicity of a chemical may be reduced or may be increased by
metabolism.

Detoxification by Metabolism

This area of biochemistry has been well studied for many years. There
are two classes of metabolism, Phase 1 and 2. Phase 1 are mainly
oxidative, reductive, or hydrolytic processes and produce substrates for
Phase 2 conjugations.

Oxidative Phase 1 reactions are usually carried out by the cytochrome
P-450 mono-oxygenase system of enzymes. Recent work indicates that
this is a family of enzymes (proteins) which can be shown by analytical
and immunological techniques to be different proteins. They are unusual
in that, unlike many other enzymes, they accept a wide range of
substrates of differing chemical structures. The enzymes in many tissues
are located in the microsomal fraction (endoplasmic reticulum) of
lipoprotein membranes in which the cytochrome P-450 is incorporated. In
general, lipophilic chemicals are accepted substrates and their oxidation
yields compounds which are more water soluble. Microsomal oxidations
include aromatic, aliphatic, alicyclic, and heterocyclic hydroxylations,
and the removal of alkyl groups attached to nitrogen or sulphur. The final
products of many of these reactions are suitable for conjugation (Phase 2)
with glucuronic acid, sulphate, glutathione, and glycine. These products
are water-soluble and are either excreted in urine or are further
metabolised to equally water-soluble compounds prior to excretion.
Other Phase 1 reactions include amine oxidase and hydrolytic reactions
catalysed by esterases, amidases, *etc.*

Although the foregoing is a short appraisal of a large field, it serves to
illustrate the principles involved. There are in most animals many systems
which serve to protect them from the deleterious actions of chemicals by
converting them rapidly to less toxic substances, followed by excretion.
Although these systems exist in all mammals, their efficacy against a
particular compound varies between different species. If large differences
in toxicity are found for a chemical in several species of mammal,
provided absorption occurs, the first hypothesis should be that the
compound is detoxified at different rates. Thus, for example, the oral
LD_{50}s (mg kg^{-1} body weight) for chlorfenvinphos [O,O-diethyl-O-2-
chloro-1-(2,4-dichlorophenyl)vinyl phosphate] are for the rat 10, mouse
100, rabbit 500, and dog 12,000. These differences are mainly accounted
for by differences in the rate of removal of one ethyl group by
microsomal oxidation (Donninger, 1971).

Intoxification by Products of Metabolism

Many examples are now known of increases in toxicity of chemicals due
to metabolism. As the chemical mechanism of Phase 1 has been

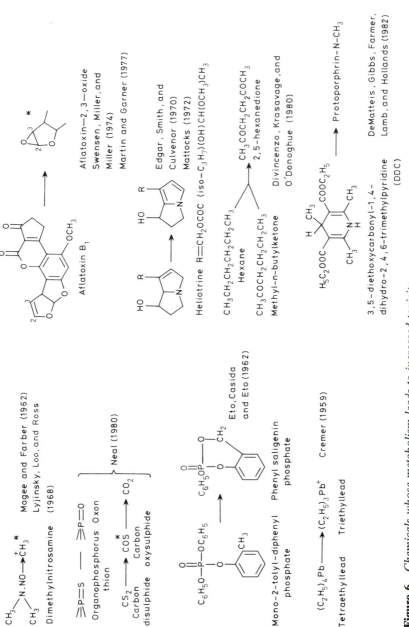

Figure 6 *Chemicals whose metabolism leads to increased toxicity*
* *These structures have not been isolated*

elucidated it is clear that, although the major throughput of products is often towards detoxification and excretion, intermediates are chemically reactive and toxic. The enzymes normally present in animals accept chemicals as substrates and metabolise them if they fit into the right conformation in the catalytic centre. It must be recognised also that the concentration or amount of reactive chemical necessary to cause toxicity if produced *in vivo* may be very small. Thus it may be calculated that out of 2 g (6 mmol) Malathion given to a 200 g rat (LD_{50}; 10 g kg^{-1}) only approximately 100 pmol of the active metabolite malaoxon will phosphorylate the acetylcholinesterase in the brain, *i.e.* approximately $2 \times 10^{-4}\%$ of the dose administered.

In examples of a metabolism which increases the toxicity of compounds (Figure 6), the site of toxicity depends on the stability of the metabolite and whether it can escape from the organ where it was produced. Dimethylnitrosamine is metabolised in the liver and causes liver necrosis. Vinyl chloride is metabolised in the liver and causes angiosarcoma in the liver. Carbon disulphide is metabolised in the liver and causes hydropic degeneration and destruction of the cytochrome P-450 system in the liver. Tetraethyllead is metabolised in the liver and causes lesions in the brain. Organophosphorus insecticides containing thiono phosphorus groups (P=S) are oxidised to the oxon form (P=O) and are toxic by inhibiting acetylcholinesterase in the nervous system. Diphenyl-2-tolylphosphate is metabolised in the liver and causes delayed neuropathy due to 'dying back' of long axons in the central and peripheral nervous system.

Although glutathione is a powerful protectant of cells from attack by electrophilic chemicals, reactive oxygen species, *etc.*, there are now examples of glutathione adducts which are themselves toxic. If potential reactive groupings are retained in the adduct then further metabolism may yield products which are selectively toxic. Thus, hexachloro-1,3-butadiene reacts with glutathione in the liver (catalysed by glutathione *S*-transferase), is excreted *via* the bile, metabolised in the gut, reabsorbed and taken up by the kidney, and metabolised by β-lyase to give a reactive compound which causes toxicity in the tubular elements of the kidney (Lock and Ishmael, 1979). For reviews in this rapidly expanding field see Dekant *et al.* (1992) and van Bladeren (1988).

For a recent review of biologically active metabolites, see Gillette (1982); Anders (1985).

B Primary Reaction with Target

The identification of the primary target is an important task achieved in only a few instances. Two examples of reversible interactions and one example of covalent interaction illustrate the principles involved.

Reversible Interaction

Cyanide gains entry as the uncharged volatile hydrocyanic acid *via* the lung or as the acid formed from salts such as sodium cyanide in the stomach. Absorption and distribution is rapid. The animal dies in convulsions with the oxygen tension in venous blood almost as high as in arterial blood. These clinical signs are caused by the inhibition of cytochrome oxidase in mitochondria, the terminal step in the electron transport chain for the utilisation of oxygen for the oxidation of substrates. The affinity of cyanide for cytochrome oxidase is $3 \times 10^5 \, M^{-1}$ (Wainio and Greenlees, 1960). The circulating concentration of cyanide required to cause death is approximately $5 \, \mu g \, ml^{-1}$ $(1.8 \times 10^{-4} \, m)$ (Aldridge and Lovatt Evans, 1946). Assuming a uniform distribution of cyanide in brain and plasma, this concentration would produce 98% inhibition of cytochrome oxidase. The interaction of cyanide with cytochrome oxidase is reversible and, if death does not occur, recovery is rapid due to the excretion of HCN by the lung and, more importantly, by its conversion in the liver to thiocyanate by the enzyme rhodanase.

Carbon monoxide is absorbed from the lung rapidly and binds to haemoglobin. The affinity of carbon monoxide for haemoglobin is 250 times that of oxygen so that exposure to 0.08% CO in air (20% O_2) will yield a 50:50 ratio $CO:O_2$ bound to the protein. The oxygen in haemoglobin is replaced by carbon monoxide and, when 50% saturation is exceeded, death may occur. Comparison of men whose haemoglobin is half-saturated with carbon monoxide (50% CO + 50% O_2) with those with 50% oxyhaemoglobin (either through haemorrhage or reduced oxygen tension) is instructive. The latter are little affected but the former are near collapse. The explanation of this lies with the properties of haemoglobin. Four molecules of oxygen combine with the four sub-units in haemoglobin. As each molecule of oxygen combines with each sub-unit, it induces an allosteric change in the remaining sub-unit so that the affinity of successive sites for oxygen are increased. Thus, under normal conditions, rapid saturation of haemoglobin with oxygen during passage through the lung is ensured. For the release of oxygen during circulation through the tissues, the reverse process takes place. Release of oxygen bound to one sub-unit with low affinity leads to the reverse allosteric change so that the affinity of one of the remaining oxygens is reduced, and so on. Carbon monoxide binds tightly to haemoglobin and also mimics oxygen, increasing the affinity of the remaining sites for oxygen. Thus, with two sites occupied by carbon monoxide (50% saturation), the oxygen on to the remaining two sites is bound so tightly that it cannot be released to the tissues (Roughton and Root, 1944). The high affinity of carbon monoxide also explains why recovery by its exhalation is so slow.

In these two examples, HCN and CO, the capacity of the tissues to utilise oxygen is impaired, but brought about by different mechanisms. Symptoms mainly originate from the central nervous system, even though in only one case (HCN) is a primary biochemical interaction present in the tissue affected. The onset of, and recovery from, symptoms is dominated by the pharmacokinetics and/or metabolism of the chemical itself.

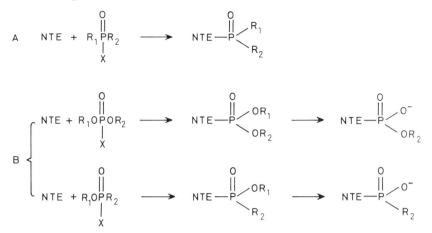

Figure 7 *Initiation of delayed neuropathy by organophosphorus compounds.*
   A. Delayed neuropathy not produced by phosphinates. Phosphinylated neurotoxic esterase cannot lose a group (R_1) because the phosphorus-carbon bond is stable.
   B. Delayed neuropathy produced by phosphates and phosphonates. Phosphorylated and phosphonylated neurotoxic esterase can lose a group (R_1) because the phosphorous-oxygen or the carbon-oxygen bounds can be broken (the NTE/ageing hypothesis)

Covalent Interaction

Certain organophosphorus compounds, while causing few or early symptoms due to inhibition of cholinesterase, cause a condition of delayed neuropathy in which symptoms of unsteadiness, progressing to paralysis, do not appear until 10–14 days after exposure. Man is a sensitive species for this condition and the chicken is the best experimental animal.

The primary target has now been identified as an esterase present in the nervous system and tightly-bound to membranes (Johnson, 1975a and b; 1982). It has been called 'neurotoxic esterase' or 'neuropathy target esterase' (NTE).

The essential reactions (Figure 7) for those compounds causing delayed neuropathy is phosphorylation or phosphonylation of the esterase followed by the release of one of the groups attached to the phosphorus (this process is called ageing because it is often a slow process). This group becomes attached to the protein, in contrast to the ageing of other phosphorylated or phosphonylated esterases when the group is released into the water. The significance of this unique behaviour of 'neurotoxic esterase' is not understood.

It is known that the reaction of di-iso-propylphosphorofluoridate (DFP), with 'neurotoxic esterase' followed by ageing takes place very rapidly (probably within 1 hour) and the excess DFP is detoxified. Signs of delayed neuropathy do not appear for 10–14 days. Therefore this chronic condition (often permanent paralysis of the limbs) is initiated by a rapid chemical reaction. The nature of the biological cascade initiated by this reaction, which leads to the clinical symptoms, is not known. Presumably by synthesis of new protein, neurotoxic esterase activity returns over the 10–14 days before symptoms appear. The 'dying back' of long axons in the peripheral nerves can be repaired, but in the spinal cord they cannot.

This example illustrates the principle that chronic diseases can be initiated by a chemical reaction which is complete long before evidence of disease is apparent. Presumably biological systems are switched on or off, and the changes, once initiated, proceed and cannot be prevented.

Recent research has provided two observations which are difficult to interpret at present and may require some modification of the 'NTE/ ageing hypothesis' (Lotti *et al.*, 1991; Johnson and Safi, 1992).

C Biochemical, Physiological, or Morphological Consequences of the Primary Reaction with Target(s)

In this phase of toxicity we are concerned with the consequences of interaction of the toxic chemical with the target (see Figure 4). Many changes may be induced; some of these lead to the symptoms, signs, or syndrome of poisoning. This phase is, therefore, the link defined in biochemical, physiological, or morphological terms between the primary chemical interaction with the target and the changes in the function of the whole organism. As illustrated in Figure 4, some reactions of chemicals with macromolecules may take place which have either no biological consequences or induce changes which do not lead to toxicity.

Organophosphorus compounds and carbamates inhibit acetylcholinesterase by phosphorylating the active centre. This inhibition of the enzyme in the nervous system is causally related to increases in the concentration of acetylcholine and this increase is in turn related to the symptoms and eventual death (often by respiratory failure). Acetylcholinesterase, with identical properties to that in the nervous system, is present on the membranes of erythrocytes. For most purposes inhibition of this enzyme in the blood is an excellent monitor for exposure and the degree of inhibition has the same relation to symptoms as the inhibition of the enzyme in the brain. Thus, although inhibition of acetylcholinesterase in human erythrocytes causes no toxicity, its inhibition by most anti-cholinesterases is a mirror of changes in the activity of the enzyme in the nervous system (Vandekar, 1980).

On chronic administration to man, cadmium accumulates in the liver and kidney. Continuous exposure to small concentrations will result in larger concentrations in these tissues because the half-life for cadmium is 16–30 years (Figure 8). The rather simple arithmetic calculations (Figure 8), using values for intake of cadmium and its half-life *in vivo,* show that concentrations in the kidney and the daily excretion can be predicted. This is a highly simplified calculation and concerns only one compart-

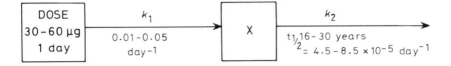

X is concentration in kidney

Kidney is 280g(70 kg man)$^{-1}$

Kidney contains $1/3$ of the body burden

<u>Calculated expected mean concentration of</u>

<u>cadmium in the kidney.</u>

$$\frac{45 \times 0.03}{6.5 \times 10^{-5} \times 3 \times 280} = 25 \text{ μg g}^{-1}$$

Published values 20–30 μg g^{-1}

<u>Expected daily excretion of cadmium</u>

$25 \times 280 \times 3 \times 6 \times 10^{-5} = 1.3$ μg day^{-1}

Figure 8 *Relationships between intake and output of cadmium and concentrations in the kidney*
(For a more extensive discussion consult Webb, 1979; Shank and Vetter, 1981)

ment, the kidney. Other, much more complicated systems have been derived (Shank and Vetter, 1981).

Cadmium is attached to metallothionein in both the liver and kidney, and cadmium complexed in this manner does not cause tubular necrosis unless the concentration in the kidney cortex exceeds 300–400 μg g^{-1} (*i.e.* overall average concentration 200 μg g^{-1}). Identification of the binding protein has allowed the development of immunoassay techniques for the determination of cadmium thionein in the urine (Tohyama, Shaikh, Nogawa, Kobayashi, and Houda, 1982; Vander Mallie and Garvey, 1980; Brady and Kofka, 1979; Chang, Vander Mallie, and Garvey, 1979). It is now established that when cadmium causes necrosis in the kidney, proteins, including cadmium thionein, are excreted in the urine.

The determination of soluble cytoplasmic enzymes released into the circulation from dying cells, *e.g.* lactic dehydrogenase, sorbitol dehydrogenase, or the transaminases can be used for the detection of cell necrosis.

The importance of measurements of indicators causally related or proportional to the disease process cannot be over-emphasised. Dose–response relationships can be compared with the reaction with the primary target (if this is known) or with the dose–response of the whole animal. Comparison of the relationships in different species with, in addition, the pharmacokinetic behaviour of the chemical, can provide valuable information and may lead to the possibility of measurement of such processes in human tissue and thereby allow an estimate of risk of exposure to man.

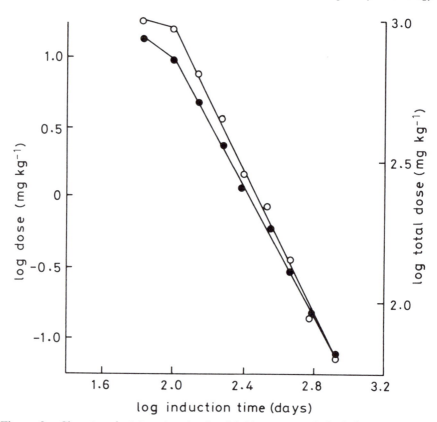

Figure 9 *Chronic administration in the drinking water of diethylnitrosamine to rats. During the course of the experiment in the seven groups receiving the highest doses, all rats developed tumours. In the two groups receiving 0.15 and 0.075 mg kg^{-1} day^{-1} 27/30 and 5/7 developed tumours by the end of the experiment*

7   Dose–Response Relationships

When making decisions about the safe use of a compound, it is vital to have information on whether there is a dose threshold, below which no toxicity is found, even with prolonged and continuous exposure.

Chemical carcinogenesis presents this problem in its most acute form, if the concept that a single lesion (one hit) on DNA is all that is necessary to initiate the growth of a tumour is accepted. With the increasing information of the many steps which may be involved in the carcinogenic process, the hypothesis is perhaps less certain; nevertheless, results are available which indicate that a safe dose of some chemicals, even if it exists, must be very small (Figure 9; Schmähl, 1979; Hoel, Kaplan, and Anderson, 1983).

Knowledge of the primary target sometimes allows a fundamental rather than an empirical approach to the solution of the problem as to whether a threshold dose exists. Measurement of the degree of reaction

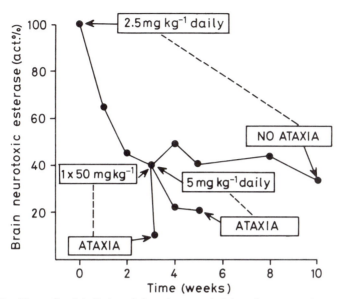

Figure 10 *Mono-2-tolyl diphenylphosphate and delayed neuropathy in the hen. The solid lines indicate the activity of brain neurotoxic esterase as a percentage of control and the dashed lines the relation between dose and clinical condition. Similar results are reported for spinal cord.*
(Reproduced by permission from Lotti and Johnson, 1980, *Arch. Toxicol.*, **45,** 263–271.)

of the chemical with the target indicates, under all circumstances, the amount of chemical delivered to its site of action. For example, it was known that a single dose of DFP caused a 70–80% inhibition, by phosphorylation, of the 'neurotoxic esterase' in the nervous system of chickens and that this was followed by the development of permanent neuropathy. It was not known if repeated exposure, to a smaller dose leading to a maintained but lower inhibition, would gradually build up damage until clinical symptoms would appear. The experiment shown in Figure 10 demonstrates that, in the chicken, inhibition may be maintained at 50% for 10 weeks with no signs of neuropathy (Lotti and Johnson, 1980). Increase in the dosage administered led to symptoms, showing that the animals do not become tolerant. This experiment answers the question in a fundamental way because delivery of the toxic chemical to the target has been established—inhibition of the neurotoxic esterase shows that the target has been reached. It seems, therefore, that 50% inhibition of neurotoxic esterase by the mechanism shown in Figure 7 can be tolerated indefinitely. To transfer this decision to man requires assumptions about the relationship between the degree of inhibition and toxicity, but it seems very likely that a threshold does exist in man as it does in the hen. Opportunities may be found to acquire information in man after accidental exposure since the same or a similar esterase to 'neurotoxic esterase' is found in lymphocytes.

For an extensive discussion of thresholds and the principles involved in their determination, see Aldridge (1985). One way of bridging the animal-man gap has recently been demonstrated (Lotti and Johnson, 1978). Both acetylcholinesterase and neurotoxic esterase are inhibited *in vitro* by many organophosphorus compounds and they must be close to each other in the cell since they sediment in the same subcellular fraction. It is impossible to predict even the acute toxicity from the rate constants for the reaction with acetylcholinesterase because the rates of disposal by metabolism may exert a dominating influence (*cf.* p. 67). If, however, the compound is delivered to the two enzymes, acetylcholinesterase and neurotoxic esterase, at the same concentration, then the following relationship should hold.

$$\frac{LD_{50}}{ED_{50}} = \frac{D_{50}(AChE)}{D_{50}(NTE)} = \frac{k_a(NTE)}{k_a(AChE)} = \frac{I_{50}(AChE)}{I_{50}(NTE)}$$

when: LD_{50} is the dose for 50% lethality (acute toxicity).

ED_{50} is the effective dose to produce delayed neuropathy in 50% of the animals.

$D_{50}(AChE)$ and $D_{50}(NTE)$ is the dose required *in vivo* to cause 50% inhibition of acetylcholinesterase and neurotoxic esterase, respectively.

$k_a(NTE)$ and $k_a(AChE)$ is the bimolecular rate constant for inhibition *in vitro* of neurotoxic esterase and acetylcholinesterase, respectively.

$I_{50}(AChE)$ and $I_{50}(NTE)$ is the concentration necessary for 50% inhibition *in vitro* (both under the same defined conditions) of acetylcholinesterase and neurotoxic esterase, respectively.

It has been shown that this relationship does hold in chickens (Lotti and Johnson, 1978). Thus it is possible to determine the ratio of the *in vitro* inhibitory potency for direct inhibitors of these enzymes using postmortem human tissue. From this ratio we can predict the potential for causing delayed neuropathy compared with acute toxicity. As a rough guide, when a ratio greater than 0.05 is obtained, the compound should be suspect.

8 Measurement of Received Dose of Reactive Chemicals

The dose of a chemical to an animal or man is usually expressed relative to body weight. For experimental studies this is an accurate statement although there is often much doubt about the amount of a toxic chemical that is absorbed by different routes of administration. Measurement of exposure of man in a working environment is extremely difficult; analyses of the concentration in the air are difficult to relate to intake by individual workers.

A new approach has been made for electrophilic chemicals which may be genotoxic. Measurement of the extent of the reaction of such electrophiles with haemoglobin *in vivo* allows the internal exposure to be derived (Ehrenberg and Husain, 1981). Haemoglobin is a suitable protein because it is available in high concentration and has a relatively long life *in vivo*. Thus measurement of, for example, alkylated residues in haemoglobin can provide an integral of the dose received over a relatively long previous exposure. For epidemiological studies, for risk assessment, and for the identification of dangerous working practices, this is valuable since the measurement is made on individual workers and is not some vague assessment of average exposure.

Analytical techniques are developing so fast that detection of a low degree of reaction with, for example, cysteine, histidine, valine, or lysine are becoming possible (Farmer, Gorf, and Bailey, 1982; Farmer, Bailey, Lamb, and Connors, 1980). From the known reaction rates with these residues, determined *in vitro* (biomolecular rate constants), the integrated dose (concentration × time) can be readily calculated.

Since reaction with DNA is directly related to the dose of some carcinogens (Weiland and Neumann, 1978; Gangler and Neumann, 1979; Gangler, Neumann, Saubner, and Scribner, 1979) it seems likely that is will be possible for some compounds to calibrate the reaction of haemoglobin with that of DNA.

There are, of course, many problems to solve, and particularly for chemicals which are bioactivated to electrophiles, but this work illustrates the use of modern developments in analysis (separative method, mass spectrometry, *etc.*) to move forward to more reliable estimates of exposure (biological monitoring). It is an accepted principle that the toxicity of a chemical will be directly related to the concentration that reaches the target. The approaches described above are leading towards an estimate of that concentration. It is also important to appreciate that this advance is due to the detailed study of reactions with a macro-molecule which has no direct relevance to the toxicity of the chemical.

9 Conclusions

The ideal of the toxicologist is to be able to predict the toxicity of a chemical by looking at its chemical structure. Knowledge of the biochemical principles of the many factors influencing toxicity and the mechanisms whereby the deleterious biological consequences are initiated allow a much more certain prediction of the type(s) of toxicity expected from a particular molecule. Quantitative prediction of toxicity (expected dose–response) is still difficult. But those working in biological research should not despair; prediction of the expected rate of a chemical reaction from the structure of the reactants is just as difficult.

In this chapter examples of biochemical principles, some accepted and some less well established, have been discussed. Perhaps the most important biochemical principle is that we must strive to ask molecular questions about toxicity (see DeMatteis and Lock, 1987). The answers, as they move from suggestions to hypotheses and concepts, will provide the framework for the assessment of hazard and the safe use of chemicals.

References

Albert, A. (1985). 'Selective Toxicity'. The physicochemical basis of therapy. Chapman and Hall, London, pp. 1–750.

Aldridge, W. N. (1980). The need to understand mechanisms. *In*: 'The Scientific Basis of Toxicity Assessment', *Ed.* H. Witschi. Elsevier/North-Holland Biomedical Press, Amsterdam. pp. 305–319.

Aldridge, W. N. (1981). Mechanisms of toxicity. New concepts are required in toxicology. *Trends Pharmacol. Sci.*, **2**, 228–231.

Aldridge, W. N. (1985). The biological basis of thresholds. *Annu. Rev. Pharmacol. Toxicol.*, **26**, 39–58.

Aldridge, W. N. (1987). Selective toxicity and specific interactions. *In*: 'Seminars of Toxicity Mechanisms' Vol. 1. *Eds* K. Hashimoto and M. Minami. Center of Academic Publications Japan, Tokyo, pp. 9–27.

Aldridge, W. N. and Lovatt Evans, C. (1946). The physiological effects and fate of cyanogen chloride. *Q. J. Exp. Physiol.*, **33**, 241–266.

Aldridge, W. N., Miles, J. W., Mount, D. L., and Verschoyle, R. D. (1979). The toxicological properties of impurities in malathion. *Arch. Toxicol.*, **42**, 95–106.

Aldridge, W. N. and Reiner, E. (1971). 'Enzyme Inhibitors as Substrates'. Interaction of esterases with esters of organophosphorus and carbonic acids. North Holland Publishing Co., Amsterdam, pp. 1–328.

Aldridge, W. N., Street, B. W., and Skilleter, D. N. (1977). Oxidative phosphorylation. Halide dependent and halide independent effects of tri-organotin and triorganolead compounds on mitochondrial function. *Biochem. J.*, **168**, 353–364.

Aldridge, W. N., Verschoyle, R. D., and Peal, J. A. (1984). *O,S,S*-Trimethyl-phosphorodithioate and *O,S,S*-Triethylphosphorothioate: pharmacokinetics in rats and effect of pretreatment with compounds affecting the drug processing systems. *Pestic. Biochem. Physiol.*, **21**, 265–274.

Anders, M. W. (*Ed.*) (1985). 'Bioactivation of foreign compounds'. Academic Press, Orlando, pp. 1–555.

Barnes, J. M. and Magee, P. N. (1954). Some toxic properties of dimethylnitros-amine. *Br. J. Ind. Med.*, **11**, 167–174.

Biempica, L., Morecki, R., Wu, C. H., Giambrone, M. A., and Rojkind, A. (1980). Immunochemical localisation of type B collagen. *Am. J. Pathol.*, **98**, 591–602.

Brady, F. O. and Kafka, R. L. (1979). Radioimmunoassay of rat liver metallothionein. *Anal. Biochem.*, **98**, 89–94.

Brown, A. W., Aldridge, W. N., Street, B. W., and Verschoyle, R. D. (1979). The behavioural and neuropathological sequelae of intoxication by trimethyl tin compounds in the rat. *Am. J. Pathol.*, **97**, 59–82.

Cameron, G. R. and Karunaratne, W. A. E. (1936). Carbon tetrachloride cirrhosis in relation to liver regeneration. *J. Pathol. Bacteriol.*, **42**, 1–21.

Chang, C. C., Vander Mallie, R. J., and Garvey, J. S. (1980). A radio immunoassay for human metallothionin. *Toxicol Appl. Pharmacol.*, **55**, 94–102.

Cremer, J. E. (1959). Biochemical studies on the toxicity of tetraethyllead and other organolead compounds. *Br. J. Ind. Med.*, **16**, 191–199.

Dekant, W., Vamvakas, S., and Anders, M. W. (1992). The kidney as a target for xenobiotics bioactivated by glutathione conjugation. *In*: 'Tissue specific toxicity: biochemical mechanisms', *Eds* W. Dekant and H.-G. Neumann. Academic Press, London, pp. 163–194.

DeMatteis, F., Gibbs, A. H., Farmer, P. B., Lamb, J. H., and Hollands, C. (1982). Liver Haem as a target for drug toxicity. In: 'Advances in Pharmacology and Therapeutics II', Vol. 5. *Eds* H. Yoshida, Y. Hagihara, and S. Ebashi. Pergamon Press, Oxford, pp. 131–138.

DeMatteis, F. and Lock, E. A. (*Eds*) (1987). 'Selectivity and Molecular Mechanisms of Toxicology'. Macmillan Press, Ltd., Basingstoke, UK, pp. 1–302.

Divincenzo, G. D., Krasavage, W. J., and O'Donoghue, J. L. (1980). Role of metabolism in hexacarbon neuropathy. *In*: 'The Scientific Basis of Toxicity Assessment', *Ed.* M. Witschi. Elsevier North-Holland Biomedical Press, Amsterdam, pp. 183–200.

Donninger, C. (1971). Species specificity of phosphate triester anticholinesterases. *Bull. W.H.O.*, **44**, 265–268.

Edgar, J. A., Smith, L. N., and Culvenor, C. C. J. (1970). Metabolic conversion of heliotridine-based pyrrolizidine alkaloids to dehydroheliotridine. *Mol. Pharmacol.*, **6**, 402–406.

Ehrenberg, L. and Husain, S. (1981). Genetic toxicity of some important epoxides. *Mutat. Res.*, **86**, 1–113.

Eto, M., Casida, J. E. and Eto, T. (1962). Hydroxylation and cyclisation reaction involved in the metabolism of tri-*o*-cresyl phosphate. *Biochem. Pharmacol.*, **11**, 337–352.

Farmer, P. B., Gorf, S. M., and Bailey, E. (1982). Determination of hydroxypropyl histidine in haemoglobin as a measure of exposure to propylene oxide using high resolution gas chromatography-mass spectrometry. *Biomed. Mass Spec.*, **9**, 69–71.

Farmer, P. B., Bailey, E., Lamb, J. H., and Connors, T. A. (1980). Approach to the quantitation of alkylated amino acids in haemoglobin by gas chromatography-mass spectrometry. *Biomed. Mass Spec.*, **7**, 41–46.

Gangler, B. J. and Neumann, H. G. (1979). The binding of metabolites formed from aminostilbene derivatives to nucleic acids in the liver of rats. *Chem Biol. Interact.*, **24**, 355–372.

Gangler, B. J., Neumann, H. G., Saubuer, N. K., and Scribner, J. D. (1979). Identification of some products from the reaction of *trans*-4-aminostilbene metabolites and nucleic acids *in vivo*. *Chem. Biol. Interact.*, **27**, 335–342.

Gillette, J. R. (1982). The problem of chemically reactive metabolites. *Drug Metab. Rev.*, **13**, 941–961.

Goldstein, A., Aronow, L., and Kalman, S. M. (1974). 'Principles of Drug Action. The Basis of Pharmacology.' John Wiley and Son, New York, pp. 1–854.

Heath, D. F. and Vandekar, M. (1957). Some spontaneous reactions of *O,O*-dimethyl-*S*-ethylthioethylphosphorothiolate and related compounds in water and on storage and their effects on the toxicological properties of the compounds. *Biochem. J.*, **67**, 187–201.

Hoel, D. G., Kaplan, N. L., and Anderson, M. W. (1983). Implication of non-linear kinetics on risk estimation in carcinogenesis. *Science*, **219**, 1032–1037.

Johnson, M. K. (1975a). The delayed neuropathy caused by some organophosphorus esters: mechanism and challenge. *Crit. Rev. Toxicol.*, **3**, 289–316.

Johnson, M. K. (1975b). Organophosphorus esters causing delayed neurotoxic effects. Mechanism of action and structure/activity studies. *Arch. Toxicol.*, **34**, 259–288.

Johnson, M. K. (1982). The target for initiation of delayed neurotoxicity by organophosphorus esters: biochemical studies and toxicological applications. *Rev. Biochem. Toxicol.*, **4**, 141–212.

Johnson, M. K. and Safi, J. M. (1992). Organophosphoramidation of neuropathy target esterase (NTE) is sufficient to initiate delayed neuropathy (DN) without the necessity of an 'ageing reaction'. *Toxicologist*, **12**, Abstract 63.

Lijinsky, W., Loo, J., and Ross, A. E. (1968). Mechanism of alkylation of nucleic acids by nitrosodimethylamine. *Nature (London)*, **218**, 1174–1175.

Lock, E. A. and Ishmael, J. (1979). The acute toxic effects of hexachloro-1,3-butadiene on the rat kidney. *Arch. Toxicol.*, **43**, 47–57.

Lotti, M. and Johnson, M. K. (1978). Neurotoxicity of organophosphorus pesticides: predictions can be based on *in vitro* studies with hen and human enzymes. *Arch. Toxicol.*, **41**, 215–221.

Lotti, M. and Johnson, M. K. (1980). Repeated small doses of a neurotoxic organophosphate. Monitoring of neurotoxic esterase in brain and spinal cord. *Arch. Toxicol.*, **45**, 263–271.

Lotti, M., Caroldi, S., Capodicasa, E., and Moretto A. (1991). Promotion of organophosphorus-induced delayed polyneuropathy by phenylmethanesulfonyl fluoride. *Toxicol. Appl. Pharmacol.*, **108**, 234–241.

Magee, P. N. (1956). Toxic liver injury: the metabolism of dimethylnitrosamine. *Biochem. J.*, **64**, 676–682.

Magee, P. N. and Barnes, J. M. (1956). The production of malignant primary hepatic tumours in the rat by feeding dimethylnitrosamine. *Br. J. Cancer*, **10**, 114–122.

Magee, P. N. and Farber, E. (1962). Toxic liver injury and carcinogenicity: methylation of rat liver nucleic acids by dimethylnitrosamine *in vivo*. *Biochem. J.*, **83**, 114–124.

Magos, L. (1981). Metabolic factors in the distribution and half time of mercury after exposure to different mercurials. *In*: 'Industrial and Environmental Xenobiotics', *Eds* I. Gut, M. Cikrt, and G. L. Plaa. Springer-Verlag, Berlin, pp. 1–15.

Magos, L. (1982). Mercury induced nephrotoxicity. *In*: 'Nephrotoxicity, Assessment and Pathogenesis,' *Eds* P. H. Bach, F. W. Bonner, J. W. Bridges, and E. A. Lock. John Wiley and Sons, Chichester, pp. 325–337.

Mahler, W. (1954). The ionisation of certain bis-phenols. *J. Am. Chem. Soc.*, **76**, 3920–3921.

Martin, C. N. and Garner, R. C. (1977). Aflatoxin B_1-oxide generated by chemical or enzymic oxidation of aflatoxin B_1 causes guanine substitution in nucleic acids. *Nature (London)*, **267**, 863–865.

Mattocks, A. R. (1972). Toxicity and metabolism of Seneccio Alkaloids. *In*: 'Phytochemical Ecology', *Ed*. J. B. Harbourne. Academic Press, New York, pp. 179–200.

Neal, R. A. (1980). Metabolism and mechanisms of toxicity of compounds containing thiono-sulphur. *In*: 'The Scientific Basis of Toxicity Assessment', *Ed.* H. Witschi. Elsevier/North-Holland Biomedical Press, Amsterdam, pp. 241–250.

Rojkind, M. and Martinez-Palomo, A. (1976). Increase in type I and type III collagens in human alcoholic liver cirrhosis. *Proc. Natl. Acad. Sci.*, **73**, 539–543.

Roughton, F. J. W. and Darling, R. C. (1944). The effect of carbon monoxide on the oxygen dissociation curve. *Am. J. Physiol.*, **141**, 17–31.

Scatchard, G. (1949). The attractions of proteins for small molecules and ions. *Ann. N.Y. Acad. Sci.*, **51**, 660–672.

Schmähl, D. (1979). Problems of dose–response studies in chemical carcinogenesis with special references to *N*-nitroso compounds. *CRC Crit. Rev. Toxicol.*, **6**, 257–281.

Seyer, J. M. (1980). Interstitial collagen polymorphism in rat liver with CCl_4-induced cirrhosis. *Biochim. Biophys. Acta*, **629**, 490–498.

Shank, K. E. and Vetter, R. J. (1981). Model description of cadmium transport in a mammalian system. *In*: 'Cadmium in the Environment', *Ed.* J. O. Nriagu. John Wiley and Son, New York, pp. 583–616.

Smith, L. L. and Wyatt, I. (1981). The accumulation of putrescine into slices of rat lung and brain and its relationship to the accumulation of paraquat. *Biochem. Pharmacol.*, **30**, 1053–1058.

Swann, P. F. and McLean, A. E. M. (1968). The effect of diet on the toxic and carcinogenic action of dimethylnitrosamine. *Biochem. J.*, **107**, 14–15P.

Swensen, D. H., Miller, E. C., and Miller, J. A. (1974). Aflatoxin B_1-2,3-oxide: evidence for its formation in rat liver *in vivo* and by human liver microsomes *in vitro*. *Biochem. Biophys. Res. Commun.*, **60**, 1036–1043.

Tobias, R. S. (1966). σ-Bonded organometallic cations in aqueous solutions and crystals. *Organomet. Chem. Rev.*, **1**, 93–129.

Tohyama, C., Shaikh, Z. A., Nogawa, K., Kobayashi, E., and Houda, R. (1982). Urinary metallothionein as a new index of renal dysfunction in 'Itai-Itai' disease patients and other Japanese women environmentally exposed to cadmium. *Arch. Toxicol.*, **50**, 159–166.

Van Bladeren, P. J. (1988). Formation of toxic metabolites from drugs and other xenobiotics by glutathione conjugation. *Trends Pharmacol. Sci.*, **9**, 295–299.

Vandekar, M. (1980). Minimizing occupational exposure to pesticides: cholinesterase determination and organophosphorus poisoning. *Residue Rev.*, **75**, 67–79.

Vandekar, M. and Heath, D. F. (1957). The reactivation of cholinesterase after inhibition *in vivo* by some dimethyl phosphate esters. *Biochem. J.*, **67**, 202–208.

Vander Mallie, R. J. and Garvey, J. S. (1979). Radioimmunoassay of metallothioneins. *J. Biol. Chem.*, **254**, 8416–8421.

Wainio, W. W. and Greenlees, J. (1960). Complexes of cytochrome C oxidase with cyanide and carbon monoxide. *Arch. Biochem. Biophys.*, **90**, 18–21.

Webb, M. (*Ed.*) (1979). 'The Chemistry, Biochemistry and Biology of Cadmium'. Elsevier/North Holland Biomedical Press, Amsterdam, pp. 1–465.

Weiland, E. and Neumann, H. G. (1978). Methaemoglobin formation and binding to blood constituents as indicators for the formation, availability and reactivity of activated metabolites derived from *trans*-4-stilbene and related aromatic amines. *Arch. Toxicol.*, **40**, 17–35.

Wyman, J. (1964). Linked functions and reciprocal effects in haemoglobin: a second look. *Adv. Protein Chem.*, **19**, 223–286.

CHAPTER 6

Animal Husbandry

A. B. G. LANSDOWN

1 Introduction

Advances in medical research and in predictive safety evaluation will continue to be dependent upon experiments in living animals for the foreseeable future, despite considerable efforts to develop alternative studies. Although a vast range of species and strains of laboratory animal is available to the experimental toxicologist, the greater proportion of controlled, reliable studies conducted for legislative purposes have employed a comparatively narrow range of species. As a consequence, the biology and animal husbandry for such animals as the Sprague-Dawley rat, New Zealand white rabbit, or the beagle dog is well known and their basic requirements under laboratory conditions appreciated.

In most aspects of experimental toxicology where sub-human species are employed as 'models' for studying human disease processes or in predicting the human response to toxic chemicals in the environment, the accent will be on good laboratory practice and reproducibility of results. As will be readily appreciated, both parameters are entirely dependent upon appropriate technician training, correct experimental procedure, and management of animal facilities and husbandry. Experience has shown that subtle variations in the conditions in animal holding rooms, diets, caging, and bedding, and inconsistencies in the genotype and microbiological status of the test animals are potentially responsible for interlaboratory differences in study results.

Animal husbandry is an immense subject and whereas ideas on 'what is acceptable' will vary widely according to the country and the legislative requirements in force at the time, the experimentalist has the ultimate responsibility for satisfying such criteria in his animals of genetic purity, microbiological cleanliness, and dietary adequacy. In Great Britain, the Universities Federation for Animal Welfare (UFAW), the Royal Society and the Medical Research Council (MRC) have set out guidelines for animal husbandry, their recommendations being updated as new information comes to hand. Practical guidelines on the use of living animals in research by the Biological Council (1984) are useful reading for the new research worker.

Clearly, the species and strain of laboratory animal selected for a particular project will vary according to the scientific rationale of the work and experience gained in previous studies. Animal models and their suitability in studying environmental toxicology, immunogenicity, carcinogenicity, reproductive toxicity, and teratogenicity will be discussed in the relevant chapters elsewhere in this volume. It is the purpose of the present review to discuss some specific and more general aspects of the husbandry of laboratory animals as they relate to experimental research.

2   The Research Animal

Experimental toxicology and other forms of research employing laboratory animals embodies scientific and humanitarian considerations, aspects of animal welfare, and legal obligations (Biological Council, 1984). In the light of intensive research into the replacement of animals in predictive studies, the experimentalist is faced with such questions as: is this test animal suitable for the work I have in mind, are the experiments ethical, and what is the likelihood of me achieving my desired intention? (Remfrey, 1984). The International Council for Laboratory Animal Science (ICLAS) (1978) has noted that too many scientific investigations fail to provide sufficient information on the animals used and on the husbandry and experimental techniques. Clearly, such information is indispensable for the correct and optimal interpretation of the results, and will be a prerequisite for the repetition of the work by other investigators, as is so often the case where the results of studies are contentious.

The policy of using the so-called 'defined' animal for research purposes is becoming more evident as legislative authorities recognise the importance of *control* in regulatory studies. As will be emphasised later, the microbiologically clean and genetically defined laboratory animal, maintained in a controlled environment by trained technicians supervised by a competent curator, is a very superior animal (Festing, 1981). Such species compare favourably with the pure chemical demanded by most research chemists!

A   Legal Aspects

The Cruelty to Animals Act was introduced into Great Britain in 1876 to control experiments on living animals which were calculated to cause pain (Advisory Committee on Animal Experiments, 1981). That Act has been in force, largely unchanged until the issue of the 1986 Animal (Scientific Procedures) Act. Internationally, the USA, Canada, and most countries of the European Economic Community, have legislation which regulates animal experimentation. These regulations vary in their details but in most cases, control the premises in which the work is conducted, the training and the competence of the scientists and technicians, and the source, transportation, and husbandry of the experimental animals (Ray and Scott, 1973). It is clear that present legislation governing the use of animals in some countries is broad and extends to farm animals and domestic species. Occasionally, as in the case of certain exotic species, special legislation such as the Dangerous Wild Animals Act (1976) (Cooper, 1978) is passed. Conceivably, such legislation would apply to most species of non-human primate used experimentally.

In Great Britain, between the issue of the 1876 Act and the recommendations of the 1986 Act, there had been a marked surge in the use of animals in predictive safety evaluation of medicines and proprietary products. To some, this excessive testing of pharmaceutical and related products represents unnecessary use of animals (Remfrey, 1976). This led to a strong desire to reduce, refine, and possibly replace the use of animal species in some experiments.

This thinking has been partly achieved in the 1986 Act, which in addition to safeguarding the use of experimental animals, seeks to monitor the rationale of the experimental procedure, the technical competence of the investigator, and the husbandry, source of supply and transportation of the animals to be used. The Act identifies the 'protected animal' as all living vertebrates (other than humans) and embraces all experiments employing foetal, larval, or embryonic forms which have attained specific levels of development (Lansdown, 1992). An experiment conducted within a designated establishment will be a 'regulated procedure'. For present purposes, it is of interest to note that the 1986 Act requires that animals used for regulated procedures should be supplied by designated breeding establishments approved by the Home Office.

In the United States, a vast amount of experimental toxicology is conducted under the auspices of the Department of Health and Human Resources, and the National Toxicology Program established in 1978 to co-ordinate and strengthen research and testing of a wide range of substances present in the environment, food, and consumer products. They have developed an integrated array of tests tailored to study harmful effects for each substance, whilst at the same time attempting to validate the tests and improve their sensitivity and reproducibility.

In the United States, much current toxicology involving animals derives from the Food and Drug Administration (FDA) who stipulate that the source of supply, species, and strain of experimental animal will be recorded in all protocols in accordance with their codes of Good Laboratory Practice (US FDA, 1978a). The FDA administer the USA Food, Drugs and Cosmetics Act, the Public Health Service Act, and Medical Device Amendments of 1976. Experimental laboratories are inspected regularly and the conditions in animal rooms, food storage, and experimental equipment checked. Special procedures are required in respect of quarantine, isolation, parasite control and vaccination, and experimental procedures. Such details as the source of animals used, identification, allocation to experimental groups, housing (including cage size, bedding, and number of animals per cage), environmental control, and diet are recorded (US FDA, 1978b).

B Animal Supply

In 1947 the MRC established the Laboratory Animals Bureau in Great Britain to monitor and control the quality and supply of laboratory animals (MRC, 1974). A census conducted at the time showed that animals were supplied by:
  (a) Breeders who sold their own animals direct to laboratories
  (b) Dealers who obtained animals from diverse sources and then resold them for research purposes.
Although this last situation may still obtain in some countries, accreditation schemes are more widely adopted nowadays for scientific reasons as well as for the health and safety of the staff concerned (especially those working with dogs, cats, and exotic species).

The production and supply of laboratory animals under an accreditation scheme aims to protect an experimentalist from zoonotic diseases and safeguard his in-house breeding and experimental animals. Accreditation schemes monitor the health status and genetic purity of laboratory animals. The first aspect relates to the hygiene and cleanliness of breeding stocks, weaning, and adult animals pending their issue to laboratories. Genetic purity will be of immense importance in research programmes employing highly inbred strains exhibiting enzymic defects or morphological abnormalities as in the scoliotic mouse (Ky/Ky) or Brattleboro mutant of the Long Evans rat (vasopressin deficient) (Valtin, 1967).

The various species/strains of laboratory animals available for research studies include:
  (1) Those bred under conventional conditions, *e.g.* rat, mouse, rabbit, guinea pig, and cat
  (2) Larger species bred under more expansive conditions and possibly requiring less rigidly controlled conditions, *e.g.* dog, ferret, sheep, and pig
  (3) Exotic species obtained from the wild state or bred in captivity in tropical countries in which they are indigenous, *e.g.* macaques, baboons, and other monkeys.
Limited success has been achieved in breeding marmosets in captivity in Great Britain and Europe but these are expensive and difficult to obtain.

The broad aim of any accreditation scheme will be to raise the quality of laboratory animals. Two main schemes are available to suppliers, namely a control scheme applicable to those species classified in the first category above, and secondly, a scheme which will be more relevant to breeders and suppliers of animals under groups 2 and 3. The first of these requires that establishments be approved by the Animals Inspectorate of the Home Office; whereas the second will require approval by the Home Office and/or the Ministry of Agriculture, Fisheries and Food. Particular problems do arise in the importation of animals from overseas where quarantine regulations apply (see below).

Nowadays, recognised suppliers of laboratory animals commonly provide certification of microbiological cleanliness and genetical purity for their animals, particularly in the case of mice, rats, rabbits, and guinea pigs which are intensively reared under defined barrier maintained conditions. Genetical purity will be monitored periodically by recognised techniques of serum protein assay, skin grafting, immunological markers, and breeding performance. True inbred animals will be isogenic and thus identical at more than 99% of their genetic loci segregated in the original colony. They will be histocompatable and will not reject skin grafts from other animals of the same strain (Festing, 1987). They are essential in immunological experiments.

Suppliers of larger laboratory species will normally provide quality assurance data and certification of good health, vaccinations (distemper, parvovirus, *etc.* for dogs, and leukopenia in cats), and deworming schedules. Veterinary certification of fitness to travel is normally required for sheep, pigs, and goats. Monkeys and animals imorted from overseas and potential cariers of zoonotic infections can be quarantined as required, and screened for infections where a scientist does not have the expertise or facilities available.

C Transportation

The principle objective in the transportation of animals is that they should arrive at their destination in a similar condition to that at their departure (Clough and Townsend, 1987). They should be subjected to minimal stress and provided with caging, boxes, *etc.* which allow acceptable environmental conditions, sufficient space for movement, and access to food and water. Consideration should be given to any special needs that a species may have in the course of a journey. In the case of larger animals, it may be necessary to obtain veterinary advice before moving an animal. When an animal has been subjected to a scientific procedure, veterinary certification of fitness for transportation is obligatory in Great Britain.

The welfare of animals in transit is subject to continual discussion and governmental legislation (News and Reports, 1986; Gibson, Paterson, and McGonville, 1986). Although much debate surrounds the transportation of larger species—horses, cattle, pigs, and poultry, many of the recommendations apply to smaller species. Animal transportation is discussed in detail and guidelines provided by the British Veterinary Association and UFAW. It is noteworthy that, in Great Britain at least, probably as many as 95% of all laboratory animals are transported from breeders and suppliers to laboratories in appropriately ventilated and equipped vehicles, as a service.

3 Experimental Animal Facilities

Each species of laboratory animal has been derived from the wild state at some time and as such will retain some biological features of its wild ancestors. The laboratory environment is far removed from the wild state but good laboratory design and husbandry should attempt to mimic natural conditions as far as is practicable. Rats and mice, for example, are crepuscular species and tend to feed, mate, and move about during twilight or dark hours. Animals exposed to bright light develop retinopathy. On the other hand, larger species including dogs and monkeys are more active and require suitable exercise facility.

A Environment

Experimental protocols these days should specify the conditions of temperature, lighting, heating, and humidity under which animals will be maintained during breeding and experimental procedures. Recommendations for common laboratory species are summarised in Table 1. It will be appreciated that deviations from these conditions may constitute a stress situation and contribute to alterations in feeding, drinking, reproductive behaviour, and responses to toxicological situations. Consistency in environmental conditions is a prerequisite in obtaining consistent and reproducible results.

Laboratory animals should be allowed sufficient oxygen and not be

Table 1 *Environmental Recommendations for Common Species of Laboratory Animals*

	Room temperature (°C)	Humidity (%)	Lighting
Mouse	20–22	50–60	12 h light : 12 h dark
Rat	20–22	50–60	12 h light : 12 h dark
Hamster	21–22	50–60	14 h light : 10 h dark
Gerbil	20–24	50–60	14 h light : 10 h dark
Rabbit	16–20	50–60	14 h light : 10 h dark
Guinea pig	18–22	45–70	12–16 h light day^{-1}
Dog	15–21	50–60	12 h light : 12 h dark
Cat	21–23	50–60	12 h light : 12 h dark
Pig	16–18	60–70	12 h light : 12 h dark
Non-human primate:			
marmoset	20–25	40–60	12 h light : 12 h dark
macaque	24	40–60	12 h light : 12 h dark
baboon	24	40–60	12 h light : 12 h dark

* Recommendations apply to adult animals. Breeding animals may require special environments.

exposed to toxic or excessive levels of carbon dioxide, ammonia (from degenerating excreta), or other pollutants. Air in animal rooms should be filtered to remove microbiological contamination. Ventilation is thus an important aspect of an animal house, at least 15–20 complete changes of air per hour are recommended for fully stocked small animal rooms (Clough, 1984). Additionally, the air in the immediate proximity of animals should be regularly replenished; small animals like rats and mice spend much of their time grouped in darker corners of cages and may be subjected to higher concentrations of carbon dioxide and ammonia.

B Caging and Accommodation

Accommodation and caging for laboratory animals varies greatly from one laboratory to another. The European Convention for the Protection of Vertebrate Animals used for Experimental and other Scientific Purposes (Council of Europe, 1986) has drawn up guidelines on this subject but its recommendations are not mandatory and are used with discretion. More recently, guidelines compiled by the Royal Society and UFAW (1987) should enable scientists to achieve desirable standards of animal welfare and husbandry.

By the European Convention (Council of Europe, 1986), any animal used or intended for use in a procedure shall be provided with accommodation, an environment, freedom of movement, food, water, and care appropriate to its well being. These will vary according to the age and anticipated use of an animal and the level of microbiological risks involved. Broadly, the size, shape, and fitting of pens, cages, *etc.* will be designed to accommodate the physiological and behavioural needs of a species. It should allow space for expected body growth. Recommended cage sizes are summarised in Table 2.

For small mammals, cages are constructed of galvanised iron, wood,

Table 2 *Guidelines on Housing Laboratory Animals in Conventional Conditions*

	Cage Height	Floor area per animal	Animal weight range
Mouse	12 cm	60–100 sq. cm	>30 g
Rat	18–20 cm	100–400 sq. cm	50–>550 g
Guinea pig	20–23 cm	200–750 sq. cm	150–>650 g
Hamster	15 cm	89–154 sq. cm	60–>120 g
Rabbits	40–45 cm	2000–6000 sq. cm	2–6 kg
Cats	50–80 cm	0.5–0.75 sq. m.	>3 kg
Dogs	1.5–2.0 m	4.5–8 sq. m.	5–>35 kg
Non-human primate:			
marmoset	80 cm	0.25 sq. m.	0.025–0.65 kg
macaques	100–200 cm	0.6–2.8 sq. m.	6–>6 kg

* Figures quoted are for animals caged in groups except in the case of dogs, rabbits, and macaques.

stainless steel, and various plastics. Whatever the material used, it is imperative that it be durable, readily sterilised, and not harmful to the animals. Catches, locks, and restraints will be adequate to contain the animals without risk to limbs or tails. Similar precautions are adopted in the selection of grids used as cage floors. It is not uncommon in long term studies, that large rodents develop ulcerative foot lesions through contact with galvanised or rough metal grids.

Larger animals will be confined in pens, these will normally have hard, non-slip floors which are readily flushed and sterilised. Where metal grids are used in indoor accommodation, these should not cause injury to an animal's feet.

C Bedding

A variety of bedding materials including sawdust, straw, sand, and synthetic materials is available for laboratory animals. To limit micro-biological contamination, the materials selected should be sterile and provide insulation, but be without injurious effect for the animals and technicians responsible for their husbandry.

Bedding materials may facilitate the spread of infections within an animal house. Sterilisation by formalin fumigation, hot air exposure, irradiation, and autoclaving is recommended (Spiegel, 1965). Under all circumstances, soiled bedding, faecal material, and animal waste should be incinerated.

D Water

Water is a further source of infection in an animal house and should be regularly monitored for impurities. Good laboratory practice requires that water used in drinking water bottles should be checked and if mains sources are defective, other sources should be used. Water bottles will be regularly sterilised and checked for leaks.

Chlorination is widely used to sterilise water supplies. It is simple and effective, and readily controlled. Cats will accept water containing $0.2-1.0\,\mathrm{mg\,l^{-1}}$ free chlorine without health risk. Most species of laboratory animal survive well on mains water supplies, although high levels of lead or other heavy metals may be detrimental to health.

4 Diet and Nutrition

Growth, maturation, and maintenance of normal body function in man and laboratory animals depends upon the availability of sufficient supplies of food, and nutritional sufficiency. Ill-health may result from generalised food deprivation or inadequacies in any of essential carbohydrates, fats, amino-acids and proteins, vitamins and minerals. The literature on this subject is vast but it is clear that different species of laboratory animals

Table 3 *Composition of Diets for Common Species of Laboratory Animals*

	Rat/Mouse		Rabbit		Dog		G/pig	Cat	Primate
	b	m	b	m	b	m			
Ash (%)	5.5	5.7	7.2	6.4	11.3	9.5	7	1.8	6.1
PROTEIN	22	16.4	16.4	13.6	23.4	18.4	17.4	10.3	18.5
FAT/OIL	3.7	2.9	1.9	2.7	8.2	8.5	3.4	4.5	3.9
CARBOHYDRATE	48	56	50	54	4.7	52	50	8.4	54
STARCH	41	38	22	29	40	43	27	?	34
FIBRE	2.1	5.6	13.2	10.8	2.3	3	8.8	?	4.3
CALCIUM	1.05	0.89	1.0	0.75	1.1	0.96	1	0.5	0.81
POTASSIUM	1.00	0.63	0.77	0.91	0.59	0.47	1.10	0.3	0.93
SODIUM	0.29	0.28	0.31	0.36	0.49	0.37	0.47	0.2	0.36
PHOSPHATE	0.70	0.80	0.71	0.60	0.84	0.71	0.78	0.3	0.71
MAGNESIUM	0.15	0.16	0.17	0.24	0.15	0.14	0.42	?	0.28
CHLORIDE	0.40	0.49	0.45	0.59	0.54	0.45	0.73	0.2	0.56
MANGANESE (mg)	50	77	60	0.81	33	26	63	?	57
COPPER	14	13	14	14.7	11	9	98	?	11
IRON	175	156	185	115	163	140	202	?	91
ZINC	60	70	54	74	83	69	71	?	71
IODINE	0.25	0.28	0.16	0.16	2.8	2.2	1	?	2
Vitamin A (IU kg^{-1})	12000	15000	14000	20000	27000	16000	29000	13500	42000
D3 (IU kg^{-1})	1200	2400	1400	1400	5500	3200	1400	440	4000
C (mg kg^{-1})	—	—	—	—	—	—	3000	—	3000
E	40	91	6.4	73	90	50	72	40	90
B1	5.5	13	9	13	6	5	23	2.5	14
B2	6	14	12	15	7	5	31	5.5	15
B6	6	16	10	12	7	6	24	2.2	35
Niacin	50	42	38	37	38	31	77	8.8	151
Pantothenic acid	17	22	17	22	26	23	29	5.5	25
Vitamin B12	0.015	0.083	0.03	0.031	0.047	0.033	0.036	0.022	0.017
Biotin	0.2	0.27	0.30	0.29	0.32	0.17	0.30	?	0.77
Vitamin K3	0.35	150	15	15	4.8	3.8	?	?	15
Choline	1100	1100	1130	1045	1500	1400	1340	880	1410

vary in their basic requirements, a feature which becomes evident when one examines the commercial diets available (Table 3).

Food deprivation and nutritional inadequacy can have varying effects according to the age and physiological state of an animal. For example, protein deprivation in a pregnant animal is a cause of maternal ill-health and may result in foetal wastage and the birth of smaller than normal offspring having an increased frequency of congenital cardiovascular, immunological, and neurological abnormality (Widdowson, 1965; Lansdown, 1977). In long term feeding trials, a 25% reduction in food intake will reduce the growth rate in rodents but tends to enhance their longevity and resistance to spontaneous tumours (Tucker, 1979; Berg, 1960). In contrast, overfeeding can lead to obesity, inactivity, and reduced life span.

On occasions, laboratory animal diets may be nutritionally adequate but animals exhibit evidence of malnutrition. This may arise in the case of infection, genetical change, or metabolic defect which render the animals incapable of digesting and assimilating dietary constituents. Thus mice infected with Coxsackie B viruses became subject to exocrine pancreatic insufficiency and incapable of digesting dietary protein (Lansdown, 1976; 1975). Alternatively, animals treated with lead or cadmium compounds tend to become deficient in zinc, copper, or iron, the heavy metal ions inhibiting essential enzyme systems and metabolic pathways for the trace metals (Landsown, 1983).

With improving standards of animal health and husbandry, there are noticeable improvements in the quality of animal foods available. These days, considerable attention is paid to the purity of nutritional constituents (especially materials obtained from natural resources), freedom from infective organisms, storage, and product stability or shelf-life. The diets available are formulated to satisfy the nutritional requirements of a species, and are mostly packed in sealed bags indicating the date and shelf-life. Where perishable foods are necessary, these will be stored in refrigerators or cold rooms (Royal Society-UFAW, 1987). Otherwise animal foods will be stored in dry, well ventilated facilities, free from vermin and other pests.

From an experimental view, it will be appreciated that deviations and inconsistencies in animal diets, particularly for rodents, can be a cause of irregularities in serum enzyme levels, haematological profiles, and liver and kidney function tests as are performed as a vital part of toxicological studies (Clapp, 1980).

A Breeding and Reproduction

The first thing that influences the growth and development of an animal is its size at birth and the number of surviving young in a litter, features that invariably relate to a mother's nutritional state (Widdowson, 1965). In species like the rat, mouse, and pig in which the gestational length is not appreciably altered by litter size, birth weight is a measure of the food available to each offspring (McCance, 1962). The small immature neonate has a greater surface area in proportion to its body weight than at later stages, and consequently loses more heat energy. Its food requirement is thus proportionately greater than that of larger litter mates, but frequently its ability to compete for available food is less. The low survival rate of the runt piglet is well documented.

Dietary requirements for the various species of laboratory animals during reproductive and breeding phases is a large subject and the reader is referred to reviews and proceedings of specialised symposia (Albanese, 1973; Hammell *et al.* 1976; Jones, 1976). Hurley (1977) for example, has conducted many studies to examine the importance of various dietary constituents in rodent pregnancies with particular reference to their effect on the foetus. Although her work relates exclusively to the rat, it is clear that imbalances in any essential nutrient will be detrimental in the foetal development of any animal, or humans (Lansdown, 1983). Vitamin deficiencies and prenatal growth defects are well understood generally, congenital neurological, optic, and skeletal defects resulting from excess vitamin A in early pregnancy, skeletal and cardiovascular deformities in hypervitaminosis D, and foetal wastage and infertility associated with vitamin E deficiency are recorded (Lansdown, 1983).

Cattle fed sweet clover containing the vitamin K antagonist, dicoumarol, developed symptoms of vitamin K deficiency and calves born to them died with haemorrhage. In other species, hypovitaminosis K is a known cause of foetal wastage, blindness, and mental retardation.

Nutritional inadequacy will have a detrimental influence on any animal in pregnancy and during lactation, the effect depending upon the species and the stage in the reproductive cycle. In addition to its effect on maternal health, undernutrition will often retard DNA-synthesis and RNA and protein replication in the offspring and may predispose it to diseases later in life (Roeder, 1973). Other effects may include adverse changes in germ cell formation, reproductive behaviour, conception, gestational development, parturition, and post-natal development (Widdowson, 1965). The lactating ability of a dam will relate to her age and physiological state, but will be a reflection on her nutritional state (Dean, 1951). Restriction of food after weaning may impair body growth, but if not excessive, an infant may accomplish normal growth and development with improved nutrition later. Detailed nutritional requirements for breeding laboratory animals are discussed by UFAW (1987).

B Maintenance

Whilst the nutritional requirements of most laboratory animals are similar, reflecting basic similarities in metabolic pathways, the sources from which these nutrients are obtained varies (Ford, 1987). Natural diets relate closely with an animal's eating habits and growth patterns in the wild state. These have become adapted and refined under laboratory conditions such that, with rodents at least, good correlations exist between food consumption and growth rate. Each strain of laboratory mouse or rat exhibits a well defined growth curve, which can serve as a reliable guide to an animal's dietary state and health status.

For larger species of laboratory animals, laboratory diets more closely resemble natural rations. Thus monkeys will receive fruit, nuts, and vegetables freshly as a normal part of their diet. These are supplemented with mineral and vitamin mixes. In dog and cat husbandry, fish or meat rations will be supplemented with commercial food products where necessary. In all cases, it will be important that these commercial rations are designed to provide all nutrients not available in natural foods.

The guinea pig is unusual in that it is incapable of synthesising ascorbic acid as obtained from green vegetables, and without this vitamin they die. Commercial rations provide about $3000\,\mathrm{mg\,kg^{-1}}$ vitamin C, thus alleviating the need to provide green vegetables in the laboratory (Table 3). The cat has an unusually high requirement for vitamin A compared with most other species. New world monkeys including tamarins and marmosets require dietary supplements of Vitamin D_3.

In summary, the nutritional requirements for laboratory animals are ideally tailored to the individual demands of the species and its age and physiological state.

C Feeding Patterns and Dietary Manipulation

Rats, mice, rabbits, guinea pigs, and most small mammals will be allowed diet *ad libitum* under laboratory conditions unless the experimental protocol dictates otherwise. However, it is common practice with larger animals to restrict feeding to once or twice daily, allowing food intake to be controlled. Under certain experimental conditions, it may be necessary to limit food intake or vary the composition of the diet. Thus 'pair-feeding' is a common practice in experimental toxicology. Where a particular treatment results in a diminution in food consumption in experimental animals, the food intake of control animals is reduced to eliminate this variable. In other situations, it is quite usual to remove food (but not water) from experimental animals for up to 24 hours before operations or terminal sacrifice. In rats and mice, food withdrawal for up to 24 hours is not excessively stressful.

It is usual in toxicological studies to vary the constitution of laboratory animal diets, either by replacing a natural constituent with a synthetic one, or by placing test substances in the food. Either way, the change may impair the acceptability of the ration, and its palatability or nutritive value. These variables are best explored in preliminary experiments. A further possibility to be considered is that alterations in the diet may influence the intestinal microflora, mucosal physiology of absorptive capacity of the gut. Thus alcohol is high in calorific value, it is readily absorbed in the rodent intestine, but it impairs the regenerative activity of the intesinal mucosa, induces changes in enzyme systems and reduces the absorptive capacity of the tissue in the long term (Lansdown, 1987).

5 Animal Maintenance and Breeding

A Rats and Mice

The vast proportion of toxicological experiments are conducted using rats or mice. These species are relatively inexpensive to maintain, they are economical on space and their biology and husbandry is well known. Although at least 300 different strains of inbred or outbred rodents are available for investigations, the majority of studies reported employ a relatively narrow range of common strains such as the CBA, TO, Balb/c, or C3H mouse, or the Sprague-Dawley, Wistar, or Fischer F344 rat. These strains breed well under laboratory conditions.

Solid-walled plastic cages with stainless steel mesh lids are preferred to older forms of rodent cages since they are secure, readily sterilised, and convenient to house on wall or free-standing racking. Current trends are towards standardisation of rodent cages (see Royal Society-UFAW guidelines, 1987). As a general rule, rodent cages will be cleaned at least twice weekly, sterilised, and animals placed in fresh cages with clean bedding as appropriate. Dead animals will be removed early as rodents are prone to cannibalism. Loss of a dead animal through cannibalism can frustrate toxicological experiments where *post-mortems* will be conducted routinely.

Rodents are social animals and where possible should be housed in groups of 2–5. However, it may be necessary to house singly in metabolic or dietary consumption studies. In animal husbandry, the psychological influence of this and other stresses are of interest (Lawlor, 1984). Some rodent strains are appreciably more nervous than others, particularly inbred strains. Irregularities in animal husbandry here may have marked influences on behavioural patterns in experimental procedures (Levine, 1960; Wells, 1975).

In another context, rodent housing has become the object of ethogram studies. Species exhibit stereotyped movements associated with survival within defined surroundings (Barnett, 1963). Locomotion, respiration, feeding, nesting, fighting, and reproductive activity have been monitored as a means of selecting suitable cages. They may be useful in comparing the behavioural characteristics of different strains of rats and mice under defined sets of conditions.

In rodent breeding, it is common to place animals in cages equipped with stainless steel grid bottoms. Oestrus cycles vary from 4 to 6 days in rats and mice and, in laboratory breeding, it is usual to arrange breeding groups of 3 female animals with one male. Male animals placed in close proximity to the female animals but in separate cages for 24 hours, will be introduced into the main group at approximately 1600 hours on the first day and removed at 0900 h on the second. Successful mating is indicated by the presence of sperm or copulation plugs in the vaginal orifice. Females will then be separated to breeding cages containing nesting materials and maintained in isolation through pregnancy, through lactation, to weaning. The grid bottomed cages used in mating prevent the loss of the vaginal plugs. In our laboratory, infant rats and mice will be separated from mothers at 3 to 4 weeks *post-partum*. It will be important in rodent breeding to keep accurate records to enable the selection of breeding stock and to identify downward trends in breeding performance. (Weihe, 1987; Cunliffe-Beamer and Les, 1987).

B   Rabbits

It is more usual to maintain rabbits singly under laboratory conditions using specially designated rooms. Rabbits are unusually sensitive to changes in environmental conditions, noise, lighting, husbandry, and even technicians. When animals are maintained on a large scale, it is important that their husbandry and maintenance be conducted quietly and methodically. As for rodents, a wide variety of rabbit caging is available commercially, usually constructed of metal. Recommended sizes vary from 40–45 cm height and 2000–6000 sq. cm floor area as a minimum for rabbits weighing from 2 to over 6 kilogrammes, (*ca.* 1000 sq. cm floor kg body weight^{-1}). It is usual for animals to be housed on wire mesh grids (16 mm mesh) allowing faecal material and urine to be disposed of in trays containing sawdust, sand, or other absorbent substance at regular intervals. Where animals are maintained in large numbers, hygiene is greatly improved when banks of cages are equipped with automatic watering and cleaning devices.

Rabbits are obligate vegetarians and as such their incisors are subject to continual growth. This will require regular checks by trained technicians who will trim their teeth from time to time to avoid excessive over-growth. Nails will also require clipping periodically. A further problem encountered in rabbit colonies is bezoar or hair ball formation. The pelage is generally thicker than in most other laboratory species and hair invariably contaminates food hoppers. When this is consumed, hair balls develop in the oesophagus leading to obstruction and death.

As rabbits are temperamental animals, they may be difficult to breed in the conventional laboratory even though the rabbit is a recognised species for reproductive and teratological studies (Gibson, Staples, and Newberne, 1966). Unlike rats and mice, rabbits ovulate at copulation, the egg is well developed and post-fertilisation development is rapid (Adams, 1970). Pregnancy is 31–32 days and litter sizes vary from 2–12 depending upon the age and strain of rabbit studied.

Conditions allowed for breeding rabbits are such as to allow a nesting compartment, draft free environment, and freedom from disturbance. After mating, the females are placed in cages with nest boxes containing straw or nesting materials 3–4 days pre-term. Care is always necessary in rabbit breeding that the young are not disturbed, and caging used is such that the very immature young do not fall out of nest boxes or become trampled upon by the mother. Newborn and young rabbits are also very sensitive to hypothermia.

C Dogs

Dogs are extensively used in laboratory investigations these days and are often selected as non-rodent species in regulatory toxicology studies. Being a larger and more active species than most others used in experimental work, dog pens are equipped with exercise areas. These may be indoors or outdoors, depending on the situation of laboratories and available space, but should exceed 4.5 sq. m per animal. Either way, sleeping accommodation should be dry, clean, and suitably ventilated. Commonly, dogs are housed singly but with an opportunity to see other animals. It is usual for dog areas to be delineated with metal bars and with solid floors that enable regular flushing and cleansing. Straw and woodshavings will be supplied as bedding material as an alternative to a constructed bed of wood, plastic, or other material. Accommodation for laboratory dogs in Great Britain and some other countries is legally controlled (Institute for Laboratory Animal Resources, 1978; Canadian Council for Animal Care, 1980–1984; Research Defense Society, 1979).

D Primate Species

Primate species used in laboratory studies present a number of specialised features in management not encountered with other species. They are all regarded as exotic species and with few exceptions are bred and obtained from sources in tropical or sub-tropical countries. As such, they may be subject to zoonotic infections transmissible to man. Not surprisingly, these animals are the object of numerous managerial publications which detail safety measures and specific recommendations on husbandry (Royal Society-UFAW, 1987; Medical Research Council, 1985; Medical Research Council, 1976). Apart from ethical considerations, it is important that an experimenter gains a good understanding of the biological and behavioural needs of the species he wishes to use.

Current guidelines (Royal Society, 1987) insist that no monkey should be housed in a cage which is less than twice an animals crown-rump length in height. Monkeys are sociable species and usually benefit from being housed so that they can see other animals. Except with marmosets, it was not usual to house monkeys other than singly, in purpose designed metal caging equipped with automatic water devices, grid floors to avoid accumulation of faecal material, and a crush-back to facilitate husbandry, anaesthesia, and dosing. Recently however, there have been moves in Great Britain to persuade laboratories to provide exercise facilities and multiple housing for large primates as a form of 'environmental enrichment'. Automatic cleaning of dropping trays is of obvious benefit as a time-consuming and safety measure.

In most laboratories, work with primate species is normally undertaken by technicians who are experienced in handling them and who are aware of the safety requirements. Thus, in my laboratory, all staff entering monkey rooms will wear safety clothing and observe the guidelines set out by the MRC Simian Virus Committee (Vizoso, 1971–1972).

E Other Species

Cats, guinea pigs, pigs, and occasionally gerbils, hamsters, and sheep are used as experimental animals. Of these species, cats only may require special considerations in management and husbandry. Animals will be purpose bred and of minimal disease status, maintained in a controlled environment with solid floors, and with adult males and queens separated. They are social animals and may be housed in groups of 5–20 with sufficient space to allow 0.5–0.75 sq. m per animal. Like dogs, cats do need some facility to exercise as well as for clawing (Royal Society, 1987) and climbing. Dirt trays are placed in indoor cat compounds. Cats do need constant temperature under laboratory conditions, and floors heated to 20°C are recommended. Technicians involved in cat husbandry will be aware of the scrupulous cleanliness required at all times.

Guinea pigs are frequently used in immunological studies. They are not usually bred in experimental laboratories because of their long gestation and economical factors, but animals obtained from commercial suppliers present few special difficulties in management or husbandry. They are housed singly or in single sex groups, allowing up to 750 sq. cm per animal depending upon its age and weight.

Pigs, sheep, and occasionally goats will be used in some experimental studies for surgery or serum production. As farm animals, they may be housed in the open or in indoor accommodation. In the latter case, grid floors are preferred for ease of cleaning and hygiene. Areas of 2–7.5 sq. m are allowed, depending upon an animals size. Good ventilation, automatic watering, and exercise facilities are essential for indoor accommodation. Regular weighing and food consumption monitoring provides a good indication of an animal's well-being, or otherwise.

6 Isolator Technology

In the past 25 years, isolators have become an established part of many large experimental animal facilities. They have developed from simple filtered-air animal boxes, fitted with long-sleeved rubber gloves for sterile manipulation, to the more sophisticated forms capable of holding up to 200 animals. Larger varieties are available for large-scale animal breeding. Isolator development has created an opportunity to produce large numbers of germ-free or gnotobiotic animals, animals of a hypo-immune status (the nude rats and mice), and to study the pathogenicity of hazardous infections transmissible to man, or to other animals.

Isolators available these days are mostly of the flexible-film type consisting of an envelope of plastic supported on a rigid metal frame. Various isolators exist but they are essentially of two main types. There are those designed for ease of operation, as in caesarian derivation of the germ-free stock, microbiological sampling and isolations, and tumour implantations. Secondly, there are the larger cuboidal or elongate isolators which are preferred for animal breeding. This type will be fitted with several pairs of long sleeved rubber gloves. Animal cages will be mounted on racks to economise on isolator space. All modern isolators are fitted with controlled air filtration, environmental monitoring devices, internal illumination, and reliable power units. Air inlets function under negative pressure, whereas exhausts will operate under positive flow and be designed as to prevent a backflow of air should there be a power failure.

Operation of isolators and the derivation of gnotobiotic animals is described elsewhere (Trexler, 1987). Broadly, isolator technology requires that all materials used within the envelope will be germ-free (except where specified organisms are introduced for study). This will include the animals and all material involved in their husbandry and experimental procedures. Thus food, bedding, instruments, and other materials will be sterilised and prepacked before introduction to isolators *via* dunk tanks or double door chambers. Hypochlorite or peracetic acid are often used in the sterilisation process.

Commonly, gnotobiotic animals will be moved from one isolator to another using 'transporter isolators' which couple to the double-door portals on the main isolators. Sterile animal cages can then be passed directly without exposure to the open environment.

Derivation of gnotobiotic stock is by caesarian section using sterile procedures at near term. The pregnant uterus will be removed from the mother under aseptic conditions and introduced into the isolator *via* a sterile bath. Freshly isolated young will be fostered by a gnotobiotic dam until weaning. A second method of deriving germ-free animals involves the transfer of fertilised eggs or preimplantation embryos to a suitably prepared recipient dam. Once a nucleus of breeding stock has been established in the isolator, successive generations can be established in other isolators or elsewhere.

Once a germ-free colony of animals has been established, it will be essential to monitor the animals and the isolator environment periodically for evidence of breakdown in the conditions. In event of contamination, an isolator may require complete sterilisation and restocking with freshly derived animals. In extreme cases, envelopes will be renewed. Most isolators have a life of about 5 years before envelopes require renewal.

7   Quality Assurance

The need to maintain animals in good quality should be the aim of all experimentalists. Not only will it be desirable to achieve microbiological cleanliness, but quality control is essential in obtaining reproducible results. Quality assurance is essential in the case of large animal breeders aiming to supply animals of high standard, but it will be important in the preservation of in-house stocks from microbiological and genetical contamination. Contamination may occur through aerobic means, by food and water, and through contact with other animals, or even humans. Of particular concern will be organisms 'imported' with animals arriving from overseas. In Great Britain at least, rigid quarantine requirements are enforced. Genetic contamination is less of a problem for most experimental toxicologists using highly outbred strains. However, to an immunologist involved in transplant studies, and experimenters requiring inbred strains, genetic 'drift' is a problem.

A   Quarantine and Isolation

Quarantine, by definition, is the compulsory maintenance of animals newly imported into a country in isolation for a minimum period of 6 months (Medical Research Council, 1976). In Great Britain, the accommodation is approved by the Ministry of Agriculture Fisheries and Food acting under the Rabies Order (1974). Although quarantine applies to all species, it is of greatest importance in the case of non-human primates. The MRC Committee on Simian Viruses (1985) considers the isolation period as a 'conditioning phase' during which animals will be observed by a veterinarian. Rhesus monkeys, for example, should provide at least three negative tuberculin tests and reactions for anergic infections.

To maintain the quality of animals in a laboratory, a curator may insist that animals arriving in his laboratory, albeit from his own country, will be isolated from in-house stocks. Some microbiological sampling is essential. Quarantine rooms will be physically isolated from all other animal holding rooms and, as far as possible, staff working there will not move to other premises without a complete change of clothing. All quarantine rooms should be subject to strict barrier conditions. Where a laboratory does not have a suitable facility for quarantine, negative pressure isolators are a convenient alternative.

B Microbiological Monitoring

In larger animal houses, regular microbiological screening of the air in animal rooms, filters, bedding, and animals is a routine procedure. It will form an essential part of an animal house engaged in breeding. This monitoring will involve checks for viruses, bacteria, helminth infections, and ectoparasites.

In the smaller laboratory, microbiological monitoring may be conducted using sentinel animals which will be killed periodically and blood, urine, faecal, and maybe throat swabs sampled. Larger animals will be screened for infections when they arrive. Where worm burdens are identified, these will be treated before animals are committed to experimental rooms.

C Air Filters

In purpose designed animal rooms, quarantine rooms, and isolators, ventilation will be through filters which are sufficient to exclude contamination from the surrounding environment. These filters will be changed regularly and checked microbiologically.

D Barrier Maintenance

In most modern laboratories, animals are bred and maintained under barrier-managed conditions. These are designed as to allow high quality specified pathogen free (SPF) animals to be available as recommended by the MRC Laboratory Animal Centre guidelines (Townsend, 1969). The form that the barrier takes will vary between laboratories but usually is such as to limit the passage of personnel, and escaped rodents to contaminate the premises. Scrupulous cleanliness will obtain throughout and rooms will be equipped with devices to exclude insects.

E   Containment of Infection

However well an animal house is constructed, occasions will arise when systems for preventing infection break down. Decontamination measures in respect of Ectromelia, *Shigella sp., Salmonella sp. etc.,* will vary, but in event of a serious infection, all animals in a room will be killed and the room fumigated. Formaldehyde used to be the fumigant of choice, but as its toxicity has become widely publicised recently, less harmful agents like Virkon (R&S. Biotech. Finedon, Northants) are preferred. Eaton (1987) indicated that once the nature and gravity of an infection is known, the curator is faced with the task of identifying the source and taking appropriate preventive measures.

Latent infections will occasionally present difficulties in quality control. These are a particular problem in laboratory monkeys such as Cynomolgus, Rhesus, and Vervet breeds. Animals may exhibit negligible serum titres for *Herpesvirus simiae* (B-virus), hepatitis virus, or measles during quarantining or under conventional laboratory conditions, but if the animals are subjected to immunosuppressive treatments or even stress, latent viral infections in spinal ganglia become active, and present a very serious risk to handlers (Perkins and O'Donoghue, 1969; Vizoso, 1971–1972; Prior *et al.,* 1964; World Health Organisation, 1971).

8   Experimental Procedures

A   Dosing

Though it is possible to use any route or method of administration, the use of injection is limited by the local tissue reactions provoked, and the duration of such treatments may be limited. It is, however, possible to carry out repeated and long term studies by subcutaneous injection if the sites used are varied. This still depends, however, on the rapid clearance of the material from the site of injection and usually will be confined to water soluble chemicals not causing a local reaction.

Oral administration is by far the most common method for long term studies. This may take the form of daily gastric intubation (gavage) or incorporation of the material into the diet. It is possible to use daily intubation even for life-time experiments and although it is time consuming and expensive, where the procedure is carried out by suitably experienced and sensitive operators there need be no unwarranted stress to the animals or loss of experimental subjects due to technical errors. It is, of course, essential that control animals are subject to the same procedures. The method has the advantage of accuracy of dosage.

Mixture of the test chemical with the diet is the simplest and most common method of administration, especially where large numbers of animals and long periods of treatment are involved. It results in some loss of accuracy of individual dosing but in most studies this is acceptable. Individual housing with frequent monitoring of body weight and food intake (at the extreme, daily) allows retrospective monitoring of the dosage but this is seldom absolutely necessary.

When giving a chemical in the diet it is important to bear in mind the changing ratio of food intake and body weight in the young animal. If a fixed concentration of test material is fed to rats there may be a 3–4 fold reduction in dosage (mg kg body weight^{-1}) between post-weaning and 3–6 months later. The worst of these variations can be avoided by appropriate modification of the dietary concentration.

In order to ensure adequate control of dosage during dietary administration, analytical investigations are important to ensure that:

The dietary mixture is homogeneous

The dietary mixture is stable over the period required (*e.g.* there is no settling of the test component or loss through volatility, and that there is no chemical interaction between the diet and the test compound)

There is no selection by the animals, to avoid or preferentially select the test material from the diet

Each batch of mixture prepared is of the required concentration.

The monitoring of the intake from the diet requires an accurate measurement of food intake and is an essential part of any good experimental design. As well as monitoring the dosage of the test material, the recorded body weights and food intakes can be used as guides to the condition of the animal. Changes of the food conversion ratio whereby less weight is gained for the same food consumed may be a useful measure of the toxic effect of a chemical.

In some instances, the presence of the test chemical reduces the palatability of the diet. The test animals then fail to gain weight normally because they eat less. In these circumstances it will be necessary to conduct a 'paired feeding assay' in which the control animals are started on trial one day later than the test animals, and are provided with precisely the weight of diet consumed by the test animals the day before, less, if necessary, the weight of the test material (usually negligible). It may then be possible to compare the weight gained in relation to the diet consumed. An adverse effect on palatability is not regarded as a toxic effect.

The administration of the test substance in the drinking water (usually provided *ad libitum*) adds to the complexity of the experiment in that both food and water consumption need to be measured.

B Examination of Animals during Experimentation

Animals will be examined periodically throughout any experimental procedure, and especially in toxicological studies. Records will include:

(a) Alterations in general appearance, physical and behavioural faculties, feeding and water consumption

(b) Appearance of eyes (with or without ophthalmoscope)
(c) Skin and hair form
(d) Mouth, teeth, and throat (especially in larger species)
(e) Presence of palpable or overt masses (? tumours) and other lesions
(f) Evidence of infections, abscesses, sores, *etc.*

Additionally, experimental protocols and good laboratory practice will normally specify sampling of blood, urine, and faeces to evaluate the influence of test procedures. It is usual to collect samples at specific times in the day to avoid changes due to diurnal rhythms

Blood sampling will depend upon the species involved. Thus for rats and mice, bleeding from the tail vein, heart, or retro-orbital sinus is usual. The posterior ear vein is the usual route in the rabbit. Cardiac puncture is used in guinea pig bleeding. For larger species, cat, dog, pig, goat *etc.*, blood sampling for an experienced technician will usually be by the jugular or femoral vein, or occasionally a more distal limb vein. The marmoset is not an easy animal to bleed, but the femoral vein is most suitable, with the animal being restrained in a body harness.

C Animal Identification and Randomisation

In good laboratory practice, it is usual to assign animals to experimental groups by randomisation or by computerised methods. Such randomisation demands an unequivocal identification of each using colour coding, tattooing, or by the use of neck or foot tags. It may be convenient in large scale studies to use a colour coding scheme to mark individual diet or treatment groups.

9 Records

In animal breeding and experimentation, a vast amount of numerical and qualitative data will be accumulated. This will necessitate accurate recording, assimilation, and storage pending report preparation. Most larger laboratories and contract research companies will computerise this information. Computer prints are admissible for legislative purposes.

In the foregoing, I have alluded to breeding and experimental protocols. The former will be compiled to demonstrate:
(a) Progress and fluctuations in the breeding patterns of particular strains and species
(b) Fecundity and the value of individual animals or pairs in a breeding programme
(c) Rates of mortality and survival in various strains/species
(d) Seasonal variations in breeding performance.

This information is then used in the selection of breeding stock, forward planning, evidence of genetic drift, breakdowns in environmental conditions, *etc.* In a toxicological programme, computerised records and statistical evaluations will be employed to identify changes in such parameters as haematological profiles, body weights, food consumption, and pathological lesions.

Certain records will be maintained by laboratories by legal obligation. These will include:

(a) Acquisition of animals, their age, sex, origin, and condition on arrival
(b) Record of animal health
(c) Dates and forms of vaccination, treatment with anthelmintic drugs, antibiotics, *etc.*
(d) Fatalities and *post-mortem* observations
(e) Emergencies—injuries to animals requiring veterinary supervision.

10   Discussion

The 1986 Animals (Scientific Procedures) Act in Great Britain provides a realistic framework within which to use and protect all vertebrate animals employed in regulated experiments. It sets out guidelines for the correct treatment of various species and the responsibilities of those who handle them. It is recognised that in a chapter of the present length, it is not possible to discuss, other than briefly, the various aspects of husbandry, nutrition, and quality assurance as expected under this new Act, the reader is referred to publications by the UFAW, WHO (1971) and Canadian Council on Animal Care (1980–1984) for more detailed reading.

It is evident from the vast number of excellent publications issued nowadays, that animal welfare, quality control, and monitoring of infections and disease are essential parts of experimental toxicology and animal research. The introduction of new technology in isolators, quarantine facilities, and animal house design, coupled with higher levels of technician training and expertise, enable us to perform our work with improved accuracy and safety.

References

Adams, C. A. (1970). Ageing and reproduction in the female animal with particular reference to the rabbit. *J. Reprod. Fertil., Suppl.,* **12,** 1.

Advisory Committee on Animal Experiments (1981). Report to the Secretary of State on the framework of legislation to replace the Cruelty to Animals Act 1876. Home Office, HMSO, London.

Albanese, A. A. (1973). The effect of maternal nutrition on the development of the offspring. *Nutr. Rep. Int.,* **7,** 243.

Barnett, S. A. (1963). 'A Study of Behaviour', Methuen, London.

Berg, B. N. (1960). Nutrition and longevity in the rat. *J. Nutr.,* **71,** 242.

Biological Council (1984). Guidelines on the use of living animals in scientific investigations. London.

Canadian Council for Animal Care (1980–1984). Guide to the care and use of laboratory animals. Ottawa.

Clapp, M. J. L. (1980). The effect of diet on some parameters measured in toxicological studies in the rat. *Lab. Anim.*, **14**, 253.

Clarke, H. E., Coates, M. E., Eva, J. K., Ford, D. J., Milner, C. K., O'Donoghue, P. N., Scott, R. P., and Ward, R. J. (1977). Dietary standards of laboratory animals. *In*: 'Report of the Laboratory Animals Centre, Diets Advisory Committee', *Lab. Anim.*, **11**, 1.

Clough, G. (1984). Environmental factors in relation to the comfort and well-being of laboratory rats and mice. *In*: 'Proceedings of Joint LASA-UFAW Symposium: Standards in Laboratory Animal Management'. Pt. 1, p. 5.

Clough, G. and Townsend, G. H. (1987). *In*: 'The Care and Management of Laboratory Animals'. *Ed.* T. Poole, Longman Scientific and Technical, Bath, p. 159.

Cooper, M. E. (1978). The Dangerous Wild Animals Act, 1976. *Vet. Rec.*, **102**, 475.

Council of Europe (1986). European Convention for the Protection of Vertebrate Animals for Experimental and other Scientific Purposes. Strasbourg, HMSO, London.

Cuncliffe-Beamer, Th. and Les, E. P. (1987). The laboratory mouse. *In*: 'The Care and Management of Laboratory Animals'. *Ed.* T. Poole, Longman Scientific and Technical, Bath, p. 275.

Dean, R. F. A. (1951). The size of a baby at birth and the yield of breast milk. *In*: 'Studies on Undernutrition, Wuppertal, 1946–9'. MRC Special Report, No. 275, 346, HMSO, London.

Eaton, P. (1987). Hygiene in the animal house. *In*: 'The Care and Management of Laboratory Animals'. *Ed.* T. Poole, Longman Scientific and Technical, Bath, p. 275.

Festing, M. F. W. (1981). The 'defined' animal and the reduction of animal use. *In*: 'Animals in Research: New Perspectives in Animal Experimentation'. *Ed.* D. Sperlinger, J. Wiley and Sons Ltd., Chichester, p. 285.

Festing, M. F. W. (1987). Introduction to animal genetics. *In*: 'The Care and Management of Laboratory Animals', *Ed.* T. Poole, Longman Scientific and Technical, Bath, p. 58.

Ford, D. J. (1987) Nutrition and feeding. *In*: 'The Care and Management of Laboratory Animals', *Ed.* T. Poole, Longman Scientific and Technical, Bath, p. 35.

Gibson, J. P., Staples, R. E., and Newberne, J. W. (1966). The use of the rabbit in teratology studies. *Toxicol. Appl. Pharmacol.*, **9**, 398.

Gibson, T. E., Paterson, D. A., and McGonville, G. (*Eds*) (1986). 'The Welfare of Animals in Transit', Proceedings of the Animal Welfare Foundation's Third Symposium. British Veterinary Association.

Hammell, D. L., Kratzer, D. D., Cromwell, G. L., and Hays, V. W. (1976). Effect of protein malnutrition of the sow on reproductive performance and on post-natal learning and performance in the offspring. *J. Anim. Sci.*, **43**, 589.

Hurley, L. S. (1977). Nutritional deficiencies and excesses. *In*: 'The Handbook of Teratology', *Eds.* J. G. Wilson and F. C. Fraser, Plenum Press, New York, **1**, 261.

Institute of Laboratory Animal Resources (1978). Guide for the use and care of laboratory animals, US Department of Health, Education and Welfare, National Institutes of Health, Bethesda, Maryland.

International Council for Laboratory Animal Science (1978). ICLAS recommendations for the specification of animals, the husbandry and techniques in animal experimentation. *ICLAS Bull.*, **42**, 4.

Jones, D. G. (1976) The vulnerability of the developing brain to undernutrition. *Sci. Prog. Oxford*, **63**, 483.

Lansdown, A. B. G. (1975). Pathological changes in pregnant mice infected with Coxsackievirus B3 and given dietary casein hydrolysate supplement. *Br. J. Exp. Pathol.*, **56**, 373.

Lansdown, A. B. G. (1976) Pathological changes in the pancreas of mice following infection with Coxsackie B viruses. *Br. J. Exp. Pathol.*, **57**, 331.

Lansdown, A. B. G. (1977) Histological observations on thymic development in foetal and newborn mammals subject to interauterine growth retardation. *Biol. Neonat.*, **31**, 252.

Lansdown, A. B. G. (1982). Teratogenicity and reduced fertility resulting from factors present in food. In: 'Toxic Hazards in Food', *Eds*, D. M. Conning and A. B. G. Lansdown, Croom-Helm, Beckenham, Kent, p. 73.

Lansdown, A. B. G. (1987). Alteration in the crypt cell population in the small intestine as an early toxic response to sub-acute ethanol administration. *Arch. Toxicol.*, **59**, 448.

Lansdown, A. B. G. (1992). The protected animal: an overview of the guidelines controlling the health and well-being of laboratory animals in the United Kingdom. Proceedings of the 8th Congress of the Federation of Asian Veterinary Associations, Manila, Philippines.

Lawlor, M. (1984). Behavioural approaches to rodent management *In*: 'Standards in Laboratory Animal Management', UFAW, **1**, 40.

Levine, S. (1960). Stimulation in infancy. *Sci. Am.*, May, 81.

McCance, R. A. (1962). Food, growth, and time. *Lancet*, **2**, 621.

Medical Research Council (1974). The Accreditation and Recognition Schemes for Suppliers of Laboratory Animals. Manual Series No 1.

Medical Research Council (1976). 'Laboratory Non-Human Primates for Bio-Medical Research in the United Kingdom'. MRC, London.

Medical Research Council (1985). The Management of Simians in Relation to Infective Hazards to Staff, London.

News and Reports (1986). Welfare Problems in todays Noah's Arks. *Vet. Rec.*, **119**, 561.

Perkins, F. T. and O'Donoghue, P. N. (1969). Hazards In Handling Simians, Laboratory Animals Handbook, No. 4, Laboratory Animals Ltd., London.

Prior, J. E., Saur, R. M., and Fegley, H. C. (1964). Zoonoses associated with laboratory monkeys. *Lab. Anim. Care*, **14**, 48.

Ray, P. M. and Scott, W. M. (1973). Animal Welfare Legislation in the E. E. C. *Br. Vet. J.*, **129**, 194.

Remfrey, J. (1976). The control of animal experimentation. Das Tier im Experiment. *Ed.* Herausgegeben Von Wolf H. Weihe, Verlag Hans Huber, Bern.

Remfrey, J. (1984) Protecting experimental animals. *Nature (London)*, **312**, 191.

Research Defense Society (1979). Notes on the Law Relating to Experiments on Animals in Great Britain, London. p. 15.

Roeder, L. M. (1973). Effect of level of nutrition on the rate of cell proliferation and of RNA and protein synthesis in the rat. *Nutr. Rep. Int.* **7**, 271.

Royal Society-Universities Federation for Animal Welfare (1987). Laboratory

Animals and their use for Scientific Purposes. I. Housing and Care. Wembley Press, London.

Spiegel, A. (1965). Sterilisation of food, bedding and other materials. *Food Cosmet. Toxicol.*, **3**, 57.

Townsend, G. H. (1969). The grading of commercially available laboratory animals. *Vet. Rec.*, **85**, 225.

Trexler, P. C. (1987). Animals of defined microbiological status. *In*: 'The Care and Management of Laboratory Animals', *Ed.* T. Poole, Longman Scientific and Technical, Bath, p. 85.

Tucker, M. J. (1979). The effect of long-term food restriction on tumours in rodents. *Int. J. Cancer*, **23**, 803.

Universities Federation for Animal Welfare (1980). Standards in Laboratory Animal Care, Parts I and II, UFAW, London.

Universities Federation for Animal Welfare (1987). 'The Care and Management of Laboratory Animals', *Ed.* T. Poole, Longman Scientific and Technical, Bath.

US Food and Drug Administration, (1978a). Laboratory Practice Regulations, Department of Health, Education and Welfare. Washington, D.C.

US Food and Drug Administration (1978b). Compliance Program Guidance Manual, Washington D.C., October 23rd.

Valtin, H. (1967). Hereditary hypothalmic diabetes insipidus in rats (Brattleboro strain). *Am. J. Med.*, **42**, 814.

Vizoso, A. (1971–1972) Progress Report, 1967–1970. Annex to MRC Report, London.

Weihe, W. H. (1987). The laboratory rat. *In*: 'The Care and Management of Laboratory Animals'. *Ed.* T. Poole, Longman Scientific and Technical, Bath.

Wells, P. A. (1975). The influence of early handling on the temporal sequence of activity and exploratory behaviour of the rat. PhD Thesis, University of London.

Widdowson, E. M. (1965). Factors affecting the growth rate of laboratory animals. *Food Cosmet. Toxicol.*, **3**, 721.

World Health Organisation (1971). Health aspects of the supply and use of non-human primates for biomedical purposes. WHO Technical Report series, No. 470, Geneva.

CHAPTER 7

Inhalation Toxicology

R. A. RILEY AND D. M. CONNING

1 Introduction

There are vast numbers of compounds in use throughout the world that are capable of being inhaled into the respiratory system. The investigation of the adverse effects of such airborne materials constitutes inhalation toxicology. This branch of toxicology is, more than any other, dominated by the technology of exposure and by the special morphology of the target organ (Leong, 1981; Salem, 1986). As a consequence, much effort is needed to ensure that test exposures are consistent in physical format and concentration, and that these parameters are sufficiently controlled to ensure adequate penetration of the lung. The biological consequences are very different where the test material is deposited at the bronchiolar level, for example, rather than the alveolar. If penetration to the alveoli is achieved, absorption may occur, in which case the toxicological evaluation is akin to any other study of systemic toxicity with one notable, and sometimes important, distinction. This is that absorption is effectively into the systemic circulation and not *via* the liver as occurs with most (but not all) materials administered orally. Thus, the major detoxifying organ may be effectively bypassed for a limited period of time and this may have consequences not identifiable following ingestion.

There is another aspect to inhalation toxicology that warrants special attention. This is the close relationship with occupational hygiene. Many of the techniques used to measure atmospheric concentrations are common to both activities and, theoretically, it should be possible to compare the human and animal reactions more closely for a given level of exposure. In practice such comparisons are rarely achieved, usually because the levels of, for example, industrial exposure are orders of magnitude less than those generated in the laboratory.

This chapter deals with the technical requirements to ensure consistent exposure of experimental animals to materials administered by inhalation, the aim being usually to achieve alveolar penetration. We do not dwell here on the range of pulmonary reactions that may occur.

2 Physical Classification of Airborne Pollutants

The many thousands of compounds which have the potential to become airborne, and therefore be inhaled, may be broadly classified according to their physical properties, although obviously there will be some overlap between groups. At the molecular level materials may be *gases*, or substances in the gaseous state at normal temperature and pressure, or they may be *vapours*, where the compound exists as a liquid or solid but is translated at a steady rate into the surrounding environment. Particulate materials, either liquid or solid, when dispersed in air are termed *aerosols* and are produced when the compound is reduced to a particle size that allows it to become airborne. Atmospheres consisting of solid particulate materials may be generated by grinding, crushing, impaction, or degradation or may be naturally occurring, *e.g.* pollens. *Fume*, usually an oxide, is produced as the result of heating the material to such a temperature that a gas is evolved which then recondenses to solid particles of a size less than 1 micron. Fumes are generated, for example, during welding processes.

Advances in technology have increased the number and amount of chemicals in the atmosphere. It is estimated that there are over 50,000 different compounds in use today and this is increasing by around 1000 per annum. The health effects of inhalation of some of these chemicals can be predicted to some extent by experimental investigation.

In order to simulate environmental conditions, a special technology has evolved relating to the design and operation of inhalation chambers and the generation and characterisation of aerosols and vapours, *etc.*

The costs associated with evaluation of toxicity by the inhalation route are considerably higher than for studies using other routes of exposure, essentially because of the cost of the specialised equipment. It is estimated that an inhalation study costs two to three times that of a comparable study by ingestion.

The procedures are those suitable for laboratories concerned with the toxicity of chemicals which may contaminate industrial or agricultural atmospheres or which may give rise to a community air pollution problem. They may also find application in the study of the effects of volatile or airborne drugs, of smokes, and of gases used in chemical warfare.

3 Methodology of Inhalation Toxicology

The principle is to establish a dynamic environment in which the test system, usually an experimental animal, is located in an airstream that contains a fixed concentration of the test material in suitable format. Static systems, in which the animal is placed in a container with a fixed volume of air with the requisite concentration, are rarely used.

Where it is important to exclude exposure by ingestion, it is necessary to arrange for individual nasal exposure only, to obviate grooming. Where this is less important, groups of animals may be housed in a chamber through which air is passed. The chamber must be of a design that ensures uniformity of distribution of the test material in the air stream, with the avoidance of turbulence which would allow local variation in concentration. The air flow must be such as to ensure the levels of water vapour emanating from the animals, of temperature, and of oxygen are maintained within fairly narrow confines, depending on the number of animals in the chamber. These constraints are more easily controlled for gases and solid-state particles than for aerosols.

In studies that are generally used are those of a basic cuboidal design (Drew and Laskin, 1975; Hinners *et al.*, 1968; Hemenway *et al.*, 1982; Laskin and Drew, 1970; Leach *et al.*, 1959; Hemenway and MacAskill, 1982; Beethe *et al.*, 1979; Doe and Tinston, 1981; Macfarland, 1983). However, the ideal chamber to ensure uniform distribution is essentially circular in design. It should be constructed of materials which will not react chemically with the test material, should allow good visibility for observation of the animals, and should be easily decontaminated.

In studies involving small groups of animals a suitably modified desiccator may conveniently be used. The animals, usually rats, mice, or hamsters, are housed within the chamber supported on a metal grid through which excreta may pass. Stainless-steel partitions can be inserted to segregate the animals if required. Atmospheres are introduced to the chamber down a central column under positive pressure. They then rise through the areas where the animals are housed and finally exit through holes positioned radially around the central input column. These output holes also allow the insertion of probes for analysis of the test atmosphere and the measurement of temperature and relative humidity. Because these chambers operate under positive pressure they must, for the safety of the investigators, sit within a ventilated space such as a fume cupboard. According to good laboratory practice for inhalation toxicology, the total space occupied by the animals in any chamber should not exceed 5% of its internal capacity. Consequently, chambers of this design limit the number of animals which may be exposed to a maximum weight of 1 kg.

Until recently, exposure chambers of cuboidal design capable of housing larger groups of animals have been fitted with pyramidal top and base units, and constructed of materials ranging from perspex to stainless steel shell with glass windows. However, the chambers recently developed at BIBRA (Riley, 1986) are constructed entirely out of borosilicate glass (Figure 1). The wide range of glass components available make construction of exposure chambers extremely flexible, both in terms of chamber volume and in the addition of external components which may be required for a particular study. Complete chambers are essentially circular in shape and therefore provide excellent aerodynamic properties.

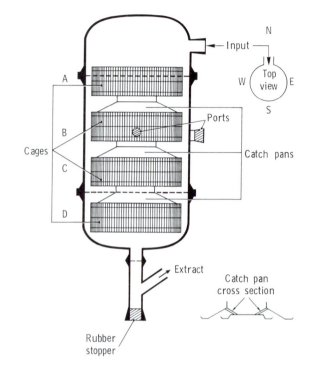

Figure 1   *Diagrammatic representation of a typical exposure chamber*

Because they are constructed of glass they afford good all-round visibility, they are inert to the atmospheres passing through them, and they are easily cleaned. Atmospheres are introduced to the chambers through a side-arm at the top, and this causes a swirling motion to ensure complete uniformity of concentration throughout the chamber. After passing through the chamber the atmospheres are exhausted under negative pressure *via* in-line filtration systems. By careful off-balancing of the exhaust and input air, a slight negative pressure of around 1 ml water gauge may be set and maintained; a water manometer connected to the chamber gives a visible indication of this pressure. All joints between components of the chamber are ground glass and consequently leaks either inwards or outwards are minimised. Holes are provided around the chamber to enable probes to be inserted to measure temperature, relative humidity, and pressure and also to enable atmospheres to be drawn to a suitable analytical system.

Animals are exposed in specially designed stainless-steel wire mesh cages, each having the capacity to house 10 animals individually. The cage units are supported on a metal frame and the chambers currently in use each have the capacity to hold six of these units. Catch pans may be inserted to prevent excreta falling onto animals positioned in lower tiers. Where nasal exposure is required the individual animals are restrained in tubes which allow only the nose or head to protrude. These tubes may be plugged directly into the side of the exposure chamber.

The generation of test atmospheres of a calculated concentration is determined by the introduction of known amounts of the test material into a constant volume of air over a given period. For gases this is readily achieved by using calibrated rotameters. The generation of vapours from solvents may be approached by numerous methods, including bubbling air through the liquid, diffusion of head space into the air stream, or by atomisation of the liquid at a constant rate (Gage, 1970a and b). The preferred method is atomisation, which can operate for lengthy periods and avoids the problems that can arise if the test material contains impurities and is unstable to continual aeration. The sample is introduced into an atomiser from a syringe connected to an infusion pump. The syringe size, injection rate, the air volume through the atomiser, and the density and molecular weight of the material determine the actual generated concentration.

Special difficulties attend the generation of atmospheres containing liquid or solid aerosols. In addition to a known and constant concentration, the particle size distribution is important, as this will determine the regions of the respiratory tract in which the aerosol is deposited, and thus influence toxicity. Particles greater than 10 microns in aerodynamic diameter tend to be trapped in the nasal cavity and main bronchi (Hatch and Gross, 1964). They may cause local damage or may be absorbed through the mucosa. Particles between 10 μm and 5 μm achieve varying degrees of penetration of the bronchial tree, the smallest likely to impact in the terminal bronchioles. Particles less than 1 μm to 5 μm in diameter will impact in the alveoli either by turbulence and velocity or by Brownian movement. Particles smaller than 0.1 μm usually will not impact but be carried out of the lung during exhalation.

The establishment of a known, constant concentration of a dust or mist within a narrow size range presents technical difficulties. For solid aerosols a generator has been developed in which the powder is compressed into a pellet in a precision bore glass tube (Figure 2). When installed in the generator the pellet is pushed at a pre-determined rate into contact with a continually rotating knife edge. This action loosens the top surface of the pellet and a negative pressure air stream removes the particles. Before introduction into the animal chamber, the aerosol is firstly passed through a cyclone which removes any particles greater than 30 μm and also assists in the separation of agglomerates which may have formed during pelleting.

Atmospheric concentrations of liquid aerosols are somewhat easier to prepare. Conventional atomisers are employed but these give a wide range of droplet size. Again, as in the solid particle generation system, coarse particles are removed by introduction into a cyclone (Gage, 1968). The coarse particles are collected by impinging on the cyclone's outer surface whereas the finer 'respirable' particles remain in the air stream which is directed into the exposure chamber.

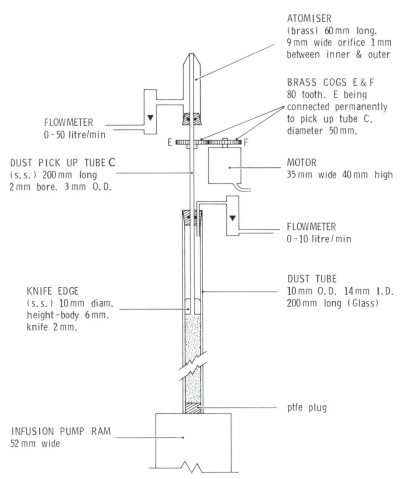

ATOMISER
(brass) 60 mm long.
9 mm wide orifice 1 mm
between inner & outer

BRASS COGS E & F
80 tooth. E being
connected permanently
to pick up tube C.
diameter 50 mm.

FLOWMETER
0 - 50 litre/min

DUST PICK UP TUBE C
(s.s.) 200 mm long
2 mm bore. 3 mm O.D.

MOTOR
35 mm wide 40 mm high

FLOWMETER
0 - 10 litre / min

DUST TUBE
10 mm O.D. 14 mm I.D.
200 mm long (Glass)

KNIFE EDGE
(s.s.) 10 mm diam.
height - body 6 mm.
knife 2 mm.

ptfe plug

INFUSION PUMP RAM
52 mm wide

Figure 2 *Diagrammatical representation of dust feed mechanisms*

It is excellent practice, and in most cases is now mandatory, to check the concentration of the test atmosphere within the breathing zones of the exposed animals. The monitoring of gases and vapours in general present little difficulty—gas chromatography, infrared spectrometry, mass spectrometry, and various specific analysers are readily available. However, it is preferable to check the analyser reading against a suitable chemical method of analysis at least once per day. This will ensure that the chosen instrument is standardised for measurement of the specific atmosphere.

With aerosols the analysis is more complex as not only should the total weight concentration be determined but more essentially the particle size distribution, for reasons mentioned above. The total mass concentration may be determined by drawing a measured volume through a membrane filter and by analysing the amount collected either by direct weighing or

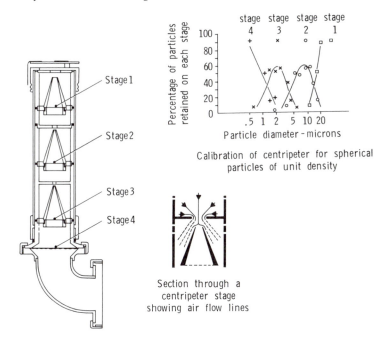

Calibration of centripeter for spherical
particles of unit density

Section through a
centripeter stage
showing air flow lines

Figure 3 *A typical cascade impactor*

preferably by a specific chemical determination. The particle size distribution should be measured using analysers which take into account the aerodynamic properties of the material. For this reason it is desirable to use a device such as a Cascade Impactor (Figure 3). This instrument employs various stages in which jets are provided in decreasing diameters (Hounam and Sherwood, 1965). As air is drawn through these stages at a constant rate, the jet velocity increases. This causes particles of decreasing size to be impacted on collection media positioned below each descending stage in the instruments. Analysis of the amount collected is again either by weight difference, or by a suitable chemical method.

When an atmosphere is introduced into a chamber, there is a delay until the calculated concentration is reached. If good mixing occurs the concentration c_t at time t is given by the following expression, (Silver, 1946):

$$t = \frac{V}{n} \log_e \left(\frac{c}{c - c_t} \right)$$

where c is the calculated concentration, n is the air flow in litres per minute, and V is the chamber volume in litres.

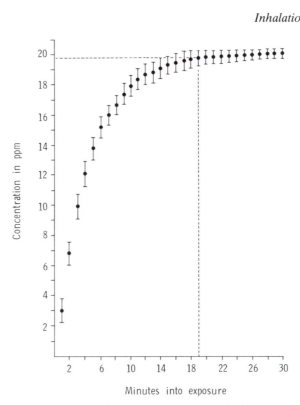

Figure 4 *Means and standard deviations of nitrogen dioxide atmospheres from 40 sample points at one minute intervals*

In simplified form the time taken to reach 99% of the desired concentration is given by:

$$t_{99} = \frac{4.605 \times V}{n}$$

Figure 4 shows the rise to equilibrium of nitrogen dioxide atmospheres in a typical BIBRA chamber. The mean values at 1 minute intervals are identical to the theoretical values calculated from the above equation, and therefore an animal may be placed in any chamber position and will receive the same concentration × time product of inhaled material as the remainder of the exposed group.

4 Lung Clearance Mechanisms

The bronchiolar tree is equipped with ciliated cells that move the layer of mucus, lining the lumen, continuously towards the epiglottis. Material that is deposited on the bronchial and bronchiolar surface is cleared from the lung by this means. Material deposited on the terminal bronchioles is usually cleared within twenty-four hours.

Material that is deposited within the alveoli is cleared by macrophage activity and transported to the local lymphatics where it is rapidly conveyed to the bronchial lymph nodes. Inert particles may be cleared from the alveoli in about five days. Where the particles are cytotoxic (*e.g.* silica) clearance may take many days.

The rate of clearance may affect local toxicity quite considerably. Numerous sensory endings are located at the surface of the respiratory tract, from the nose to the alveolar level. They are easily stimulated by a variety of chemicals and their stimulation is followed by a variety of reflex responses. Such responses will result in bronchoconstriction which will, in turn, further impede both inhalation and clearance. The differential control of airway potency by such mechanisms is not fully understood but where it is necessary to measure such effects, an estimate of respiratory and expiratory resistance is required. In man and larger animals this can be done with flow meters that measure the velocity of air moving into and out of the lungs. For the measurement of respiratory patterns in smaller mammals, a pressure sensor and plethysmograph are required. Such a sensor (Figure 5) and body plethysmograph (Figure 6) have been developed at BIBRA to measure the intensity of these reflex responses. The sensor (UK Patent No. 2119932) will detect small pressure changes in a plethysmograph which result from the inspiration and expiration of the animal. These pressure changes cause a movement of a permanent magnet located on a rubber tambour and changes in magnetic flux density are detected by a fixed Hall Effect i.c. The output signals, which are directly proportional to the tidal volume and respiratory rate of the animal, may be channelled to a suitable recording system for evaluation.

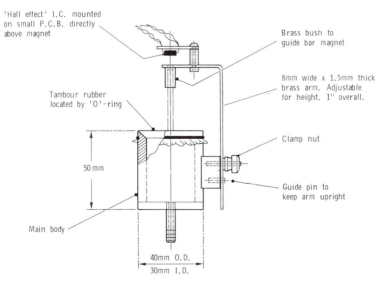

Figure 5   *The BIBRA pressure sensor*

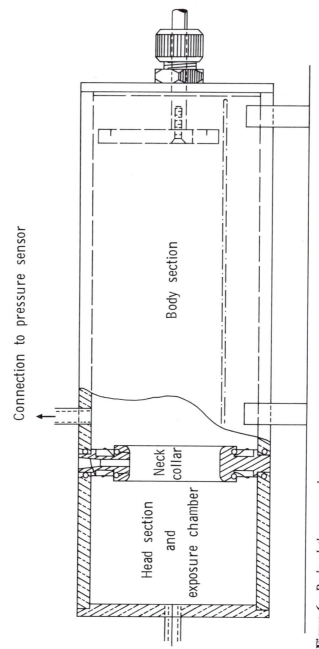

Connection to pressure sensor

Body section

Neck collar

Head section and exposure chamber

Figure 6 *Body plethysmograph*

In some circumstances, it may be necessary to more closely mimic respiratory patterns adopted by man. Essentially, man is a facultative mouth breather whereas most experimental animals, including primates, are obligatory nose breathers. In the latter, much of the inhaled material may be filtered out in the nasal cavity whereas in man the mechanism may be bypassed. In this situation the dog may be a suitable animal. Exposure can be by chamber or by a mask suitably equipped with a mouth tube or by a means of obstructing the nostrils. One problem with the dog is the well developed musculature surrounding the distal bronchi and the residual numbers of intra-alveolar ducts. These allow much greater distribution of material at the level of the alveolar spaces than occurs in man. This is of greater importance with gaseous materials where absorption may be enhanced. With particulate materials, including aerosols, alveolar impaction is still the most important mechanism.

5 Conclusion

Irrespective of the species, the main aim of inhalation studies is to generate and expose animals to consistent atmospheres over a defined period. This requires a multi-disciplinary approach and in most instances relates to the characteristics of the test article. Experiments of this nature should be conducted with constant surveillance.

References

Beethe, R. L. *et al.* (1979). Evaluation of a recently designed multi-tiered exposure chamber. *Inhal. Toxicol. Res. Inst.*, LF-67.

Doe, J. E. and Tinston, D. (1981). Novel chambers for long term inhalation studies. *In*: 'Inhalation Toxicology and Technology', *Ed.* B. K. J. Leong. Ann Arbor, Michigan. Ann Arbor Science Publishers, pp. 77–88.

Drew, R. T. and Laskin, S. (1975). Environmental Inhalation Chambers. *In*: 'Methods of Animal Experimentation', *Ed.* W. I. Gray, Vol. 4. Academic Press, New York. pp. 1–41.

Gage, J. C. (1968). Toxicity of paraquat and diquat aerosols generated by a size selective cyclone: effect of particle size distribution. *Br. J. Ind. Med.*, **16**, 11–14.

Gage, J. C. (1970a). Experimental Inhalation Toxicology. *In*: 'Methods in Toxicology', *Ed.* G. E. Paget, Blackwell Scientific Press, pp. 258–278.

Gage, J. C. (1970b). The sub-acute inhalation toxity of 109 industrial chemicals. *Br. J. Ind. Med.*, **27**, 1–18.

Hatch, T. F. and Gross, P. (1964). 'Pulmonary Deposition and Retention of Inhaled Aerosols'. Academic Press, New York.

Hemenway, D. R. *et al.* (1982). Inhalation toxicology chamber performance: a quantitative model. *Am. Ind. Hyg. Assoc. J.*, **43**, 120–127.

Hemenway, D. R. and MacAskill, S. M. (1982). Design, development and test results of a horizontal flow inhalation facility. *Am. Ind. Hyg. Assoc. J.*, **43**, 874–879.

Hinners, R. G. *et al.* (1986). Animal inhalation exposure chambers. *Arch. Environ. Health,* **16,** 194–206.

Hounam, R. F. and Sherwood, R. J. (1965). The cascade centripeter: a device for determining the concentration and size distribution of aerosols. *Am. Ind. Hyg. Assoc. J.,* **26,** 122–131.

Leach, L. J. *et al.* (1959). A multi-chamber exposure unit designed for chronic inhalation studies. *Am. Ind. Hyg. Assoc. J.,* **20,** 13–22.

Leong, B. K. J. (1980). 'Proceedings of the Inhalation Toxicology and Technology Symposium'. Kalamazoo, Michigan. Ann Arbor Scientific Publishers.

Macfarland, H. N. (1983). Designs and operational characteristics of inhalation exposure equipment: a review. *Fundam. Appl. Toxicol.,* **3,** 603–613.

Riley, R. A. (1986). A new approach to the construction of small inhalation chambers: design and evaluation. *Am. Ind. Hyg. Assoc. J.,* **47**(3), 147–151.

Salem, H. (1986). 'Inhalation Toxicology Research Methods, Applications and Evaluation'. Marcel Dekker Inc., New York and Basel.

Silver, S. D. (1946). Constant flow gassing chambers: principles influencing chamber design and operation. *J. Lab. Clin. Med.,* **31,** 1153–1161.

CHAPTER 8

Histopathology in Safety Evaluation

J. G. EVANS AND W. H. BUTLER

1 Introduction

Histopathology has long been the most important of the investigations undertaken for safety evaluation, because the morphological evidence of a pathological process is the most consistent of the changes that can be identified as the result of a toxic process (Butler, 1982). Consequently, the practice of pathology in toxicological studies has become very extensive and specialised. It is not practical in a chapter of this type to cover the field in detail, so only the important principles are described and exemplified by commonly encountered lesions. References for more extensive reading are provided, should more specialised information be required.

In essence, treatment may induce changes that would not otherwise occur or result in increased severity or in the earlier occurrence of a lesion that is normally found in the species or strain of animals being used. This is of particular importance in studies designed to detect chronic toxic effects and conducted over a substantial part of an animal's lifetime as the incidence of degenerative or neoplastic lesions is often associated with ageing (Burek, 1978). The elucidation of such problems requires careful attention to experimental detail and an awareness of the disease activities likely to be encountered. The following account will provide a basis for such expertise, particularly in respect of degenerative or infective diseases. The issues with respect to neoplastic disease are dealt with in Chapter 15.

2 The Role of Histopathology

Histopathological examinations of tissue form an important and time consuming part of many safety evaluation studies. The examination is dependent on the correct identification of lesions that may be present, and a proper appreciation of their significance. Therefore separation of spontaneous from induced pathology is of major importance in the majority of toxicological studies. It must be remembered, however, that in certain instances treatment may not result in an additional, clearly defined effect, but merely modify the severity or the extent of a disease that occurs naturally.

Spontaneous diseases in laboratory animals may be divided into two broad classes. The first are infectious diseases, resulting from viruses, bacteria, richetsiae, or myoplasm; parasitic infestation may also be included in this group. The second group are non-infectious diseases. These are often age related and may be markedly influenced by genetic factors and modified by dietary, hormonal, and animal husbandry (Butler, 1982).

Infectious diseases are largely controlled, in modern animal units, by good husbandry and management practice. Animals are now available that are free from defined parasites and micro-organisms so that the diseases that these agents cause do not form a major problem in histological examination of tissue from toxicological studies (Clough, 1987). The range of infectious disease in laboratory animals has been extensively documented (see bibliography), and will not be discussed in detail. A number, however, which are of specific historical interest still give rise to sporadic outbreaks of disease and these will be briefly mentioned. The non-infectious diseases, in contrast, form an important group for the pathologists practicing in toxicology. With this in mind particular emphasis will be placed on non-neoplastic diseases as they are commonly seen in the laboratory rat and mouse.

3   Examination of Animals and Tissue Preparations

It cannot be emphasised too strongly that a proper systematic examination of animals at necropsy with an accurate description of all lesions is an essential prequisite for a proper histopathological assessment. Failure to do so at this stage may seriously compromise any conclusions that are drawn from a study, or indeed may completely invalidate them. For this reason staff performing these duties should be properly trained, and supervised at the necropsy by the pathologist assigned to the study. Selection of tissue is generally specified by the protocol and, unless otherwise stated, covers samples of all tissues forming the animal's body.

In particular, if electron microscopic examination is envisaged it is important to design the protocol to ensure the optimum conditions for such an examination. This may require the use of additional animals for fixation by perfusion to obtain optimal results. Unless sufficient care is taken over sampling and preservation of tissue for this purpose, any conclusions may be strictly limited (Butler, 1982).

To prevent autolytic changes, tissues are generally 'fixed' to inhibit degradative enzyme activity. Commonly used fixatives for both electron and light microscopy are solutions of various aldehydes: formaldehyde for light microscopy and glutaraldehyde for electron microscopy. The fixatives themselves are made up as weak solutions (2–4%), buffered at the pH and osmolality of body fluids (Hayat, 1981). To cut thin sections of 'fixed' tissue it must be sufficiently rigid and so the tissue is dehydrated through graded alcohols and then infiltrated and embedded in a suitable material so that it can be cut. In the case of light microscopy this is paraffin wax; and for electron microscopy an epoxy resin or araldite, or possibly a mixture of the two, is used. Sections for light microscopy are generally cut at $5\,\mu$m on a microtome fitted with a steel knife. For electron microscopy, a diamond or glass knife is used and sections cut at approximately 50 nm.

Sections in their natural state are colourless and show little contrast. In order to differentiate the various cytological components, the tissue has to be stained. In the case of electron microscopy the stains used are heavy metals, such as lead or uranium salts, that bind differentially to the various organelles and thus inhibit the passage of electrons to varying degrees (Glauert, 1974). For light microscopy a wide variety of natural and synthetic dyes are used. These are broadly classified as basic or acidic dyes. The former are charged positively at the pH of the staining solution and bind to acidic groups bearing negative charges, and the latter are negatively charged and behave in the opposite manner. A typical basic dye is toluidine blue which stains RNA and DNA. An example of an acidic dye is acid fuchsin which stains the proteins of the cytoplasmic matrix, mitochrondria, and smooth endoplasmic reticulum. Commonly used dyes in routine histology are eosin (E) and haematoxylin (H). Cells stained by H and E show a blue nucleus and a pink cytoplasm that may contain clumps of blue or basophilic material (Bancroft and Stevens, 1977). In addition, a variety of stains have been developed to demonstrate, in a more or less specific manner, various structural components of the tissue. Examples of these are Van Gieson stain for connective tissue, Feulgen's method for DNA, and Periodic Schiff stain for glycol groups in sugars. The last two methods fall more appropriately into the realm of histochemistry and may require special methods of fixation. Histochemical methods also include the demonstration of enzyme activity in tissue sections. The aim of enzyme histochemistry is to demonstrate the presence of specific enzymes at their site of activity in life. To a certain extent these two requirements, that of localisation and maintenance of activity, are mutually exclusive. Some enzymes diffuse rapidly from their initial site while others are particularly susceptible to the inhibitory effects of fixation. For the latter, including majority of oxidative enzymes, the tissue is rapidly frozen and sections cut in a cryostat at $-20°$C. The section is then incubated to demonstrate the activity of the enzyme prior to fixation. Others, such as the hydrolytic enzymes, may be subjected to controlled fixation in aldehydes at 4°C for a few hours, following which the tissue is frozen and the sections cut.

The general principles of enzyme histochemistry are the same for all enzymes and are typified by the method for glucose-6-phosphatase. This enzyme catalyses the hydrolysis of glucose-6-phosphate in the following way:

$$glucose\text{-}6\text{-}phosphatase \rightarrow glucose + phosphate$$

The method is dependent on the capture of the phosphate group by lead ions that are included in the incubation medium. Lead phosphate precipitates as a colourless deposit which is then treated with hydrogen sulphide to form lead sulphide. This appears as a black or brown deposit (Bancroft and Stevens, 1977).

Detailed accounts of histological and histochemical techniques may be obtained from a number of texts referenced in the bibliography.

Immunocytochemical methods have become standard in routine human diagnostic pathology using both immunofluorescent and immuno-peroxidase techniques. The same techniques have been applied in safety assessment but are not, at present, widely used (True, 1990).

Advances in molecular biology have led to the development of methods that allow the *in situ* analysis of nucleic acids. Both messenger ribonucleic acid and genomic deoxyribonucleic acid may be identified using these techniques. Duplex strands of nucleic acid are separated by gentle heating in the presence of labelled complimentary strands of nucleic acid. On cooling, and in the right conditions, the labelled probe will hybridise with the target sequence and can be visualised by the appropriate method. The introduction of non-radiolabelled probes has meant that these techniques may be used for routine diagnostic purposes. For example, biotin labelled probes may be detected immunologically or by highly specific binding of avidin or streptavidin. The latter may be labelled with a fluorochrome, with colloidal gold, or with an enzyme so that they be seen with the light or electron microscope (Burns and Magee, 1992).

The techniques of molecular biology that are applicable in diagnostic histopathology are being continually upgraded, and to an extent simpli-fied, and so may be used for investigative purposes in toxicological pathology.

4   General Histopathological Terms

The structures of the developed body are derived from three germ layers in the embryo: the ectoderm, mesoderm, and endoderm. The ectoderm forms the skin and its appendage and much of the nervous system. The mesoderm forms the supporting structures of the body—bone, muscle, the connective tissues, and the vasculature and blood components. The endoderm forms the gut and associated glands such as the liver, salivary glands, and pancreas.

Clearly organs such as the liver, formed typically from endoderm, have within their substance blood vessels and connective tissue that are of mesodermal origin and thus the whole organ is of mixed embryonic origin. In the case of the liver, the hepatocytes are arranged in a glandular pattern around the vascular tree (Rappaport, Borowy, Lougheed, and Lotto, 1954). The liver is the organ that has been most extensively studied following both acute and chronic injury and many of the concepts of cellular injury have been developed from the examination of the reaction of hepatocytes (Farber and Fisher, 1979). It must be emphasised that the response of cells to injury is limited and generally recognised as a consequence of a disturbance in the ionic and water balance within the cell. In addition, there may be the accumulation of metabolic products such as free lipid or the breakdown products of phospholipid membranes. These changes are often referred to as cellular degenerations, and in the case of water balance are recognised in classical pathology by varying degrees of cell swelling (La Via and Hill, 1971). In the least severe of these, the cell loses its normal texture and has a faint granular appearance. When the fluid accumulation is more pronounced, discrete, fine droplets may be seen in the cell, while in the most severe form extensive large vacuoles are formed displacing much of the normal contents of the cell to the periphery.

It was once considered that these changes represented stages that would eventually lead to the death of the cell, although this is now not generally held. Indeed, cells exhibiting even the most advanced of these changes are not inevitably committed to die, in fact, cell death may occur without the cell passing through any of the stages associated with excess water accumulation (Farber, 1982).

Dead or dying cells are recognised in living tissue by a series of cytoplasmic and nuclear changes. Tissue recognised as such is said to be necrotic and must be distinguished from the general autolytic changes that occur after generalised body death. In the cytoplasm there is loss of detail often associated with increased eosinophilia and shrinkage of the cell. The nucleus may appear contracted and dense (pyknotic) or may be broken into several pieces (karryorhexsis).

In addition to these direct manifestations of cellular injury, the body may mount a series of tissue changes that are designed to limit the extent of the injury or that may precede the stage of tissue repair. This response, termed inflammation, is largely vascular and involves both the blood vessels themselves and the cellular elements of the blood. Inflammation has, by convention, been termed acute or chronic in respect to the time for which the inflammatory change has been present. In addition, certain histological features are associated with either an acute or chronic reaction, although they are not totally exclusive. Thus acute inflammation is largely vascular with the accumulation of fluid and neutrophils at the site. Fluid accumulation is not a prominent feature of chronic inflammation, the main histological feature being the presence of mononuclear cells (macrophages and lymphocytes) and with the proliferation of fibrous tissue and the formation of new blood vessels. Chronic inflammation is also often associated with various stages of repair and adaptation to tissue damage (Thomson, 1978).

The mechanism of the inflammatory reaction is complex and dependent on the release of chemical mediators, both by damaged cells and by the cells of the inflammatory response itself. The latter may indeed be central to the maintenance of long-term recalcitrant chronic inflammatory changes (Taussig, 1984).

5 Non-neoplastic Histopathological Changes in Laboratory Animals

A wide variety of bacterial, virological, and parasitological diseases have been described in laboratory animals. The majority of these are rarely seen in practice as they have been eliminated by good animal husbandry in well-maintained animal houses. This process begins at the phase of animal production with breeding from specific, pathogen-free stock. This is continued by barrier techniques in the animal house. In addition, many of the diseases of the larger laboratory animals, such as the cat and dog, are well controlled by vaccination policies and the common spontaneous diseases of these species such as distemper, canine viral hepatitis, leptospirosis, feline panleukopenia, and feline infectious peritonitis are now more correctly in the realm of clinical veterinary pathology and should be of little consequence in experimental toxicology. For the majority of this section, therefore, the histopathological changes that may confuse toxicological investigation will be confined to non-infectious diseases. A number of infectious diseases, however, have been central to the development of good laboratory practice and have in the past led to conflicting findings. Among these are Tyzzer's disease and mouse hepatitis. The former is caused by the bacterium *Bacillus piliformis,* and the latter occurs as a result of a virus infection. Both agents cause a focal necrotising hepatitis and, although the former was first reported in the mouse, it is known now to affect a wide range of species (Ganaway, Allen, and Moore, 1971).

Other examples are Sendai virus, which in mice has been associated with a necrotising bronchiolitis that may progress to a low-grade purulent pneumonia; and Sialodacroadenitis, which is caused by another virus and which results in necrosis within the salivary glands (Parker and Richter, 1982; Jacoby, Bhatt, and Jonas, 1979).

While the infectious diseases listed may cause problems from time to time, the most confusing histopathological findings in laboratory animals fall within the group of diseases generally described as constitutional and are largely degenerative in nature and often coincident with old age. In some instances these may affect almost 100% of animals and to be so severe as to cause their death. Thus, recognition and diagnoses of these diseases may be of considerable importance in for example, carcinogenicity studies as a judgement may be required as to the probable cause of the animal's death. In contrast, others may affect only a small proportion of animals and are probably of little consequence in affecting the overall mortality rate, although the incidence may be affected by treatment.

Among the former is nephrosis, or chronic glomerulonephrosis, in the aged rat. This is a severe degenerative disease of the kidney characterised by glomerulonephrosis, the formation of hyaline casts, tubular atrophy, and fibrosis (Gray, 1977). Associated with these structural changes are severe metabolic disturbances, particularly of phosphorus and calcium. Retention of phosphorus and loss of calcium through the kidney results in hyperplasia of the parathyroid gland, which raises the calcium levels in the serum and results in the deposition of calcium at various sites in the body including the walls of blood vessels, the mucosa of the gastro-intestinal tract, the lung, the heart, and the kidney itself. Other forms of calcium deposits, not associated with chronic renal disease, may also be encountered. One that may confuse toxicological interpretation is nephro-calcinosis, which is characterised by deposition of accretions of calcium in kidney tubules at the cortico-medullary junction. In some instances only an occasional tubule may be affected and the amount of calcium deposited is small. In other cases the deposit of calcium may be extensive, resulting in necrosis with further deposits of calcium in the dead tubular cells (dystrophilic calcification). The pathogenesis of this lesion is poorly understood. The initial lesion may occur at a site distant from the calcium deposit while the dietary balance of calcium, phos-phorus, and magnesium, and the hormonal status of the animal play an important contributory role (Glaister, 1986).

Cardiovascular disease is also common in old rats. This may be associated with the chronic glomerulonephrosis previously described, and result in extensive deposition of calcium salts in blood vessel walls and in the heart. Areas of degeneration, necrosis, and fibrosis may also be seen within the myocardium and this may eventually lead to myocardial fibrosis. The latter may also occur in the absence of other degenerative diseases and in this instance is predominantly in the immediate sub-endocardial region and is characterised by the replacement of myocardial fibres by an orderly arrangement of fibrous tissue. Again, the pathogene-sis of the lesion is not understood. It has been suggested that the myocardial necrosis results from a viral infection of anoxia, which induces an abnormal response to physiological demand. Certain drugs, such as isoproternol, can cause a similar pathological effect (Anvers and Cohen, 1979). Myxoid degeneration of the cardiac valves may also be seen in old rats and an association between this condition and nephrosis has been made.

Other parenchymatous organs in the rat are also susceptible to various degenerative effects in old age. Thus, in the thyroid there may be varying degrees of colloid loss and in some instances fibrosis. In addition, a diffuse and nodular hyperplasia of the C-cells is a feature that may be confused with a benign tumour. The adrenal gland may also show areas of hyperplasia and hypertrophy; the affected cells often being filled with lipid (Burek, 1978). Testicular atrophy is occasionally seen, both in young and old animals. There may be complete loss of germinal epithelium or merely a relative decrease in cell numbers with the formation of multinucleate giant cells from spermatids. Atrophy of the ovary is a prominent feature in old female rats and may be of two morphological forms: either the ovary appears shrunken and cystic with loss of follicular tissue or there is overgrowth of granulosa-thecal cells which may mimic a benign tumour of the granulosa cells (Glaister, 1986).

Mammary glands in old animals may occasionally show mild acinar hyperplasia, sometimes with duct ectasia and sometimes with more pronounced cystic development.

Paralysis of the hind limbs of ageing rats has been described in several strains. Microscopically this is accompanied by demyelination and axon degeneration and distention of the axon sheaths in the white matter of the cord and the peripheral nerves. There is an associated atrophy of the lumber and hind limb musculature. This condition is particularly prominent in male (WAG × BN)F$_1$ rats over three years of age (Anvers and Cohen, 1979).

Lymphoid tissue shows marked atrophic changes with age and reduction of germinal follicles is often associated with an abundance of haemosiderin-containing macrophages. Some nodes may show cystic dilatation of sinusoids and occasionally mesenteric lymph nodes may be grossly distended. Splenic atrophy may mimic angioma in that loss of lymphoid tissue leaves little other than the supporting structure. The thymus also shows an age associated atrophy. This is characterised by a loss of lymphocytes and a proliferation of the epithelial component. In the most advanced form, little other than a fibro-vascular stroma remains, often heavily infiltrated by adipose tissue.

Old mice may show similar degenerative conditions. In addition, certain strains (*e.g.* strain A) are susceptible to the development of deposits of amyloid in various tissues. Sites include the kidney, the gastro-intestinal mucosa, and the heart. Amyloid itself is a complex glycoprotein that has a characteristic periodicity when examined in the electron microscope. Amyloid stains with Congo Red and shows characteristic red-green birefringence under the polarising microsome. In the kidney the deposit is most pronounced in the glomerular tuft, and in the most severe form there may be extensive loss of protein through the kidney with eventual death of the animal (Conner, Conner, Fox, and Rogers, 1983).

The mouse shows marked strain and sex differences in histological structure and disease patterns. Reference to the literature should be made, as only a few specific examples will be mentioned here. In the kidney of female animals, Bowman's capsule has a flattened squamous epithelium, whereas in the sexually mature male, the epithelium is cuboidal in form. In mature males the terminal duct of the submaxillary gland is formed of columnar cells filled with intensely eosinophilic granules, while that of the female is formed of low-cuboidal cells lacking granules. The adrenal gland of female mice contains more lipid than that of the male, giving it a paler gross appearance. In the adrenal gland of young mice there is a layer of cells contained between the cortex and medulla and is called the X zone. This layer disappears rapidly in males at the time of sexual maturity, and more slowly at the time of the first pregnancy in females. If pregnancy does not intervene, the zone may persist throughout the life of female animals (Dunn, 1965).

Small cystic structures are commonly found in the thyroid and pituitary of old mice, while ectopic thymic tissue may be seen in the thyroid gland of certain strains (*e.g.* BALB/c). The adrenal gland may show proliferation in the subcapsular region of spindle cells. These are most common in female animals and may be increased by the presence of tumours. Proliferation of lymphoreticular tissue is also prominent in aged mice and great care is required in distinguishing benign proliferations from those associated with malignancy.

The conditions described represent some of the common spontaneous diseases of laboratory rats and mice. Many of them have underlying genetic mechanisms, although it is the interaction of genetic and environmental factors that determine their expression. It should be noted that changes in dietary or husbandry procedures can markedly alter the pattern of disease, irrespective of experimental treatment (Tucker, 1979). Indeed, even if there is a treatment-related effect, this may operate through an indirect rather than a direct mechanism. Thus, the induction of strain may result in considerable changes in the hormonal status of the animal which may markedly affect the pattern of disease seen.

References

Anvers, M. R. and Cohen, B. J. (1979). Lesions associated with aging. *In* 'The Laboratory Rat' *Eds* H. J. Baker, J. R. Lindsey, and S. H. Weisbroth, Vol. 1, p. 377. Academic Press, New York.

Bancroft, J. D. and Stevens, A. (*Eds*) (1977). 'Theory and Practice of Histological Techniques'. Churchill-Livingstone, Edinburgh, London, and New York.

Burek, J. D. (1978). 'Pathology of Ageing Rats'. CRC Press, West Palm Beach.

Burns, J. and McGee, O'D. (1992). Tissue nucleic acid analysis. *In* 'Oxford Textbook of Pathology' *Eds* O'D. McGee, P. G. Isaacson, and N. G. Wright, pp. 2284–2288. Oxford University Press, Oxford, New York, and Tokyo.

Butler, W. H. (1981). Histological control of spontaneous animal pathology. *In* 'Animals in Toxicological Research' *Eds* I. Bartošetc, A. Guaitani, and E. Pacei, p. 71. Raven Press, New York.

Clough, C. (1987). Quality in laboratory animals. *In* 'Laboratory Animals' *Ed.* A. A. Tuffery, p. 79, A Wiley-Interscience publication, Chichester, New York.

Conner, M. W., Conner, B. H., Fox, J. G. and Rogers, A. E. (1983). Spontaneous amyloidosis in outbred CD-1 mice. *Surv. Synth. Path. Res.*, **1**, 67.

Dunn, T. B. (1965). Spontaneous lesions of mice. *In* 'The Pathology of Laboratory Animals' *Eds* W. E. Ribelin and J. R. McCoy, p. 303. Charles C. Thomas Publisher, Springfield, Illinois, USA.

Farber, E. and Fisher, M. M. (1979). 'Toxic Injury of the Liver'. Part A. Marcel Decker, New York and Basel.

Farber, J. L. (1982). Membrane injury and calcium homeostatis in the pathogenesis of coagulative necrosis. *Lab. Invest.*, **47**, 114.

Ganaway, J. R., Allen, A. M. and Moore, T. D. (1971). Tyzzer's disease. *Am. J. Pathol.*, **64**, 717.

Glaister, J. (1986). 'Principles of Toxicological Pathology'. Taylor and Francis, London and Philadelphia.

Glauert, A. M. (1974). 'Practical Methods in Electron Microscopy' Vol. 3 part 1. Fixation, dehydration, and embedding of biological specimens. North Holland Publishing Co., Amsterdam, Oxford.

Gray, J. E. (1977). Chronic progressive nephrosis in albino rat. *CRC Crit. Rev. Toxicol.*, **5**, 115.

Hayat, M. A. (1981). 'Fixation for Electron Microsocopy'. Academic Press. New York, London.

Jacoby, R. O., Batt, P. N. and Jonas, A. M. (1979). Viral diseases. *In* 'The Laboratory Rat' *Eds* H. J. Baker, J. R. Lindsey, and S. H. Weisbroth Vol. 1, p. 272 Academic Press, New York.

LaVia, M. F. and Hill, R. B., Jr. (1975). 'Principles of Pathobiology'. 2nd edn. University Press, London, Oxford.

Parker, J. C. and Richter, C. D. (1982). Viral diseases of the respiratory system. *In* 'The Mouse in Biomedical Research' *Eds* H. L. Foster, J. D. Small, and J. G. Fox, Vol. 1, p. 28. Academic Press, New York.

Rappaport, A. M., Borowy, Z. J., Lougheed, W. M. and Lotto, W. N. (1954). Subdivision of the hexagonal liver lobules into a structural and functional units. *Anat. Rec.*, **119**, 11.

Taussig, M. J. (1984). 'Processes in Pathology and Microbiology'. 2nd edn. Blackwell Scientific Publications. Oxford, London.

Thomson, R. G. (1978). 'General Veterinary Pathology'. W. B. Saunders Co. Philadelphia, London, Toronto.

True, L. D. (1990). 'Atlas of diagnostic immunohistopathology'. *Ed.* J. P. Lippincott, Gower Medical, New York and London.

Tucker, M. J. (1979). The effect of long-term food restriction on tumours in rodents. *Int. J. Cancer*, **23**, 803.

Bibliography

Histological Methods

'Manual of Histological Techniques'. J. D. Bancroft and H. C. Cook, Churchill-Livingstone, Edinburgh, London, Melbourne, and New York, 1984.

'The Theory and Practice of Histological Techniques'. *Eds*. J. D. Bancroft, and A. Stevens. Churchill-Livingstone, Edinburgh, London, and New York, 1977.

General Pathology

'An Introduction to General Pathology'. W. G. Spector, Churchill-Livingstone, Edinburgh, London, New York 1977.

'Processes in Pathology'. M. J. Taussig, Blackwell Scientific Publications, Oxford, London, Edinburgh, 1979.

Veterinary and Laboratory Animal Pathology

'Pathology of Laboratory Animals', Vol. I and II. *eds* K. Benirschhe, F. M. Garner, and J. C. Jones. Springer-Verlag, New York, Heidelberg, Berlin, 1978.

'Pathology of Ageing Rats'. J. D. Burek, CRC Press, Inc. 1978.

'Pathology of Laboratory Animals'. *Eds* E. Cotchin, and F. J. C. Roe. Blackwell Scientific Publications, Oxford and Edinburgh, 1967.

'Laboratory Animal Medicine'. *Eds* J. G. Fox, B. J. Cohen, and F. M. Loew. Academic Press Inc. (Harcourt Brace Jovanovich, Publishers) New York, London. 1984.

'Principles of Toxicological Pathology'. J. R. Glaister, Taylor and Francis. London and Philadelphia, 1986.

'Rat histopathology. A Glossary for use in toxicity and carcinogenicity studies'. P. Greaves and J. M. Faccini, Elsevier Scientific, Amsterdam, 1984.

'Veterinary Pathology'. H. A. Smith and T. C. Jones, Lea and Feiberger, Philadelphia.

CHAPTER 9

The Metabolism and Disposition of Xenobiotics

S. D. GANGOLLI AND J. C. PHILLIPS

Introduction

Metabolism can refer to the sum of the chemical reactions that serve to maintain life or to the chemical transformations effected by living systems on foreign chemicals (Hayes, 1975). Pharmacokinetics is the discipline which quantifies, as a function of time, the absorption, distribution, metabolism, and excretion of chemicals in the body (Watanabe, Ramsey, and Gehring, 1980). The WHO Scientific Group, in their recommendations on the procedures for investigating intentional and unintentional food additives (WHO Scientific Group, 1967), recognised the importance of metabolic and pharmacokinetic data in the safety evaluation of food chemicals. This group recommended that, whenever possible, toxicity studies should be conducted on an animal species most similar to man with regard to its metabolic, biochemical, and toxicological characteristics in relation to the test substance. Furthermore, the group considered that at a relatively early stage there was a need to obtain information on the metabolic fate of the test compound in experimental animals and in man.

A similar view was expressed by the Scientific Committee of the Food Safety Council in their proposed Safety Decision Tree Scheme for food safety assessment (Food Safety Council, 1978). The Committee recommended that studies on metabolism and pharmacokinetics should be undertaken immediately after acute toxicity has been determined, and that information on the metabolic disposition of a compound should be used both in the choice of animal model and in the definition of dose regimes and other conditions relevant to the conduct of long-term animal studies. The objectives of metabolic and pharmacokinetic studies as defined by the Committee may be summarised as follows:

(1) To obtain information on the absorption, biotransformation, disposition, and excretion of a test substance after single or multiple doses in experimental animals and in man.

(2) To obtain a quantitative measure of these events at different dose levels of the compound and to elucidate the conditions likely to stress or overload the metabolic and excretory capacity of the experimental animal.

In this chapter, we aim to provide a general review of metabolic and excretory processes in animals and man. The sequence of the various sections in this chapter is:

(1) Sites of absorption
(2) Distribution
(3) Metabolic reactions
(4) Sites of metabolism
(5) Bioactivation and detoxification processes
(6) Factors influencing metabolism and toxicity

In the following chapter, a brief review of the theoretical considerations relevant to understanding the change in concentration of a compound in blood and tissues following administration is given and practical methods and techniques in metabolic studies discussed.

1 Sites of Absorption

The main sites of exposure and routes of entry of foreign chemicals into the human body are the skin, lungs, and the gastro-intestinal tract. Although there are special anatomical and physiological features characterising and regulating the entry of substances from each of these sites, there are certain common mechanisms applicable for the absorption of chemicals across membranes and into the systemic circulation. These 'uptake' mechanisms fall broadly into two categories: (1) passive transport, and (2) specialised transport systems.

A Passive Transport Systems

Two mechanisms mediate the passive transport of chemicals across membranes. The first is simple diffusion and applies to very small, hydrophilic, non-ionic compounds and gases capable of penetrating through waterfilled pores (about 4 Å radius) on the surface of membranes. With passive transport, molecules pass across a concentration gradient and the phenomenon is described by Fick's first law of diffusion. At dynamic equilibrium, the equation as applied to transport across a single phase membrane (Davson, 1943) is:

$$dD/dt = \frac{KA(C_o - C_i)}{X}$$

where dD/dt is the rate of transport of a compound across a membrane, K is a constant, A is the cross-sectional area of membrane exposed to the compound, C_o and C_i are the concentrations of the compound on the outside and inside of the membrane respectively, and X is the thickness of the membrane.

Examples of compounds that are absorbed by simple diffusion are the gases (*e.g.* CO, CO_2, HCN, NO_2) in the lung, the nerve gas, sarin, in the skin, and urea in the gastro-intestinal tract.

The second mechanism involved in passive transport relates to the ability of a compound to diffuse through both the aqueous and lipid components of membranes. The degree of lipophilicity possessed by a compound determines the facility with which it will be absorbed, and therefore the relationship between the partition coefficient (P) of a compound and its rate of permeation through membranes has been extensively studied (see, for example review by Houston and Wood, 1980). When applied to intestinal transport, the absorption rate constant of a compound (K_a) is related to $\log P$ by:

$$\log K_a = -a(\log P)^2 + b \log P + c$$

where a, b, and c are the regression coefficients (Hansch and Clayton, 1973).

The measurement of partition coefficients by the somewhat tedious and time-consuming two-solvent distribution technique (using n-octanol and water) has been supplemented by the more elegant thin-layer chromatography method developed by Biagi, Barbaro, Gamba, and Guerra (1969), and the more recent high performance liquid chromatography method of Unger, Cook, and Hollenberg (1978).

The lipophilicity of a compound is influenced by its degree of ionisation in solution. This depends on the pK_a of the compound and the pH of the medium, the relationship being described by the Henderson–Hasselbalch equations for acids and bases:

$$\text{for an acid: } pK_a - \text{pH} = \log(c_u/c_i)$$
$$\text{for a base: } pK_a - \text{pH} = \log(c_i/c_u)$$

where c_i and c_u are the concentrations of the ionised and unionised compound, respectively.

Thus the degree of ionisation of weakly acidic organic compounds, such as salicylic acid, benzoic acid, and the barbituric acids, is least in the acidic conditions prevailing in the normal stomach and hence they are readily absorbed from this site. Conversely, weakly basic compounds (*e.g.* alkaloids, paracetamol, *etc.*) are more readily absorbed from the small intestine. However, the lipophilicity, and hence the absorption, of strongly ionised compounds can be increased by interactions with counter-ions. Compounds capable of such ion-pair formation include tetracyclins, quaternary ammonium compounds, and sulphonic acids.

B Specialised Transport Systems

The two major processes classified as specialised transport are active transport and carrier-mediated facilitated diffusion. Active transport processes, involving uptake of compounds against a concentration gradient using energy-dependent enzymic reactions, are largely responsible for the absorption of essential components such as sugars, amino acids, and nucleic acids. The translocation of these substances across membranes involves their interaction with appropriate cell surface receptors and the process, which requires the expenditure of energy, is governed by the kinetics of enzymic reactions. These active transport processes are rate limiting with respect to availability of substrate, receptor sites, and ATP or other related sources of energy, and can be inhibited by competitive substrates and metabolic inhibitors. Thus, for example, the active transport of glucose is inhibited by the non-nutritive analogue deoxyglucose, acting as a competitive substrate, and the uptake of potassium by the sodium/potassium ATPase pump mechanism can be reduced by the presence of rubidium (Tobin, Akera, Han, and Brody, 1974). Metabolic inhibitors, such as ouabain and tetrodotoxin can inhibit active transport by interfering with Na/K ATPase or sodium uptake (Ullrick, Capasso, Rumrich, and Sato, 1978; Kao, 1981).

Facilitated transport is the other specialised process in which carrier-mediated mechanisms are involved. This process is not energy dependent and does not operate against a concentration gradient. Examples of such processes are the phlorizin-sensitive transport of sugars across membranes and the transport of calcium ions by the calcium-binding protein, calmodulin. The latter process can be inhibited by a number of organic bases, such as the phenothiazines, acting as competitive substrates (Weiss and Levin, 1978).

Other mechanisms of absorption include pinocytosis and phagocytosis, whereby inert or high molecular weight substances are translocated intact across membranes. Examples of such processes are the persorption of macromolecular substances, such as degraded carrageenan from the small intestinal lumen of the guinea pig (Sharratt, Grasso, Carpanini, and Gangolli, 1970) and the uptake of horse-radish peroxidase from the rat gut (Walker, Isselbacher, and Block, 1972).

Skin

Absorption of both lipophilic and hydrophilic compounds through the intact skin is mediated primarily by passive diffusion mechanisms, compounds being taken up principally by transepidermal diffusion. However, diffusion *via* hair follicles and sebaceous glands can also occur. The principles and methods for studying percutaneous absorption have been reviewed by Barry (1983) and the 'state-of-the-art' in predicting skin absorption recently summarised (Scott, Guy, and Hadgraft, 1990).

Lung

In the lung the mechanism for the absorption of inhaled gases is by passive diffusion. Toxicants absorbed into the circulation from the lungs include carbon monoxide, oxides of nitrogen and sulphur, and vapours of industrial solvents such as chlorinated hydrocarbons. Particles of $1\ \mu m$ diameter and below can penetrate the alveoli and gain access to the circulation by phagocytosis.

Gastro-intestinal tract

Both passive and active carrier-mediated transport mechanisms effect the absorption of compounds from the gastro-intestinal tract. Simple diffusion allows the penetration of non-ionic hydrophilic substances such as urea, and of anionic substances with hydration radii of less than 2.9Å, such as Br^-, Cl^- and NO_3^-. Cations, on the other hand, require a specialised carrier mediated transport system for absorption from the gastro-intestinal tract. Lipophilic foreign compounds that are absorbed by passive diffusion include drugs (barbiturates, paracetamol, *etc.*), food additives (butylated hydroxyanisole and other antioxidants, benzoic acid, and flavouring esters) and environmental chemicals (phthalic acid esters and other industrial solvents and plasticisers, halogenated pesticides, *etc.*). Examples of compounds absorbed by active transport include 5-fluorouracil and some serine and threonine derivatives of nitrogen mustards (Schanker and Jeffrey, 1961). In addition some dipeptides, tripeptides, and oligopeptides are also actively absorbed from the small intestine (Matthews, 1975; Das and Radhakrishnan, 1976).

The neonatal absorption of intact proteins and particulate material is well documented. Although macromolecular absorption supposedly ceases on the maturation of the epithelial cells, there is evidence showing that adult animals continue to absorb biologically significant quantities of large molecules and particulates (Le Fevre and Joel, 1977), probably by pinocytosis. Examples of such compounds include colloidal silver, polystyrene latex, asbestos fibres, and starch particles.

2   Distribution

Following the absorption of a compound into the systemic circulation, it may be distributed throughout the body to tissues, prior to metabolism and excretion. Unless tissue distribution is very rapid, the blood can be considered to be the 'central compartment' in a kinetic description of the uptake, distribution, and excretion of a compound from the body (see the following chapter). The rate of removal of a compound from the blood is dependent on a number of factors, including blood flow through the various organs and the extent of binding of the compound to tissue and/or plasma proteins. The pharmacokinetic consequences of drug-protein interactions have been discussed (Jusko and Gretch, 1976). The plasma concentration of a compound does not, therefore, necessarily reflect the concentration in any particular tissue, and in many instances, tissues have been found to have a specific affinity for a compound. This may explain in part tissue specific injury, and frequently results in prolonged retention of compounds in the body. Thus, paraquat accumulates preferentially in the lungs, and at early times after administration, the concentration in this organ increases as the plasma concentration falls (Smith *et al.*, 1978). Similarly, lipophilic compounds such as hexabromobiphenyl, DDT, and dioxins accumulate in skin and adipose tissue, but not in muscle and liver (Tuey and Matthews, 1981).

3 Metabolic Reactions—Bioactivation and Detoxication Processes

The metabolism and biotransformation of foreign compounds in the body may be effected by a variety of enzymic processes, which can be categorised as follows:
(1) Hydrolysis
(2) Oxidation
(3) Reduction
(4) Conjugation
(5) Miscellaneous (including synthesis)

Williams (1959) proposed that these five categories could be simplified into two types of reactions, namely Phase I reactions (hydrolysis, oxidation, and reduction) and Phase II reactions (conjugation and synthesis). In general, Phase I reactions convert compounds into substrates suitable for Phase II metabolism. Some workers consider the further metabolism of glutathione conjugates to be a third phase of metabolism (Phase III).

Each of these categories of metabolic reactions will be described briefly and their role in the bioactivation and detoxification of compounds discussed.

A Hydrolysis

Hydrolases (EC. 3xxx) constitute a heterogeneous group of enzymes generally of low specificity, capable of the hydrolysis of carboxylic acid esters, amides and lactones, organophosphate and sulphate esters, saccharides, glycosides and glucuronides, peptides, and related structures of both endogeneous and exogeneous origin. Examples of some of the exogenous compounds used as drugs, food chemicals, pesticides, and industrial solvents capable of being hydrolysed in the body are shown in Table 1. Recently it has been shown that there are multiple isozymes of hepatic carboxylesterase (EC 3.1.1.1), many of which have been characterised (see *e.g.* Hosokawa *et al.*, 1990).

Table 1 *Examples of some foreign chemicals capable of being metabolized by enzymatic hydrolysis in the body*

Use	Chemical class	Example
Drug	Carboxylic acid ester	Acetylsalicylic acid
		Pethidine, atropine
	Carboxylic acid amide	Phenacetin, penicillin,
		chloramphenicol
	Glycoside	Digitalis alkaloids
Food additive		
Flavouring agents	Carboxylic acid ester	Ethyl acetate
Emulsifiers	Fatty acid esters	Acetylated monoglycerides
Preservatives	Carboxylic acid ester	Hydroxybenzoates, gallates
Sweetener	Dipeptide	Aspartame
Agrochemicals		
Herbicides and	Carboxylic acid esters	Pyrethroids, carbamates
insecticides	and amides	Carbanilates,
Industrial Chemicals		
Solvents and	Carboxylic acid esters	Dialkyl phthalates, adipates
plasticizers		

The hydrolysis of various esters by serum albumin has also been reported (Kuruno, Kondo, and Ikeda, 1983).

B Oxidation

Mixed-function Oxidase System

A group of enzymes which are associated principally with the membranous components of the endoplasmic reticulum and collectively referred to as the microsomal mixed function oxidase (MFO) system, constitute the most important group of enzymes mediating the oxidation of both endogenous and foreign compounds. A wide variety of foreign compounds including drugs, food additives, and industrial chemicals are metabolised by the MFO system and examples of some of the oxidation reactions and substrates are shown in Table 2. Essential requirements for MFO activity are molecular oxygen, NADPH, the flavoprotein reductase, and the haemoprotein 'cytochrome P-450'. The range of enzymic reactions, the mechanisms involved, and other related topics pertaining to the MFO system have been extensively reviewed [see, for examples, Jakoby, Bend, and Caldwell (1982); Sato and Omura (1978); Testa and Jenner (1976); Ullrich, Roots, Hildebrandt, Estabrook, and Conney (1977); and more recently Schuster (1989)]. Although a detailed discussion of the MFO system is beyond the scope of this chapter, certain important aspects of the system will be mentioned briefly. These are:
(1) The essential components of the enzyme complex

(2) Characteristics of the substrate/cytochrome P-450 interaction
(3) The cytochrome P-450 superfamily
(4) Inducers and inhibitors of enzymic activity.

Following the observation that MFO activity was present in the microsomal fraction of tissue homogenates and that the enzyme activities were greater in the smooth endoplasmic reticulum, various isolation procedures were developed to investigate the nature of the complex. The solubilisation, separation, and reconstitution procedures pioneered and developed by Coon and his co-workers (Coon, Autor, Boyer, Lode, and Strobel, 1973) have shown that three distinct components are essential for microsomal MFO activity. These are: lipids (phosphatidylcholine), NADPH-cytochrome P-450 (cytochrome c) reductase, and the terminal oxidase cytochrome P-450. Historically, substrates of the MFO system were divided broadly into two groups based on their effects on the spectral properties of cytochrome P-450 preparations (Remmer, Shenkman, Estabrook, *et al.*, 1966). Type I substrates, exemplified by hexobarbital, interact with the oxidised haemoprotein to give a spectral peak at 385–390 nm and a trough at 418–427 nm. Type II substrates (*e.g.* aniline) exhibit an interaction spectra with a peak at 425–435 nm and trough at 395–405 nm. Some substrates (*e.g.* phenacetin) also contain chemical structures associated with both type I and type II compounds and consequently interact with cytochrome P-450 to give a reverse type I (or modified type II) spectrum (*i.e.* peak at 420 nm and trough at 392 nm). This is a concentration-dependent effect, probably caused by the interaction of the substrate with binding sites different from type I sites. There is generally a good agreement between the spectral dissociation constant values (K_s) for type I substrate-binding spectra and K_m values for enzyme activity indicating that the former probably manifests a true reflection of the enzyme substrate complex. On the other hand, the correlation between K_s and K_m values for type II substrates is extremely poor. The binding of substrates to cytochrome P-450, largely determined by the lipophilicity of the compound, is accompanied by changes in protein conformation and spin state of the iron (Shichi, Kumaki, and Nebert, 1978). Thus, type I binding results in the transformation of a low spin haem-iron to a high spin state, whereas type II and reverse type I bindings result in the formation of a low spin cytochrome P-450 accompanied by the direct co-ordination of the substrate with the haem-iron.

The 'Cytochromes P-450' are now recognised to comprise a superfamily of haemoproteins with an identical prosthetic group, but different apoprotein structures, which are responsible for the different substrate specificities. Currently at least 27 families with up to eight subfamilies and up to 23 individual forms have been identified (Nebert *et al.*, 1991; Soucek and Gut, 1992). Many of these families are present in all mammals, and although the greatest number of forms have been characterised in the rat, human orthologues have been found for a substantial number of them.

Table 2 *Enzymic Metabolism of Foreign Compounds*

Reaction Category [Enzyme System]	Specific biotransformation	Chemicals
OXIDATION [Microsomal mixed-function oxidase]	Aromatic hydroxylation Aromatic epoxidation	Benzene, polycyclic aromatic hydrocarbons acetanilide
	Aliphatic epoxidation: Aliphatic hydroxylation: O-dealkylation: S-dealkylation: S-oxidation: N-oxidation N-hydroxylation de-sulphuration: de-halogenation: N-dealkylation:	Olefins, vinyl chloride, allyl alcohol Alkanes, fatty acids Codeine, phenacetin, Methyl mercaptan, methylisothiourea Carbon disulphide, thioacetamide, parathion Hydroxylamine, 2-acetylamino fluorene Dimethylaniline, morphine, imipramine Parathion Halothane, DDT, dichloromethane t-Amines, amphetamine, dialkylnitrosamines
OXIDATION [Alcohol dehydrogenase] [Aldehyde dehydrogenase, Aldehyde oxidase Xanthine oxidase]	Alcohol → Aldehyde: Aldehyde → Acid	Ethanol, allyl alcohol, methoxyethanol Formaldehyde, tolualdehyde
[Monoamine oxidase, Microsomal mixed-function amine oxidase]	Amines → Aldehyde:	Tryptamine, tyramine, benzylamine
REDUCTION [Alcohol dehydrogenase] [Aromatic aldehyde-ketone reductase (NADPH-dependent)]	Carbonyl reduction	Cyclohexanone Benzaldehyde p-Nitroacetophenone Chloral hydrate

Enzyme	Reaction	Examples
[Microsomal mixed-function epoxide reductase]	Polycyclic arene oxide reduction	Benzo(a)pyrene-4,5-oxide
[Microsomal and bacterial azo reductase]	Azo reduction to primary amines	Prontosil, butter-yellow and other azo-colours azobenzene
[Microsomal and bacterial nitro-reductase] [xanthine oxidase]	Nitro aromatic compounds amines	Nitrobenzene chloramphenicol niridazole, nitrazepam
[Microsomal N-oxidoreductase]	N-oxide → tertiary amine	Nicotine-N-oxide, indacine-N-oxide
[Microsomal dehalogenase]	Aryl and alkylhalide dehalogenation	DDT, trichlorofluoromethane
HYDROLYSIS [Carboxylic acid esterase]	Ester hydrolysis	Ethyl acetate, malathion, propyl gallate, acetylsalicylic acid
[Amidases] [Microsomal epoxide hydrolase]	Amide hydrolysis Epoxide → Diol	Penicillin dimethoate Styrene oxide, benzo(a) pyrene-4,5-oxide
[Glycosidases-glucuronidase glycosidases] [Aryl sulphatase]	Glucuronide hydrolysis Glycoside hydrolysis Aryl sulphate hydrolysis	Phenolphthalein glucuronide Cyasin, amygdalin, thioglycosides Steriod sulphates

Table 2 (*continued*)

Reaction Category [Enzyme System]	Specific biotransformation	Chemicals
CONJUGATION [UDP-glucuronyl transferase]	Glucuronic acid conjugation with:	
	aryl and alkyl alcohols	Phenol, propane-1, 2-diol, 7-hydroxycoumarin, sterols
	primary and secondary amines	Aniline, sulphonamides, cyproheptadine *N*-hydroxyarylamines
	sulphydric compounds	Thiophenols
	carboxylic acids	Phenylacetic acid, other arylacetic acids, carbamic acids
[Acyl-CoA: amino acid *N*-acyl transferases]	Amino acid (*e.g.* glycine) conjugation with carboxylic acids	Benzoic acid, salicylic acid, cinnamic acid

[Sulphotransferases]	Sulphate conjugation with: phenols, aromatic amines, alkanols	Phenol, 2-acetylaminofluorene, oestradiol and other sterols
[Acetyl CoA: arylamine transferases]	Acetylation of aromatic amines	Isonicotinic acid hydrazide, sulphanilamide
[Methyl transferases] [Phosphotransferases]	O- and N-Methylation reaction Phosphate conjugation with phenols	Catechol, imidazole, histamine, Phenol
[Glutathione-S-transferases]	Glutathione conjugation with: arene and alkene oxides and epoxides	Phenanthrene-9,10-oxide, styrene-7,8-oxide, benzo(a)pyrene-4,5-oxide
[–glutamyl trans peptidase + cysteinyl glycinase + N-acetylase]	Aryl and alkyl halides, nitrates	Chloracetonitrite 1,2-dichloro-4-nitrobenzene methyl bromide

The early approaches to characterisation of cytochrome P-450, namely affinity and ion-exchange chromatography (see, *e.g.* Gibson and Schenkman, 1978; Ryan *et al.,* 1979) have now largely been replaced by gene isolation and directed expression in suitable cell lines (see Gonzalez, 1988, 1990). Since each P-450 gene nearly always produces a single protein, the current nomenclature for the gene, transcript and product are the same. Thus in rat and human, italicised CYP1A1 indicates the gene and non-italicised CYP1A1 both the corresponding mRNA and protein.

A wide variety of compounds of differing chemical structures have been found to induce the MFO system, with at least six different classes of inducers now recognised (Soucek & Gut, 1992). These are barbiturates (*e.g.* phenobarbitone), polycyclic aromatic hydrocarbons (*e.g.* 3-methylcholanthrene), steroids (*e.g.* dexamethasone), simple aliphatic compounds (*e.g.* acetone), hypolipidaemic drugs (*e.g.* chlofibrate), and macrolide antibiotics (*e.g.* rifampicin). Possible mechanisms involved in the induction process include transcriptional activation, mRNA stabilisation and protein stabilisation (Gonzalez, 1988; 1990; Kimura *et al.,* 1986; Song *et al.,* 1989). The cellular expression of polycyclic hydrocarbon-inducible P-450s is mediated by a cytosolic receptor, the *Ah* receptor, which interacts with the inducer, is translocated to the nucleus, and stimulates translation by interaction with nuclear DNA. Receptor mechanisms may also be involved in the induction of other subfamilies, although no receptor for phenobarbitone has been found (Murray and Reidy, 1990).

Many inhibitors of drug metabolism have been identified, although few are specific for particular subfamilies (Testa and Jenner, 1981; Murray, 1987). Inhibitors may exert their effect by interfering with substrate binding or electron transport and others may require metabolic activation for their activity. The major families/subfamilies of cytochrome P-450, with some known inducers and inhibitors are summarised in Table 3.

Mono-oxygenase activity is also found in other liver cell organelles. The outer membrane of mitochondria contain cytochrome P-450, but little cytochrome c-reductase; reduction being mediated by NADH-cytochrome b_5-reductase (Raw, 1978). Nuclear mono-oxygenase activity is thought to be responsible for the toxification of many genotoxic compounds including 2-acetylaminofluorene and aflatoxin B_1 (Kawajiri, Yonekawa, Hara, and Tagashira, 1979; Guengerich, 1979). Activity at this site is inducible, and results in qualitatively similar patterns of metabolism to that of the microsomal enzyme.

Other Oxidative Reactions

These reactions include the NAD-linked dehydrogenases responsible for the oxidation of ethyl alcohol and other primary and secondary alkanols to aldehydes (alcohol dehydrogenase) and the oxidation of aldehydes to acids by aldehyde dehydrogenases. Other enzyme systems known to mediate these reactions include catalase for methanol oxidation, and

Table 3 *The cytochrome P-450 superfamily—some inducer, inhibitors, and substrates*

Subfamily/ protein	Inducer[a]	Inhibitors[b]	Substrates
1A			
1A1	3MC; TCDD; βNF; ARO	αNF; 9-HE	} 7-Ethoxresorufin
1A2	ISO; 3MC; TCDD; βNF	αNF; 9-HE	} Caffeine; acetanilide
2A			
2A1	PB; βNF; 3MC; PCN	—	Coumarin, testosterone
2A2	Non-inducible	—	
2B			
2B1	PB; ethanol; DEX	SB; SKF-525A	Pentoxyresorufin,
2B2	PB, ARO; PCN; DEX	—	Benzphetamine
2B4	PB	—	
2D			
2D1	Non-inducible	Ajamalicine; quinidine	Debrisoquine
2E			
2E1	Ethanol, DMSO Isoniazid	Diallyl sulphide	Dimethylnitrosamine 4 Nitrophenol
3A			
3A1	PCN; PB; DEX	Erythromycin	Testosterone
3A2	PB	—	Testosterone
3A4	PB; DEX	17α-Ethynyl- estradiol	Nifedipine
3A6	RIF	—	—
4A			
4A1	CLOF	Terminal acetylinic fatty acids	Lauric acid

[a] 3MC = 3-methylcholanthrene;   TCDD = 2,3,7,8-tetrachlorobenzodioxin; βNF = β-naphthoflavone;   ISO = isosafrole;   PB = phenobarbitone;   ARO = Aroclor   1254; DMSO = dimethylsulphoxide;   PCN = pregnenolone-16α-carbonitrile;   DEX = dexamethasone; RIF = rifampicin; CLOF = clofibrate.
[b] α-NF = α-naphthoflavone; 9-HE = 9-hydroxyellipticine; SB = secobarbital.
Data from Murray and Reidy (1990); Soucek and Gut (1992).

aldehyde oxidase and xanthine oxidase for the oxidation of aldehydes to acids (McMahon, 1971).

Other oxidation reactions of importance in the metabolism of foreign compounds include the oxidative deamination of amines by either the monoamine oxidase system or by a microsomal FAD-containing mono-oxygenase enzyme (Ziegler, 1980). The oxidation of nitrogen in organic molecules has recently been reviewed (Gorrod and Damani, 1985). A number of novel oxidative reactions have also been identified. These include the co-oxidation of polycyclic aromatic hydrocarbons and aromatic amines by prostaglandin hydroperoxidase (Eling, Boyd, Reed, Mason, and Sivarajah, 1983) and various peroxidative activities mediated by cytochrome P-450 (Testa, 1987).

C   Reduction

The two main classes of chemicals that are enzymically reduced in the body are those containing either carbonyl groups or nitrogen functional groups. The oxido-reductive interconversion of primary and secondary alcohols to aldehydes and ketones, respectively, is mediated by cytosolic NAD-linked alcohol dehydrogenase. At physiological pH, the reaction equilibrium favours reduction. However, although ingested ketones, such as cyclohexanone and hexobarbitone are reduced to the corresponding secondary alcohol, reactive aldehydes are usually further oxidised to acids. Examples of aldehydes reduced *in vivo,* however, include phthalaldehydic acid (Shiobara, 1977) and 2[1'-phenoxy-2'chloro]acetaldehyde (Osterloh, Karakaya, and Carter, 1980). Additionally, cytosol from a number of tissues contains at least three NADPH-dependent aldehyde reductases (M.W. range 34,000–36,000) capable of the reduction of many aromatic ketones and aldehydes (Wermuth, Münch, and Von Wartburg, 1977; Sawada, Hara, Kato, and Nakayama, 1979). An interesting feature of these enzymes is their stereo-selectivity. Thus, for example, *R*-warfarin is reduced almost exclusively to *R,S*-warfarin alcohol in humans, only a trace of the *R,R*-alcohol being formed.

Aromatic nitro- and azo-compounds and tertiary amine *N*-oxides are three important groups of nitrogen containing compounds metabolised by reduction. Aromatic nitro compounds are reduced to amines, *via* nitroso and hydroxylamine derivatives, by mammalian hepatic microsomal NADPH-linked nitro-reductase activity. This enzyme activity, inducible by phenobarbitone and DDT treatment, and inhibited by SKF-525A, suggests the obligatory involvement of cytochrome P-450. A number of other proteins possess nitro-reductase activity. These include NADPH-cytochrome c reductase, DT-diaphorase, xanthine oxidase, aldehyde oxidase, and lipoyl dehydrogenase (Wang, Behrens, Ichikawa, and Bryan 1974; Hewick, 1982). Aldehyde oxidase, for example, is a cytosolic molybdenum-containing enzyme, which catalyses the reduction of compounds such as nitrofurazone, *N*-nitrosodiphenylamine, and various sulphoxides (Tatsumi, Yamada, and Kitamura, 1983).

An hepatic azo-reductase enzyme capable of metabolising aromatic azo compounds to their component amines, probably *via* an intermediate hydrazo-derivative, is found in both microsomal and cytosol fractions. The properties of the microsomal enzyme, such as inhibition by oxygen and inducibility by typical inducers of cytochrome P-450, suggests that this enzyme is associated with the cytochrome P-450 dependent MFO system. The cytosolic enzyme from rat liver has been purified and shown to be a flavoprotein of molecular weight 52,000, containing two molecules of FAD per enzyme molecule (Huang, Miwa, and Lu, 1979). The enzyme activity is inhibited by dicoumarol and markedly induced by 3-methylcholanthrene. NADPH and NADH can serve as co-factors. The electron transport properties, spectral characteristics and mechanism of inhibition by dicoumarol suggests that this enzyme is identical with DT-diaphorase (Huang, Miwa, and Lu, 1979).

   N-Oxide reduction to tertiary amines may be carried out enzymatically or non-enzymatically, depending on the substrate, site of metabolism, and the animal species. An hepatic microsomal NADPH-linked cytochrome P-450 dependent enzyme, inducible by phenobarbital, catalyses the reduction of indicine N-oxide to indicine (Powis and Wincentsen, 1980). In contrast, a cytosolic enzyme, probably a flavoprotein, requiring NADPH in preference to NADH, mediates the reduction of nicotine 1^1-N-oxide to nicotine. Haemoglobin non-enzymatically catalyses the reduction of a number of N-oxide compounds including those of trimethylamine, nicotinamide, imipramine, and dimethylaminoazobenzene, to their corresponding amines (Bickel, 1969). In the process, haemoglobin is oxidised to methaemoglobin. Other nitrogen containing compounds known to be biologically reduced include aliphatic nitro, nitroso, hydroxylamine, oxime, and hydroxamic acids.

   Reduction of carbon-carbon double bonds has not been widely noted, however the saturation of the α,β-unsaturated ketone LY140091 by a cystolic enzyme was recently reported (Lindstrom and Whitaker, 1984).

D Conjugation

These reactions constitute a diverse group of enzyme-mediated processes whereby foreign compounds or their metabolites are linked to endogenous compounds to form more polar water-soluble products capable of being readily excreted from the body in the bile or the urine, depending on overall molecular weight. The major conjugation reactions lead to the formation of glucuronides, mercapturic acids, sulphates, and amino acid and acetyl derivatives. Less common conjugation reactions include glycosylation, methylation, fatty acid conjugation, and phosphorylation.

Glucuronide conjugates

UDP-glucuronyltransferase (UGT) catalyses the transfer of glucuronic acid from UDP-glucuronic acid (UDPGA) to endogenous or foreign substrates to form O(ether)-, O(ester)- N-, S-, and less commonly C-glucuronides. Exogenous substrates include alcohols, phenols, carboxylic acids, hydroxylamines, aromatic amines, and sulphydryl compounds. In the conjugation reactions involving these functional groups, the $C-1$ carbon atom of the glucuronic acid moiety of UDPGA is activated to facilitate nucleophilic attack by the oxygen, nitrogen, carbon, or sulphur atom of the substrate. The glucuronide derivative thus fromed has the β-configuration, and UDP is the leaving group.

The UGTs are a group of isoenzymes of 50–60 kDa localised primarily in the hepatic endoplasmic reticulum and nuclear envelope (Burchell and Coughtrie, 1992). In human liver, nine enzymes in two sub-families have been identified, based on sequence identity. The four isoenzymes in sub-family 1 catalyse the glucuronidation of phenols and bilirubins, but generally not steroids or bile acids, and the five isoenzymes of sub-family 2 catalyse the glucuronidation of steroids, bile acids, and some xenobiotics.

UGT enzyme activity can be enhanced *in vitro* by the addition of detergents such as Triton X-100 and Lubrol or organic solvents (*e.g.* n-hexane, chloroform, and ether). Treatment of animals *in vivo* with phenobarbital, 3-methylcholanthrene, DDT, chlorobiphenyls, or TCDD leads to the induction of hepatic microsomal glucuronyl-transferase activity. On the other hand, inhibition is effected by treatment with hydrazines, harmol, and by the major metabolite of hydantoin, 5-(*p*-hydroxyphenyl)-5-phenylhydrantoin (Mulder, 1974; Batt, Ziegler, and Siest, 1975). *In vitro* inhibition of enzyme activity is effected by the addition of UDP-galactose, UDP-xylose, galactose, galactosamine (probably at the level of UDP-glucose dehydrogenase) and by piperonyl butoxide (Hanninen and Marniemi, 1970; Lucier, McDaniel, and Matthews, 1971).

Mercapturic acid conjugates

The formation of mercapturic acids constitutes the end product of a series of biochemical reactions initiated by the conjugation of electrophilic, hydrophobic substrates with the tripeptide glutathione. Substrates for this conjugation reaction, which is mediated by glutathione *S*-transferases, (Habig, Pabst, and Jakoby, 1974) include alkyl and aryl halides, alkylmethane-sulphonates, dialkylphosphoric acid triesters, epoxides, alkenes, organic thiocyanate, and triazines. The sequence of enzymic reactions leading to the formation of mercapturic acids is exemplified by the metabolism of benzyl chloride to benzylmercapturic acid (Figure 1).

Glutathione *S*-transferase (GST: EC 2.5.1.18) is a family of enzymes of overlapping substrate specificity (Vos and Van Bladeren, 1990), each enzyme species being a dimer differing in sub-unit composition. The isoenzymes are mainly present in cytosol, although a microsomal form has been identified. Elution chromatography of the transferases from rat liver cytosol on carboxymethylcelluloses has revealed up to 12 different isozymes (Stockman, Beckett, and Hayes, 1985; Mannervik, 1987), which along with the enzymes from mouse and human liver cytosol are assigned to one of three classes (α, μ and π) based on structural and catalytic properties (see Table 4).

Individual isoenzymes of glutathione *S*-transferase are inducible by a

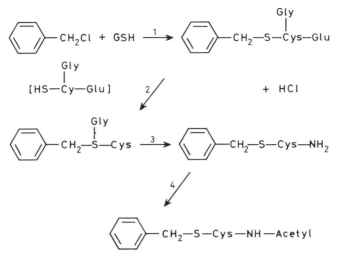

Benzylmercapturic acid

Figure 1 *Conjugation of benzyl chloride with glutathione. The enzymes catalysing the reactions are*: (1) *Glutathione S-transferase*, (2) *γ-Glutamyl transferase*, (3) *Peptidase*, (4) *Acetyl-CoA acetyl transferase*.

variety of chemicals (Mannervik, Alin, Guthenberg, Jensen, and War-holm, 1985), principally by translational activation of the corresponding genes (Vos and Van Bladeren, 1990). Induction by phenobarbitone, 3-methylcholanthrene, and benzo(a)pyrene has been reported by Clifton and Kaplowitz (1978), and polychlorinated biphenyls and TCDD have also been shown to induce this enzyme activity in rat liver cytosol (Kohli, Mukhtar, Bend, Albro, and McKinney, 1979; Baars, Jansen, and

Table 4 *Classification of rat cytosolic glutathione S-transferases*

Class		
Alpha (α)	*Mu (μ)*	*Pi (π)*
1.1	3.3	7.7
1.2	3.4	
2.2	4.4	
8.8	3.6	
	4.6	
	6.6	

Isoenzyme 5.5 not yet classified.
Mouse isoenzymes—class Alpha N4.4
                clas Mu N1.1; C1.1; C2.2; D1.1
                class Pi N3.3.
Data from Vos and Van Bladeren (1990).

Breimer, 1978). Other compounds known to be inducers of GST activities in hepatic and extrahepatic tissues include the antioxidants butylated hydroxyanisole (BHA) and ethoxyquin (Benson, Batzinger, Ou, *et al.*, 1978) and *trans*-stilbene oxide (Guthenberg, Morgenstern, DePierre, and Mannervik, 1980). A wide range of inhibitors of GSTs are also known (Mannervik and Danielson, 1988) many of which have differential effects on the various isoenzymes. Interestingly, the herbicides 2,4-D and 2,5,4-T, which are strongly inhibitory to μ enzymes, appear to act by a mechanism not asociated with binding to the active site (Dierickx, 1983). *In vivo,* inhibition may also result from the depletion of GSH by treatment with compounds such as diethylmaleate, and by thiol blocking agents (*e.g.* *p*-hydroxymercuribenzoate) and alkylating agents (Askelöf, Guthenberg, Jakobson, and Mannervik, 1975). Other compounds possessing a carbonyl (or similar) group with an electrophilic character which are capable of reacting non-enzymatically with glutathione and other thiols, include the 2-methylmercaptotriazine herbicide, Cyanatryn (Bedford, Crawford, and Hutson, 1975).

Sulphate conjugates

A variety of sulphotransferases present in mammalian liver cytosol catalyse the formation of *O*-sulphate esters of simple phenols, other aromatic alcohols, and steroids (androstenolone and estrone). Additionally, arylamines are metabolised by sulphate conjugation to sulphamates. The donor of the sulphate group is 3'-phosphoadenosine-5'-phosphosulphate (PAPS) formed from ATP and sulphate ions by the action of three interlinked enzymic reactions, namely ATP-sulphurylase, ADP-sulphurylase, and APS-kinase. Thus, the depletion of the intracellular ATP pool by mitochondrial inhibitors, *e.g.* rotenone, can reduce the availability of the co-factor (PAPS) and thereby inhibit the phosphotransferase activities. Other inhibitors of these enzymes include pentachlorophenol and 2,6-dichloro-4-nitrophenol which act as 'dead end' competitive substrates. The sulphotransferase have been reviewed in detail by Jakoby, Sekura, Lyon, Marcus, and Wang (1980).

Other Conjugation Reactions: Amino Acid Conjugates

Glycine is the amino acid most commonly used for the conjugation of aromatic acids in mammals. In the typical sequence of biochemical processes involved in the conjugation of benzoic acid, for example, the first step is the esterification of the acid with CoA in the inner mitochondrial matrix, to form benzoyl CoA. This reaction, mediated by butyryl-CoA synthetase and requiring ATP, proceeds *via* the formation of the intermediate AMP derivative (C_6H_5CO-AMP) which then reacts with CoA to produce benzoyl CoA. Subsequently the acyl CoA combines with glycine, catalysed by benzoyl transferase to form hippuric acid. 'Benzoyl transferase' and 'phenylacyl transferase' both of which are acyl-CoA: amino acid *N*-acyl transferases have been isolated from bovine-liver mitochondria. The former enzyme is specific for benzoyl-CoA, salicyl-CoA and certain short linear and branch-chain fatty acyl-CoA's, whereas the latter specifically utilises the CoA esters of phenylacetic acid and indole-3-acetic acid. These related enzymes, separated by SDS disc-electrophoresis and by other procedures, have been shown to be single polypeptide chains (molecular weight approximately 33 kDa). In addition, there appears to be a third enzyme in rat liver specific for bile acid-CoA derivatives which is capable of forming bile acid conjugates with both glycine and taurine (Killenberg and Jordan, 1978).

Taurine and glutamine conjugates have been shown to be formed with certain arylacetic acids, including 1- and 2-naphthylacetic acids, hydratropic acid, and 3-phenoxybenzoic acid in mammals (Idle, Millburn, and Williams, 1978). Other amino acids participating in conjugation reactions include histidine, serine, and alanine in bats, ornithine in reptiles and certain birds, and arginine in Arachnida. In addition, conjugation with peptides (other than GSH) has been reported (*e.g.* with phenothiazine in calves: Waring and Mitchell, 1985).

Acetyl conjugates

Liver cytosol contains a group of enzymes, the acetyl CoA: *N*-acetyltransferases, which catalyse the acetylation of aliphatic and aromatic amines including drugs such as isoniazid, dapsone, and the sulphonamides. Animal studies have shown a heterogeneity and polymorphism in the *N*-acetyltransferases with respect to tissue distribution, substrate specificity and pH-activity characteristics. One enzyme, found primarily in the liver and gut, catalyses the acetylation of a wide range of compounds and is considered to be the enzyme responsible for genetic polymorphism. Another acetyltransferase found in extrahepatic sites is active towards *p*-aminobenzoic acid and has a limited substrate specificity.

In marked contrast to the other conjugation reactions, acetylation decreases the hydrophilicity of the compound.

Methyl conjugates

Enzymic methylation of *O*-, *N*- and *S*- atoms in a wide range of substrates is effected by the corresponding methyltransferases using *S*-adenosylmethionine as the methyl donor.

Soluble (*i.e.* cytosolic) catechol-, hydroxyindole-, and di-iodotyrosine-*O*-methyl transferases have been identified as separate enzymes. In contrast, phenol-*O*-methyltransferase is a microsomal enzyme. Substrates for the *N*-methylation reaction include primary, secondary, and tertiary amines and azaheterocyclics, and for *S*-methylation include diethyl dithiocarbamate and carbon disulphide.

Phosphate conjugates

Rat liver mitochondria contain an ATP-dependent enzyme capable of phosphorylating 1-aminopropan-2-ol. Other examples of enzyme mediated phosphorylation reactions are the formation of 1-naphthyl dihydrogen phosphate from 1-naphthol and of di-(2-amino-1-naphthyl) hydrogen phosphate from 2-naphthylamine (Binning, Darby, Heenan, and Smith, 1967; Boyland, Kinder, and Manson, 1961; Troll, Tessler, and Nelson, 1963). However, little is known of the enzymes mediating these reactions and the extent to which they are involved in the metabolism of foreign compounds.

Novel conjugates

Many novel conjugate reactions, including those with endogenous chemicals, such as cholesterol, fatty acids, and phospholipids have been reported (see reviews by Paulson *et al.,* 1986 and Testa, 1987).

4   Sites of Metabolism

Before discussing the biotransformation and disposition of foreign compound with respect to specific tissues, the organisation and functions of various intracellular organelles involved in metabolic processes at the cellular level will be outlined. The liver cell is used as an illustrative example, and the various sub-cellular components are shown in the electron micrograph (Figure 2). A detailed discussion of the methods for the isolation and purification of these organelles is beyond the scope of this chapter, and many texts describing the relevant methods for disruption and fractionation (by centrifugation) have appeared (see, for example, Birnie, 1972; Snell and Mullock, 1987). A brief outline of the structure and function of the various organelles follows.

A   Intracellular Organelles

Nucleus

The various components of the cell nucleus are contained within a double membrane. The outer one is part of the cell's endoplasmic reticulum, and carries a number of ribosomes. The inner membrane, which is separated from the outer by the perinuclear space, is lined by the lamina densa, and the whole structure contains a number of pores. Within the membranes is the finely dispersed euchromatin, the fibrous hetero- and nucleolar-associated chromatin, the nucleolus, and a number of other bodies of

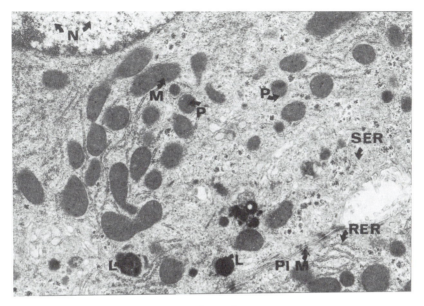

Figure 2

granular appearance. (Bouteille, Laval, and Dupuy-Coin, 1974). A wide range of enzyme activities have been found associated with the nucleus, apart from those involved in RNA and DNA synthesis and processing. These include methylases, acetylases, proteases, kinases, deacetylases, steroid dehydrogenase, cytochrome oxidase, glycosyl transferases, ATPase, carboxylesterases, phosphatases, mixed-function oxidases, and 5'-nucleotidase (Olson and Busch, 1974).

Mitochondria

The mitochondria, of which there are about 400 in a liver cell, contain the enzymes responsible for oxidative phosphorylation and related energy transfer processes, and are the site of production of high-energy phosphates (*e.g.* ATP) in the cell. These rod-shaped bodies, with sizes up to 5 μ by 0.7 μ, are bounded by a double membrane. The outer, limiting membrane encloses a folded, inner membrane which gives rise to the cristae. Thus, the mitochondrion consists of four compartments in which enzymes can be located, namely the outer membrane, the inter-membrane space, the inner membrane including the cristae, and the matrix enclosed by the inner membrane's surface. Enzyme activities associated with the outer membrane include monoamine oxidase, NADH-cytochrome b_5 reductase, acyl transferases, phosphotransferases and kinases, and cytochrome P-450-dependent monooxygenases. NADH-linked dehydrogenases, succinate dehydrogenase, fatty acyl transferases, and cytochrome P-450 are associated with the inner membrane and citric acid

Table 5 *Tissue localisation of some enzymes mediating xenobiotic transformation reactions*

Tissue	Hydrolysis	Oxidation	Reduction	Conjugation
Skin	Carboxylic acid esterase EH	MFO system *e.g.* AHH; ethoxycoumarin *O*-deethylase; Steroid dehydrogenase, MAO	Ketosteroid reductase Progesterone desaturase	UDP-GT Sulphotransferase COMT
Lung	EH	MFO system *e.g.* AHH, *N*-and *O*-demethylases *N*-oxidase and hydroxylase	*N*-oxide reductase	UDP-GT Phenol sulphotransferase
Gastro-intestinal tract Small intestine	Non-specific carboxylic acid esterase	MFO system *e.g.* AHH, *N*- and *O*-deethylase biphenyl hydroxylase	'*N*-hydroxy-AAF reductase'	UDP-GT GSH-T *N*-acetyl transferase Sulphotransferase Glycine conjugase
Microflora	Carboxylic acid esterase Sulphate ester hydrolase Amide hydrolase Glucuronidase Glucosidase Sulphamatase	*N*- and *O*-dealkylase Dehalogenase Deaminase	Nitro-reductase Azo-reductase *N*-oxide reductase Epoxide reductase Aldehyde reductase Aromatase	Methyl and acetyl transferase

Organ	Hydrolases/Esterases	Oxidases	Reductases	Transferases
Liver	Carboxylic acid esterase Amidase EH	MFO system *e.g.* aromatic and aliphatic hydroxylase; N-, O- and S-dealkylases: epoxidase; sulphoxidase; N-hydroxylase, *etc.* ADH and Ald DH Catalase, MAO	Azo-reductase Nitro-reductase N-oxide reductase Ketoreductase	UDP-GT; GSH-T; Sulphotransferase Acetyl and methyl transferases
Kidney	Glucuronidase Aryl sulphatase Carboxylic acid esterase Amidase EH	MFO system *e.g.* aromatic and aliphatic hydroxylase N-dealkylase; Fatty acid hydroxylase ADH		UDP-GT; GSH-T Sulphotransferase Amino-acid conjugase Acetyl transferase
Adrenals		MFO system *e.g.* steroid hydroxylase		UDP-GT GSH-T Sulphotransferase
Testes	Glucuronidase Aryl sulphatase Carboxylic acid esterase	MFO system *e.g.* steroid hydroxylase	Aromatase	UDP-GT GSH-T Sulphotransferases

Abbreviations: EH—Epoxide hydratase; MFO—Mixed function oxidase; UDP-GT—Uridine diphosphoglucuronsyltransferase GSH-T—Glutathione-S-transferase; COMT—Catechol-*O*-methyl transferase; ADH—alcohol dehydrogenase Ald.DH—Aldehyde dehydrogenase; MAO—Monoamine oxidase; AAF—*N*-acetylaminofluorene cycle enzymes, transaminases, kinases, and NADP-linked de-hydrogenases are among the enzymes found in the matrix. Adenylate kinase and nucleoside diphosphokinase have been found in the inter-membrane space. Other enzyme activities found in mitochondria include esterases and superoxide dismutase. An extensive discussion of the nature and distribution of enzymes in the mitochondrion can be found in the review of Ernster and Kylenstierna (1969).

Microsomes

The microsomes are a heterogeneous mixture of membrane vesicles derived from the endoplasmic reticulum (e.r.). The separation of microsomal preparations into rough (ribosome-containing) and smooth (ribosome-free) fractions by density gradient centrifugation procedures, has led to the study of enzyme topology in these membranes. Rough e.r., which is composed of thin lamellae made up of two opposing membranes, often occurring in more or less parallel stacks, contains a high concentra-tion of electron transport proteins, glucose-6-phosphatase, and ATPase activity. Smooth e.r. is made up of a maze of fine tubules, the membranes being in intimate contact with the mitochondria (Claude, 1969). It contains high levels of cytochromes b_5 and P-450 (De Pierre and Erster, 1977; Bergman and Dallner, 1976) and is generally more active in metabolising xenobiotics than the rough e.r. (Holtzman, Gram, Gigon, and Gillette, 1968; Gram, Schroeder, Davis, Reagan, and Guarino, 1971). Other enzyme activities associated with microsomes include monoamine oxidase, NAD(P)H-cytochrome c reductase, nucleoside diphosphatase, esterase, aldolase, and glutamine synthetase (Amar-Costesec, Beaufay, Feytmans, Thines-Sepoux, and Berthet, 1969).

Peroxisomes

Peroxisomes are widely distributed in animal cells, and consist of a single limiting membrane enclosing an electron-dense granular matrix and a crystalloid core. They can be differentiated from lysozomes by the absence of other matrix features such as vacuoles, lipid droplets, or membranous structures. Catalase is the principal marker enzyme for peroxisomes, however other oxidases are present including D-amino acid oxidase, urate and glyoxylate oxidase. Other enzmes found in peroxi-somes include carnitine acyl transferase, NAD(P)-dependent de-hydrogenases, thiolase, enoyl-CoA hydratase and fatty acyl CoA synthe-tase (Reddy and Lalwani, 1983).

Lysozomes

Lysozomes are present in the cytoplasm of the majority of mammalian cells, with the exception of the erythrocytes. They appear as 'dense bodies' in electron micrographs, consisting of a fine, granular matrix enclosed by a single membrane. The matrix often contains other structures such as osmiophilic droplets, vesicles and membrane-like material. Lysozomes contain a wide range of hydrolytic enzymes including acid phosphatase, DNAase and RNAase, esterases, aryl sulphatases, glycosidases, and lipases (for a detailed discussion of lysozomal enzymes, see Tappel, 1969).

Plasma membranes

The plasma membrane of a liver cell is that part of the surface of the hepatocyte which is in contact with other cells. The fraction as prepared by discontinuous density gradient centrifugation (Lansing, Belkhode, Lynch, and Lieberman, 1967), contains desmosome-rich trilamellar structures and circular vesicles. As well as 5'-nucleotidase, the widely accepted marker enzyme for plasma membranes, these structures also contain acid and alkaline phosphatase activity, phosphodiesterase and nucleoside triphosphate pyrophosphohydrolase. In addition, the plasma membrane contains a large number of receptor sites for the active uptake of essential nutrients, and hormones, including those for amino acids, sugars, fats, glucagon, and prolactin.

The main mammalian tissues involved in the metabolism of foreign compounds are shown in Table 5, along with representative examples of the major enyme categories known to be present at these sites. Although enzymes such as the mixed-function oxidases and conjugates are generally widely distributed in most tissues, certain important biochemical and morphological features distinguish and characterise the range of biotransformation reactions effected in each of the organs. A brief description of these features, as they relate to the biotransformation and disposition of foreign compounds, follows.

B   Skin

This tissue constitutes approximately 10% of the normal body weight of mammals, and is thus one of the largest organs of the body. It consists of two distinct components; the outer and thinner epidermis and the thicker, underlying dermis. In addition, there are several structures within the skin, such as sweat glands, sebaceous glands, and hair follicles. The skin is involved in several important physiological and biochemical functions, including the regulation of body temperature, of water loss and retention, of sebum production *via* androgens, and of carbohydrate and lipid metabolism *via* insulin. The skin is also one of the most active tissues in the body for protein synthesis (Freedberg, 1972). In addition, the skin is capable of activating or inactivating several steroidal hormones, such as the androgen dehydroepiandrosterone, which is converted first to testosterone and subsequently to 5α-dihydrotestosterone.

Steroid metabolism in the skin has been reviewed comprehensively by Hsia (1971) and Rongone (1977).

The presence of haem-proteins in the skin has been known for some time, although cytochrome P-450 was first demonstrated in rat skin microsomes less than 20 years ago (Bickers, Kappas, and Alvares, 1974). The amount of microsomal protein in weanling rat skin is small $(3-5 \, mg \, g^{-1}$ wet tissues) compared with the liver $(20-25 \, mg \, g^{-1}$ wet tissue) (Bickers and Kappas, 1980) and MFO enzyme activities are also low (Willemsens, Vanden Bossche, and Lavrijsen, 1989). The activities are inducible by polycyclic hydrocarbons, PCBs, TCDD, and other compounds known to be inducers of P4501A1 in the liver. The most extensively studied xenobiotic metabolism is that of the microsomal MFO-associated AHH activity, reported to be present predominantly in the epidermal layer (Thompson and Slaga, 1976). Other MFO activities found in the skin include aniline hydroxylase and 7-ethoxycoumarin deethylase (Pannatier, Jenner, Testa, and Etter, 1978; Pohl, Philpot, and Fouts, 1976). Due to technical difficulties in the preparation of cutaneous tissues for metabolic studies, very few definitive experiments have been carried out to delineate the substrate specificities of the enzyme activities associated with the MFO system in the skin.

C Lung

The mammalian lung contains over 40 different cell types, although only four are unique to the lung (Sorokin, 1970). These four types of pulmonary epithelial cells, concentrated in the broncho-alveolar region and known to contain enzyme systems capable of metabolising foreign chemicals are: (1) non-ciliated bronchiolar cells (Clara cells); (2) squamous alveolar cells (membranous pneumocytes: type I cells); (3) great alveolar cells (granular pneumocytes type II cells); and (4) pulmonary alveolar macrophages. The Clara cells, dome-shaped in form and containing lipid droplets in their apices, have an abundant complement of smooth endoplasmic reticulum. Type I cells have a flattened shape and attenuated cytoplasm with limited intracellular organelles, whereas type II cells are cuboidal in shape, and rich in rough endoplasmic reticulum, Golgi apparatus, lysosomes, and lamellar bodies. Type II cells eventually mature into type I cells over a period of approximately 3 months. Although type I cells are only half as numerous as type II cells, they cover about 25 times as much of the alveolar surface as the latter cell type. The pulmonary alveolar macrophages, present in the alveolar spaces, are particularly rich in lysosomes containing hydrolases.

The main locations for the enzymes of the MFO complex in this organ are the type II pneumocytes and the Clara cells. The wide variety of substrates metabolised in the lung has been reviewed by Gram (1980). Whereas phenobarbitone treatment has no significant inductive effect on pulmonary MFO enzymes (Uehleke, 1968), polycyclic aromatic hydro-carbons, naphthoflavone and phenothiazine derivatives have a marked inducing capability. Several P450s have been isolated from lung tissue, including forms with similar properties to those of forms 4A1 and 4A2 (Murray and Reidy, 1990).

D   Gastro-intestinal Tract

Small intestine

The gastro-intestinal tract was originally thought to act only in the absorption of foreign compounds, but it is now recognised that the epithelium is an important site for the metabolism of ingested com-pounds. Mixed-function oxidase enzyme activity, which is present along the length of the intestine, is most active in the duodenum and decreases towards the caecum. Activity also varies at different levels of the mucosa cells, with the lowest activity found in the crypts and the highest in the villous tip cells (Hoensch, Woo, and Schmid, 1975). Although the specific activities of MFO enzymes in the small intestine are low compared to the liver (see review by Houston and Wood, 1980) the extensive surface area provided by the mucosal cells and their rapid turnover renders this tissue quantitatively important. In contrast to the relatively low uninduced Phase I (oxidative) reaction capability, this organ shows high glucur-onyltransferase activity. The mucosal cells respond to most of the model inducers of the MFO system (Wollenberg and Ullrich, 1977).

Microflora

The gastro-intestinal tract in mammals contains a wide variety of micro-organisms. Over 400 species of bacteria have been identified and the main groups found in man, namely the bacteridaceae, propionobac-teridaceae, lactobacillaceae, enterobacteriaceae, micrococcaceae, and yeasts are anaerobic organisms. There are large interspecies differences in the numbers and types of micro-organisms present in the gut and also in their distribution along the gastrointestinal tract (Rowland and Walker, 1983; Savage, 1977).

In view of the anaerobic character of the resident microflora and the low redox potential prevailing in the caecum and colon, reductive reactions constitute the major biotransformations effected on foreign chemicals. These reactions include the reduction of nitrate, nitro-, and azo-compounds (see Table 5). The other major class of metabolic reactions carried out by the gut microflora is the enzymic hydrolysis of esters, amides, peptides, and glucuronides. The gastro-intestinal hydroly-sis of glucuronides excreted *via* the bile and the subsequent intestinal reabsorption of the metabolite formed leads to entero-hepatic circulation and the possibility of second-pass hepatic metabolism. Entero-hepatic circulation has been found to be of significance in the metabolism of stilboestrol, butylated hydroxytoluene, and various food flavours, including linalool.

A number of factors have been found to modify the enzymic profile of gut microflora. For example, treatment of rats with cyclamate for prolonged periods leads to adaptive changes in the microflora resulting in an increase in the metabolism of this compound by the induction of gut flora sulphamatase activity (Renwick, 1986). On the other hand, chronic exposure of rats to metronidazole leads to the inhibition of its metabolism by the gut flora (Eakins, Conroy, Searle, Slater, and Willson, 1976). The metabolic capabilities of microflora in various species has been reviewed by Scheline (1973; 1980) and by Rowland, Mallett and Wise (1985).

E   Liver

A major and possibly the most intensively studied site for the metabolism of foreign compounds is the liver. The relative uniformity in the morphology of the tissue, the abundance of metabolising enzyme activities and the ease of preparation for assay has, in part, been responsible for the wide ranging biochemical and histological investigations on this organ. The brief description of the morphology of the liver is intended to provide an appreciation of the tissue distribution of enzyme activities.

There is a growing body of evidence that the original description of the liver in terms of a hexagonal configuration of liver cells proposed by Kiernan in 1833, does not equate with the current concepts of the functional unit of the liver, namely the acinus, first described by Rappaport and his co-workers. This unit is seen as a parenchymal mass consisting of a terminal portal tract, hepatic arteriole, bile ductule, and lymph vessels, contained between two or more central veins. Thus, the classical description of midzonal periportal, focal, and centrilobular regions of the liver correspond to zones 1, 2, and 3 respectively in the Rappaport description. (For a more detailed discussion of this scheme, see Rappaport, 1969). Recent histochemical and electron microscopic studies have demonstrated that enzyme distribution within the liver lobule is not uniform. The activity of respiratory enzymes is particularly high in zone 1, whereas NADP-dependent MFO enzyme activities are concentrated in zone 3. Furthermore, immunohistochemical and microspectrophotometric studies have shown that phenobarbital-inducible cytochrome P-450 is predominantly in zone 3 whereas the 3-methylcholanthrene-inducible cytochrome P-448 is more evenly distributed throughout the liver lobule. (Baron, Redick, Kapke, and Guengerich, 1981).

The MFO enzyme system is most generally studied in the microsomal fraction, although cytochrome P-450 has also been detected in hepatic, mitochondrial, and nuclear fractions by ESR and microspectrophotometry. The mitochondrial haemoprotein and associated AHH activity are inducible by 3-methylcholanthrene, and the reaction was more rapid in the presence of NADH than with NADPH. The specific content of cytochrome P-450 in the rat hepatic nuclear fraction is about one-ninth of that in the microsomes, and the NADPH-cytochrome c reductase specific activity about 45%. (Sikstrom, Lanoix, and Bergeron, 1976). Rigorous purification procedures have established that these MFO activities are genuinely associated with hepatic mitochondrial and nuclear fractions and not due to adventitious microsomal contamination.

F Kidney

The kidney contains many of the enzyme systems capable of metabolising foreign compounds found in the liver, but the complex morphology and biochemical functions of the former makes it difficult to compare the tissue distribution and role of metabolising enzymes in the two organs. The two major anatomical regions of the kidney are the outer cortex and the medulla, the former being the major component. The cortex is composed mainly of proximal tubule cells and also of glomeruli, distal tubules, collecting ducts, blood vessels, and connective tissue. The endoplasmic reticulum in the tubule cells is structurally similar to that found in liver parenchymal cells, but constitutes only a small fraction of the total tubule volume. The luminal surface of the tubular cells is lined with a brush border consisting of numerous invaginated villi. The basal surface is characterised by frequent infolding, within which lie rod-shaped mitochondria. The medulla includes the papilla containing the loops of Henle, the collecting ducts, the descending and ascending vasa recta and their associated capillary plexus and interstitial tissue. (For a more detailed discussion of kidney structure and function, see Brenner and Rector, 1976.)

The renal cytochrome P-450 system, located mainly in the cortex, has many similarities but also certain differences to the hepatic system. These differences reside in the absorption maximum of the CO-cytochrome P-450 interaction spectrum, substrate specificities, and inducibility. The kidney cytochrome P-450 system, like that in the liver, is also involved in the metabolism of endogeneous substrates, having an essential role in fatty acid hydroxylation, prostaglandin synthesis and vitamin D metabolism.

The kidney cortex contains at least two forms of cytochrome P-450, one inducible by treatment with polycyclic aromatic hydrocarbons and the other inducible by dietary lauric acid. The latter form is relatively specific for fatty acid ω-hydroxylation. The renal MFO system metabolises substrates at rates generally lower than those found with the liver enzyme and the array of substrates metabolised is also much restricted. (Jones, Orrhenius, and Jakobson, 1980).

G   Other Tissues

Other organs of importance in the metabolism of foreign compounds include the adrenals, gonads, and placenta.

The adrenal cortex, in common with other steroid-producing tissues, contains an abundance of cytochrome P-450, located both in the microsomes and in the mitochondria. The former is required for the 17α- and 21-hydroxylation of steroids, whereas the latter mediates the 11β- and 18-hydroxylation and the side-chain cleavage of cholesterol. The mitochondrial steroid hydroxylation system has been resolved into three components—an NADPH-dependent flavoprotein (adrenodoxin reductase), a non-haem iron protein (adrenodoxin), and cytochrome P-450. There is little information on the chemical inducibility of adrenal cytochrome P-450; one report claims that phenobarbitone and 3-methylcholanthrene treatment are without effect (Feuer, Sosa-Lucero, Lumb, and Moddel, 1971).

Both the testes and the ovaries contain relatively low levels of MFO enzymes. However, the close proximity of these enzymes to the site of germ cell production confer particular toxicological relevance to their metabolic capabilities. The placenta contains a wide range of enzymes mediating the metabolism of foreign compounds, including the MFO system, conjugases, reductases, and hydrolases (for reviews, see Juchau, 1980; Heinrichs and Juchau, 1980).

5   Bioactivation and Detoxification Processes

The metabolic biotransformation of foreign compounds leading either to the generation of biological reactive metabolites (and thus to toxicity), or to the formation of normal physiological constituents or biologically inert excretory products, is depicted schematically in Figure 3.

In this scheme, five main types of metabolic events and the consequent biological effects may be identified. These are as follows:

Type 1: Compound 'X', toxic *per se* is metabolised *via* Pathway A to yield a non-toxic end product.

Type 2: Compound 'X', non-toxic *per se* is metabolised by a Pathway B reaction process to form a biologically reactive, and therefore, toxic metabolite.

Type 3: Compound 'X', non toxic *per se*, is metabolised by a Pathway B reaction process to a reactive (toxic) product which in turn is metabolised by a Pathway A process to yield a non-toxic product.

Type 4: Compound 'X', non-toxic *per se*, is metabolised as in type 3 reaction. The non-toxic product at step 2 is further metabolised by a type B reaction to yield a toxic species.

Type 5: Compound 'X', non-toxic *per se*, can be metabolised either by a type A reaction or by a type B reaction. Thus, in contrast to the

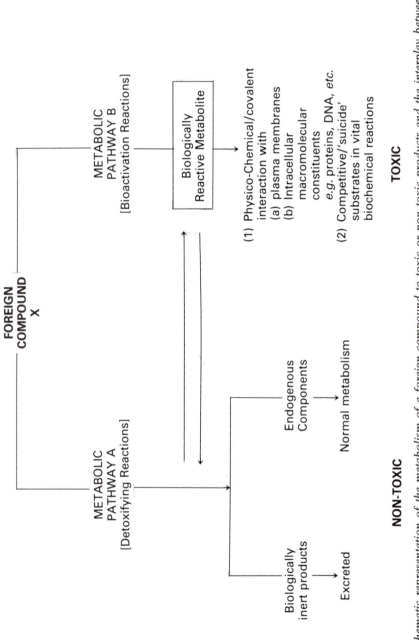

Figure 3 Schematic representation of the metabolism of a foreign compound to toxic or non-toxic products and the interplay between detoxifying and bioactivation processes in determining the eventual biological fate of a compound.

Table 6 *Some examples of biotransformation reactions on foreign compounds leading to toxic or non-toxic metabolites*

Type	Compound X	Pathways	Products
(1)	SO$_2$ Phenol Ethylene oxide	**A** Sulphite oxidase UDP-GT EH	Sulphate *O*-glucuronide Ethylene glycol
(2)	Chloroform Di-(2-ethylhexyl)phthalate Red 2G Ethylene dibromide	**B** MFO non-specific esterase azo reductase GSH-T	Phosgene Mono-ester Aniline Thiiranium ion
(3)	Allyl alcohol	**B** ADH **A** GSH-T	Acrolein→ Mercapturic acid

	Bromobenzene	MFO GSH-T
		   or EH
		   or UDP-GT
			Bromobenzene oxide
			→Mercapturic acid
			or -diol, or -glucuronide

	B	A	B	
	MFO	EH	MFO	
(4) Benzo(a)pyrene (BP)	MFO	Sulphotransferase	MFO	BP → epoxide → -diol → -diol epoxide
2-Acetylaminofluorene (2-AAF)			non-enzymic rearrangement	AAF → N-oxide → N,O-sulphate → -sulphonate
		A ‖ B	B	
(5)				
Aniline	MFO C-hydroxylase	‖	MFO N-hydroxylase	p-Aminophenol ‖ Phenylhydroxylamine
Phenacetin	UDP-GT or Sulphotransferase	‖	MFO	-glucuronide or sulphate ‖ quinone?

UDP-GT: Uridine diphosphoglucuronosyltransferase; GSH-T: Glutathione-S-transferase; EH: Epoxide hydratase; MFO: Mixed-function oxidase; ADH: Alcohol dehydrogenase.

situation with types 3 and 4 reactions where the metabolic processes operate in tandem, in type 5 reactions the two metabolic processes act simultaneously and therefore the biological effect is determined by the relative metabolic contribution of two processes.

It should be noted that Pathways A and B may each consist of a number of enzymic reactions acting on the compound or its metabolites. Examples of some of the foreign compounds metabolised by these various types of reactions are shown in Table 6. The picture that emerges from a survey of the complex and dynamic interplay of enzymic reactions participating in the metabolic disposition of foreign compounds, is that, in the main, enzymic reactions are toxicologically 'neutral', in that they cannot be categorised exclusively as bioactivating or detoxifying processes. Thus, for example, the MFO enzyme system can participate in both bioactivation and detoxification reactions and conjugation reactions, generally considered to be a detoxification step, can yield toxic end products, such as glutathione or sulphate conjugates (for reviews, see Van Bladeren, Bruggeman, Jongen, Scheffer, and Temmink, 1987; Mulder, Meerman, and van der Goorbergh, 1986).

Two further factors have a determining influence on the metabolism and biological fate of a compound. The first is the influence of the kinetics of the enzymic processes involved, and the second is the relationship of the various organs and intracellular organelles involved in the bioactivation and detoxification of the compound, with respect to the target site. Examples of the latter include the interactive metabolic transformation of nitrobenzene by the liver and the gastro-intestinal microflora (Levin and Dent, 1982) and of 1,3-hexachlorobutadiene by the liver and kidney (Nash, King, Locke, and Green, 1984). Other important determinants in the toxicology of a compound are the route of exposure of the species to the compound and the pharmacokinetics following absorption. The various factors known to influence pharmacokinetics and metabolism of compounds will be discussed in the next section.

6   Factors Influencing Metabolism and Pharmacokinetics

Differences in the biological fate of foreign compounds observed both between and also within animal species can be ascribed to the following:
(1) Differences in rates of absorption and transfer across cell membranes.
(2) Differences in rates of metabolism.
(3) Differences in routes of metabolism.
A number of factors contribute to these differences in xenobiotic metabolism. They include the age, sex, and strain of the animal model, the nutritional status of the animal, and other dietary factors. An

appreciation of the influence of these various factors in modifying the metabolism and pharmacokinetics of a compound in an animal model is essential for the understanding of the underlying toxic mechanisms. A brief description of these factors follows.

Rates of Absorption

In an earlier section, a brief description of the influence of molecular size, lipophilicity, and degree of ionisation on the absorption of foreign chemicals from various sites of exposure in the body was presented. In addition to these physico-chemical properties, other factors of significance in this context are the physical form of the compound, the dose level administered and the frequency of exposure to the foreign compound.

Whereas inert particulate material is unlikely to be absorbed into the body, except for the minute amounts persorbed across the intestinal lumen by pinocytosis, the uptake from the skin of absorbable compounds can be accelerated by various means. These include the presence of solvents, such as dimethylsulphoxide, and increased hydration of the stratum corneum by a vehicle.

Both the dosage and frequency of exposure to a foreign compound can affect its absorption, particularly when absorption is carrier mediated and therefore potentially saturable.

Species differences in the skin absorption rates of chemicals have been observed. Thus, whereas cutaneous permeability in the guinea pig, pig, and monkey is similar to man, the skin of the rat and rabbit is much more, and that of the cat much less permeable, compared with man (Scala, McOsker, and Reller 1968; Coulston and Serrone, 1969; Wester and Maibach, 1977).

Enterohepatic Circulation

Differences in extent of enterohepatic circulation (EHC), which is the reabsorption of a compound excreted into the bile from the gut and returned to the liver via the portal blood, are also known to affect the pharmacokinetics and biological effect of xenobiotics. Since one consequence of EHC is the maintenance of a disproportionately high concentration of compound in liver and gut for a longer period than would occur in its absence, exposure of the animal to potentially toxic compounds is increased. Thus, for example, species differences in the toxicity of indomethacin to the gastric mucosa are directly related to the extent of EHC (Duggan, Hooke, Noll, and Kwan, 1975).

Rates of Metabolism: Inter-species Differences

There is extensive literature on the quantitative differences observed in the metabolic rates of chemicals in various animal species (see review by

Table 7 *Species differences in the rates of metabolism of foreign compounds*

Species	Circulating half-life (min)		
	Hexobarbital	Antipyrine	Aniline
Man	360	600	—
Dog	260	107	167
Rat	140	141	71
Mouse	19	11	35
Aromatisation of quinic acid (% conversion in 72 hr)			
Man	60		
Old World Monkey	40–50		
Dog	0		
Rodent (Rat, Mouse)	0–5		

Hathway, Vol. 3, 4, 5, and 6). Some examples of species differences in the rate of metabolism of various drugs are shown in Table 7. It is clear that the mouse is more efficient than man in the metabolism of hexobarbital and antipyrine, for example, and that man converts quinic acid to benzoic acid more rapidly than the old world monkey or the dog. Other examples of species related differences in the rates of metabolism of chemicals include the 2-hydroxylation of biphenyl in the mouse (high) and in the rat (low), and the glucuronidation of monobutylphthalate in the hamster (high) and in the rat (low).

Rates of metabolism: Intra-species Differences

The metabolism of foreign compounds within a particular species may be modified by a number of factors, including the age, sex, strain, nutritional status and diet of the animals. A brief note on each of these factors follows:

Age. The activities of enzymes associated with the hepatic MFO complex are generally extremely low in the developing foetus and newborn animal. Liver size and the specific activities of the MFO enzymes progressively increase with age until puberty, as do the enzymes involved in conjugation reactions (Dutton, 1978). However, phenol sulphotransferase activity is present in foetal liver at an earlier stage than glucuronyltransferase activity.

The human foetus has appreciably higher levels of MFO enzyme activities compared with its animal counterpart, but less than in human adults. However, the glucuronidation ability in the human foetus is low. Hence, factors leading to the increased synthesis of bilirubin without a concomitant increase in UDP glucuronyltransferase activity can result in jaundice in premature human babies. In contrast to glucuronidation, sulphation of some drugs (*e.g.* paracetamol) is more rapid in the foetus than in the adult (Levy, Khanna, Soda, Tsuzuki, and Stern 1975).

Sex. In general, male rats and humans, but not male mice, tend to metabolise compounds more rapidly than females. Thus, for example, glyceryl guaiacolate ether, a centrally acting muscle relaxant, has a shorter circulating half-life in male rats compared to female animals because of the higher *O*-demethylase activity in the males (Giri, 1973). Similarly, the synthetic androgen, trengestone, is metabolised more readily to the 16α-hydroxy-derivative by male rat liver slices than by female preparations and, although a sex-difference was not seen with human liver slices, marked differences in the excretion of unchanged compound after oral administration to humans have been reported (Breuer, Kime, and Knuppen, 1973). The exhalation of acetaldehyde following the i.p. administration of ethanol in C57BL/6J mice is approximately 5–6 times greater in male animals than in females (Redmond and Cohen, 1972) and castration of the male mice abolishes this difference. Increased xenobiotic metabolising activity in males is also exemplified by the relative resistance of male rats to the toxicity of the organophosphorus insecticide, parathion.

Strain Differences. The literature is replete with examples of intra-species variation in the metabolism and disposition of foreign chemicals being ascribable to genetically determined strain differences (for reviews, see for example Hathway, Vol. 1–6; Nebert, Robinson, Niwa, Kumaki, and Poland, 1975). The induction of arylhydrocarbon hydroxylase (AHH) activity by 3-methylcholanthrene or β-naphthoflavone occurring in genetically responsive (C57BL/6J and C3H/He) strains of mice but not in the DBA/2 strain is due to an autosomal dominant trait capable of being segregated in heterozygous offspring. The presence or the absence of AHH induction has been found to correlate with the hepatic *N*-hydroxylation of 2-acetylaminofluorene in C57BL/6N and DBA/2N strains (Thorgeirrson, Felton, and Nebert, 1975). Other examples of strain differences in metabolism include the 7-hydroxylation of coumarin in mice (Lush and Andrew, 1978), the 4- and 6- hydroxylation of debrisoquine in inbred strains of rat (Al-Dabbagh, Idle, and Smith, 1981) and the *N*-acetylation of *p*-aminobenzoic acid and other arylamines in rats and mice (Tannen and Webber, 1979). In humans, examples of racial metabolic differences include the acetylation of drugs such as dapsone and sulphamethazine and the hydroxylation of debrisoquine. A recent example of genetic polymorphism in drug oxidation relates to the anti-angina compound perhexiline (Gould, Amoah, and Parke 1986; Cooper, Evans, and Whibley, 1984). For this drug, impaired oxidation results in increased toxicity.

Dietary Constituents and Nutritional Status. There has been a growing recognition of the importance of both macro- and micro-components in the diet and of nutritional status in influencing the metabolism and pharmacokinetics of foreign compounds since the early observations by Mueller and Miller (1950) and Brown, Miller, and Miller (1954) showing that dietary riboflavin modified hepatic azoreductase activity towards the carcinogenic azo compound 4-dimethylaminoazobenzene in the rat and mouse. Subsequent reports have shown that the biotransformation of drugs can be significantly affected by the intake of carbohydrates, fat, protein, vitamins, and minerals. Many reviews on the extensive and growing literature on this subject have been published including those by Campbell and Hayes (1974); Parke and Ioannides (1981); and Hayes and Campbell (1980).

The high intake of glucose and other carbohydrates in experimental animals and in man has been found to decrease haem synthesis and the activities of enzymes associated with the hepatic MFO system. Consequences of excess carbohydrate intake include prolonged phenobarbital hypnosis (Strother, Throckmorton, and Herzer 1971) and depressed hepatic dimethylnitrosamine demethylase activity (Venkatesan, Arcos, and Argus, 1970).

Animals maintained on diets totally deficient in lipids have markedly depressed hepatic MFO enzyme activities. The progressive addition of saturated and unsaturated fats to the diet leads to the restoration of these enzyme activities to normal levels. In particular a high intake of unsaturated fats, *e.g.* herring oil rich in linoleic acid, has been found to 'permit' the maximum induction of hepatic cytochrome P-450 and the MFO system (Marshall and McLean, 1971). Cholesterol has also been found to exert a stimulatory effect on liver MFO enzymes (Lambert and Wills, 1977).

Protein deficiency has been shown to decrease the content of phosphatidylcholine in cytochrome P-450 and to decrease the activities of NADPH cytochrome P-450 reductase and of numerous MFO enzymes (Campbell and Hayes, 1976). On the other hand, the excessive intake of protein increases hepatic MFO activity in the rat and leads to a significant reduction in the circulating half-life of antipyrine and theophylline in man (Kappas, Anderson, Conney, and Alvares, 1976).

The inadequate intake of micronutrients such as ascorbic acid, riboflavin, tocopherol, calcium, copper, magnesium, selenium, and zinc has been found generally to decrease the activities of the MFO enzyme system and other xenobiotic metabolising enzymes (see, for example, Gibson and Skett, 1986). This may be as a result of an inhibition in the synthesis of the enzyme or essential co-factors. Other elements such as chromium, manganese, nickel, cadmium, and cobalt have been found to decrease cytochrome P-450 content by inhibiting the δ-ALA synthetase activity and inducing haem oxygenase activity.

Food deprivation has been found to alter markedly the hepatic MFO system. The effect is sex dependent in that whereas in the male rat some of the enzymes capable of being stimulated by androgens are depressed, in female animals a slight increase in enzymic activity is observed (Kato

Table 8 *Species differences in the routes of metabolism of foreign compounds*

Compound	Principal Metabolic Reaction	Species	Reference
	Phase I reactions		
Coumarin	7-hydroxylation	Man, Baboon	1
	3-hydroxylation	Rat, Guinea-pig, Rabbit	
BHT	Oxidation of -OH	Rat, Man	
	Oxidation of t-butyl	Man	
2-AAF	N-oxidation	Rat, Mouse, Dog	3
	C-hydroxylation	Guinea-pig, Lemming	
Cyclohexylamine	C-Hydroxylation	Rat	4
	Deamination	Man	
	Phase II reactions		
4-Chlorophenylacetic acid	Glycine conjugation	Rat	5
	Glutamine conjugation	Man	
Phenol	Sulphate conjugation	Cat	6
	Glucuronidation	Pig	
	Sulphation/glucuronidation	Rat	

Ref. (1) Cohen (1979)
    (2) Daniel, Gage, and Jones (1968)
    (3) Lotlikar, Enomoto, Miller, and Miller, (1967)
    (4) Renwick (1986)
    (5) James, Smith, and Williams (1972)
    (6) Capel, French, Milburn, Smith, and Williams (1972)

and Gillette, 1965; Gram, Guarino, Schroeder, Davis, Reagan, and Gillette, 1970).

Routes of Metabolism—Species Differences

Qualitative differences in the pathways involved in the metabolism of foreign compounds in various animal species have been found, and a number of examples of species differences in either phase 1 or phase 2 metabolic processes are shown in Table 8. These differences have been related to the varying susceptibility of the species to the biological effect of the particular compound. Although many attempts have been made to predict which route of metabolism would be favoured in a species or genus derived from structural considerations, no clear predictive framework has emerged (see, for example, the studies on conjugation of arylacetic acids by Caldwell, 1981).

A number of enzymes involved in the metabolism of foreign compounds have been found exclusively in man and other primates. These include the enzymes involved in the N^1-glucuronidation of methoxysulphonamides, and the O-methylation of 3,5-di-iodo-4-hydroxybenzoic acid.

Concluding Remarks

This review has attempted to present a general perspective of some of the important factors and sites in the body involved in the absorption, metabolism, and biological disposition of foreign compounds.

The rationale for this scheme of presentation was to enable the reader to glean an appreciation and recognition of the inter-relationship of the various facets involved in determining the metabolic fate of foreign compounds in experimental animals. Such an understanding could usefully form the basis for the interpretation of metabolic data and the application of the information in the design and conduct of animal toxicity studies.

It must be reiterated that each of the factors mentioned constitutes a complex topic in its own right. Active and vigorous research is still proceeding in order to explain the role of these various factors in xenobiotic metabolism. At this stage, the state-of-the-art, while illuminating important aspects of the subject, awaits further development to enable a realistic assessment to be made of the biological effects of chemicals to man.

References

Al-Dabbagh, S. G., Idle, J. R., and Smith, R. L. (1981). Animal modelling of human polymorphic drug oxidation—the metabolism of debrisoquine and phenacetin in rat inbred strains. *J. Pharm. Pharmacol.*, **33,** 161–164.

Amar-Costesec, A., Beaufay, H., Feytmans E., Thines-Sempoux, D., and Berthet J. (1969). Subfractionation of rat liver microsomes. *In*: 'Microsomes and Drug Oxidations', *Eds* J. R. Gillette, A. H. Conney, G. J. Cosmides, R. W. Estabrook, J. R. Fouts, and G. J. Mannering. Academic Press NY., pp. 41–58.

Askelöf, P., Guthenberg, C., Jakobson, I., and Mannervik, B. (1975). Purification and characterisation of two glutathione S-aryltransferase activities from rat liver. *Biochem. J.*, **147,** 513–522.

Baars, A. J., Jansen, M., and Breimer, D. D. (1978). The influence of phenobarbital, 3-methylcholanthrene and 2,3,7,8-tetrachlorodibenzo-*p*-dioxin on glutathione S-transferase activity of rat liver cytosol. *Biochem. Pharmacol.*, **27,** 2487–2494.

Baron, J., Redick, J. A., Kapke, G. F., and Guengerich, F. P. (1981). An immunohisto-chemical study on the localisation and distribution of phenobarbital- and 3-methylcholanthrene-inducible cytochrome P-450 within the livers of untreated rats. *J. Biol. Chem.*, **256,** 5931–5937.

Barry, B. W. (1983). Dermatological Formulations—Percutaneous Absorption Drugs and the Pharmaceutical Sciences, Vol. 18. Marcel Dekker Inc., New York.

Batt, A-M., Ziegler, J-M., and Siest, G. (1975). Competitive inhibition of glucuronidation by *p*-hydroxyphenyl hydantoin. *Biochem. Pharmacol.*, **24**, 152–154.

Bedford, C. T., Crawford, M. J., and Hutson, D. H. (1975). Sulphoxidation of cyanatryn, a 2-mercapto-*sym*-triazine herbicide, by rat liver microsomes. *Chemosphere*, **4**, 311–316.

Benson, A. M., Batzinger, R. P., Ou, S.-Y. L., Bueding, E., Cha, Y.-N., and Talalay, P. (1978). Elevation of hepatic glutathione S-transferase activities and protection against mutagenic metabolites of benzo(a)pyrene by dietary antioxidants. *Cancer Res.*, **38**, 4486–4495.

Bergman, A. and Dallner, G. (1976). Properties of a rat liver smooth microsomal subfraction not aggregated by Mg^{2+}. *Life Sci.*, **18**, 1083–1090.

Biagi, G. L., Barbaro, A. M., Gamba, M. F., and Guerra, M. C. (1969). Partition data of penicillins determined by means of reversed-phase thin-layer chromatography. *J. Chromatogr.*, **41**, 371–379.

Bickel, M. H. (1969). The pharmacology and biochemistry of *N*-oxides. *Pharmacol. Rev.*, **211**, 325–354.

Bickers, D. R. and Kappas, A. (1980). The skin as a site of chemical metabolism. *In*: 'Extrahepatic Metabolism of Drugs and Other Foreign Compounds'. *Ed.* T. E. Gram, Ch. 8, MTP Press Ltd, London U.K. pp. 295–318.

Bickers, D. R., Kappas, A., and Alvares, D. P. (1974). Differences in inducibility of hepatic and cutaneous drug metabolising enzymes and cytochrome P-450 by polychlorinated biphenyls and 1,1,1-trichloro-2,2-bis(*p*-chlorophenyl)ethane (DDT). *J. Pharmacol. Exp. Ther.*, **188**, 300–309.

Binning, A., Darby, F. J., Heenan, M. P., and Smith, J. N. (1967). The conjugation of phenols with phosphate in grass grubs and flies. *Biochem. J.*, **103**, 42–48.

Birnie, G. D. (1972). 'Subcellular components. Preparation and fractionation'. 2nd Edn, Butterworths/University Park Press.

Bouteille, M., Laval, M., and Dupuy-Coin, A. M. (1974). Localisation of nuclear functions as revealed by ultrastructural autoradiography and cytochemistry. *In*: 'The Cell Nucleus'. Vol I, Ch 1. *Ed.* H. Busch, Academic Press, New York, pp. 3–71.

Boyland, E., Kinder, C. H., and Manson, D. (1961). The biochemistry of aromatic amines. 8, Synthesis and detection of di-(2-amino-1-naphthyl) hydrogen phosphate, a metabolite of 2-naphthylamine in dogs. *Biochem. J.*, **78**, 175–179.

Brenner, B. M. and Rector, F. C. (1976). 'The Kidney'. W. B. Saunders, Philadelphia, USA.

Breuer, H., Kime, D. E., and Knuppen, R. (1973). Metabolism of 6-chloro-9β-10-pregna-1,4,6-triene-3,20-dione in rat, rabbit, monkey, and man. *Acta Endocrinol. (Copenhagen)*, **74**, 127–143.

Brown, R. R., Miller, J. A., and Miller, E. C. (1954). The metabolism of methylated azo-dyes. IV Dietary factors enhancing demethylation *in vitro*. *J. Biol. Chem.*, **209**, 211–222.

Burchell, B. and Coughtrie, M. W. H. (1992). UDP-glucuronosyl-transferases. *In*: 'Pharmacogenetics of Drug Metabolism'. *Ed.* W. Kalow. Pergamon Press Inc., New York, pp. 195–225.

Caldwell, J. (1981). Current status of attempts to predict species differences in drug metabolism. *Drug Metab. Rev.*, **12**, 221–237.

Campbell, T. C. and Hayes, J. R. (1974). Role of nutrition in the drug metabolising enzyme system. *Pharmacol. Rev.*, **26,** 171–197.

Campbell, T. C. and Hayes, J. R. (1976). The effect of quantity and quality of dietary protein on drug metabolism. *Fed. Proc. Fed. Am. Soc. Exp. Biol.*, **35,** 2470–2474.

Capel, I. D., French, M. R., Millburn, P., Smith, R. L., and Williams, R. T. (1972). The fate of [^{14}C]phenol in various species. *Xenobiotica*, **2,** 25–34.

Claude, A (1969). Microsomes, endoplasmic reticulum and interactions of cytoplasmic membranes. *In*: 'Microsomes and Drug Oxidations'. *Eds* J R. Gillette, A. H. Conney, G. J. Cosmides, R. W. Estabrook, J. R. Fouts, and G. J. Mannering. Academic Press, NY, pp. 1–39.

Clifton, G. and Kaplowitz, N. (1978). Effect of dietary phenobarbital, 3,4-benzo(a)pyrene and 3-methylcholanthrene on hepatic, intestinal, and renal glutathione S-transferase in the rat. *Biochem. Pharmacol.*, **27,** 1284–1287.

Cohen, A. J. (1979). Critical review of the toxicology of coumarin with special reference to interspecies differences in metabolism and hepatotoxic response and their significance to man. *Food Cosmet. Toxicol.*, **17,** 277–289.

Coon, M. J., Autor, A. P., Boyer, R. F., Lode, E. T., and Strobel, H. (1973). *In*: 'Oxidases and Related Redox Systems'. *Eds* T. E. King, H. S. Mason, and M. Morrison. University Park Press, Baltimore, p. 529.

Cooper, R. G., Evans, D. A. P., and Whibley, E. J. (1984) Polymorphic hydroxylation of perhexiline maleate in man. *J. Med. Genet.*, **21,** 27–33.

Coulston, F. and Serrone, D. M. (1969). The comparative approach to the role of non-human primates in evaluation of drug toxicity in man. *Ann. N. Y. Acad. Sci.*, **162,** 681–704.

Daniel, J. W., Gage, J. C., and Jones, D. I. (1968). The metabolism of 3,5-di-tert-butyl-4-hydroxytoluene in the rat and in man. *Biochem. J.*, **106,** 783–790.

Das, M. and Radhakrishnan, A. N. (1976). Role of peptidase and peptide transport in the intestinal absorption of proteins. *World Rev. Nutr. Diet*, **24,** 58–87.

Davson, H, (1943). *In*: 'Permeability of Natural Membranes'. *Eds* H. Davson and J. F. Danielli. Cambridge University Press, p. 11.

DePierre, J. W. and Ernster, L. (1977). Enzyme topology of intracellular membranes. *Annu. Rev. Biochem.*, **46,** 201–262.

Dierickx, P. J. (1983). Interaction of chlorophenoxyalkyl acid herbicides with rat liver glutathione *S*-transferases. *Food Chem. Toxicol.*, **21,** 575–579.

Duggan, D. E., Hooke, K. F., Noll, R. M., and Kwan, K. C. (1975). Enterohepatic circulation of indomethacin and its role in intestinal irritation *Biochem. Pharmacol.*, **24,** 1749–1754.

Dutton, G. J. (1978) Developmental aspects of drug conjugation, with special references to glucuronidation *Annu. Rev. Pharm. Toxicol.*, **18,** 17–36.

Eakins, M. N., Conroy, P. J., Searle, A. J., Slater, T. F., and Willson, R. L. (1976). Metronidazole (Flagyl) a radio-sensitizer of possible clinical use in cancer chemotherapy: some biochemical and pharmacological consideration. *Biochem. Pharmacol.*, **25,** 1151–1156.

Eling, T., Boyd, J., Reed, G., Mason, R., and Sivarajah, K. (1983). Xenobiotic metabolism by prostaglandin endoperoxide synthetase. *Drug Metab. Rev.*, **14,** 1023–1053.

Ernster, L. and Kylenstierna, B. (1969). Mitochondrial membranes—structure composition and function. *In*: 'Mitochondria—Structure and Function', *Eds* L. Ernster and Z. Drahota, FEBS Symposium 17, pp. 5–31.

Feuer, G., Sosa-Lucero, J. C., Lumb, G., and Moddel, G. (1971). Failure of various drugs to induce drug-metabolising enzymes in extrahepatic tissues of the rat. *Toxicol. Appl. Pharmacol.*, **19**, 579–589.

Food Safety Council (1978). Proposed System for Food Safety Assessment. *Food. Cosmet. Toxicol.*, **16** (Suppl 2), 1–36.

Freedberg, I. M. (1972). Pathways and controls of epithelial protein synthesis. *J. Invest. Dermatol.*, **59**, 56–65.

Gibson, G. G. and Schenkman, J. B. (1978). Purification and properties of cytochrome P-450 obtained from liver microsomes of untreated rats by lauric acid affinity chromatography. *J. Biol. Chem.*, **253**, 5957–5963.

Gibson, G. G. and Skett, P. (1986). *In*: 'Introduction to Drug Metabolism'. Chapman and Hall, Ch. 5.

Giri, S. N. (1973). The pharmacological action and *O*-demethylation of glyceryl guaiacolate ether in male and female rats. *Toxicol. Appl. Pharmacol.*, **24**, 513–518.

Gonzalez, F. J. (1988). The molecular biology of P450s. *Pharmacol. Rev.*, **40**, 243–288.

Gonzalez, F. J. (1990). Molecular genetics of the P450 superfamily. *Pharmacol. Ther.*, **45**, 1–38.

Gorrod, J. W. and Damani, L. A. (1985) 'Biological Oxidation of Nitrogen in Organic Molecules'. Ellis Horword Ltd., Chichester, UK.

Gould, B. J., Amoah, A. G. B., and Parke, D. V. (1986). Stereoselective pharmacokinetics of perhexiline. *Xenobiotica*, **16**, 491–502.

Gram, T. E. (1980). The metabolism of xenobiotics by mammalian lung. *In*: 'Extrahepatic Metabolism of Drugs and Other Foreign Compounds'. *Ed.* T. E. Gram. Ch. 3, MTP Press Ltd., London. pp. 159–209.

Gram, T. E. (1982). 'Extrahepatic Metabolism of Drugs and Other Foreign Compounds'. MTP Press Ltd., London.

Gram, T. E., Schroeder, D. H., Davis, D. C., Reagan, R. L., and Guarino, A. M. (1971). Further studies on the submicrosomal distribution of drug metabolising components in liver. Localisation in fractions of smooth microsomes. *Biochem. Pharmacol.*, **20**, 2885–2893.

Gram, T. E., Guarino, A. M., Schroeder, D. H., Davis, D. C., Reagan, R. L., and Gillette, J. R. (1970). The effect of starvation on the kinetics of drug oxidation by hepatic microsomal enzymes from male and female rats. *J. Pharmacol. Exp. Ther.*, **175**, 12–21.

Guengerich, F. P. (1979). Similarities of nuclear and microsomal cytochromes P-450 in the *in vitro* activation of aflatoxin B_1. *Biochem. Pharmacol.*, **28**, 2883–2890.

Guthenberg, C., Morgenstern, R., DePierre, J. W., and Mannervik, B. (1980). Induction of glutathione S-transferases A, B and C in rat liver cytosol by *trans*-stilbene oxide. *Biochem. Biophys. Acta*, **631**, 1–10.

Habig, W. H., Pabst, M. J., and Jakoby, W. B. (1974). Glutathione S-transferases; the first enzymic step in mercapturic acid formation. *J. Biol. Chem.*, **249**, 7130–7139.

Hanninen, O. and Marniemi, J. (1970). Inhibition of glucuronide synthesis by physiological metabolites in liver slices. *FEBS Lett.*, **6**, 177–181.

Hansch, C. and Clayton, J. M. (1973). Lipophilic character and biological activity of drugs. II, The parabolic case. *J. Pharm. Sci.*, **62**, 1–21.

Hathway, D. E. (1970–1981). Specialist Periodical Reports, Vol. 1–6 'Foreign Compound Metabolism in Mammals'. The Chemical Society. London.

Hayes, J. R. and Campbell, T. C. (1980). Nutrition as a modifier of chemical carcinogenesis. *In*: 'Carcinogenesis' Vol. 15, 'Modifiers of Chemical Carcinogenesis' *Ed.* J. J. Slaga, Raven Press, New York, Chapter 11, pp. 207–241.

Hayes, W. J. (1975). 'Toxicology of Pesticides'. Williams & Wilkins Co. Baltimore.

Heinrichs, W. L. and Juchau, M. R. (1980). Extrahepatic drug metabolism: The Gonads, *In*: 'Extrahepatic Metabolism of Drugs and other Foreign Compounds' *Ed.* T. E. Gram, Ch 9. MTP Press Ltd., London. pp. 319–332.

Hewick, D. S. (1982). Reductive metabolism of nitrogen containing functional groups. *In*: 'Enzymic Basis of Detoxification: Metabolism of Functional Groups'. *Eds* W. B. Jackoby, J. R. Bend, and J. Caldwell. Academic Press, New York. pp. 151–170.

Hoensch, H., Woo, C. H., and Schmid, R. (1975). Cytochrome P-450 and drug metabolism in intestinal villous and crypt cells of rats: Effect of dietary iron. *Biochem. Biophys. Res. Commun.*, **65**, 399–405.

Holtzman, J. L., Gram, T. E., Gigon, P. L., and Gillette, J. R. (1968). The distribution of the components of mixed-function oxidase between the rough and the smooth endoplasmic reticulum of liver cells. *Biochem J.*, **110**, 407–412.

Hosokawa, M., Maki, T., and Satoh, T. (1990). Characterisation of molecular species of liver microsomal carboxylesterases of several animal species and humans. *Arch. Biochem. Biophys.*, **277**, 219–227.

Houston, J. B. and Wood, S. G. (1980). Gastro-intestinal absorption of drugs and other xenobiotics. *In*: 'Progress in Drug Metabolism', Vol. 4. *Eds* J. W. Bridges and L. F. Chasseaud. J. Wiley and Sons, pp. 57–129.

Hsia, S. L. (1971). Steroid metabolism in human skin. *Mod. Trends Dermatol.*, **4**, 69–88.

Huang, M-T., Miwa, G. T., and Lu, A. Y. H. (1979). Rat liver cytosolic azo-reductase. Purification and characterisation. *J. Biol. Chem.*, **254**, 3930–3934.

Idle, J. R., Millburn, P., and Williams, R. T. (1978). Taurine conjugates as metabolites of arylacetic acids in the ferret. *Xenobiotica*, **8**, 253–264.

Jakoby, W. B., Bend J. R., and Caldwell, J. (1982). 'Metabolic Basis of Detoxification: Metabolism of Functional Groups'. Academic Press, New York.

Jakoby, W. B., Sekura, R. D., Lyon, E. S., Marcus, C. J., and Wang, J-L. (1980). Sulphotransferases. *In*: 'Enzymic Basis of Detoxification', *Ed.* W. B. Jackoby. Academic Press, New York. Vol. II, pp. 199–228.

James, M. O., Smith, R. L., and Williams, R. T. (1972). The conjugation of 4-chloro- and 4-nitro-phenylacetic acid in man, monkey and rat. *Xenobiotica*, **2**, 499–506.

Jones, D. P., Orrhenius, S., and Jakobson, S. W. (1980). Cytochrome P. 450-linked mono-oxygenase systems in the kidney. *In*: 'Extrahepatic Metabolism of Drugs and Foreign Compounds', *Ed.* T. E. Gram. Ch. 2, MTP Press Ltd., London. pp. 123–158.

Juchau, M. R. (1980). Extrahepatic drug biotransformation in the placenta. *In*: 'Extrahepatic Metabolism of Drugs and Foreign Compounds'. *Ed.* T. E. Gram, Ch. 4, MTP Press Ltd. London. pp. 211–238.

Jusko, W. J. and Gretch, M. (1976). Plasma and tissue protein binding of drugs in pharmacokinetics. *Drug Metabol. Rev.*, **5**, 43–140.

Kao, Y. (1981). Tetrodotoxin, Saxitoxin, Chiriquitoxin. New persepctives on ionic channels. *Fed. Proc., Fed. Am. Soc. Exp. Biol.*, **40**, 30–35.

Kappas, A., Anderson, K. E., Conney, A. H., and Alvares, P. (1976). Influence of dietary protein and carbohydrate on antipyrene and theophylline metabolism in man. *Clin. Pharmacol. Ther.*, **20**, 643–653.

Kato, R. and Gillette, J. R. (1965). Sex differences in the effects of abnormal physiological states on the metabolism of drugs by rat liver microsomes. *J. Pharmacol. Exp. Ther.*, **150**, 285–291.

Kawajiri, K., Yonekawa, H., Hara, E., and Tagashira. Y. (1979). Activation of 2-acetylaminofluorene in the nuclei of rat liver. *Cancer Res.*, **39**, 1089–1093.

Killenberg, P. G. and Jordan, J. T. (1978). Purification and characterisation of bile acid-CoA: amino acid N-acetyltransferase from rat liver. *J. Biol. Chem.*, **253**, 1005–1010.

Kimura, S., Gonzalez, F. J., and Nebert, D. W. (1986). Tissue-specific expression of the mouse dioxin-inducible P_1-450 and P_3-450 genes: differential transcriptional activation and mRNA stability in liver and extrahepatic tissues. *Mol. Cell Biol.*, **6**, 1471–1477.

Kohli, K. K., Mukhtar, H., Bend, J. R., Albro, P. W., and McKinney, J. D.. (1979). Biochemical effects of pure isomers of hexachlorobiphenyl-Hepatic microsomal epoxide hydrase and cytosol glutathione-S-transferase activities in the rat. *Biochem. Pharmacol.*, **28**, 1444–1446.

Kuruno, Y., Kondo, T. and Ikeda, K. (1983). Esterase-like activity of human serum albumin: enantioselectivity in the burst phase of reaction with *p*-nitrophenyl-methoxyphenyl acetate. *Arch. Biochem. Biophys.*, **227**, 339–341.

Lambert, L. and Wills, E. D. (1977). The effect of dietary and lipid peroxide sterols and oxidised sterols on cytochrome P-450 and oxidative demethylation in the endoplasmic reticulum. *Biochem. Pharmacol.*, **26**, 1417–1421.

Lansing, A. I., Belkhode, M. L., Lynch, W. E., and Lieberman, I. (1967). Enzymes of plasma membrane of liver. *J. Biol. Chem.*, **242**, 1772–1775.

LeFevre, M. E. and Joel, D. D. (1977). Mini review. Intestinal absorption of particulate matter. *Life Sci.*, **21**, 1403–1408.

Levin, A. A. and Dent, J. G. (1982). Comparison of the metabolism of nitrobenzene by hepatic microsomes and caecal microflora from Fischer F344 rats *in vitro* and the relative importance of each *in vivo*. *Drug Metab. Dispos.*, **10**, 450–454.

Levy, G., Khanna, N. N., Soda, D. M., Tsuzuki, O., and Stern, L. (1975). Pharmacokinetics of acetaminophen in the human neonate. Formation of acetaminophen glucuronide and sulfate in relation to plasma bilirubin concentration and D-glucaric acid excretion. *Pediatrics*, **55**, 818–825.

Lindstrom, T. D. and Whitaker, G. W. (1984). Saturation of α,β-unsaturated ketone: a novel xenobiotic biotransformation in mammals. *Xenobiotica*, **14**, 503–508.

Lotlikar, P. D., Enomoto, M., Miller, J. A., and Miller C. (1967). Species variation in the *N*- and ring hydroxylation of 2-acetylaminofluorene and effects of 3-methylcholanthrene treatment. *Proc. Soc. Exp. Biol. Med.*, **125**, 341–346.

Lucier, G. W., McDaniel, O. S., and Matthews, H. B. (1971). Microsomal rat liver UDP Glucuronyltransferase: Effects of piperonyl butoxide and other factors on enzyme activity. *Arch. Biochem. Biophys.*, **145**, 520–530.

Lush, I. E. and Andrews, K. M. (1978). Genetic variation between mice in the metabolism of coumarin and its derivatives. *Genet. Res.*, **31**, 177–186.

Mannervik, B. (1987). Glutathione transferase: *In*: 'Drug Metabolism—from Molecules to Man' *Eds* D. J. Benford, J. W. Bridges, and G. G. Gibson. Taylor and Francis Ltd., London, pp. 30–39.

Mannervik, B., Alin, P., Guthenberg, C., Jensson, H., and Warholm. M. (1985). *In*: 'Microsomes and Drug Oxidation' *Eds* A. R. Boobis, J. Caldwell, F. De Matteis, and C. Elcombe. Taylor and Francis, London, p. 221.

Mannervik, B. and Danielson, U. H. (1988). Glutathione transferases. Structure and catalytic activity. *CRC Crit. Rev. Biochem.,* **23**, 283–336.

Marshall, W. J. and McLean, A. E. M. (1971). A requirement for dietary lipids for induction of cytochrome P-450 by phenobarbitone in rat liver microsomal fraction. *Biochem. J.,* **122**, 569–573.

Matthews, D. M. (1975) Intestinal absorption of peptides. *Physiol. Rev.,* **55**, 537–608.

McMahon, R. E. (1971). Enzymic oxidation and reduction of alcohols, aldehydes and ketones. *Handb. Exp. Pharmacol.,* **28**, 500–517.

Mueller, G. C. and Miller, J. A. (1950). The reductive cleavage of 4-dimethyl-aminoazobenzene by rat liver: Reactivation of carbon dioxide treated homogenates by riboflavin-adenine dinucleotide *J. Biol. Chem.,* **185**, 145–154.

Mulder, G. J. (1974). On non-specific inhibition of rat liver microsomal UDP-glucuronyltransferase by some drugs. *Biochem. Pharmacol.,* **23**, 1283–1291.

Mulder, G. J., Meerman, J. H. H. and van der Goorbergh A. M. (1986). *In*: 'Xenobiotic Conjugation Chemistry' *Eds*. G. D. Paulson, J. Caldwell, D. H. Hutson, and J. J. Menn. Am. Chem. Soc., Washington, p. 282.

Murray, M. (1987). Mechanisms of the inhibition of cytochrome P-450-mediated drug oxidation by therapeutic agents. *Drug Metab. Rev.,* **18**, 55–81.

Murray, M. and Reidy, G. F. (1990). Selectivity in the inhibition of mammalian cytochromes P-450 by chemical agents. *Pharmacol. Rev.,* **42**, 85–101.

Nash, J. A., King, L. J., Locke, E. A., and Green, T. (1984). The metabolism and disposition of hexachloro-1,3-butadiene in the rat and its relevance to nephrotoxicity. *Toxicol. Appl. Pharmacol.,* **73**, 124–137.

Nebert, D. W., Robinson, J. R., Niwa, A., Kumaki, K., and Poland, A. P. (1975). Genetic expression of aryl hydrocarbon hydroxylase activity in the mouse. *Cell Physiol.,* **85**, 393–414.

Nebert, D. W., Nelson, D. R., Coon, M. J., Estabrook, R. W., Feyereisen, R., *et al.* (1991). The P450 superfamily: Update on new sequences, gene mapping and recommended nomenclature. *DNA Cell Biol.,* **10**, 1–14.

Olson, M. O. J. and Busch, H. (1974). Nuclear proteins. *In*: 'The Cell Nucleus', Vol. III, Ch. 6, *Ed.* H. Busch Academic Press, New York, pp. 211–268.

Osterloh, J. D., Karakaya, A., and Carter, D. E. (1980). Isolation and identification of the polar metabolites of chlorpheniramine in the dog. *Drug Metab. Dispos.,* **8**, 12–14.

Pannatier, A., Jenner, P., Testa, B., and Etter, J. C. (1978). The skin as a drug-metabolising organ. *Drug Metab. Rev.,* **8**, 319–343.

Parke, D. V. and Ioannides, C. (1981). The role of nutrition in toxicology. *Annu. Rev. Nutr.,* **1**, 207–234.

Paulson, G. D., Caldwell, J., Hutson, D. H., and Menn, J. J. (1986). Xenobiotic Conjugation Chemistry, ACS Symposium Series 299, Washington DC, USA.

Pohl, R. J., Philpot, R. N., and Fouts, J. R. (1976). Cytochrome P-450 content and mixed function oxidase activity in microsomes isolated from mouse skin. *Drug Metab. Dispos.*, **4,** 442–450.

Powis, G., and Wincentsen, L. (1980). Pyridine nucleotide co-factor requirements of indicine-*N*-oxide reduction by hepatic microsomal cytochrome P-450. *Biochem. Pharmacol.*, **29,** 347–351.

Rappaport, A. M. (1969). Anatomic considerations: *In*: 'Diseases of the Liver'. *Ed.* L. Schiff. 3rd Edn., J. B. Lippincott Co, Philadelphia. pp. 1–49.

Raw, I. (1978). Cytochrome P-450 in the liver mitochondrial outer membrane of 20-methylcholanthrene or Aroclor treated rabbits. *Biochem. Biophys. Res. Commun.*, **81,** 1294–1297.

Reddy, J. K. and Lalwani, N. D. (1983). Carcinogenesis by hepatic peroxisome proliferators: Evaluation of the risk of hypolipidemic drugs and industrial plasticizers to humans. *Crit. Rev. Toxicol.*, **12,** 1–58.

Redmond, G. P. and Cohen, G. (1972). Sex difference in acetaldehyde exhalation following ethanol administration in C57BL mice. *Nature (London)*, **236,** 117.

Remmer, H., Schenkman, J., Estabrook, R. W., Sasame, H., Gillette, J., Narasimhula, S., Cooper D. Y., and Rosenthal, O. (1966). Drug interaction with hepatic microsomal cytochrome. *Mol. Pharmacol.*, **2,** 187–190.

Renwick, A. G. (1986). The metabolism of intense sweetners. *Xenobiotica*, **16,** 1057–1071.

Rongone, E. L. (1977). Cutaneous metabolism. *In*: Adv. Mod. Toxicol. Vol. 4: 'Dermatotoxicology and Pharmacology'. *Eds* R. N. Margulli, and H. J. Marbach. John Wiley and Sons, New York.

Rowland, I. R. and Walker, R. (1983). The gastrointestinal tract in food toxicology In: 'Toxic Hazards in Food'. *Eds* D. M. Conning and A. B. G. Lansdown. Ch. 6. Croom Helm, London and Canberra. pp. 183–274.

Rowland, I. R., Mallett, A. K., and Wise, A. (1985). The effect of diet on the mammalian gut flora and its metabolic activity. *CRC Crit. Rev. Toxicol.*, **16,** 31–103.

Ryan, D. E., Thomas, P. E., Korzeniowski, D., and Levin, W. (1979). Separation and characterisation of highly purified forms of liver microsomal cytochrome P-450 from rats treated with polychlorinated biphenyls, phenobarbital, and 3-methylcholanthrene. *J. Biol. Chem.*, **254,** 1365–1374.

Sato, R. and Omura, T. (1978). 'Cytochrome P-450'. Academic Press, New York.

Savage, D. C. (1977). Interactions between the host and its microbes. *In*: 'Microbial Ecology of the Gut'. *Eds* R. T. J. Clark and T. Bauchop. Academic Press, London. pp. 277–310.

Sawada, H., Hara, A., Kato, F. and Nakayama, T. (1979). Purification and properties of reductases for aromatic aldehydes and ketones from guinea-pig liver. *J. Biochem. (Tokyo)*, **86,** 871–881.

Scala, J., McOsker, D. E., and Reller, H. H. (1968). The percutaneous absorption of ionic surfactants. *J. Invest. Dermatol.*, **50,** 371–379.

Schanker, L. S. and Jeffrey, J. J. (1961). Active transport of foreign pyrimidines across the intestinal epithelium. *Nature (London)*, **190,** 727–728.

Scheline, R. R. (1973). Metabolism of foreign compounds by gastrointestinal micro-organisms. *Pharmacol. Rev.*, **25,** 451–523.

Scheline, R. R., (1986). *In*: 'Extrahepatic Metabolism of drugs and Other

Foreign Compounds'. *Ed.* T. E. Gram, MTP Press Ltd., London, pp. 551–580.

Schuster, I. (1989). Cytochrome P-450 Biochemistry and Biophysics. Taylor and Francis, London.

Scott, R. C., Guy, R. H., and Hadgraft, J. (1990). Prediction of Percutaneous Penetration. Methods, Measurement, Modelling. IBC Technical Services Ltd., London.

Sharratt, M., Grasso, P., Carpanini, F. M. B., and Gangolli, S. D. (1970). Carrageenan ulceration as a model for human ulcerative colitis. *Lancet,* **2,** 932.

Shichi, H., Kumaki, K., and Nebert, D. W. (1978). Circular dichroism studies on the binding of type I substrates and reverse type I compounds to rabbit liver microsomal cytochrome P-450. *Chem. Biol. Interact.,* **20,** 133–148.

Shiobara, Y., (1977)., The effect of carboxyl substituents on the metabolism of aromatic aldehydes. *Xenobiotica,* **7,** 457–468.

Sikstrom, R., Lanoix, J., and Bergeron, J. J. (1976). An enzymic analysis of a nuclear envelope fraction. *Biochem. Biophys. Acta,* **448,** 88–102.

Smith, L. L. *et al.* (1978). In: 'Industrial and Environmental Xenobiotics'. *Eds* J. R. Fouts and I. Gut. Excerpta Medica, Amsterdam, pp. 135–140.

Snell, K. and Mullock, B. (1987) 'Biochemical Toxicology, a Practical Approach' IRL Press, Oxford.

Song, B.-J., Veech, R. L., Park, S. S., Gelboin, H. V., and Gonzalez, F. J. (1989). Induction of rat hepatic *N*-nitrosodimethylamine demethylase is due to protein stabilization. *J. Biol. Chem.,* **264,** 3568–3572.

Sorokin, S. P. (1970). The cells of the lung. *In*: 'Morphology of Experimental Respiratory Carcinogenesis'. *Eds* P. Nettesheim, M. R. Hanna, and J. W. Deatherage U.S. Atomic Energy Commission.

Soucek, P. and Gut, I. (1992). Cytochromes P-450 in rats: structures, functions, properties, and relevant human forms. *Xenobiotica,* **22,** 83–103.

Stockman, P. K., Beckett, G. J., and Hayes, J. R. (1985). Identification of a basic hybrid glutathione S-transferase from human liver. *Biochem. J.,* **227,** 457–465.

Strother, A., Throckmorton, J. K., and Herzer, C. (1971). The influence of high sugar consumption by mice on the duration of action of barbiturates and *in vitro* metabolism of barbiturates, aniline and *p*-nitroanisole. *J. Pharmacol. Exp. Med.,* **179,** 490–498.

Tannen, R. H. and Webber, W. W. (1979). Rodent models of the human isoniazid-acetylator polymorphism. *Drug Metab. Dispos.,* **7,** 274–279.

Tappel, A. L. (1969). Lysozomal enzymes and other components. In: 'Lysozomes in Biology and Pathology', 2. *Eds* J. T. Dingle and H. B. Fell. Frontiers of Biology 14B, Ch. 9, North Holland Publishing Co., Amsterdam, pp. 209–244.

Tatsumi, K., Yamada, H., and Kitamura, S. (1983) Reductive metabolism of *N*-nitrosodiphenylamine to the corresponding hydrazine derivative. *Arch. Biochem. Biophys.,* **226,** 174–181.

Testa, B. (1987). Recently discovered routes of metabolism. *In*: 'Drug Metabolism—from Molecules to Man'. *Eds* D. J. Benford, J. W. Bridges, and G. G. Gibson. Taylor and Francis, London pp. 563–580.

Testa, B. and Jenner, P. (1976). 'Drug Metabolism: Chemical and Biochemical Aspects'. Marcel Dekker Inc., New York.

Testa, B. and Jenner, P. (1981). Inhibitors of cytochrome P-450's and their mechanism of action. *Drug Metab. Rev.,* **12,** 1–117.

Thompson, S. and Slaga, T. J. (1976). Mouse epidermal aryl hydrocarbon hydroxylase. *J. Invest. Dermatol.,* **66,** 108–111.

Thorgeirrson, S. S., Felton, J. S. and Nebert, D. W. (1975). Genetic differences in the aromatic hydrocarbon inducible *N*-hydroxylation of 2-acetylaminofluorene and acetaminophen-produced hepatotoxicity in mice. *Mol. Pharmacol.,* **11,** 159–165.

Tobin, T., Akera, T., Han, C. S., and Brody, T. M. (1974). Lithium and rubidium interactions with sodium- and potassium-dependent adenosine triphosphatase. A molecular basis for the pharmacological actions of these ions. *Mol. Pharmacol.,* **10,** 501–508.

Troll, W., Tessler, A. N., and Nelson, N. (1963) Bis(2-amino-1-naphthyl) phosphate, a metabolite of β-naphthylamine in human urine. *J. Urol. Nephrol.,* **89,** 626–627.

Tuey, D. B. and Matthews, H. B. (1981). Distribution and excretion of 2,2′,4,4′,5,5′-hexabromobiphenyl in rats and man: Pharmacokinetic model predictions. *Toxicol. Appl. Pharmacol.,* **53,** 420–431.

Uehleke, H. (1968). Extrahepatic microsomal drug metabolism. *In:* 'Sensitization to Drugs'. Proc. Tenth European Soc. for the Study of Drug Toxicity Excerpta Media Intn. Congress Series 181, Oxford pp. 94–100.

Ullrich, V., Roots, A., Hildebrandt, A., Estabrook, R. W., and Conney, A. H. (1977). 'Microsomes and Drug Oxidation'. Pergamon Press, Oxford.

Ullrick, K. J., Capasso, G., Rumrich, G., and Sato, K. (1978). Effect of *p*-chloromercuribenzoate (pCMB), ouabain, 4-acetamido-4′-isocyanoto-stilbene-2,2′-disulphonic acid (SITS) on proximal tubular transport processes. *In:* 'Membrane Toxicity' *Eds* M. W. Miller and A. E. Shamoo. Plenum Press, NY. pp. 3–10.

Unger, S. H., Cook, J. R. and Hollenberg, J. S. (1978). Simple procedure for determining octanol-aqueous partition, distribution and ionisation coefficients by reversed-phase high-pressure liquid chromatography. *J. Pharm. Sci.,* **67,** 1364–1367.

Van Bladeren, P. J., Bruggeman, I. M., Jongen, W. M. F., Scheffer, A. G., and Temmink, J. H. M. (1987). The role of conjugating enzymes in toxic metabolite formation. In: 'Drug Metabolism—From Molecules to Man'. *Eds* D. J. Benford, J. W. Bridges, and G. G. Gibson. Taylor and Francis, London, pp. 151–170.

Venkatesan, N., Arcos, J. C., and Argus, M. F. (1970). Amino acid induction and carbohydrate repression of dimethylnitrosamine demethylase in rat liver. *Cancer Res.,* **30,** 2563–2567.

Vos, R. M. E. and Van Bladeren, P. J. (1990). Glutathione *S*-transferases in relation to their role in the biotransformation of xenobiotics. *Chem. Biol. Interact.,* **75,** 241–265.

Walker, W. A., Isselbacher, K. J., and Bloch, K. J. (1972). Intestinal uptake of macromolecules: Effect of oral immunisation. *Science,* **177,** 608–610.

Wang, C. Y., Behrens, B. C., Ichikawa, M., and Bryan, G. T. (1974). Nitroreduction of 5-nitrofuran derivatives by rat liver xanthine oxidase and reduced nicotinamide adenine dinucleotide phosphate-cytochrome c reductase. *Biochem. Pharmacol.,* **23,** 3395–3404.

Waring, R. H. and Mitchell, S. C. (1985). The metabolism of phenothiazine in the neonatal calf: identification of drug-polypeptide conjugates from urine. *Xenobiotica*, **15**, 459–468.

Watanabe, P. G., Ramsey, J. C., and Gehring P. J. (1980) Pharmacokinetics and metabolism of industrial chemicals. *In*: 'Progress in Drug Metabolism', Vol. 5. *Eds* J. W. Bridges and L. F. Chasseaud. J. Wiley and Sons, pp. 311–343.

Weiss, B. and Levin, R. M. (1978). Mechanism for selectively inhibiting the activation of cyclic nucleotide phosphodiesterase and adenylate cyclase by antipsychotic agents. *Advan. Cyclic Nucleotide Res.*, **9**, 285–303.

Wermuth, B., Münch, J. D. B., and von Wartburg, J. P. (1977). Purification and properties of NADPH-dependent aldehyde reductase from human liver. *J. Biol. Chem.*, **252**, 3821–3828.

Wester, R. C. and Maibach, H. I. (1977). Percutaneous absorption in man and animals: a perspective. *In*: 'Cutaneous Toxicity', *Eds* V. A. Drill and P. Lazar. Academic Press, New York.

WHO Scientific Group (1967). Procedures for Investigating Intentional and Unintentional Food Additives. *WHO Tech Rep. Ser.*, **348**, WHO, Geneva.

Williams, R. T. (1959) 'Detoxification Mechanisms', 2nd Edn. Chapman and Hall Ltd., London.

Willemsens, G., Vanden Bossche, H., and Lavrijsen, K. (1989). Cytochrome P-450-dependent metabolism of retinonic acid in epidermal microsomes from neonatal rats. Effects of azole antifungals. *In*: 'Cytochrome P450: Biochemistry and Biophysics' *Ed.* I. Schuster. Taylor and Francis, London, pp. 767–770.

Wollenberg D. and Ullrich, V. (1977). Characterisation of the drug mono-oxygenase system in mouse small intestine. *In*: 'Microsomes and Drug Oxidation'. *Eds* V. Ullrich, *et al.*, Pergamon Press, New York. pp. 675–679.

Ziegler, D. M. (1980). Microsomal flavin-containing mono-oxygenases. *In*: 'Enzymatic Basis of Detoxification' Vol. I. *Ed.* W. B. Jakoby. Academic Press, New York. Vol. I, pp. 201–230.

CHAPTER 10

Theory and Practice in Metabolic Studies

J. C. PHILLIPS AND S. D. GANGOLLI

Introduction

In the design and conduct of experimental studies on the metabolic fate and biological disposition of a compound, two important aspects merit consideration. These are the underlying theoretical principles and the relevant practical techniques. A brief description of the salient features of these two aspects follows:

1 Theoretical Considerations

The biological effect that a foreign compound has on a living organism is a function of the concentration of that compound at the target site and the duration of exposure. Pharmacokinetics, which is the discipline that quantifies with respect to time, the uptake, distribution, metabolism, and elimination of compounds, therefore provides important information with which to assess the potential hazard of the compound. Understanding the pharmacokinetic characteristics of a compound over a wide dose range, after single and repeated exposure, and between species, greatly facilitates the extrapolation of toxicity data from various test species to man. The importance of pharmacokinetic studies in toxicology has been the subject of many recent reviews (see for example, Gehring, Watanabe, and Blau, 1976; Withey, 1977; Watanabe, Ramsey, and Gehring, 1980).

A detailed discussion of the analysis of pharmacokinetics data and the assumptions inherent in the various mathematical models is beyond the scope of this chapter and the reader is referred to the many standard texts (*e.g.* Goldstein, Aronow, and Kalman, 1974; Niazi, 1979; Gibaldi and Perrier, 1975). However, some of the basic concepts will be outlined, and the various pharmacokinetic parameters that can be derived will be discussed.

The most common approach to pharmacokinetic characterisation is to depict the body as a system of 'compartments', although it must be remembered that these compartments do not necessarily correspond to anatomical or physiological entities. The one-compartment model describes the body as a single, homogeneous unit and is most useful for the analysis of blood concentration–time data or urinary excretion–time data when the administered compound is rapidly distributed in the body. If, as is frequently the case, the administered compound takes some appreciable time to distribute within the body, a two-compartment (or multi-compartment) model must be used. A two-compartment model consists of a 'central compartment', which contains blood plasma and (in the absence of data to the contrary) the metabolically active tissues such as the liver and kidneys, connected to a 'peripheral compartment'. The choice of model depends on a number of factors, including the number of tissues or fluids being sampled and the frequency of sampling.

Many pharmacokinetic processes, for example the rate of absorption, elimination, or biotransformation of a compound, are satisfactorily described by first-order (or linear) kinetics. Under these conditions the rate of the process is proportional to the concentration of the compound. In its simplest form, 'linear' kinetics obeys the equation:

$$\frac{-\mathrm{d}c}{\mathrm{d}t} = kc \tag{1}$$

where c is the concentration of compound at the time t, and k is the rate constant.

If, however, one of the pharmacokinetic processes is saturable, as in the case of carrier-mediated uptake or enzyme-mediated metabolism, the system may obey non-linear Michaelis–Menten kinetics. In this situation

$$\frac{-\mathrm{d}c}{\mathrm{d}t} = \frac{V_\mathrm{m}c}{K_\mathrm{m} + c} \tag{2}$$

where V_m and K_m are the maximum rate and Michaelis constant of the process.

At very low concentrations of compound, as is frequently the case in 'real-life' situations, c is much less than K_m, so that first-order kinetics apply. In contrast, where c is very large, as in maximum tolerated dose experiments, the kinetics of the system approximate to zero-order.

A   One-compartment Model

If a plot of the logarithm of the plasma concentration of a compound, given by intravenous injection, against time can be fitted to a single straight line, the body is acting as a one-compartment system (Figure 1). The slope of the line is equal to $-k_\mathrm{el}/2.303$, where k_el is the apparent

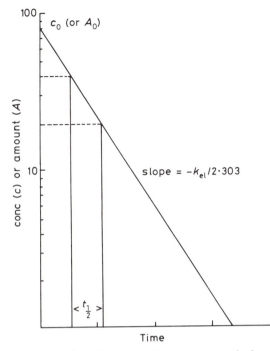

Figure 1 *Blood concentration–time curve for a compound after i.v. injection into a one-compartment system. Elimination is apparent first-order with a rate constant* k_{el}

first-order elimination rate constant. k_{el} is also equal to $0.693/t_{1/2}$ where $t_{1/2}$ is the elimination half-life. The elimination half-life is independent of the concentration of the compound and is constant for the whole period of elimination of the compound (if $t_{1/2}$ varies with dose, it is likely that either the system is obeying zero-order kinetics, and/or that absorption is prolonged).

The volume of distribution of the compound (V_d) is a measure of the apparent space within the body available to contain the compound.

Thus:

$$V_d = \text{Amount of drug in body}/c \qquad (3)$$

where c = concentration in the fluid measured. For this example,

$$V_d = \text{i.v. dose } (D)/\text{concentration at time zero } (c_0)$$

c_0 is the intercept of the plasma concentration–time curve on the concentration axis. An alternative method for obtaining V_d is to use the relationship:

$$V_d = D/(AUC_{0 \to \infty} \times k_{el}) \qquad (4)$$

where $AUC_{0 \to \infty}$ is the area under the plasma concentration–time curve, from time zero to infinity. The area can be estimated by a variety of methods, including the 'cut and weight' method, the trapezoidal rule, or the interpolation method (Corney and Heath, 1970). This method is applicable to estimations for non-i.v. administered compounds, and also to some multi-compartment linear system.

The rate constant of urinary excretion of a compound (k_{ex}) is the difference between the overall elimination rate constant (k_{el}) and the sum of the individual rate constants for elimination by all extra-renal pathways (*e.g.* biotransformation, biliary excretion). A semi-log plot of urinary excretion rate of unmetabolised compound against time should yield a straight line with a slope of $-k_{el}/2.303$ and an intercept of $K_{ex} \cdot V_d \cdot c_0$. It is important to remember that the urinary excretion rates determined experimentally are not instantaneous rates, but averages over a finite time period. However, the average rate closely approximates to the instantaneous rate at the mid-point of the collection period, providing this is no longer than the half-life of the compound. An alternative method for calculating k_{ex} and the problems of extra-renal elimination and further biotransformation of primary metabolites are discussed by Levy and Gibaldi (1975). For some drugs, k_{ex} is proportional to kidney function as reflected in creatinine or inulin clearance.

The kinetics of urinary excretion may also be characterised by renal clearance value (Cl_R). This is the rate of urinary excretion of the compound divided by the plasma concentration. Thus:

$$Cl_R = (dA_u/dt)/c \tag{5}$$

In practice, this is determined by dividing the urinary excretion rate at the mid-point of the collection period by the plasma concentration at this time. However, a more satisfactory approach is to plot urinary excretion rate against plasma concentration at the collection mid-point, the slope of the resultant straight line being Cl_R.

As by definition,

$$(dA_u/dt) = k_{ex}A \tag{6}$$

and $A = V_d c$ from Equation (3), substitution in Equation (5) shows that renal clearance also equals $k_{ex} \cdot V_d$.

The clearance ratio of a compound is defined as the ratio of its renal clearance to that of inulin. Since inulin clearance is a measurement of glomerular filtration rate, a clearance ratio of greater than one indicates excretion, in part, by renal tubular secretion. If tubular reabsorption is appreciable and/or glomerular filtration is incomplete (as in the case of a drug bound to protein), the clearance ratio will be less than unity.

Whole body clearance (Cl_b) is analogous to renal clearance and defined as:

$$Cl_b = k_{el} V_d \tag{7}$$

Whole body clearance is the sum of the individual clearances by the various tissues of the body (for example liver, lung, *etc.*).

Another way of defining clearance, which is independent of any physiological model, is in terms of blood flow through an organ (Q) and the concentration of the compound in the blood entering and leaving the organ. Thus the rate of elimination of the compound from the circulation by an organ at a steady state, is the difference between the amount entering the organ (Qc_{in}) and the amount leaving (Qc_{out}). From Equation (5),

$$Cl_{organ} = \frac{Qc_{in} - Qc_{out}}{c_{in}}$$

or $Cl_{organ} = Q\,E_{organ}$, where E_{organ} is called the extraction ratio.

The extent of absorption of a compound from the gastro-intestinal tract can be determined from plasma concentration–time curves following oral and i.v. administration of the same dose of compound. Thus:

$$\text{Apparent availability} = \frac{AUC_{0 \to \infty}\,(\text{Oral})}{AUC_{0 \to \infty}\,(\text{i.v})} \tag{8}$$

The fraction absorbed (F) expressed as a percentage is referred to as the 'bioavailability'; it is only equal to the apparent availability if the volume of distribution and elimination rate constant of the compound are independent of the route of administration.

The rate of absorption of a compound can also be determined from plasma concentration–time curves. A typical curve, showing an initial lag phase (Figure 2) can be described by the equation:

$$c = \frac{k_{ab}FD[e^{-k_{el} \cdot t} - e^{-k_{ab} \cdot t}]}{[k_{ab} - k_{el}]V_{d}} \tag{9}$$

where c is the concentration, F is the fraction of the administered dose D absorbed, V_{d}, is the apparent volume of distribution, t is the time and k_{ab} and k_{el} are the rate constants for absorption and elimination respectively.

If $k_{ab} > k_{el}$, a plot of log c against t will eventually have a slope of $-k_{el}/2.303$. Extrapolating this line to $t = 0$ and using the method of residuals, the plot of log residual against t has a slope of $-k_{ab}/2.303$. If $k_{ab} < k_{el}$, the reverse situation obtains, with the slope of the concentration–time plot equal to $-k_{ab}/2.303$ and the slope of the residuals plot equal to $-k_{el}/2.303$. This is sometimes called a 'flip-flop' model. The details of the mathematical models and alternative methods for calculating absorption rates are described by Levy and Gibaldi (1975) as are the limitations of this approach, which applies only if the compound is absorbed and eliminated by first order processes.

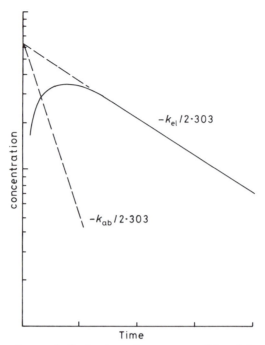

Figure 2 *Absorption and elimination of a compound involving two consecutive first-order processes. The absorption rate constant* k_{ab} *is determined by the method of residuals*

The pharmacokinetics of a compound following repeated, rather than single, administration is clearly of great importance in toxicological studies. If the plasma $t_{1/2}$ is small compared with the interval between doses, the bulk of the compound may be eliminated during this period and, thus, the situation after multiple doses will be very similar to that after a single dose. However, if the $t_{1/2}$ is long, the compound will accumulate in the body as each new dose is added to the residue of the previous dose, eventually reaching a steady state (plateau). The mathematical treatment of repetitive dose kinetics has been discussed by many authors, including Dost (1968), Levy and Gibaldi (1975), and Wagner, Northam, Alway, and Carpenter (1965).

The change in concentration of compound in plasma following repeated i.v. administration is described by the equation

$$c = \frac{D(1 - e^{-nk_{el}\tau}] e^{-k_{el}t}}{V_d(1 - e^{-k_{el}\tau})} \tag{10}$$

where D, V_d and k_{el} are as in Equation (9) above, c is the concentration in plasma of the compound during the dosing interval, n is the number of dose, τ the time interval between doses and t the time since the last dose.

The average plasma concentration at the plateau ($\bar{c}_\infty$) is given by:

$$\bar{c}_\infty = \frac{F \cdot D}{V_d k_{el} \tau} \tag{11}$$

where F = the bioavailability (Equation 8).

This equation is valid both for i.v. administration (where $F = 1$) and the administration by other routes ($F \leqslant 1$). The average concentration $\bar{c}_\infty$, is not directly related to either the minimum (c_∞) min or the maximum (c_∞) max concentration achieved at steady state. These are given by:

$$(c_\infty)\,\text{min} = \frac{D}{V_d}\left(\frac{1}{1 - e^{-k_{el}\tau}}\right) e^{-k_{el}\tau} \quad \text{and} \quad (c_\infty)\,\text{max} = (c_\infty)\,\text{min} \cdot e^{k_{el}\tau}$$

As the time taken for the plateau concentration to be reached is often quite long, one practical application of this type of kinetic analysis is the calculation of a 'loading dose' of compound that will rapidly produce any desired plateau plasma concentration with any dosing interval and maintenance dose. The average body burden of compound ($\bar{X}_\infty$) at the steady state is given by:

$$\bar{X}_\infty = \bar{c}_\infty \cdot V_d \tag{12}$$

B Multi-compartment Models

A compound that is injected intravenously into the body usually takes some time to distribute and therefore the plasma concentration–time curve is not mono-exponential. If the curve can be resolved into two linear components (biexponential) it may be described by the equation:

$$c = Ae^{-\alpha t} + Be^{-\beta t} \tag{13}$$

and justify the representation of the body as an open two-compartment linear system. It is usually assumed that the compound is eliminated exclusively from the central compartment, and although elimination from both compartments, or exclusively from the peripheral compartment, is possible, the three models are not distinguishable on the basis of simple plasma or urine elimination data. During the initial distributive phase, the rate of decline of the plasma concentration of the compound is greater than in the post-distributive phase. If the two curves are resolved by 'feathering', the slopes of the two lines are $-\alpha/2.303$ and $-\beta/2.303$ respectively and the intercepts on the concentration axis are A and B (Figure 3).

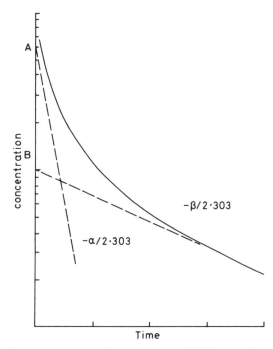

Figure 3 *Blood concentration of a compound following i.v. injection in a two-compartment system. A and α are determined by 'feathering' the curve*

It should be noted that the timing of blood sampling dictates the number of compartments to which the data can be best fitted. For example, the kinetics of calcium metabolism in man and the rat can be fitted to a two compartment model if blood is sampled between 2 and 72 hours following an i.v. dose of $^{45}CaCl_2$ or to a four-compartment model if sampled between 0 and 72 hours (Bronner and Lemaire, 1969). The rate constants for the movement of compound between the central and peripheral compartment (K_{12} and K_{21}) and for elimination from the central compartment (K_{10}) are given by:

$$K_{21} = \frac{A\beta + B\alpha}{A + B} \qquad K_{12} = \alpha + \beta - K_{21} - K_{10} \quad \text{and} \quad K_{10} = \alpha\beta/K_{21} \quad (14)$$

(Mayersohn and Gibaldi, 1971).

It should be noted that the elimination rate constant (K_{10}) is not the same as the terminal half-life, as in the one-compartment model. Determination of these rate constants allows an assessment to be made of the effect of factors such as age, sex, and genetic influences on the disposition of a compound. A knowledge of K_{12} allows the calculation of the amount of compound in the peripheral compartment (A_p) after i.v.

administration of dose D, using the equation:

$$A_p = \frac{D \cdot K_{12}}{\beta - \alpha} (e^{-\alpha t} - e^{-\beta t}) \tag{15}$$

Although this may be useful in rationalising pharmacological effects with tissue levels of compound it should be remembered that A_p may not accurately reflect the actual amount of compound in a particular 'peripheral' tissue. The apparent volume of distribution of a compound in the two-compartment model is given by:

$$V_d = \frac{D}{AUC_{0 \to \infty})\beta} \tag{16}$$

This equation applies regardless of route of administration, but can be simplified to:

$$V_d = \frac{D}{\left[\dfrac{A}{\alpha} + \dfrac{B}{\beta}\right]\beta} \tag{17}$$

for i.v. administration only.

By analogy with the single-compartment situation, the fraction of an oral dose absorbed is given by:

$$F = \frac{(AUC_{0 \to \infty}) \text{ oral} \times \beta \text{ oral}}{(AUC_{0 \to \infty}) \text{ i.v.} \times \beta \text{ i.v.}}$$

Whole body clearance in this model is defined as $V_d\beta$ and is equal to the clearance from the central compartment; it can be calculated from Equation (16). As in the one-compartment model, average steady state concentrations after repetitive dosing are given by:

$$c = \frac{F \cdot D}{V_d K_{10} \tau} \tag{18}$$

providing elimination is from the central compartment [k_{el} in Equation (11) is equivalent to K_{10} in Equation (18)].

C Non-linear Pharmacokinetics

As mentioned earlier, many biological processes involving enzymes or carrier-systems are saturable and are thus best described by Michaelis–Menten equations. There are two limiting cases in *in vivo* situations; K_m much larger than c and c much greater than K_m. In the former situation,

Equation (2) reduces to:

$$\frac{-dc}{dt} = \frac{(V_m)}{(K_m)} c \tag{19}$$

which has the same form as the first order rate equation, with V_m/K_m equivalent to k_{el}. In the latter case, Equation (2) reduces to

$$\frac{-dc}{dt} = V_m \tag{20}$$

where the rate is constant and independent of c. The biotransformation of ethanol (Lundquist and Wolthers, 1958) and salicylate (Levy, 1965) approach the conditions described in Equation 20.

For a compound eliminated by a single capacity-limited process, the plasma concentration–time curve in the post-absorptive, post-distributive phase can be used to estimate the *in vivo* apparent K_m and V_m values. Plotting the rate of change of $c(\Delta c/\Delta t)$ as a function of c at the mid point of the sampling interval using a linearised version of the Michaelis–Menten equation:

$$\frac{c}{(\Delta c/\Delta t)} = \frac{K_m}{V_m} + \frac{c}{V_m} \tag{21}$$

yields a straight line with a slope of $1/V_m$ and an intercept of K_m/V_m. The change in tissue concentration of a compound and urinary excretion of chemical and metabolites can also be investigated in non-linear systems (see for example, Gibaldi and Perrier, 1975).

2   Practical Considerations in Metabolic Studies

The following practical considerations need to be taken into account in the design and conduct of metabolic studies:
   (1) The chemistry of the test compound
   (2) The choice of animal model
   (3) The choice of appropriate experimental techniques.

A   Test Compound

The importance of establishing the chemical identity and purity of the test compound, and the nature and levels of accompanying contaminants prior to embarking on metabolic studies, is self evident. Ideally, metabolic studies should be carried out on chemically defined compounds, free of impurities. As analytical methodology and preparative techniques improve, this objective is being achieved. However, difficulties arise in the case of natural products and certain categories of synthetic compounds which are manufactured to comply with technical specifications, tailored specifically to meet a technological end use. Analytical studies on synthetic materials, such as azo-dyes, flavouring agents, emulsifiers, and industrial solvents, frequently show them to be complex mixtures of parent compound and subsidiary constituents. Thus, metabolic studies may either be undertaken using the purified major component, as in the recent study with the azo-dye amaranth (Phillips, Bex, Mendis, Walters, and Gaunt, 1987), or with materials representative of the commercially used products such as the studies with the azo-dye Brown HT (Phillips, Mendis, and Gaunt, 1987) and emulsifier YN (Phillips, Gaunt, and Gangolli, 1975).

In the case of naturally occurring products, the problems associated with obtaining a pure compound or even material of standard composition are formidable. Many of these substances, such as gums, resins, waxes, and essential oils, are not only complex, poorly defined mixtures, but also subject to variations in chemical composition due to geographical, seasonal, and soil factors.

In addition to data on the chemical composition of a test compound, it is essential to obtain, prior to embarking on metabolic studies, information on the physico-chemical properties of the compound. The information of particular relevance in this context is that relating to the stability and reactivity of the test compound. This should include data on the compound's volatility, its likelihood to undergo spontaneous degradation or polymerisation, and its likely reactivity with the vehicle used for administration or with dietary constituents.

An essential consideration in conducting metabolic studies is the availability of appropriate methods for the isolation and determination of the parent compound and its metabolites in biological samples. As human contact with foreign compounds generally occurs at low levels of exposure, and metabolic studies are carried out at comparable dose levels, it is important to establish that the methodologies for 'clean-up' procedures and analytical determinations are adequately sensitive and specific for the purpose. These investigations may be carried out on the compound *per se,* or on radiolabelled or stable isotope-labelled preparations.

B   Choice of Animal Model

Mention was made in the previous chapter of some of the factors responsible for the variability in the absorption, metabolism, and disposition of foreign compounds in experimental animals and that these differences could be attributed in part to the species, strain, sex, and age of the animal. Clearly the influence of these factors needs to be considered in the choice of an animal model for metabolic studies and, subsequently, for the conduct of toxicological studies to generate data for the evaluation of the safety of the compound in man. However, the seemingly simple problem of choosing an animal model capable of metabolising a compound in manner similar to man is fraught with difficulties. (For a discussion of the criteria for species similarities, see Ruelius, 1975). Variability in the metabolism and pharmacokinetics of compounds in humans can be as wide as in experimental animals. In addition to marked differences due to age and sex, genetic polymorphism is an important determinant for metabolic differences in man. Thus, for example, the differences in the hydroxylation of the alicyclic antihypertensive drug, debrisoquine, in humans, and in the oxidation of other compounds including bufuralol, nortriptyline, phenacetin, and phenformin (Davis and Boobis, 1983) have been ascribed to genetic polymorphism. Since the ideal situation, that of having identified an entirely comparable and practical animal model, is unlikely to be achieved, a working compromise based on a sound understanding of the similarities and differences in the metabolism and disposition of a compound in man and in an experimental animal forms the basis for the selection of a suitable animal species. Difficulties in extrapolating experimental data obtained in animals to man have been discussed recently by Garattini (1985).

C   Experimental Investigations in Metabolic Studies

The experimental animal species commonly used in metabolic studies are rodents (rats, mice, guinea-pigs, and hamsters), canines (dog), felines (cat and civet), rabbits, and non-human primates.

The *in vivo* and *in vitro* experimental procedures generally used for the studies include:

(1) Whole animal studies
(2) Isolated perfused organ technique
(3) Cells in culture
(4) Preparations of tissues and subcellular fractions.

The experimental protocol for a particular metabolic study depends on the nature of the compound and the specific information required. A brief outline description of the various experimental procedures follows.

Whole Animal Studies

Investigations in the intact animal may be categorised as either non-invasive procedures or terminal studies, *i.e.* studies where the animal is killed following treatment for examination of the internal organs.

Non-invasive procedures are widely used for investigating metabolism and pharmacokinetics of drugs and food additives in man and also in large experimental animals (*e.g.* cat, dog). Briefly, the procedure involves the collection of serial samples of blood, urine, faeces, saliva, or respired air following the administration of the test compound, and the subsequent analysis of the samples for the parent compound and/or metabolites. Thus the pharmacokinetics of antipyrine, hexabarbitone, and phenylbutazone in man can be determined by the sequential analysis of blood samples. In the case of antipyrine, the salivary levels of the drug have been found to reflect accurately the blood levels, and this method has been widely used for assessing hepatic microsomal mixed function oxidase activity in humans (Fraser, Mucklow, Murray, and Davis, 1976).

The '$^{14}CO_2$ breath test' has been used for measuring the demethylation of ^{14}C-labelled antipyrine in man (Hepner and Vessell, 1975). This method has also found application in the metabolism of ^{14}C-labelled glycodiazine, diazepam, and caffeine and other compounds metabolised by enzymic demethylation, deacetylation, and decarboxylation reactions (see review by Bircher and Preisig, 1981). A variation of this procedure is the use of ^{13}C-labelled compounds, whereby the amount of $^{13}CO_2$ and other stable isotope-labelled metabolites are measured in the exhaled air and in body fluids by mass spectrometry or Fourier transform NMR spectroscopy. This method has been used in investigations on the metabolism of chloroform in man (Fry, Taylor, and Hathway, 1972) and the use of stable isotopes in pharmacological research has been reviewed (Baillie, 1981; Hawkins, 1977).

In 'terminal' *in vivo* studies, the analysis of respired air, body fluids (including bile), and excreta is accompanied by the examination of internal organs for residues of the test compound, its metabolites, and products formed by the incorporation or interaction of the metabolites with endogenous components. Radiolabelled compounds are widely used in these studies (see Table 1 for list of radionuclides frequently used in whole-animal studies) and the vast literature on foreign compound metabolism is based largely on this method. Recently, attention has focused on the kinetics of the formation, disposition, and removal of products formed by the covalent binding of electrophilic metabolites of test compounds with endogenous macromolecules such as DNA, RNA, and protein (Pohl and Branchflower, 1981). Examples of such studies include investigations into the turnover of hepatic DNA and protein and their adducts following exposure of rats and mice to vinyl chloride and vinylidene chloride (Reitz, Watanabe, McKenna, Quast, and Gehring 1980).

Whole-body autoradiographic procedures provide another method for demonstrating the tissue distribution of radiolabelled compounds in experimental animals. The techniques employed have been described in detail by Rogers (1967), Ullberg and Larsson (1981), and more recently by Bénard, Burgat, and Rico (1985). The general method consists of administering to experimental animals the radiolabelled test compound, and then at timed intervals killing and rapidly freezing the animals. Serial sections of the carcass are cut using a freezing microtome, and the sections removed using adhesive transparent tape. The sections adhering to the tapes are dried while frozen and the radioactivity visualised by exposure to *X*-ray film. At the end of the exposure period, the sections may be fixed and stained. This technique provides a rapid survey of radioactivity in many organs and has the advantage that loss or

Table 1 *Some radionuclides used in whole animal metabolic studies*

Element	Radioisotope	Particle emission*	Use†
Hydrogen	^{3}H	β; 18; 12.3 yr	A, B
Carbon	^{11}C	β^+; 970; 20 min	B
	^{14}C	β; 156; 5730 yr	A, B
Fluorine	^{18}F	EC, β^+; 640; 1.8 hr	A
Sodium	^{22}Na	EC, β^+; 540; 2.6 yr	A
	^{24}Na	β; 1390; 15 hr	B
Phosphorus	^{32}P	β: 1710; 14.3d	A, B
	^{33}P	β: 248; 25d	A
Sulphur	^{35}S	β: 167; 87d	A, B
Chlorine	^{36}Cl	β: 709; 307,000 yr	A, B
Calcium	^{45}Ca	β: 254; 164d	A, B
	^{47}Ca	β: 690,2000; 4.7d	B
Chromium	^{51}Cr	EC; 28d	A, B
Iron	^{55}Fe	EC; 2.7 yr	B
	^{59}Fe	β: 270,460; 45d	A, B
Zinc	^{65}Zn	EC, β^+; 325; 243d	B
Bromine	^{82}Br	β; 440; 36 hr	A
Cadmium	^{109}Cd	EC; 462d	A, B
Iodine	^{125}I	EC; 60d	A, B
	^{129}I	β; 150; 1.6×10^7 yr	
	^{131}I	β: 335,608; 8.1d	A, B
Gold	^{198}Au	β: 960; 2.7d	A, B
Mercury	^{203}Hg	β: 212; 47d	A, B

* Principle emission; E_{max} (KeV); $t_{1/2}$ EC = electron capture
† A = Audoradiography; B = Whole Body

translocation of soluble compounds is kept to a minimum. The whole-body autoradiography method has been widely used in the mouse and rat; the availability of suitable freezing microtomes being the constraining factor in using larger animals. The disadvantages of the technique are that large amounts of radiolabelled compound are required, that the time factor involved in developing and processing the *X*-ray films precludes the generation of rapid results, and the results are at best semi-quantitative, giving little information on the nature of the radiolabelled material in tissues. Although the standard method cannot be used for volatile compounds, since the autoradiographs are obtained from dried sections, low temperature autoradiography, in which cut sections of the frozen ($-80°$C) carcass are exposed to *X*-ray film has been successfully used for inhalation anaesthetics (Cohen and Hood, 1969) and volatile nitrosamines (Brittebo and Tjälve, 1982).

Isolated Perfused Organ Technique

This procedure has been used for metabolic studies on the liver, kidney, lung and small intestines. The experimental techniques have been described in detail (Colowick and Kaplan, 1981; Ross, 1972).

In studies on the perfused liver, the animal species commonly employed is the rat. The organ can be completely isolated from the animal or left *in situ* in the killed animal. Metabolic studies using this technique have been carried out on a wide range of compounds including phenacetin, paracetamol, suphanilamide, and isoniazid. This technique has the advantage that the intact architecture of the liver is retained and the flow rate and the composition of the perfusion medium can be controlled. Furthermore, serial samples of the perfusate and bile can be obtained. Additionally, test compounds and inhibitors can be introduced into the perfusion medium at concentrations that would be toxic in the intact animal. Disadvantages of the isolated liver system include the difficulty of obtaining identical liver preparations at the same time, and the relatively short period of time over which these preparations are viable (about 6–8 hours).

The isolated perfused kidney method has been used for the study of the transport and metabolism of endogenous compounds, such as choline and vitamin D, and of foreign chemicals, *e.g.* isoproterenol, salicylates, and paracetamol. The rat kidney is the one most extensively used, although studies with dog and rabbit kidneys have also been reported. The use of the isolated perfused kidney in drug disposition studies has been reviewed by Bekersky (1983).

The use of the isolated perfused lung, generally obtained from small laboratory animals, (*e.g.* rats, guinea-pigs, and rabbits), provides a versatile technique for biochemical and metabolic studies. The test compound can be readily introduced directly into the pulmonary artery or into the trachea by inhalation, and serial blood samples can be obtained during the course of the investigation. The principal limitation of this technique is that experiments are restricted to a duration of about 4 hours.

The importance of the small intestine as a major site of xenobiotic metabolism has led to the application of the perfusion technique to this organ. The procedure is conducted either by the *en bloc* removal of the entire small intestine from the animal followed by vascular perfusion with heparinised or defibrinated blood in a recycling system or the autoperfusion of a segment of the small intestine *in situ*. The former approach requires considerable surgical skill and the use of glucocorticoid and norepinephrine in the perfusate to compensate for denervation. The animal generally used is the rat.

One of the principal advantages common to all of the isolated perfused organ systems, is the ability to study the metabolism of the test compound in isolation from the contributory role of other sites in the body.

Isolated Cells in Culture

The isolated cell suspension and cell culture systems have been extensively used in recent years for the study of xenobiotic metabolism. The mammalian cell types commonly used are hepatocytes, kidney cells, pulmonary cells, enterocytes, and nerve cells. Descriptions of the methods for the preparation of cultures and application of the techniques are given fully in Colowick and Kaplan (1981b).

Inevitably, isolated hepatocytes constitute the most widely used cell type. These cells retain in culture their ability to carry out the wide range of phase I and phase II metabolic transformations. Furthermore, the levels of cofactors required for these reactions are intracellularly generated and maintained in a manner analogous to the situation that obtains in the intact animal. Thus, sequential phase I and phase II biotransformation reactions on a substrate can be carried out by freshly isolated hepatocytes at rates comparable to the whole organ. The application of this technique for the study of the metabolism of a wide variety of chemicals including biphenyl, norbenzphetamine and aminopyrine has been extensively reviewed (Fry, 1982; Fry and Bridges, 1977; Fry, 1983).

Suspensions of intact, fully functional, isolated kidney cells have been found to have widespread utility in biochemical research (*e.g.* Bach, Kettley, Ahmed, and Dixit, 1986) and to a lesser extent in metabolic studies. The most complete data are available for the renal metabolism of paracetamol (Jones, Sundby, Ormstad, and Orrenius, 1979); however, phase I metabolism of benzo(a)pyrene, 7-ethoxycoumarin, and biphenyl has been investigated, as well as the phase II metabolism of 4-methylumbelliferone and benzoic acid (Fry, Wiebkin, Kao, Jones, Gwynn, and Bridges, 1975). Isolated kidney cells offer an ideal model for the study of the relationship of cellular transport to metabolic processes. Thus, kidney cell preparations active in amino acid uptake can be used for the study of GSH synthesis and utilisation, *e.g.* in the paracetamol conjugation reaction (Moldéus, Anderson, Norling, and Ormstad, 1980).

The development of procedures for the isolation of specialised cell types from the lung and the maintenance of these cells in culture has provided the means for the investigation and elucidation of cell-specific xenobiotic metabolism. Thus, for example, alveolar type II cells and non-ciliated bronchiolar epithelial cells (Clara cells) have been separated by centrifugal elutriation from rabbit and rat lung and the cytochrome P-450 dependent MFO system investigated (Philpot, Anderson, and Eling, 1977; Jones and Fouts, 1980).

Suspensions of enterocytes obtained from the small intestines of rat and guinea-pig and, maintained in culture, have been used to study the metabolism of benzo(a)pyrene, ethylmorphine, biphenyl, and other chemicals (Pinkus, 1981). Although the overall levels of enzyme activities in the enterocytes are low, most of the enzymes associated with the hepatic MFO system have been found in the gut epithelium. The main disadvantage of the isolated enterocyte system is the limited viability of the preparations.

The mammalian nerve cell preparation best suited for culturing purposes is that obtained from embryonic tissue, although recently Smith and McInnes (1986) have succeeded in maintaining cells from adult mouse dorsal root ganglia in culture. Two types of culture can be used: the first consists of 'mixed' cultures containing neurons and a variety of glial cells, ependymal cells, and fibroblasts, whereas the second approach attempts to separate the different cell types before culturing in order to achieve pure (mostly neural or glial) cultures. The former procedure allows the normal physiological interactions between the different cell types to be studied, and its uses have been reviewed recently by Spencer, Crain, Bornstein, Peterson, and Van de Water (1986); the latter procedure enables the interaction between a compound and a specific cell type to be investigated. There appears to be little published information on metabolic studies of foreign compounds by these preparations.

Testicular cells in culture have been used extensively to study the mechanisms of toxicity of various toxins, including phthalate esters and glycol ethers (Gray and Beamond, 1984; Gray, Moss, Creasy, and Gangolli, 1985). Individual cells types including Sertoli cells and spermatocytes have been used, and more recently co-culture systems have been developed in which, for example, Sertoli and germ cells are cultured together to provide a more realistic model of the effect of toxins *in vivo* with respect to cell–cell interactions. The development of compartmental apparatus may further advance understanding of cell–cell interactions, and permit investigations into the role of metabolic activation in this system.

Whole Tissue and Intracellular Homogenates

Whole tissue homogenates and subcellular fractions obtained by differential centrifugation procedures constitute the most widely used methods for studies on the metabolism of foreign compounds. The organ of pre-eminence in these studies is the liver, and the literature on methodology, enzyme kinetics, cofactor requirements, and effects of inducers and inhibitors on metabolic reactions in respect of these systems is extensive. Excellent symposia proceedings and reviews on this subject appear regularly (*e.g.* 'Microsome and Drug Oxidation' meetings).

Detailed description of the preparation procedures and enzyme assay methods are available (Brodie and Gillette, 1971; Snell and Mullock, 1987). The use of these *in vitro* preparations and the many refinements introduced, such as purified enzyme fractions, have provided an invaluable tool for elucidating the biochemical mechanisms responsible for the transformation of foreign chemicals. However, the limitations of the system need to be taken into account in arriving at a realistic assessment of the results. Metabolic changes in *in vitro* systems are subject to variation due to preparation procedures, ionic strength and pH of the medium, levels of cofactors and nutrients, and to end-product inhibition or activation. In addition, the results of *in vitro* enzyme induction and inhibition studies require to be interpreted with caution. Thus, for example, whereas *in vitro* 3,3,3-trichloropropylene oxide and cyclohexene oxide inhibit epoxide hydratase activity, in the intact animal these compounds act as inducers of this enzyme, whilst reducing liver GSH content (Oesch, 1974; Oesch, Jerina, Daly, and Rice, 1973).

Each of the various techniques and methods briefly described in this section play an important part in providing an insight into the mechanisms mediating the metabolic transformation of compounds in the body. Each of the techniques has advantages and disadvantages, and an appreciation of these is essential for their intelligent use in generating relevant metabolic data.

References

Baillie, T. A. (1981). The use of stable isotopes in pharmacological research. *Pharmacol. Rev., 33,* 81–132.

Bach, P. H., Ketley, C. P., Ahmed, I., and Dixit, M. (1986). The mechanism of target cell injury by nephrotoxins. *Food Chem. Toxicol., 24,* 775–779.

Bekersky, I. (1983). Use of the isolated perfused kidney as a tool in drug disposition studies. *Drug Metab. Rev., 14,* 931–960.

Bénard, P., Burgat, V., and Rico, A. G. (1985). Application of whole-body autoradiography in toxicology. *Crit. Rev. Toxicol., 15,* 181–215.

Bircher, J. and Presig, R. (1981). Exhalation of isotopic CO_2. *In:* 'Methods in Enzymology', Vol. 77. *Ed.* W. B. Jakoby. Academic Press Inc. (London) Ltd., pp. 3–9,

Brittebo, E. B. and Tjälve, H. (1982). Tissue specificity of *N*-nitrosodibutylamine metabolism in Sprague–Dawley rats. *Chem. Biol. Interact., 38,* 231–242.

Brodie, B. B. and Gillette, J. R. (1971). *In:* 'Handbook of Experimental Pharmacology', Vol. 28., 'Concepts in Biochemical Pharmacology', Pt. 2. Springer Verlag, Berlin.

Bronner, F. and Lemaire, R. (1969). Comparison of calcium kinetics in man and the rat. *Calcif. Tissue Res., 3,* 238–248.

Cohen, E. N. and Hood, N. (1969). Application of low temperature autoradiography to studies of the uptake and metabolism of voltaile anaesthetics in the mouse. I. Chloroform. *Anaesthesiology, 30,* 306.

Corney, P. L. and Heath, D. F. (1970). A simple way of estimating turnover rates from specific activity–time curves. *J. Appl. Physiol., 28,* 672–674.

Colowick, S. P. and Kaplan, N. O. (1981a). Organ perfusion: *In:* 'Methods in Enzymology', Vol. 77, Pt. B. *Ed.* W. B. Jakoby. Academic Press Inc. (London) Ltd. pp. 81–129.

Colowick, S. P. and Kaplan, N. O. (1981b). Cells. *In:* 'Methods in Enzymology', Vol. 77, Pt C. *Ed.* W. B. Jakoby. Academic Press Inc. (London) Ltd., pp. 130–168.

Davis, D. S. and Boobis, A. R. (1983). Drug metabolism and the safety evaluation of drugs and chemicals. *In:* 'Animals in Scientific Research: An Effective Substitute for Man?' *Ed.* P. Turner. Macmillan Press, London, pp. 69–80.

Dost, F. H. (1968). 'Grundlagen der Pharmakokinetics'. Thieme, Stuttgart.

Fraser, H. S., Mucklow, J. C., Murray, S., and Davis, D. S. (1976). Assessment of antipyrine kinetics by measurement in saliva. *Br. J. Clin. Pharmacol.*, **3**, 321–325.

Fry, B. J., Taylor, T., and Hathway, D. E. (1972). Pulmonary elimination of chloroform and its metabolites in man. *Arch. Int. Pharmacodyn. Ther.*, **196**, 98–111.

Fry, J. R. (1982). *In*: 'Reviews on Drug Metabolism and Drug Interactions' *Eds* A. H. Beckett and J. W. Gorrod. Freund Publishing House, Tel Aviv.

Fry, J. R. (1983). A review of the value of isolated hepatocyte systems in xenobiotic metabolism and toxicity studies. *In*: 'Animals in Scientific Research: An Effective Substitute for Man', *Ed*. P. Turner. Macmillan Press, pp. 81–90.

Fry, J. R. and Bridges, J. W. (1977). The metabolism of xenobiotics in cell suspensions and cell cultures. *In*: 'Progress in Drug Metabolism', Vol. 2. *Eds* J. W. Bridges and L. F. Chasseaud. John Wiley and Sons, pp. 71–118.

Fry, J. R., Weibkin, P., Kao, J., Jones, C. A., Gwynn, J., and Bridges, J. W. (1978). A comparison of drug metabolising capability in isolated viable rat hepatocytes and renal tubule fragments. *Xenobiotica*, **8**, 113–120.

Garattini, S. (1985). Toxic effects of chemicals: difficulties in extrapolating data from animals to man. *Crit. Rev. Toxicol.*, **16**, 1–29.

Gehring, P. J., Watanabe, P. G., and Blau, G. E. (1976). Pharmacokinetic studies in evaluation of the toxicological and environmental hazards of chemicals. *In*: 'New Concepts of Safety Evaluation', *Eds* M. A. Mehlman, R. E. Shapiro, and H. Blumenthal. Hemisphere Publishing Corp., Washington DC, pp. 195–270.

Gibaldi, M. and Perrier, D. (1975). Pharmacokinetics. *In*: 'Drugs and the Pharmaceutical Sciences', Vol. 1. *Ed*. J. Swarbrick, Marcel Dekker, New York.

Goldstein, A., Aronow, L., and Kalman, S. M. (1974). 'Principles of Drug Action: The Basis of Pharmacology', 2nd Edn., John Wiley and Sons, New York.

Gray, T. J. B. and Beamand, J. A. (1984). Effect of some phthalate esters and other testicular toxins on primary cultures of testicular cells. *Food Chem. Toxicol.*, **22**, 123–131.

Gray, T. J. B., Moss, E. J., Creasy, D. M., and Gangolli, S. D. (1985). Studies on the toxicity of some glycol ethers and alkoxyacetic acids in primary testicular cell cultures. *Toxicol. Appl. Pharmacol.*, **79**, 490–501.

Hawkins, D. R. (1977). The role of stable isotopes in drug metabolism. *In*: 'Progress in Drug Metabolism', Vol. 2., *Eds* J. W. Bridges and L. F. Chasseaud, John Wiley and Sons, Chichester, pp. 163–218.

Hepner, G. W. and Vesell, E. S. (1975). Quantitative assessment of hepatic function by breath analysis after oral administration of (^{14}C)-aminopyrine *Ann. Intern. Med.*, **83**, 632–638.

Jones, D. P., Sundby, G. B., Ormstad, K., and Orrenius, S. (1979). Use of isolated kidney cells for study of drug metabolism. *Biochem. Pharmacol.*, **28**, 929–935.

Jones, K. G. and Fouts, J. R. (1980). *Pharmacologist*, **22**, 277.

Levy, G. (1965). Pharmacokinetics of salicylate elimination in man. *J. Pharm. Sci.*, **54**, 959–967.

Levy, G. and Gibaldi, M. (1975). Pharmacokinetics. *In*: 'Concepts in Biochemical Pharmacology', Pt 3. *Eds* J. R. Gillette and J. R. Mitchell, Ch. 59, Springer Verlag, Berlin. pp. 1–34.

Lundquist, F. and Wolthers, H. (1958). The kinetics of alcohol elimination in man. *Acta Pharmacol. (KbH)*, **14**, 265–289.

Mayersohn, M. and Gibaldi, M. (1971). Mathematical models in pharmacokinetics. Solution of the two compartment open model. *Am. J. Pharm. Educ.*, **35**, 19–28.

Moldéus, P., Andersson, B., Norling, A., and Ormstad, K. (1980). Effect of chronic ethanol administration on drug metabolism in isolated hepatocytes with emphasis on paracetamol activation. *Biochem. Pharmacol.*, **29**, 1741–1745.

Niazi, S. (1979). 'Textbook of Biopharmaceutics and Chemical Pharmacokinetics'. Appleton, Century, Croft, New York.

Oesch, F. (1974). Purification and specificity of a human microsomal epoxide hydratase. *Biochem. J.*, **139**, 77–88.

Oesch, F., Jerina, D. M., Daly, J. W., and Rice, J. M. (1973). Induction activation and inhibition of epoxide hydrase: An anomolous prevention of chlorobenzene-induced hepatotoxicity by an inhibitor of epoxide hydrase. *Chem. Biol. Interact.*, **6**, 189–202.

Phillips, J. C., Bex, C. S., Mendis, D., Walters, D. G., and Gangolli, S. D. (1987). The metabolic disposition of ^{14}C-labelled amaranth in the rat, mouse, and guinea-pig. *Food Chem. Toxicol.*, **25**, 947–954.

Phillips, J. C., Gaunt, I. F., and Gangolli, S. D. (1975). Studies on the metabolic fate of ^{32}P-labelled emulsifier YN in the mouse, guinea-pig, and ferret. *Food Cosmet. Toxicol.*, **13**, 23–30.

Phillips, J. C., Mendis, D., and Gaunt, I. F. (1987). The metabolic disposition of ^{14}C-labelled Brown HT in the rat, mouse, and guinea pig. *Food Chem. Toxicol.*, **25**, 1013–1019.

Philpot, R. M., Anderson, M. W. and Eling, T. E. (1977). *In*: 'Metabolic Functions of the Lung'. *Eds* Y. S. Bakhle and J. R. Vane. Marcel Dekker, Inc., New York, p. 123.

Pinkus, L. M. (1981). Separation and use of enterocytes. *In*: 'Methods in Enzymology', Vol. 77. *Ed.* W. B. Jakoby. Academic Press Inc. (London) Ltd., pp. 154–161.

Pohl, L. R. and Branchflower, R. V. (1981). Covalent binding of electrophilic metabolites to macromolecules. *In*: 'Methods in Enzymology', Vol. 77. *Ed.* W. B. Jakoby. Academic Press Inc. (London) Ltd., pp. 43–49.

Reitz, R. H., Watanabe, P. G., McKenna, M. J., Quast, J. F., and Gehring, P. J. (1980). Effects of vinylidene chloride on DNA synthesis and DNA repair in the rat and mouse: a comparative study with dimethylnitrosamine. *Toxicol. Appl. Pharmacol.*, **52**, 357–370.

Rogers, A. W. (1967). 'Techniques of autoradiography'. Elsevier Publishing Co., Amsterdam.

Ross, B. D. (1972). 'Perfusion technique in Biochemistry'. Oxford Univ. Press (Clarendon), London.

Ruelius, H. W. (1975). Interaction of drug metabolism with the other elements of drug safety evaluation. *Drug Metab. Rev.*, **4**, 115–133.

Smith, R. A. and McInnes, I. B. (1986). Phase contrast and electron microscopical observations of adult mouse DRG cells maintained in primary culture. *J. Anat.*, **145**, 1.

Snell, K. and Mullock, B. (1987). 'Biochemical Toxicology: a Practical Approach', IRL Press, Oxford.

Spencer, P. S., Crain, S. M., Bornstein, M. B., Peterson, E. R., and Van de Water, T. (1986). Chemical neurotoxicity: detection and analysis in organotypic cultures of sensory and motor systems. *Food Chem. Toxicol.*, **24,** 539–544.

Ullberg, S. and Larsson, B. (1981). Whole-body autoradiography. *In*: 'Progress in Drug Metabolism'. Vol. 6, *Eds* J. W. Bridges and L. F. Chasseaud. J. Wiley and Sons, London, pp. 249–289.

Wagner, J. G., Northam, J. I., Alway, C. D., and Carpenter, O. S. (1965). Blood levels of drug at the equilibrium state after multiple dosing. *Nature* (*London*), **207,** 1301–1302.

Watanabe, P. G., Ramsey, J. C. and Gehring, P. J. (1980). Pharmacokinetics and metabolism of industrial chemicals. *In*: 'Progress in Drug Metabolism', Vol. 5, *Eds* J. W. Bridges and L. F. Chasseaud. J. Wiley and Sons, London, pp. 311–343.

Withey, J. R. (1977). Pharmacokinetic Principles. *Proc. 1st Internat. Cong. Toxicol.*

CHAPTER 11

Immunotoxicology — Conceptual Problems

H. E. AMOS

1 Introduction

Recognition that chemicals interacting with the immune system in a 'non' immunologically specific manner might constitute a potential hazard to health, has not been universally embraced by toxicologists involved in risk assessment. The difficulty has been to convincingly establish a cause and effect relationship. Untoward responses resulting from transient perturbations in immune responsiveness may involve a considerable latent period before they become manifest. Moreover, interpretation of immunological assay systems used in safety evaluation are themselves open to criticism. Many assays depict *in vitro* phenomena, which cannot be directly related to a clinical state. For example, a reduced *in vitro* proliferative response of lymphocytes to the mitogen, phytohaemagglutinin, is frequently equated with a clinical state of immune suppression. This is quite unjustified unless account is taken of the reserve capacity of the immunological system.

There is no doubt, however, that cells involved in the immune response can be a target for toxic damage which will then compromise the functional integrity of the response. It will either become downgraded, leading to degrees of suppression, or upgraded, producing autoallergic reactions and hypersensitivity responses. The perception that modulation of the immune response by 'immunotoxicants' under controlled conditions can be a therapeutic tool has been widely acclaimed. Their value has been proven in the control of tissue rejection and improvements in delivery systems particularly the humanisation of murine monoclonal antibodies, offers considerable optimism for the future.

2 Inadvertant Immunomodulation

Of major interest to immunotoxicologists are those 'immunotoxicants' which inadvertently participate in the immunological system. They may include drugs, environmental chemicals and impurities in products manufactured by recombinant DNA, and other biotechnologies (D'Agnolo, 1983). In fact, the range of products is such that it is virtually impossible to avoid contact with at least some of them. The question must be asked therefore:

'What is the clinical significance of inadvertant exposure to immunotoxicants?'

It is well established that experimentally induced immunosuppression in animals leads to an increased tumour incidence and in some cases increased metastatic spread (Pimm and Baldwin, 1978; Baldwin, 1973). The mechanism, however, is complicated, it is not a straightforward reduction in immuno surveillance brought about by the 'immunotoxicant' targeting T lymphocytes. Studies with chemical carcinogens, for example, have shown that suppression can be manipulated so as to remove T suppressor cells which will in fact increase tumour immunity (North *et al.*, 1982).

The clinical effect of an 'immunotoxicant' which downgrades the immunological system will depend upon which cell function has been depleted. This was elegantly demonstrated by Spreafico *et al.* (1984) who defined the immunological profile in mice treated with doxorabicin (DXR) and its close chemical analogue daunrabicin (DNR). They showed that DXR given up to the LD_{10} level had no effect on macrophage function but with DNR, the multifunction macrophage was severely depressed, affecting phagacytosis, antigen expression, and non-specific direct cytotoxicity. Thus closely related chemicals subserving the same therapeutic effect can differ in their toxicity towards a particular cell type.

Spreafico *et al.* (1984) attempted to relate the susceptibility of immunologically important cells to different immunosuppressive agents and showed, for example, the sequence for corticosteroids to be:

Thymocytes (TH) > Precursor cytotoxic lymphocyte (CTLp)
> cytotoxic lymphocytes > T suppressor cells(Ts) > B cells

and for cyclophosphamide the ranking is:

Ts > TH > B > CTLp

The clinical significance of this diversity is that the overall response of the immunological system to challenge by an 'immunotoxicant' is the algebraic sum of various positive and negative cell responses. For each agent, therefore, which specifically targets immunologically important cell types, a fingerprint can be established which can then be used to predict the clinical consequence. Unfortunately the clinical consequence can be predicted only in the animal species which donated the cells. Studies with the alkyl tin series had already shown that species variation in cell responsiveness could be expected (Seinen *et al.*, 1977). In these experiments thymocytes were the main target and neonatal rats the most sensitive species, there was virtually no toxic effect on thymocytes from man.

Three main points emerge from these quoted examples which have implications for toxicologists involved in safety evaluation and risk assessment.

(a) The degree and type of immunological imbalance is dependent upon the immunopharmacological profile of chemicals.
(b) Unless a profile is defined it is difficult to predict with any degree of confidence the likely clinical outcome.
(c) Profiles tend to be species specific.

The implications clearly relate to how can toxicologists effectively screen for 'immunotoxicants', given that each one has a distinct target cell profile? Before considering proposals which have been put forward to answer the question, the issue of clinical significance needs comment. Detectable changes at a clinical level are only significant to the toxicologist if they are untoward and pose a hazard to health. There are a few examples of chemicals down regulating the immunological system in man following inadvertant exposure and these are associated with major toxicological accidents. Two of which, in particular, are illustrative.

A   Polybrominated Biphenyls in Michigan (PBBs)

In 1973 a commercial preparation containing polybrominated biphenyls was mistakenly used instead of magnesium oxide in the preparation of a feed supplement for lactating cows. Subsequently large numbers of cows and farmworkers were exposed to PBBs and sustained toxic injury, including an immunotoxic response (Carter, 1976).

PBBs are fat soluble and stored in the thymus, liver, brain, and adipose tissue where they persist for long periods of time. A longitudinal study of Michigan farm workers exposed to PBBs was undertaken firstly in 1976–77 and again in 1981–83 (Bekesi *et al.*, 1978; 1983). Changes in the immunological profiles of long term survivors were monitored. The main findings were decreased numbers of T cells, an increase in null cells, and a decreased proliferative response of lymphocytes to *in vitro* stimulation with mitogens. These changes, detectable during the first survey, were still present at the time of the second analysis.

Plasma levels of PBBs ranged from 0.6 to 70 p.p.m. and in 25 percent of the subjects followed there was no decrease in levels between the two surveys. No correlation was found between serum levels and the prevalence of clinical symptoms or immunological findings, but the PBB content of white blood cells did directly correlate with immune disfunction (Robez *et al.*, 1982).

It would appear, therefore, that continued exposure of immunologically functional cells to the immunotoxicant contributed to the persistence of the immunological defect. What remains still to be unequivocally proven is whether the degree of suppression resulting from the persistent defect will eventually lead to an increased cancer risk. The work has been extensively reviewed by Bekesi *et al.* (1987).

B Accidents Involving Exposure to Polychlorinated Dibenzo-*p*-dioxins (TCDD)

Polychlorinated dibenzo-*p*-dioxins and polychlorinated dibenzofurans are two series of tricyclic aromatic compounds with similar physical, chemical, and biological properties. They are present as contaminants in different commercial products such as phenoxy herbicides, polychlorinated biphenyls, diphenyl ether herbicides, and hexachlorophene. They have also been reported to occur in fly ash from industrial incinerated waste.

There have been at least seven major accidents involving TCDD over the past fifteen years, but the one involving inadvertent exposure at Seveso in Italy offered the best opportunity to study the immunological defect. The official conclusion to the studies was 'that no significant difference in immuno function between exposed and control children could be found' (Seveso, 1979). This finding could not be predicted on the basis of animal model studies.

It was demonstrated in the early 1970s that the immunological system was a particularly sensitive target for TCDD in laboratory animals (Vos and Moore, 1974). The cells involved were those participating in differentiation and expression of cell mediated immunity and T cell dependent antibody production, which are located within the thymus gland. It has been suggested that the action of TCDD on thymus epithelial cells, which are involved in T cell maturation, is through the expression of the *Ah* receptor (*Ah* designing the genetic locus for aryl hydrocarbon responsiveness) (Poland and Glover, 1980).

The *Ah* receptor can be detected in different tissues. Rats have a high concentration of *Ah* receptor activity expressed in the thymus, whereas in mice the concentration in the thymus is about 1/4 of that expressed in the liver (Mason and Okey, 1982). Moreover, some strains of mice [DBA/2J(D2)] are *Ah*-receptor deficient and resistant to the toxic effect of TCDD at concentrations which cause marked toxicity in sensitive strains [C57BL/6 (B6)]. Thus, the *Ah* locus segregates with TCDD induced thymus atrophy in mice (Vecchi *et al.*, 1983).

The recognition that genetic constraints can play a role in producing inter and intra species differences in response to challenge with 'immunotoxicants' might be one explanation why the official report of the Seveso incident concluded that there was no demonstrable immunological defect in man.

These two illustrative studies do not, as yet, prove that exposure to PBB, or TCDD, cause a hazard to health which is mediated through the immunological system. Time, undoubtedly, will be the final arbiter, but faced with a scenario that TCDD, for example, is an active tumour promoter in animals (Pitot *et al.*, 1980; Poland and Glover, 1982) and since the T cytotoxic effector cell is a major cell type, participating in immune surveillance against neoplasms (Herberman and Ortaldo, 1981), modulation of T lymphocyte maturation by the *Ah* receptor could be a mechanism associated with a potential carcinogenic risk of TCDD. It is, therefore, prudent for toxicologists to devise methods for identifying 'immunotoxicants' in safety evaluation assessments.

3   Relevance to Man of Immunotoxicological Models in Animals

The basis of any safety evaluation study is to model an 'effect' in animals then make judgements on the relevance of the data to man. There is, therefore, an element of empiricism in safety assessment studies, but the empirical approach has proved its value in that man has been protected from most major poisons. It is not quite so certain, however, whether confidence in animal model data can be extended to the field of immunotoxicology.

As stated in an earlier section, each immunotoxicant will have a particular cell profile and it is the summation of the 'effects' which determine the ultimate clinical response. Many of the early workers in immunotoxicology failed to appreciate this fact, and drew clinical conclusions on the basis of a single assay system. Thus, many claims for specific xenobiotics causing an immuno depressed state were premature. Sharma and Zeeman (1980) illustrated this point rather well. They studied the effect of DDT on antibody production in mice and showed that they could either increase or decrease the production by altering conditions of the model system. The main variables influencing the result were dose and dosing schedules, nature of the antigen and time between antigen challenge, and the effect studied.

Even immunotoxicity data generated in man is not without question. The Bekesi studies on PBBs quoted earlier (Bekesi *et al.*, 1978) have been challenged by Silva *et al.* (1979). These investigations were unable to find any differences in the numbers or function of various classes of lymphocytes in high or low PBB exposed individuals *versus* a control group.

Immunotoxicologists are now well aware of these pitfalls and efforts are being made to address the central issue which is to determine how loss of immune function, as measured *in vitro*, equates to a clinical effect *in vivo*. Much of the credit for this approach must go to Dean and co-workers (Dean *et al.*, 1984). They have assembled data obtained over a number of years with a variety of chemicals tested at three different dose levels in host resistant model systems. They evaluated the lack of independence of two variables—*in vitro* immune function and altered host susceptibility, using Spearman's rank correlation co-efficient.

They were able to generate data in which significant correlation was obtained between depressed cytolysis by natural killer cells (NK) and increased susceptibility to *in vivo* challenge with B16F10 melanoma cells (RhO = 0.758). Similarly, they also demonstrated a correlation between loss of cytotoxic lymphocyte (CTL) activity and increased susceptibility to challenge with PYB6 sarcoma cells.

These data go a long way towards answering the question—what degree of immune suppression is necessary to induce directly overt toxic effects? From Dean and co-workers' data it was possible to deduce that the functional reserve for the CTL model was approximately <10–50 percent, whereas with the NK system it was more like 60–70 percent. It is only when this reserve has been exhausted that the susceptibility to tumour challenge is noted.

Functional assay of this type, in which it is possible to judge the clinical implications of changes in the immunological systems, must be the basis for developing any immunotoxicity safety evaluation screen. The data base for recommending the introduction of immunotoxicity testing into routine screens for risk assessment is, however, not yet adequate. A fact fortunately recognised by most regulating agencies.

4   Hypersensitivity as an Immunotoxic Response

Difficulties associated with screening chemicals for their direct effect on cells of the immunological system are relatively minor when compared to screening chemicals for immunogenicity. This is due to our inability to control in assay systems the many variables which govern the immunological response. The demonstration that a chemical will induce antibody formation in experimental animals is not by itself indicative of a toxic response or a prediction that the compound will cause hypersensitivity tissue damage in man.

Practically all chemicals can be made to stimulate antibodies in model systems, given the right conditions. But small molecular weight chemicals are not complete antigens in their own right. They are referred to as haptens.

The hapten hypothesis is the assumption that drugs become covalently bound to a macromolecular carrier and are then able to perform the functions of antigens and induce an allergic state. It is now well recognised that the ability of low molecular weight organic chemicals to stimulate an antibody response is a direct function of their reactivity with nucleophilic groups on proteins or other macromolecules.

An aspect of research, therefore, into immunotoxic drug reactions, has concentrated on the formation of complete immunogenic units from a composition of a hapten and carrier molecule as this is the one variable which can to some extent be controlled. Once a complete antigen is formed, however, host variables will dictate the final clinical response and these are not sufficiently characterised to be taken into account in safety evaluation systems, but it should be noted that one major regulatory authority, the Ministry of Health and Welfare in Japan, does demand that all drugs be tested for their immunogenicity in experimental systems.

5   Antigen Formation with Haptens

There are a number of possible mechanisms by which haptens can form covalent linkages with macromolecules under physiological conditions. The main ones are as follows.

A   Direct Protein Reactivity

Compounds with functional groups capable of covalently linking to nucleophilic side chains of proteins in neutral conditions, without prior biotransformation or chemical modification, clearly have immunogenic possibilities. Drug designers are well aware of this and steps are taken to limit direct protein reactivity by careful molecular manipulation.

B   Structural Alteration to Instability

Degradation

The beta lactam antibiotics, such as penicillin, induce antibodies in the majority of subjects taking the drug. It has been shown that the antibodies are not directed towards penicillin itself, but to various breakdown products. The major antigenic determinant, the penicilloyl conjugate, can be formed in a number of ways, one of which is through the penicillin rearrangement product, penicillenic acid, a highly reactive anhydride capable of rapidly conjugating free lysine residues in proteins (Parker *et al.*, 1962).

The ease by which penicillin undergoes rearrangement to form a variety of antigenic determinants is not shared by the majority of drugs, yet most of our information about immunotoxic drug reaction has been gained by studying penicillin!

Polymerisation

Some small molecular weight compounds undergo polymerisation in solution to produce structures of considerable size and complexity. The net effect of such a process is to convert a non-immunogenic hapten into a molecule capable of directly stimulating antibodies. It is not known to what extent polymer formation contributes to the antibody response generated against drugs, but the potential should be recognised.

6   Metabolic Conversion and Enzyme Interactions

It is difficult to determine with any degree of certainty exactly what happens to a drug after it gets inside the body. Enzymic mechanisms exist which are aimed at converting the parent drug into polar metabolites which facilitate elimination. Some metabolic intermediates can be highly protein reactive but the evidence that such complexes stimulate antibodies is sparse. The following two examples, however, support the contention that antigens can be formed from metabolic products.

A   Practolol

The cardioselective B-receptor blocking agent, practolol, induced in a small number of patients a syndrome defined as 'oculomucutaneous'. As the name implies the main sites of tissue damage were confined to the eyes and serosal surfaces, including the peritoneum and skin.

Epidemiological studies identified two factors which suggested that the tissue damage might not be due to a direct toxic effect of the drug but might implicate an immunological element in the pathogenesis. These were:

(a) the syndrome was not dose related
(b) the incidence was low.

Although attempts to identify an antibody with specificity for the practolol hapten were unsuccessful it was possible to show antibody activity against a practolol product generated in an *in-vitro* mixed function oxygenase system using rat and hamster liver microsomes (Amos, Lake, and Atkinson, 1977; Amos, Lake, and Artis, 1978). The clinical significance of the antibody in relation to the pathogenesis of the syndrome was not established, but its demonstration did indicate that metabolic products can stimulate a specific immunological response.

B   Halothane

Halothane is a fluorinated hydrocarbon used as a volatile anaesthetic and there is a clinical association between repeated use of the drug and hepatic damage. The possibility that a hypersensitivity mechanism may be involved in the pathogenesis arose following a number of epidemiological observations.

(a) Accelerated onset of hepatic changes after more than one halothane exposure, (Inman and Mushin, 1974).
(b) Unexplained post-operative pyrexia following exposure, (Trey, Lipworth, and Chatman, 1968).
(c) Frequent past history of allergy.

The first observation that a minor metabolite may be involved in the production of hepatic necrosis was the finding that a product of halothane metabolism, *via* the major oxidative pathway was capable of rendering rabbit hepatocytes antigenic (Neuberger *et al.,* 1980). It is not clear from the work whether the metabolite is essential for the antibody specificity or whether it functions to alter self determinants, but either way the immunogenic role of the metabolite must be acknowledged.

7   Alteration of Drugs by Chemical Interaction

The proposition that environmental chemicals may convert non protein reactive compounds into ones which can covalently link, has not been fully appreciated. At present there is one example which illustrates the possibility. The topical antimicrobial agent, chlorhexidine, is a bis-biguanide structurally unsuited for creating covalent linkages. There is evidence, however, implicating chlorhexidine in IgE antibody formation and clinical anaphylaxis (Ohtoshi *et al.,* 1986). Compliance with the hapten hypothesis would demand that chlorhexidine must covalently link to proteins to stimulate the IgE response. Experimental studies showed that chlorhexidine could be converted into a protein reactive species by chlorine or hypochlorite. The amount of chlorine required to convert the $-NH-$ residues of chlorhexidine into the corresponding chloramide, NCl, was less than 1 p.p.m. in water and the conversion was very rapid in neutral conditions (Rose, personal communication). The converted N-chloro derivative bound avidly to model proteins and the conjugates were shown to induce IgE antibodies in experimental animal models (Layton, Stanworth, and Amos, 1986; 1987).

The points to note in this example are that the chlorine interaction was very rapid and required very little free chlorine; the level of 1 p.p.m. is reached in the outlet source of many domestic water supplies. Protein conjugates formed by N-chloro conversion of the hapten were immunogenic and stimulated IgE responses. If such a mechanism was reponsible for chlorhexidine antigen formation it is possible to ask might it have a more general role in immunotoxicology? Further work will tell.

8   Pseudo-allergic Reactions

A fundamental requirement for a hypersensitivity reaction is an antigen reacting with specific antibody or with specifically allergised cells. The consequences of such interactions alter effector pathways which can lead to tissue damage or, as in the case of anaphylaxis, produce potentially life threatening conditions.

The term 'pseudo-allergic' refers to reactions which mimic signs and symptoms of hypersensitivity disorders, but without involving immunological mechanisms. Thus the inducing compound does not behave as an antigen.

One of the most important pseudo-allergic reactions is with compounds which induce signs of clinical anaphylaxis, such as synacthen (Jaques and

Brugger, 1969), and intravenous anaesthetic drugs (Lorenz *et al.*, 1969). An important target of these compounds is the complement system, particularly direct fixation of C_3. Fixation of this major complement component by chemicals, without involving C_1, C_4, or C_2, will initiate fixation of the end terminal sequence of the complement cascade and generate biologically active fragments called anaphylatoxins. The fragments have the property of reacting with mast cells and liberating vasoactive amines which bring about the anaphylactic process unless tested in a cohort of genetically high responders to the specific antigen (Gleichmann *et al.*, 1989).

The implication for the immunotoxicologist is that pseudo-allergic response can respond to dose–response relationships, whereas true immunological reactions are not dose-related in the accepted manner.

9 Conclusion

The identification of chemicals which target the immunotoxicological system and alter its function is becoming a task for toxicologists involved in safety evaluation. Systems are being devised and validated which could eventually be incorporated into routine toxicology screens, but the fundamental question whether transient modulation producing either a downgrading or an upgrading of the immunological response relates to a serious health hazard, such as an increased tumour incidence, is still unresolved. Definitive data are lacking but it must be stated that the evidence which is accumulating would make it irresponsible for toxicologists to ignore the possibility of a correlation.

References

Amos, H. E., Lake, B. G., and Atkinson, H. A. C. (1977). Allergic drug reactions: An *in vitro* model using a mixed function oxidase complex to demonstrate antibodies with specificity for a practolol metabolite. *Clin. Allergy*, **7**, 423.

Amos, H. E., Lake, B. G., and Artis, J. (1978). Possible role of antibody specific for a practolol metabolite in the pathogenesis of oculomucocutaneous syndrome. *Br. Med. J.*, **1**, 402–404.

Baldwin, R. W. (1973). Immunological aspects of carcinogenesis. *Adv. Cancer Res.*, **18**, 1.

Bekesi, J. G., Holland, J. F., Anderson, H. A., Fischbein, A. S., Rom, W., Wolff, M. S., and Selikoff, I. J. (1978). Lymphocyte function of Michigan dairy farmers exposed to polybrominated biphenyls. *Science*, **199**, 1207–1209.

Bekesi, J. G., Roboz, J. P., Solomon, S., Fischbein, A. S., and Selikoff, I. J. (1983). Altered immune function in Michigan residents exposed to polybrominated biphenyls. *In*: 'Immunotoxicology', *Eds* G. G. Gibson, R. Hubbard, and D. V. Parke; pp. 181–191. Academic Press, New York.

Bekesi, J. G., Roboz, J. P., Fischbein, A. S., and Selikoff, I. J. (1987). Clinical immunology studies in individuals exposed to environmental chemicals. *In*: 'Immunotoxicology', *Eds* A. Berlin, J. Dean, M. H. Draper, E. M. B. Smith, and F. Spreafico. pp. 347–367. Martinus Nijhoff.

Carter, L. T. (1976). Michigan's PBB incident: chemical mix up leads to disaster. *Science*, **192**, 240–243.

D'Agnolo, G. (1983). The control of drugs obtained by recombinant DNA and other biotechnologies. *In*: 'Current Problems in Drug Toxicity'. *Eds* G. Zbinden, F. Cohadon, J. Y. Detaille, and G. G. Mazue. John Libby Eurotext. Paris and London. pp. 241–247.

Dean, J. H., Lloyd, D., Lauer, L. S., House, R. V., Ward, E. C., and Murrary, M. J. (1984). Experience with validation of Methodology for Immunotoxicity Assessment in Rodents. *In*: 'Immunotoxicology'. *Eds* A. Berlin, J. Dean, M. H. Draper, E. M. B. Smith, and F. Spreafico. Martinus Nijhoff. pp. 135–158.

Gleichman, E., Kimber, I., and Purchase, I. F. H. (1989). Immunotoxicology: suppressive and stimulatory effects of drugs and environmental chemicals on the immune system. *Arch. Toxicol.*, **63**, 257–273.

Herberman, R. A. and Ortaldo, J. R. (1981). Natural killer cells; their role in defence against disease. *Science*, **214**, 24–30.

Inman, W. H. W. and Mushin, W. W. (1974). Jaundice after repeated exposure to halothane: An analysis of reports to the committee on safety of medicines. *Br. Med. J.*, **1**, 5.

Jaques, R. and Brugger, M. (1969). Synthetic polypeptides related to corticotrophin acting as histamine liberators. *Pharmacology*, **2**, 361.

Layton, G. T., Stanworth, D. R. and Amos, H. E. (1986). Factors influencing the immunogenicity of the haptenic drug chlorhexidine in mice. II. The role of the carrier and adjuvants in the induction of IgE and IgG anti-hapten responses. *Immunology*, **59**, 459–465.

Layton, G. T., Stanworth, D. R., and Amos, H. E. (1987). Factors influencing the immunogenicity of the haptenic drug chlorhexidine in mice. I. Molecular requirements for the induction of IgE and IgG anti-hapten antibodies. *Mol. Immunol.*, **24**, 133–141.

Lorenz, W., Doenicke, A., Halbach, S., Krumey, I., and Werle, E. (1969). Histaminfreisetzung and magensaftsekretion bei narkosen mit propanid (Epantol). *Klin. Wochenschr.*, **47**, 154.

Mason, M. E. and Okey, A. B. (1982). Cytosolic and nuclear binding of 2,3,7,9-tetrachlorodibenzo-*p*-dioxin to the *Ab* receptor in extra hepatic tissues of rats and mice. *Eur. J. Biochem.*, **123**, 209–215.

Neuberger, J. M., Mieli-Vergani, G., Tredger, M., Davis, M., and Williams, R. (1980). Oxidative metabolism of halothane in the immuno-pathogenesis of the associated liver damage (abstract). *Gut*, **21**, 467.

North, W. J., Dye, E. S., Mills, C. D., and Chandler, J. P. (1982). Modulation of antitumour immunity; immunological approaches. Springer Seminars in *Immunobiology*, 5/2. 193–200.

Ohtoshi, T., Yamanuchi, N., Tadokoro, K., Miyachi, S., Suzuki, S., Miyamoto, T., and Muranaka, M. (1986). IgE antibody—mediated shock reaction caused by topical application of chlorhexidine. *Clin. Allergy*, **16**, 155–162.

Parker, C. W., de Weck, A. L., Kern, M., and Eisen, H. N. (1962). The preparation and some properties of acid derivatives relevant to penicillin hypersensitivity. *J. Exp. Med.*, **115**, 803–819.

Pimm, M. V. and Baldwin, R. W. (1978). Immunology and immunotherapy of experimental and clinical metastases. *In*: 'Secondary Spread of Cancer'. *Ed.* R. W. Baldwin. Academic Press, London, pp. 163–209.

Pitot, H. C., Goldsworth, T., Campbell, H. A., and Poland, A. (1980). Quantitative evaluation of the promotion by 2,3,7,8-tetrachlorodibenzo-*p*-dioxin of hepatocarcinogenesis from diethylnitrososamines. *Cancer Res.*, **40**, 3616–3620.

Poland, A. and Glover, E. (1980). 2,3,7,8-tetrachlorodibenzo-*p*-dioxin segregation of toxicity with the *Ah* locus. *Mol. Pharmacol.*, **17**, 86–94.

Poland, A. and Glover, E. (1982). Tumour promotion by TCDD in skin of HRS/J hairless mice. *Nature (London)*, **300**, 271–273.

Robez, J., Greaves, J., Holland, J. F., and Bekesi, J. G. (1982). Determination of polybrominated biphenyls in serum by negative chemical ionization mass spectrometry. *Anal. Chem.*, **54**, 1104–1108.

Rose, F. (1984). Personal communication.

Seinen, W., Vos, J. G., Van Spenje, I., Snock, M., Brando, R., and Hooykaas, H. (1977). Toxicity of organotin compounds. II. Comparative *in vivo* and *in vitro* studies with various organotin and organolead compounds in different animal species with special emphasis on lymphocyte cyctotoxicants. *Toxicol. Appl. Pharmacol.*, **42**, 197–212.

Seveso, (1979) The escape of toxic substances at the ICEMSA establishment on 10 July 1976, and the consequent dangers to health and the environment due to industrial activity. A translation by the Health and Safety Executive of the offical report of the Parliamentary Commission of Enquiry, by permission of the Parliament of the Republic of Italy.

Sharma, R. P. and Zeeman, M. G. (1980). Immunologic alterations by environmental chemicals: relevance of studying mechanisms versus effects. J. *Immunopharmacol.*, **2**, 285–307.

Silva, J., Kaufmann, C. A., Simon, D. G., Landrigan, P. J., Humphrey, H. E. B., Heath, C. W., Wilcox, K. R., van Amburg, G., Kaslow, R. A., Ringel, A., and Hoff, K. (1979). Lymphocyte function in humans exposed to polybrominated biphenyls. *J. Reticulendothel. Soc.*, **26**, 341–346.

Spreafico, F., Merendino, A., Braceschi, L., and Sozzani, S. (1987). Immunodepressive drugs as prototype immunotoxicants. *In*: 'Immunotoxicology'. *Eds* A. Berlin, J. Dean, M. H. Draper, E. M. B. Smith, and F. Spreafico. Martinus Nijhoff. pp. 192–207.

Trey, C., Lipworth, L., and Chalmers, T. C. (1986). Fulminant hepatic failure. *New Engl. J. Med.*, **279**, 798.

Vecchi, A., Sireni, M. A., Canearata, M., Recchia, M., and Garatini, S. (1983). Immunosuppressive effects of 2,3,7,8-tetrachlorodibenzo-*p*-dioxin in strains of mice with different susceptibility to induction of aryl hydrocarbon hydroxylase. *Toxicol. Appl. Pharmacol.*, **68**, 434–441.

Vos, J. G. and Moore, J. A. (1974). Suppression of cellular immunity in rats and mice by maternal treatment with 2,3,7,8-tetrachlorodibenzo-*p*-dioxin. *Int. Arch. Allergy Appl. Immunol.*, **47**, 777–794.

CHAPTER 12

Perspectives—the Evaluation of Reproductive Toxicity and Teratogenicity

A. B. G. LANSDOWN

1 Introduction

The quality of an infant at birth and its capacity to undergo further development and maturation is an expression of the viability and fertilising capability of germ cells, successful conception, and subsequent intrauterine growth. Peri- and post-natal development will usually depend upon the growth potential of the individual and its ability to cope with environmental circumstances. Toxicologists have become aware that these environmental circumstances may act at any stage from germ cell production in the male or female of the species to produce infertility, genetic damage, or failure in reproductive potential, through conception to intrauterine development and peri-natal stages, leading to structural, functional, and behavioural changes in the offspring. Intrauterine development is a highly vulnerable stage in life when growth in the offspring depends on its genotype and its response to environmental factors.

Life before birth has become a highly emotive subject in most modern societies. Declining birth rates, local increases in the birth of children with structural and functional abnormalities, and causes of peri-natal death have presented considerable epidemiological concern in recent years, and clinical research has been directed towards developing methods such as ultrasonography and biochemical tests to identify 'at-risk' pregnancies, as well as in identifying genetical or environmental causes of foetal distress. Thus, in a recent study of foetal and neonatal mortality in Curacao and the Netherlands Antilles, an analysis revealed an overall death rate of 34.2 per 1000 total births (Wildschutt *et al.*, 1987). In 28 of 223 consecutive foetal and neonatal deaths, malformation was the principle cause of fatality; others included hypertension, antepartum haemorrhage, and asphyxiation. Among the environmental causes of malformation, infection, complications associated with intrauterine growth retardation, and maternal metabolic disorders were identified. The role of infectious agents is well known and numerous excellent reviews on this subject are available (Dudgeon, 1970; Brown and Karunas, 1972; Mims, 1968; Overall, 1972; Overall and Glasgow, 1970). Low birthweight and its post-natal consequences for survival and subsequent growth are widespread international problems and a subject for intense research among epidemiologists, clinicians, and experimental toxicologists (Coid and Ramsden, 1973; Lansdown and Coid, 1974; Lansdown, 1977; Levin *et al.*, 1975; Edwards, 1969; Lechtig *et al.*, 1975). Such environmental causes as dietary inadequacy, maternal exposure to sub-toxic thresholds of chemicals in the environment, or genetical factors have been identified. Of human pregnancies recognised by a missed menstrual period, about 15% end in miscarriage, although this level may be as high as 50% if one includes pregnancies diagnosed by means of the more sensitive hormonal assays. To the Darwinist, this pre-natal loss involving progeny subject to retarded growth, genetical damage, or unspecified complications, is a form of natural selection. The implications of this concept in limiting the dissemination of such life-threatening diseases as insulin-dependent diabetes, blood group disorders, anaemias, and other conditions through the community are considerable (Valdheim 1986; 1985).

To the clinician and the toxicologist, germ cell production conception and pre- and peri-natal growth in the offspring present different challenges. In the former case, the past 25 years have seen a surge of interest in the development of *in vitro* fertilisation and ovum transplantation in an attempt to overcome the problem of reproductive failure, and the use of ultrasound, serum α-foetoprotein, and amniocentesis in the diagnosis of congenital disorders. Toxicologists in contrast, directed by the lessons learned in the thalidomide episode in the late 1950s, and subsequent catastrophes such as the Minamata situation and the inadvertent use of Agent Orange in the Vietnam war, have sought to identify chemicals and environmental hazards which are potentially harmful in the broad reproductive process in man or other species. Interestingly, the techniques developed as an integral part of clinical *in vitro* fertilisation programmes have proved particularly useful in studying the responses of pre-conception embryos to potentially toxic agents.

It is the purpose of the present review to focus on tests available in the toxicological evaluation of substances to impair processes in germ cell production, reproductive activity, and pre-natal development. As will be seen with several examples selected, many present attitudes and legislative requirements have been fashioned by real-life experiences.

2 Historical

Man has exhibited a strong interest in reproductive disorders or abnormalities in the development of members of his own race and in various species of animals since earliest times. Many illustrations exist in wall paintings and ancient manuscripts of individuals with bizarre structural deformities, dating from the times of the ancient Babylonian and Egyptian dynasties. Even through the Middle Ages and until comparatively recent times, reproductive inadequacy or deformity in the neonate was attributed to the action of demons or curses, or even to acts of God.

Perhaps the first person to apply embryological principles to the study of congenital abnormalities was the eminent physician William Harvey (1651), who is reputed to have noticed that harelip, as occurs in some deformed infants, resembles a condition seen normally at an earlier stage in development. To Harvey, this harelip condition represented a state of 'arrested growth', a hypothesis that has been employed subsequently to explain the development of such abnormalities as cleft palate, ectopia cordis, and gastrorachischisis.

In the 18th century, Etienne Geoffrey St. Hilaire (1832) and his son performed a series of classical studies using chick embryos. They demonstrated that developmental abnormalities such as spina bifida or anencephaly could be induced by shaking the eggs or partly coating the shells with an impermeable material. Later, they extended their work to study patterns of human developmental abnormality. They coined the term 'teratology' to describe the study of monstrosities and grotesque developmental anomalies.

Important work showing that developmental anomalies may be induced by environmental agents dates from the early 1940s when pre-natal infection was recognised as a cause of structural and behavioural abnormality. Thus, Gregg and workers in Australia demonstrated a correlation between rubella infection in early pregnancy and the birth of infants with congenital heart disease, deafness, cataract, and other deformities, known collectively as the 'rubella syndrome' (Gregg, 1941; Swan *et al.*, 1943). Subsequently, a wide range of viruses and other infections have been identified as potential causes of reproductive failure, infertility, or congenital deformity in humans or sub-human species (Lansdown, 1978).

Perhaps the most important single event, creating an awareness that foetal development can be influenced by xenobiotic agents administered to a pregnant mother at sensitive stages in pregnancy, was the identification of the hypnotic drug thalidomide as a cause of limb and spinal abnormalities (Lenz, 1961; McBride, 1961).

Even now, there is controversy as to how this growth impairment occurs but it is clear that the foetus is most vulnerable during the period of maximum cell division and organogenesis. Whilst this concept is generally true in respect of teratogens that pass transplacentally to accumulate in foetal tissues, work with a broader range of chemical teratogens has established that, depending upon the nature of a teratogen and its site of action–accumulation–metabolic breakdown, foetal development may be adversely influenced at any stage from conception to birth. It is evident also from recent studies that nutritional inadequacy, drugs, or exposure of pregnant mothers to adverse environmental conditions in late pregnancy may influence post-natal development and maturation (Dobbing and Sands, 1971; Dobbing, 1974). At these later stages, organogenic processes in the foetus have reached an advanced state. However, organs such as the heart, lungs, and brain are in the essential phase of 'functional development'. The foetal brain is particularly susceptible to environmental changes at this late stage in gestation and Dobbing and his colleagues have demonstrated, using suitable animal models, that nutritional deprivation or factors leading to foetal growth retardation during the stage known as the 'brain growth spurt' can precipitate irreversible damage and be manifest post-natally as mental retardation (Dobbing and Sands, 1971). The prospect of catch-up growth has been investigated in human babies and in a variety of animal models, and the science of behavioural teratology has emerged.

It would be unproductive to list the wide range of events occurring over the past 25 years from which present day thinking and the design of tests to identify environmental agents capable of impairing reproductive processes derive. It is purposeful to note that the emergence of reproductive toxicology has been greatly influenced by real-life problems as diverse as infertility and reproductive failure in men following occupational exposure to lead or chronic alcoholism, and foetal abnormality and pre-natal death resulting from the consumption of bread contaminated with mercurial fungicides. As new information comes to hand, so testing procedures and precautionary measures will be revised accordingly (Lansdown, 1987).

3   Principles of Reproductive Toxicity and Teratogenesis

The reproductive process involves germ cell production and maturation in the parental generation, fertilisation, conception, and intrauterine growth of the offspring, parturition, and development after birth. In each case, normal development depends upon the viability of the genotype and appropriate genetical instructions for morphogenic events. Constituent cells in the embryo/foetus will undergo a programmed sequence of cyto-differentiation, metabolic and functional changes, migration, and even death, in achieving normal organogenic events. Within a species these development patterns may show small variations, but for a particular strain or species they will fall within the 'expected range'. Small and subtle deviations in development are documented for humans and most strains of laboratory animals commonly employed in reproductive studies.

Reproductive toxicology embodies several basic toxicological principles but also involves some specialised approaches and expertise. The period of intrauterine growth represents a highly vulnerable stage in development when the offspring are particularly sensitive to the toxic effects of exogenous agents, nutritional changes in the mother, and adverse environmental physical factors. Maternal illness, inappropriate choice of drugs, hypoxia, and hyperthermia have been implicated variously in reproductive failure, defective foetal development, and reduced postnatal survival (Edwards, 1969; Lechtig *et al.*, 1975). Epidemiological evidence suggests that up to 80% of human conceptions abort and that at least 6% of live births are complicated by some structural or functional defect.

Estimates of the frequency of these abnormalities vary greatly on account of the wide racial, socio-economic, and geographic variation in population studies. Statistics are also complicated by difficulties arising in the definition of 'what constitutes an abnormality'. The identification of minor deviations in human development is a further problem. In early foetal mortality and stillbirth, it is quite likely that an undiagnosed deformity is present, possibly reflecting a mutagenic or functional defect.

Defects in reproductive performance may be due to genetical or environmentally induced damage in germ cell production, alteration in physiological factors, disease processes in male or female partner, or through immunological factors. In controlled laboratory experiments conducted under defined conditions, most of these variables may be eliminated such that the true influence of an environmental compound acting on a particular stage in the reproductive cycle can be assessed accurately.

Six general principles apply in teratological studies, each being supported by abundant clinical and experimental evidence (Table 1) (Wilson, 1959). As a generalisation, the susceptibility of foetal tissues to deformation or damage relates largely to the nature of the exogenous influence and its pharmacological effect on tissues differentiating at the

Table 1 *General principles of teratogenicity*

1	The susceptibility of a conceptus to a teratogenic agent depends upon its genotype and the manner in which this interacts with the environmental factor.
2	The susceptibility of a conceptus to a teratogenic agent relates closely to its development stage at the time of exposure.
3	Teratogenic influences act upon sensitive cells and tissues of the embryo or foetus by specific mechanisms to induce deviant development.
4	Deviant development may be manifest as growth retardation, structural deformity, functional disorder, or by mortality.
5	A teratogenic agent will induce abnormalities in sensitive tissues, the effect being related to the nature of the agent and the route by which exposure occurs.
6	Manifestations of deviant development increase in degree as the dose level of the teratogenic agent increases. They range from a no-effect threshold to a totally lethal effect.

time of exposure. In the foetus, each tissue system seems to exhibit a stage of critical sensitivity outside which it exhibits little or no response. For example, in the rat foetus exposed to trypan blue, dye accumulated in the lysosomes of the yolk sac endoderm resulting in nutritional insufficiency in the foetus and deformities in those organs developing at that time. At later stages, when the function of the yolk sac placenta is superceded by the chorioallantoic placenta, trypan blue exerted minimal effect on foetal development (Beck and Lloyd, 1963). Stockard (1921) recognised these critical phases of development which he indicated could occur at any stage from conception to birth and involve functional or morphological parameters. In the present discussion, it is appropriate to make a distinction between the terms 'embryo' and foetus. For present purposes, embryonic development is defined as that period during which organ development occurs and when a defect in cell function results in malformation. In the foetal stage the principle organ systems are differentiated and development failure is a potential cause of pathological damage or functional disorder.

The genetic make-up of a conceptus is the setting in which teratogenesis occurs; genetic and extrinsic factors interact to varying degrees to initiate developmental abnormality. Responses range from embryo-lethality to no-effect, depending upon the teratogen and species of animal involved. Experience with such drugs as thalidomide, salicylate, and anti-cancer agents has shown that the dose causing maternal toxicity is an unreliable guide to the level required to induce foetal deformity. Thalidomide exhibits a very low toxicity threshold in mothers, but is highly teratogenic to the foetus at low concentrations. In contrast, the teratogenic dose of sodium salicylate, in rats at least, closely approaches that causing maternal toxicity (Lansdown, 1970). Many examples exist to illustrate ways in which the mother exerts a protective effect against teratogenic influences. This is particularly true in the case of nutritional deficiencies. In the case of drugs, compounds are possibly detoxified in the maternal liver. The placenta may act as a barrier for some toxic agents and infections (influenza virus). Occasionally, as in the case of cadmium and lead ions, which are strongly teratogenic in rodents if they enter the circulation, substances ingested with the diet are poorly absorbed from the maternal gastrointestinal tract.

It may well be that for many teratogens the concentration absorbed into the maternal circulation correlates well with that seen in the foetus (Gillette, 1977). However, since drug metabolising enzyme systems are poorly developed or not present at all in the foetal liver, toxic compounds tend to concentrate in the tissues. This would account in part for the greater toxicity seen in foetal tissues compared to those of the mother. Examples also exist to show that the teratogenic potential of a compound in a particular species relates closely to the rate at which it is metabolised. Alcohol is a well known human teratogen but is appreciably less effective in the rat, probably on account of the very much faster rate of metabolism in that species (Kennedy and Persaud, 1979). Variations in the maternal metabolism of teratogens are attributable to differences in plasma-protein binding, tissue distribution patterns, and excretion (Smith and Caldwell, 1977). In the laboratory situation, these factors will normally be examined in simple preliminary studies well before teratogenicity and reproductive trials are planned.

4 Epidemiological and Experimental Aspects

Strongest evidence that a particular chemical or environmental factor is likely to prove injurious to human reproduction is provided by large scale population studies, such as those conducted in Japan and Iraq in relation to accidental mercury poisoning (Takeuchi and Matsumoto, 1969; Bakir *et al.*, 1973). Chronic maternal alcoholism (Jones *et al.*, 1973), rubella virus infection (Swan *et al.*, 1943; Gregg, 1941), ionising radiation (Hiraoka, 1961), and exposure to drugs like thalidomide (Lenz, 1961; McBride, 1961), are further examples of teratogenicity and reproductive failure where irrefutable clinical evidence exists.

Occasionally, clinical studies suggesting that a particular agent or condition is responsible for increased numbers of still-births, congenital deformities, and reproductive failures are equivocal. One such example concerned the potential teratogenic risks posed by trace levels of anaesthetic gases to pregnant nurses and other personnel working in hospital operating theatres (Corbett *et al.*, 1974; Lansdown, Pope *et al.*, 1976). These and other workers failed to identify a clear cause–effect relationship or to establish threshold levels of halothane, nitrous oxide, or other gas necessary to cause toxic symptoms. Subsequent studies in which pregnant rats were exposed to high sub-anaesthetic levels of single or combinations of more than one gas failed to confirm teratogenicity (Lansdown, Pope, *et al.*, 1976; Pope, Halsey, *et al.*, 1975; Pope, Halsey, *et al.*, 1978). Pregnant animals were exposed at periods of gestation at which maximum teratogenicity occurs with other teratogens. Where foetal growth retardation was evident, this was attributed to reduced food consumption through anaesthetic-induced drowsiness. Although it would be unwise to extrapolate these observations in terms of a no-effect in humans without further wide-ranging trials, possibly employing other species, it is conceivable that the problems identified in the human cases resulted from a complex interaction of many factors, *i.e.* stress, tiredness, *etc.* Nevertheless, as a precaution a large proportion of hospital operating theatres in the United Kingdom are fitted with scavenging devices to control air quality and to remove trace levels of anaesthetic gases.

Many examples exist of clinical evidence that a substance or environmental condition causes reproductive distress or teratogenicity where the results have been confirmed experimentally. Thalidomide, salicylates, hypoxia, hyperthermia, and various nutritional imbalances are examples. Occasionally, however, reproductive or teratogenic risks identified through accurate clinical observations have not been reproduced under experimental conditions. Perhaps the best known of these is the rubella syndrome. A large number of studies have been conducted using a wide variety of laboratory animals, including sub-human primates. Although some congenital deformities have been produced resembling those produced by rubella infection in humans, viraemia has not been recorded on many occasions and the true rubella syndrome of congenital heart disease, abortion, cataract, and other changes has not been reproduced (Sever, Meyer, *et al.*, 1966).

In the main, experiments designed to demonstrate teratogenicity or reproductive toxicity these days tend to be prospective rather than retrospective in outlook. It is hoped that, as a consequence of the experience gained in the last 25 years, the vast proportion of hazards presently in the environment or in permitted drug lists will be appreciated, and recommendations made accordingly. In the years ahead, through nationally and internationally controlled safety evaluation procedures, it is expected that most hazards among new drugs, food additives, industrial processes, and agricultural preparations will be identified before the human risk is encountered.

5 Mechanisms of Reproductive Toxicity and Foetal Abnormality

Agents in the environment may adversely influence reproduction in humans and other species by their toxic action at any time, from germ cell production and maturation in the parental generation, to post-natal growth and development in the offspring. This toxicity may involve a specific stage in the reproductive cycle or be of a more general type. For simplicity, it is preferable to consider the action of the various agents according to whether they act directly or otherwise on germ cell production and reproductive performance in male or female parent, or alternatively are toxic in some way to post-conceptional development of the progeny. It is recognised that normal development in the progeny either pre-natally or post-natally depends upon the viability of the genotype and the way in which this interacts with its environment.

A Germ Cell Production and General Reproductive Performance

A wide range of exogeneous agents including viral infections, metal ions, alcohol, anti-cancer and anti-viral drugs, pesticides, and food additives are known to impair spermatogenesis, leading to reduced reproductive capacity and maybe sterility in the long term. On occasions, defects in spermatozoal morphology have been linked with metabolic, functional, and structural abnormalities in the surviving progeny. It is also evident from clinical observations with certain drugs that induced hormonal changes in the male or female partner may impair libido, reproduction cycles, or germ cell production, leading to reproductive failure (Lansdown, 1983; Dorrington and Gore-Langford, 1981; Nicholl, 1980).

Reproductive activity may be depressed in the event of ill-health. This frequently leads to altered physiological and endocrinological activity. It occurs with infections like influenza, cytomegalovirus, and several enterovirus infections, conditions which are acknowledged causes of impaired foetal development.

B Teratogenesis

Teratogenesis in the present context is used in its narrowest sense to define the action of any xenobiotic chemical or environmental condition on structural or functional development in the embryo/foetus. This may occur by toxic action in susceptible tissue or organ systems, or through a detrimental effect upon the health and nutritional state of the pregnant

Table 2 *Levels of action of environmental agents in cells of the developing embryo/foetus*

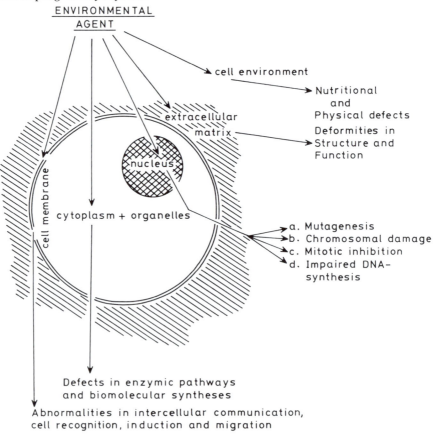

mother. By this latter mechanism, foetal growth is disturbed through imbalances in the intrauterine environment or the availability of nutrients essential for optimal growth. The mother is unable to satisfy the physiological demands of her offspring.

In view of the extreme diversity of environmental agents capable of impairing foetal growth in one or more species, it is clearly impracticable to attempt formulating a unifying hypothesis to explain the ways in which development abnormalities arise. For present purposes, a simplified scheme is proposed where specific levels of toxic action are identified. In each case, the text will be illustrated by experimental studies employing specific chemical factors or genetically determined conditions. The scheme envisaged (Table 2) recognises the cell as the main functional unit and identifies four principle routes of action (Saxen, 1976). This represents a considerable simplification of the mechanisms for abnormal development suggested by Wilson (1972), who considered up to ten possible ways in which sensitive tissues might respond to toxic insults. In the present discussion, it is recognised that cell sensitivity varies according to generative phases of the tissue and the gestational age of the individual.

The four-level model for teratogenesis allows for the fact that certain teratogens including metal ions, hypervitaminosis A, and some drugs influence tissues to different degrees according to their development state, functional condition, and activity. It offers a broad framework within which to discuss mechanisms of deviant development, avoiding many of the controversies that have arisen over what constitutes a primary route and that which is clearly of secondary importance.

The Intracellular Compartment

Genetical or chromosomal damage is an expression of changes resulting from exposure to toxic agents capable of altering DNA structure. On occasions, this may be difficult to distinguish from responses due to such factors as ageing, irradiation, or ultraviolet light, all of which are capable of mutagenicity (Kolata, 1980). Chromosomal damage is more frequently seen in patients with cancer or congenital defects but the reverse is not invariably true. The incidence of defects in the newborn attributable to chromosomal damage is an indication of clastogenicity minus the rate of intrauterine death.

In the human race, the relative contribution of genetical factors to congenital defects is not known. Approximately 5% of malformations are due to somatic mutations, 10% to chromosomal damage, and 5% to identifiable causes in the environment (Wilson, 1977). The so-called 'inborn errors of metabolism' are included in this group. The mutation is reflected often in terms of an enzymic deficiency affecting a specific metabolic pathway.

The contribution of genetic damage to teratogenesis has been discussed on many occasions (Wilson, 1977) but often the data for a particular group of chemicals are incomplete and the interpretations questionable. Studies conducted with a range of well known organochlorine and organophosphorus pesticides are an illustration (Table 3) (Sternberg, 1979). In summary, the mutagenic potential of these agents seems to be a good indication of their carcinogenic potential in the tests available but not of their teratogenic activity. DDT, for example, caused chromosomal damage in mice, and in two cell culture systems it induced dominant lethal mutations in rats and mice, *but* it was not teratogenic in either species. Strangely, DDT actually prolonged reproductive life in rats by up to five months, and protected against the teratogenic effects of salicylates, Benlate, and chloridine. Multigeneration studies in rats have demonstrated that aldrin, carbaryl, dieldrin, and malathion may lead to

Table 3 *Mutagenic potential of a range of organochlorine and organophosphorus pesticides*

Insecticide	Toxicity		
	Mutagenicity	Teratogenicity	Fertility
DDT	+[1,5], −[2,3]	−[a,b]	−[***]
Aldrin	−[3]	+[a,b]	
Dieldrin	−[1,3,5,6,]	+[a,b]	+[**]
Endrin		+[a,b]	
Kepone		+[a,b]	+[***]
Chlordane	−[5]	−[a]	−[***]
Heptachlor	−[3]		+[***], +[***]
Mirex	−[5]	+[a,b]	−[**], −[***]
Lindane	−[3]	−[a,b]	
Dichlorvos	+[1,3,8], −[5]		
Malathion	−[3]	+[a,b]	
Parathion		+[a,b]	
Carbaryl	+[1,3]	+[a,b]	+[***]
Diazinon	−[3]		
Captan	+[1,4]		
Nitrosocarbaryl	+[1,4]		

Key—1. Chromosomal damage *in vivo* or *in vitro*
  2. Unscheduled DNA synthesis
  3. AMES TEST
  4. DNA base pair substitution test
  5. Dominant lethal assay (rats or mice)
  6. Heritable translocation test
  7. Chromosomal damage in human peripheral blood culture
  8. Chromosomal damage in Drosophila sp. salivary gland
  a. Embryotoxicity
  b. Teratogenicity
  ** Oestrus changes in dogs, rats and rabbits with or without
        pregnancy failure
 *** Reduced fertility
(Reproduced by permission from Sternberg (1979) *Pharm. Therap.*, **6,** 147)

infertility when tested over three generations (Smith and Caldwell, 1977), although the experimental evidence has not been altogether convincing in every case.

More consistent evidence of teratogenicity is seen with those agents which impair DNA synthesis and mitotic activity at periods of high organogenesis. Ionising radiation is a well known cause of mutation in human patients and an acknowledged cause of congenital abnormalities (Yamazaki *et al.,* 1954; Brent, 1972; Wilson, 1964a). Whole-body radiation leads to the widespread production of free radicals and is also regarded as a cause of mutation (Sternberg, 1979). Mitotic arrest, cell death, and impaired cell proliferation occur with milder teratogenic influences like rubella infection, folate antagonists, cortisone, and agents which inhibit the mitotic spindle (the stathmokinetic agents). Impaired cell division in any tissue is a potential cause of asynchronous growth, hypoplasia, and possibly agenesis, depending upon the susceptibility of the tissue at the time of exposure.

Neural crest cells are particularly sensitive to the effects of purine and pyrimidine antimetabolites. These agents will induce exencephaly, ancephalocoele, and spina bifida if they are administered to the pregnant mother at the stage of neural fold formation in the embryo.

Defective regulatory mechanisms are a further teratogenic manifestation of intracellular toxicity. Anomalies in cell migration, proliferation, and intercellular communication occur. Naeye and Blanc (1965) concluded that hypoplastic development in the offspring of mothers infected with rubella early in pregnancy was due to a viral inhibition of cell proliferation. The organs had, in fact, failed to achieve the 'critical mass' for the particular gestational age and subsequent events were delayed or absent.

Vitamin A is an amphiphilic compound which readily penetrates the cell membrane to cause mitochondrial swelling, release of lysosomal enzymes, and inhibition of DNA synthesis. Impaired cell division and migration occurring at critical stages in the development of the cephalic mesoderm leads to derangements in the overlying neuroectoderm and the formation of exencephaly and spina bifida (Marin-Padilla, 1966). Cell death, mitotic arrest, and defective migration patterns probably underlie the micromelias, cleft palate, and cardiovascular deformities associated with cortisone, cadmium ion, salicylates, and possibly thalidomide (Saxen, 1970; Poswillo, 1975; Andrew and Zimmerman, 1971).

The several genetical deformities operating at the cellular level seem to involve deficiencies in cytoplasmic enzyme systems. Phenylketonuria, lysosomal alpha-L-iduronidase deficiency (Hurler's Syndrome), and diabetes mellitus are examples. As a consequence of the various enzyme deficiencies, some metabolites reach toxic levels, whereas others exhibit defective or alternative patterns of biodegradation. The defects are manifest as foetal mortality or as structural, functional, or behavioural abnormalities which are appreciated in the neonatal period.

The Cell Membrane

Foetal development involves a complex pattern of cell aggregation, morphogenetic movements, and cell migration, all of which depend to some degree on the biochemical characteristics of the cell membrane, and the intercellular passage of essential nutrients and inducer substances. Much of this subject is very unclear and speculative. However, some interesting genetic and non-genetic models exist to illustrate how damage induced in the cell surface can adversely affect morphogenesis and functional development.

The vertebrate limb, for example, involves extensive cellular interaction in development. The early limb bud exists as a finger-like projection of mesenchyme ensheathed in ectodermal tissue. The apical ridge of this ectoderm exhibits a profound influence on the morphogenetic patterns of skeletal structures within the mesenchyme (Cooke and Summerbell, 1980; Saunders and Fallon, 1966). Extirpation, reversal, or injury to this ridge of ectoderm has been shown to cause marked abnormalities in the limb bud (Wolpert, 1976). Although a large number of teratological agents have been shown to affect adversely the development of the limb bud, the mechanisms are probably not specific for that tissue, but form part of a more complicated condition.

Cell death is an important event in the morphogenesis of a number of structures including the limb bud, kidney, and palate. This means that any teratogen selectively altering morphogenic patterns of cell death is a potential cause of structural abnormality in that tissue. Janus Green B, for example, has been shown to inhibit the degeneration of the interdigital tissues in the development of the limb bud, syndactyly being the resulting deformity (Menkes and Deleanu, 1964). Involutionary changes characterise the development of the thymus and thyroid from branchial pouch primordia, and also the derivation of the pituitary gland from Rathke's pouch of stomodeal origin. Defects in the degeneration of these primitive tissues result in a persistence of the embryonic tissues into post-natal life as 'embryonic rests' or hamartomata. Such changes are occasionally seen as part of the normal background pathology in laboratory rodents (Lansdown and Grasso, 1971; Lansdown, 1982) but occur rarely in other species (Rest, 1987).

Morphogenetic cell death has an important implication in the normal development of the mammalian palate. Palatal primordia develop in the lateral mesenchyme of the buccal region and migrate towards the midline where fusion occurs following the degeneration of the intervening epithelial tissues. Defects in this degenerative process underlie the formation of cleft palate commonly associated with cortisone treatment in foetal mice. Cleft palate may also be attributable to defects in mucopolysaccharide synthesis and cell division (Ross and Walker, 1967). Although hypervitaminosis A is also a cause of cleft palate in various species, the biochemical mechanism is probably complex, involving changes in the synthetic capacity of mesenchymal cells and their interaction with tissues of ectodermal origin (Marin-Padilla, 1966; Kochhar, 1968).

The Synthesis and Composition of the Extracellular Matrix

Many foetal tissues are characterised by an intracellular matrix of specific chemical composition. Probably the best known example is the ground substance of cartilage and bone, which is frequently deformed or absent in foetuses from mothers exposed to teratogenic agents in early or mid-gestation. Skeletal abnormalities are seen where exogeneous agents or genetical determinants adversely influence the synthesis and degradation of mucopolysaccharides of cartilage matrix. Achondroplasia, chondrodystrophy, and chondrodysplasia are all well known conditions affecting bone development in humans, cattle, and laboratory animals (Duffell, Lansdown, and Richardson, 1985).

Teratogens may impair the biosynthesis of chondroitin sulphate and lead to defective cartilage formation and ossification patterns. An example is seen with sodium salicylate, which has been shown to inhibit the sulphation of the polysaccharide moiety in early cartilage development (Lansdown, 1970). Deviant cartilage formation with hypertrophy of chondocytes was shown to impair subsequent ossification patterns. Damage in the skull and spinal regions lead to manifestations of exencephaly and spina bifida in affected foetuses. Micromelias were also common observations.

Tetracycline, hypervitaminosis A, and cortisone exhibit a pronounced effect on skeletal development in many species (Saxen and Kaitilia, 1972; Morriss, 1972; Lansdown, 1983b). The mechanism is probably complex. However, tetracycline shows a particular affinity for calcium ions in intact foetuses and in cultured tissues. Impaired skeletal development in this case is attributed to a state of 'non-availability' of essential micronutrients. Other examples exist where ionic 'competition' is a cause of structural deformity. Collagen and elastic tissue are other examples of intercellular tissues which may be abnormal in teratogen exposed foetuses. Abnormal collagen formation is seen in cattle with the hereditary disease dermatosparaxis. Affected individuals exhibit defects in procollagen synthesis (Lenaers et al., 1974). Lathyrogenic agents, such as beta-aminopropionitrile, inhibit procollagen peptidase activity, and abnormalities involving vascular and skeletal tissues are seen experimentally. Nutritional copper deficiency is a further well known condition affecting collagen development. It is probable that copper acts in some way in the cross-linking of collagen fibres.

The Cellular Environment

In accordance with Wilson's (1959) six principles of teratogenicity, foetal development is determined by the close interaction between the foetal genotype and its environment. This environment relates not only to the intrauterine medium, which is intimately linked to the state of health of the mother, but also the intrafoetal conditions. This latter aspect reflects foetal nutrition, intercellular ionic balances, and physiological conditions.

Maternal metabolic diseases, malnutrition, and abnormal environmental conditions (hypoxia, hyperthermia, *etc.*) are well known causes of distress in human pregnancies and they have been examined experimentally in several animal models (Edwards, 1969). Often, conditions such as these are expressed at the foetal level in terms of nutritional insufficiency (Hurley and Cosens, 1974). This may arise through defects in utero-placental circulation or through the inability of the mother to metabolise nutrients essential for normal foetal growth. The influence of maternal ill-health in pregnancy is well documented in the medical press.

A further aspect of maternal ill-health which is well known is that of viral infection in pregnancy, especially with influenza, rubella, and herpes viruses. With the influenza virus at least, where infection of the conceptus has not been confirmed, foetal deformities and mortality are probably largely due to fever (hyperthermia) in the mother (Edwards, 1969). The work has been conducted in a number of experimental animal models, the structural defects seen closely resembling those seen in infected mothers. Metabolic imbalances attributable to disease processes are a further possible explanation.

Coxsackievirus infection in mice leads to a profound exocrine pancreatic insufficiency and atrophy within two days of infection, and as a consequence mothers become incapable of digesting dietary protein (Lansdown, 1975b). Labile reserves of protein accumulated in the maternal liver become depleted with the result that foetal development is retarded and mortality increased (Lansdown, 1976). The level of foetal effect seen correlated well with the stage in pregnancy at which infection occurred (Lansdown, 1975a). Subsequent studies demonstrated that this foetal distress could be prevented using immunological means or by feeding the animals with diets containing protein in a digested form (*i.e.* casein hydrolysate). Foetal development, as determined by measuring α-foetoprotein and albumin levels, was within normal limits for the species and gestational age (Lansdown, Coid, and Ramsden, 1975).

Nutritional causes of congenital deformity are well documented with reference to the serious problems encountered in developing countries. Protein and/or deficiency in specific amino acids is associated with a range of defects involving the development of skeletal, vascular, immunological, and nervous systems. Usually, these deformities form part of a more general pattern of growth retardation with reduced cell production (Naeye, Blanc, and Paul, 1973). Deficiencies in essential vitamins, minerals, fats, and amino acids occurring through starvation, metabolic deficiencies, or through the action of toxic compounds in the diet are discussed elsewhere (Lansdown, 1983b; Hurley, 1977).

6 Legislative Aspects of Reproductive Toxicity Testing

Reproductive toxicity is defined as that part of general toxicology dealing with the adverse influence of exogeneous agents upon the reproductive process (ECETOC, 1983). Teratogenicity represents a specific module within this framework and relates only to the effects of test chemicals on embryonic/foetal growth and their ability to induce malformations recognisable at birth or at later stages.

Early reproductive toxicity studies were conducted only on food additives, pesticides, and other preparations for which exposure could be expected over several generations. Studies would also have been conducted with hormonal preparations or products which might have been administered to women of child-bearing age. The test most commonly used before 1960 was the two-litter test where animals were dosed from before mating and through two complete cycles of mating, pregnancy, and lactation. Toxicity was assessed on the basis of reduced levels of conception, litter size, and post-natal viability. Surprisingly, thalidomide induced changes in all three parameters but was not suspected of being teratogenic. Retrospectively, it seems likely that its teratogenic properties were not identified in the two-litter test on account of its exceedingly low toxicity. In due course, Lenz (1961) and McBride (1961) clearly recognised the association between thalidomide therapy in women in the first trimester and the birth of babies with limb and spinal defects. Since that time, there has been a dramatic reappraisal of the safety evaluation of substances for reproductive toxicity.

The publication 'Guidelines for Reproductive Studies for the Safety Evaluation of Drugs for Human Use' by the Food and Drug Administration of the United States (FDA, 1966) is the basis of most predictive safety evaluation studies conducted these days. The recommendations are for single generation studies for the evaluation of drugs and chemicals likely to be ingested by humans over short periods, and for substances that are voided from the body within a short time. In contrast, for agents to which people might be exposed over a prolonged period, such as drugs used in chronic treatments, pesticides, and food additives, and where ingestion might extend over several generations, multigeneration studies would be conducted.

A Single Generation Studies

These tests recommended by the Food and Drug Administration (1966) are perhaps better known as the 'Three Segment Studies' on account of the specification for separate fertility studies with general reproductive performance (segment 1), embryotoxicity and teratogenicity (segment 2), and peri- and post-natal dosing (segment 3) modules.

Experiments conducted under segment 1 provide an overall insight of the potential risks accompanying a test compound without identifying either its target tissue or mechanism of action. Rats and mice are preferred for this study and they will be dosed for a sufficient time before mating to reveal adverse effects on gametogenesis, and then through mating, gestation, and post-natally until the end of lactation and weaning. Half the female animals will be killed and examined in mid-gestation to allow an investigator to assess the influence of his test compound on conception and foetal survival. Comparison of the number of corpora lutea in each ovary and the conception rate in each uterine horn provides an assessment of pre-implantation loss. It may be that, if a non-pregnancy situation is diagnosed at this stage, a study will be terminated and repeated using lower dose levels. Supplementary studies may prove useful in determining reasons for the non-pregnancy.

Segment 2 studies are more appropriately termed the teratogenicity phase, since they seek to identify the influence of a test compound upon post-conception development; they also cover the broader aspects of embryo-lethality and altered growth patterns. Rats, mice, and rabbits are used, pregnant animals being dosed throughout the period of optimal organogenesis. Experience with a wide range of known teratogens has shown that highest rates of malformations occur following dosing during this period. Dosing at earlier stages may result in increased embryonic loss. Experiments are terminated shortly before delivery and parturition, thus avoiding the loss and cannibalism of deformed or dead offspring. External examination of foetuses by hand lens, with or without transverse sectioning to detect visceral abnormalities, is mandatory. Most teratologists routinely conduct skeletal examinations in foetuses stained with alizarin red dye. On occasions, foetal tissues may be examined histologically for subtle abnormalities.

Segment 3 of the single generation studies is designed specifically to evaluate the influence of a test compound on later stages of foetal development, parturition, and early post-natal life. Again the rat, mouse, and rabbit are species of choice. Animals are dosed during the latter third of gestation and through lactation and weaning. The principal advantage of this peri- and post-natal dosing study is that an investigator is able to determine the influence of his test compound on functional development of foetal organs. The specifications allow an appreciation of changes that would not be detected under the recommendations in segments 1 or 2. Tests for behaviour or organ function will be conducted in neonates at the discretion of investigating scientists.

Where a test compound administered to a pregnant animal is excreted in the milk, segment 3 studies should provide evidence of foetal toxicity arising through transplacental exposure, and also of adverse effect through consuming 'contaminated milk'. The tests will demonstrate the influence of a test compound upon lactation and maternal behaviour.

The three segment study is criticised on account of its inadequacy in showing up the effects of test compounds on fertility and reproductive performance (Palmer, 1981). Regulatory authorities in Great Britain and in the EEC recommend that there may sometimes be an advantage in conducting additional male and female fertility studies in segment 1, with treated males being mated with untreated females and *vice versa*. However, experience with a large number of compounds has demonstrated that this measure is warranted in less than 1% of cases. The modification may allow a greater appreciation of toxicity than the guidelines set by the Food and Drug Administration permit.

B Multigeneration Studies

Multigeneration studies are not conducted with some compounds in the light of their chemical configuration and intended usage. They will be performed, however, where prolonged exposure in the human environment is envisaged (DHSS, 1982). Multigeneration studies may extend over three generations of test animals but this is time consuming and costly. These days it is more common for the tests to be conducted over two generations unless the intended use of the substance predicts that further tests are desirable.

In the evaluation of multigeneration studies, the offspring of successive generations will be examined for normality and maturation with reproductive potential. For scientific and economic reasons, mice and rats are normally selected for this type of work. A proportion of the offspring from each generation will be sacrificed (usually 50%) and examined macroscopically and histologically for abnormalities. The tests are modified by the various regulatory authorities but these modifications generally apply to the number of animals required in each test group and in the specifications for sacrifice and evaluation of toxic changes.

C Additional Studies

Guidelines published by the various regulatory authorities for reproductive toxicity and teratogenicity are specific in their requirements; that is, which species shall be used, the period of dosing, and the means of evaluation of toxic change. It will be appreciated, however, that the costs involved in conducting the entire spectrum of studies in intact animals can be prohibitively expensive. As a consequence, authorities such as ECETOC (1983) encourage the development of short term assays such as:
  (a) Male fertility and spermatozoal morphology tests
  (b) Female fertility and the influence of test substances on oestrus cycles
  (c) *In vitro* tests for embryotoxicity
  (d) Dominant lethal assay tests for mutagenicity.
Specific recommendations are not available yet for these additional tests, but there is evidence that many research groups are using them on

account of the smaller number of animals required. The tests are also preferred since they can provide specific information in a shorter time than is possible where the more conventional tests are employed.

ECETOC (1983) consider that more emphasis should be placed upon developing and validating the short term assays in reproductive toxicology. It is hoped that, in due course, the tests required by the various national and international regulatory authorities will be harmonised.

7 Experimental Considerations

A Species

Many strains and species of laboratory animal have been used in reproductive studies, but the majority of regulatory studies have been conducted in the popular strains of mouse, rat, and rabbit. More rarely, dogs, ferrets, and pigs have been employed as non-rodent species. Species selection has been largely based on the convenience of the animals under laboratory conditions, a detailed knowledge of their reproductive biology, and their sensitivity to known human teratogens like thalidomide. Although monkey species are phylogenically closer to humans, they exhibit a number of important metabolic and behavioural differences such that rarely have they been found more suitable than rabbits and rats. The latter species are considerably less expensive to use and do not require the highly specialised level of management and husbandry.

It is common practice in reproductive studies to administer test compounds intragastrically or in the diet to mimic the anticipated pattern of human exposure. Interspecies variations exist in the gastro-intestinal absorption of exogeneous materials and food substances, principally on account of differences in gastric acidity, intestinal flora, metabolising enzymes in the intestinal epithelium, and in the characteristics of the mucosal surfaces. Preliminary studies are profitably conducted to examine the pattern of intestinal absorption of test compounds before conducting more definitive tests, since it is well known with many proven teratogens that a close relationship exists between the level of compound in the maternal circulation and the foetal response. The rate at which a test compound is metabolised has a marked bearing on the levels of teratogenicity seen. Alcohol, for example, is metabolised more slowly in the human than in the rat and is appreciably more toxic. Imipramine is eliminated more slowly from the body in rabbits and rats than in humans (Harper, Palmer, and Davies, 1965).

The reproductive biology and foetal development are well known in rats and mice, making them choice species for many types of toxicity study (Wilson, 1964b). They were used in the early two-litter tests which preceded the present reproductive toxicity tests. However, their value as a sole test species is limited on account of their low sensitivity to thalidomide. Nowadays, the rabbit is preferred as the main species for most teratological tests, with rats and mice being employed as second species.

Many experimental studies have been conducted in mice but their value in teratological studies is limited in view of their known susceptibility to spontaneous cleft palate (Loevy, 1962). This feature has been used to advantage in demonstrating the interaction between genetical factors and environmental factors in abnormality (Pinsky and DiGeorge, 1965; Biddle and Fraser, 1976).

The sensitivity of the rabbit foetus to the deforming influence of thalidomide is well appreciated (Giroud, Tuchmann-Duplessis, and Mercier-Parot, 1962), but other features that make this an attractive species for reproductive toxicology include its spontaneous ovulation pattern and comparatively large size (Gibson, Staples, and Newberne, 1966). Although the rate of spontaneous abnormalities in the rabbit is slightly higher than in rodents, this is not considered to be a serious disadvantage. In each study, there will be a statistical evaluation of the number of congenital abnormalities in test and control groups (Stadler, Kessidjan, and Perraud, 1983).

The ferret has been recommended as a suitable species in teratological studies. It is a small carnivore and may be suitable where a non-rodent species is sought (Beck, 1975). However, it seems that few studies so far have been conducted using this species and there is still a paucity of background data on its general biology and susceptibility to known teratogens. On the other hand, the hamster has featured widely in experimental studies and its sensitivity to several known human teratogens is appreciated (Ferm, 1965). Although it is a convenient laboratory animal and has a comparatively short gestation (16 days), it is notoriously difficult to dose by intravenous injection.

B   Dose of Test Compound and Route of Administration

The magnitude of the dose and the route of administration of a test compound to animals in reproductive toxicity experiments can have a profound bearing on the outcome and levels of foetal abnormality. Experience has shown that these two factors do not operate in isolation, but can interact in the production of congenital abnormalities (Fraser, 1964; Mellin, 1964).

Experimental teratology and reproductive toxicity differ from most pharmacological experiments in that test compounds are usually administered at doses which are sufficiently high to produce an observable effect. Occasionally, the doses administered are unrealistically high and cannot mimic anticipated human exposure levels. In one such experiment, mice were given 1900–2500 mg kg^{-1} sodium chloride, a dose which could be expected to cause extensive ionic and metabolic imbalances in maternal and foetal tissues; but foetal abnormalities were reported (Nishimura and Miyamoto, 1969). Other experiments have sought to demonstrate the teratogenicity of caffeine in rats, but the doses administered were equivalent to that contained in 38–72 cups of coffee (Le Chat, Borlee, *et al.*, 1980; Palm, Arnold, *et al.*, 1978).

In regulatory studies, test compounds will normally be administered to test animals by the route that is most relevant in human exposure. Several examples exist to demonstrate that, whereas some substances are without obvious risk under normal conditions of exposure, they may be teratogenic under unusual circumstances. Cadmium and lead ions are not teratogenic in hamsters or rats if given by oral route, but become so if they are injected subcutaneously or intravenously (Ferm, 1969; McClain and Becker, 1975). The relevance of this observation in terms of human risk is unclear. In the case of lead ions, there is a reported ten-fold difference in the intestinal absorption in rats and in humans (Moore, Hughes, and Goldberg, 1979). Humans have been reported to absorb about 10% of an ingested dose of lead (Rabinowitz, Wetherill, and Kopple, 1976), but in their 1977 survey The Scandinavian Council for Environmental Information Workshop concluded that ". . . lead is not a teratogen in humans in the classical sense."

In conclusion, it must be emphasised that the biological properties of a test substance cannot be equated with clinical risk. It is the circumstances of exposure to the substance that are of paramount importance, that is the route, level of dose, and the time of exposure.

C The Period of Dosing

Environmental agents may impair the reproductive activity of a species or alter the developmental pattern of its progeny at any stage, from germ cell formation in the parental generation through to post-natal growth and maturation in the offspring. In the foetus there is a good correlation between the type of abnormalities seen and the period in gestation at which they are exposed to the deforming influence. For present purposes, the reproductive cycle is considered in three main phases:

(a) Gametogenesis
(b) Development in the embryo/foetus
(c) Post-natal growth and maturation.

The various regulatory authorities take these factors into account in making their specifications (Johnson and Everitt, 1984; Tuchmann-Duplessis, 1984).

Experimental studies have demonstrated that each embryonic/foetal tissue exhibits its own peak of sensitivity to deforming influences (genetical or environmental). In most species, the period of high teratogenicity tends to be during early or mid-gestation when organogenesis is optimal. In the mouse foetus, cleft palate was regularly induced following administration of cortisone to pregnant mothers at 13–14 days (Biddle and Fraser, 1976). Exposure of rats to trypan blue dye, sodium aurothiomalate, or cadmium ions produced maximum

Table 4 *Range of reproductive effects and manifestations of teratogenicity as a function of the time of exposure to toxic agents during the reproductive cycle*

State of Exposure	Response		
	Male	Female	Conceptus
Before mating	Reduced fertility, loss of libido, abnormal sperm characteristics, spermatocyte mutation	Reproductive failure, abnormal oestrus cycles, hormonal changes, defects in ovum transport, mutation	—
During pregnancy	—	Generalised maternal illness, metabolic, hormonal, and nutritional defects, circulatory and other functional changes, toxaemia	Preconception loss, stillbirth, intrauterine death, structural and functional abnormalities, biochemical changes, growth retardation, transplacental carcinogenesis, somatic mutations, post-natal distress.
Peri- and post-natal period	—	Failure in lactation, toxic metabolites in milk, prolonged gestation, deficiencies in maternal behaviour, weaning, *etc.*	Peri-natal mortality, functional and developmental abnormality, mental and behavioural deformity, carcinogenic changes with mutations, post-natal toxicity through substances present in milk

teratogenicity during the period 8–10 days of gestation when foetal nutrition is largely by way of the yolk sac placenta. In the human, maximum sensitivity to thalidomide is seen at about 37–50 days (Lenz, 1961; Lenz and Knapp, 1962). Experimental evidence shows that, in animals treated with teratogens at different stages in the reproductive cycle, the responses seen are specific for the stage of exposure (Table 4).

D   Experimental Observations

Identification of defects in reproductive performance or in foetal growth constitutes a vital part of any predictive safety evaluation study. The quality of the results relates directly to the sensitivity of the methods used and to the experience of the investigator.

For comparative purposes, the influence of environmental agents on the reproductive performance of a test species or the development of its offspring may be assessed by the following indices:

    (a) Pregnancy (*i.e.* the number of pregnancies recorded as a percentage of those exposed)
    (b) Conception rate
    (c) Foetal survival
    (d) Malformation
    (e) Parturition
    (f) Post-natal survival (1, 10, 20, 50 days, *etc.*).

These indices are valuable in comparing the toxicity of different agents (administered at comparable pharmacological doses), different dose levels to construct a dose–response relationship, and test models.

The pregnancy index will be a measure of the influence of a test compound upon germ cell production, reproductive behaviour, and conception. Defects in the male animal may be attributable to impaired spermatogenesis, sperm motility, and fertilising ability (capacitance), hormonal changes, or lowered reproductive behaviour. An analysis of spermatozoal morphology and fertilising ability may be conducted under *in vitro* conditions. Other tests envisaged will include semen viscosity, acidity, and chemical constitution. In the rabbit, sperm samples can be readily obtained using the artificial uterus technique. In small species, sperm samples are obtained from epididymal preparations.

Reproductive activity in female animals shows sensitivity to hormonal changes as well as disturbances in general health. Drug-induced oestrus changes are readily examined by vaginal smear techniques and assays of circulating hormone levels. *In vitro* fertilisation is a further potentially useful test in assessing the viability of the ovum.

A wide range of methods is available for determining foetal development in teratological studies. Selection of procedure will normally be limited by equipment available and economic factors. Commonly, an investigator will count resorption sites, dead embryos/foetuses, and stillbirths. Live progeny delivered by Caesarian section at near term are examined routinely by hand lens and gross abnormalities in the shape and proportions of the head and limbs will be noted. Transverse sections (3–5 mm) allow an assessment of visceral changes (Wilson, 1959). Skeletal abnormalities commonly occur in teratological experiments where test compounds exhibit a direct influence on the development of cartilage or bone, as well as those which induce a generalised impairment in foetal growth. Dawson's alizarin red-S technique is widely used in staining the skeletons of foetuses treated with teratogens (Dawson, 1926). Addition of counterstains, such as methyl green, methylene blue, and methyl blue, has proved useful in demonstrating cartilagenous anomalies in whole foetal mounts. Teratological effects on skeletal development may be assessed by the following:

(a) The number and extent of ossifications
(b) Observation of absent or defective ossifications
(c) Defects in the configuration of ossifications, *e.g.* wavy ribs, supplementary ribs or digits.

In mechanistic studies, sections taken by Wilson's (1959) method may be preserved in formalin or Bouin's fixative and preserved for histopathological examination. Alternatively, as in Ullberg's (1958) studies, histological preparation with subsequent autoradiography may be employed to identify sites of drug deposition/metabolism in foetuses from mothers treated with radiolabelled compounds. This specialised type of study will not be conducted routinely in many laboratories.

In peri- and post-natal experiments, toxic changes are sought in the neonates and young offspring. Methods available for this include timing of specific developmental events (*e.g.* eye opening, hair growth, *etc.*) and behavioural pharmacology. This last parameter has become a very specialised area and applies very much to the functional development of the brain. Different laboratories have developed their own particular expertise and such methods as the righting reflex, open-field exploration, movement, and response to external stimuli are adopted (Barlow and Sullivan, 1975). In many respects these studies resemble those conducted routinely in pharmacological laboratories.

8 Statistical Evaluation

Reproductive and teratological studies generate a vast amount of quantitative data. The onus will be upon the statistician to evaluate the significance in differences in such parameters as:

(a) Reduced fertility
(b) Reproductivity
(c) Conception rates
(d) Intrauterine and post-natal growth rates
(e) Changes in the sex ratio of the offspring
(f) Increases in the incidence of structural and functional defects in the offspring

(g) Viability

(h) Weaning index

(i) Patterns of post-natal growth in the offspring.

In order to obtain results that are statistically sound, it is important that sufficient numbers of animals are assigned to test and control groups using a table of random numbers. This procedure will be adopted also in culling litters to a constant size for longer term observations. Randomisation will also be used in selecting animals for mating in multigeneration studies.

In segment 1 tests, ten animals will be allocated per treatment group for each step of the experiment; that is ten for examination in mid-gestation and ten for autopsy at weaning. Experience has shown that, at best, only agents having a potent effect on reproductive processes can be detected (Palmer, 1981). It is necessary to obtain a 40–50% difference between treated and control groups to achieve significance in Fischer's Exact Test.

Palmer (1974) considered that, on the basis of several hundred tests, analysis of litters rather than individual foetuses provided the only valid sample unit for statistical purposes and that litter values are not normally distributed. Thus non-parametric modes of analysis are preferred, but they are not infallable and may be used only as a guide for making a judgement, and not the sole criterion for decision.

In assessing the significance of major malformations, a statistical analysis may have limited application. It is likely that group sizes of at least 60 animals—that is three times the number required by the Food and Drug Administration—would be necessary to establish the statistical significance of a 100-fold increase in the malformation rate in the event of a 0.1% incidence of 'background' malformations.

9   General Observations

In more than 25 years since the thalidomide tragedy, toxicological studies have assumed greater importance than at any time previously, and research and development costs have increased disproportionately. Legislative authorities throughout the world have sought to introduce, or improve where necessary, screens to identify those environmental chemicals liable to impair reproductive processes or development of the offspring, long before a human is exposed by intention or accidentally. As new information has come to hand, through experimental toxicology or as a consequence of epidemiological study, legislative requirements have become more stringent and far reaching. As in other disciplines of toxicology, it will be the responsibility of the experimentalists to modify their tests to take into account the nature of the substance they are testing and the manner in which a human being may be exposed.

The three segment study for reproductive toxicology and teratology will provide a lot of information, but it is complex and time consuming. It may be preferable to conduct simpler tests to provide answers to specific questions (Palmer, 1974). Alternatively, one could use the simple tests in the early stages of toxicity evaluation and then progress to more definitive trials. At all times, the toxicologist should be aware of the limitations of any test he conducts, and if necessary apply further tests to identify toxic effects.

Having completed necessary toxicological screens, the toxicologist will attempt to extrapolate his observations in terms of human risk. This will always be a contentious subject, but in reproductive toxicology at least, experience has shown that the studies described above using the rat, mouse, or rabbit as test species are normally adequate in revealing those environmental agents likely to be injurious in human reproduction or pregnancy. Although *in vitro* tests for studying the influence of environmental agents on embryonic or foetal growth are widely conducted in many larger laboratories at present, they are not currently accepted by most major regulatory authorities as an alternative to studies conducted in intact animals. It is likely, however, that if a test compound does provide evidence of risk in preliminary *in vitro* studies a company may decide against carrying out further studies at that stage for economic reasons.

Acknowledgement

I am grateful to Mrs Anne Murray for her assistance in preparing this chapter.

References

Andrew, F. D. and Zimmerman, E. F. (1971). *Teratology*, **4,** 31.
Bakir, F., Damluji, S. R., Amin-Zaki, L., and Murtaiha, M. (1973). *Science,* **181,** 230.
Barlow, S. and Sullivan, F. M. (1975). *In*: 'Teratology: Trends and Applications'. *Eds* C. L. Berry and D. E. Poswillo. Springer-Verlag, New York, p. 103.
Beck, F. (1975). *In*: 'New Approaches to the Evaluation of Abnormal Development'. *Eds* D. Neubert and H. J. Merker. Thieme, Stuttgart, p. 8.
Beck, F. and Lloyd, J. B. (1963). *J. Embryol. Exp. Morphol.,* **11,** 175.
Biddle, F. G. and Fraser, F. C. (1976). *Genetics,* **84,** 743.
Brent, R. L. (1972). *Davis' Gynec. Obstr.,* **2,** 1.
Brown, G. C. and Karunas, R. S. (1972). *Am. J. Epidemiol.,* **95,** 207.
Carter, C. O. (1969). *Br. Med. Bull,* **25,** 52.
Coid, C. R. and Ramsden, D. B. (1973). *Nature (London),* **241,** 460.
Cooke, J. and Summerbell, D. (1980). *Nature (London),* **287,** 697.
Corbett, T. H., Cornell, R. G., Endres, J. L., and Leiding, B. S. (1974). *Anaesthesiology,* **41,** 341.
Dawson, A. B. (1926). *Stain Technol.,* **1,** 123.
Department of Health and Social Security (1982). 'Guidelines for the Testing of Chemicals for Safety', Report No. 27, p. 33. London.

Dobbing, J. (1974). *Pediatrician*, **53**, 2.

Dobbing, J. and Sands, J. (1971). *Biol. Neonat.*, **19**, 363.

Dorrington, J. and Gore-Landford, R. E. (1981). *Nature (London)*, **290**, 600.

Dudgeon, J. A. (1970). *In*: 'Clinical Virology: The Evaluation and Management of Human Viral Infections', *Eds* Debre and Celers. Saunders, Philadelphia, p. 792.

Duffell, S. J., Lansdown, A. B. G., and Richardson, C. (1985). *Vet. Rec.*, **117**, 571.

ECETOC (1983). 'Identification and Assessment of the Effects of Chemicals on Reproduction and Development', Monograph No. 5, Brussels.

Edwards, M. J. (1969). *Teratology*, **2**, 313.

Food and Drug Administration (1966). 'Guidelines for Reproductive Studies for Safety Evaluation of Drugs for Human Use', Washington.

Ferm, V. H. (1965). *Lab Invest.*, **14**, 1500.

Ferm, V. H. (1969). *Experientia*, **25**, 56.

Fraser, F. C. (1964). *In*: 'Proceedings of the Second International Conference on Congenital Malformations', *Ed.* M. Fishbein, International Medical Congress, New York, p. 277.

Gibson, J. P., Staples, R. E., and Newberne, J. W. (1966). *Toxicol. Appl. Pharmacol.*, **9**, 398.

Gillette, J. R. (1977). *In*: 'Handbook of Teratology', *Eds* J. G. Wilson and F. C. Fraser. Plenum, New York, **3**, p. 35.

Giroud, A., Tuchmann-Duplessis, M., and Mercier-Parot, L. (1962). *C. R. Soc. Biol.*, **156**, 765.

Gregg, N. M. (1941). *Trans. Ophthal. Soc. Austr.*, **3**, 35.

Harper, K. H., Palmer, A. K., and Davies, R. E. (1965). *Arzneim-Forsch.*, **15**, 1218.

Harvey, W. (1651). *In*: 'Exercitiones et Generatione Animalium Typis du Gardiansis', Pulleyn, Coemetaria Paulino, Amsterdam.

Hiraoka, S. (1961). *Acta Anat. Nippon*, **36**, 161.

Hurely, L. S. and Cosens, G. (1974). *In*: 'Trace Element Metabolism in Animals', *Eds* W. G. Hoekstra, J. W. Suttie, H. E. Ganther, and W. Mertz. University Park Press, Baltimore, **2**, p. 516.

Johnson, M. and Everitt, B. (1984). 'Essential Reproduction'. Blackwell Scientific Publications, Oxford.

Jones, K. L., Smith, D. W., Streissguth, A. P., and Mirianthopoulos, N. C. (1973). *Lancet*, **1**, 1076.

Kennedy, L. S. and Persaud, T. V. N. (1979). *In*: 'Advances in the Study of Birth Defects'; *Ed.* T. V. N. Persaud. MTP Press, Lancaster, **2**, p. 223.

Knill-Jones, R. P., Moir, D. O., and Rodrigues, L. V. (1972) *Lancet* **1**, 1326.

Kochar, D. M. (1968). *Teratology*, **1**, 299.

Kolata, G. B. (1980). *Science*, **208**, 1240.

Lansdown, A. B. G. (1970). *Food Cosmet. Toxicol.*, **8**, 647.

Lansdown, A. B. G. (1975a). *Br. J. Exp. Pathol.*, **56**, 119.

Lansdown, A. B. G. (1975b). *Br. J. Exp. Pathol.*, **56**, 373.

Lansdown, A. B. G. (1976). *Teratology*, **13**, 291.

Lansdown, A. B. G. (1977). *Biol. Neonat.*, **31**, 252.

Lansdown, A. B. G. (1978). *In*: 'Advances in the Study of Birth Defects', *Ed.* T. V. N. Persaud, MTP Press, Lancaster, p. 292.

Lansdown, A. B. G. (1982). *Vet. Rec.*, **110**, 429.

Lansdown, A. B. G. (1983a). *Trends Pharmacol. Sci.*

Lansdown, A. B. G. (1983b). *In*: 'Toxic Hazards in Food', *Eds* D. M. Conning and A. B. G. Lansdown. Croom-Helm, London, p. 73.

Lansdown, A. B. G. (1987). *In*: 'The Future of Predictive Safety Evaluation', *Eds* A. N. Worden, D. V. Parke, and J. Marks. MTP Press, Lancaster, **2**, p. 77.

Lansdown, A. B. G. and Coid, C. R. (1974). *Br. J. Exp. Pathol.*, **55**, 101.

Lansdown, A. B. G., Coid, C. R., and Ramsden, D. B. (1975). *Nature (London)*, **254**, 599.

Lansdown, A. B. G. and Grasso, P. (1971). *J. Comp. Pathol.*, **81**, 141.

Lansdown, A. B. G., Pope, V. D. B., Halsey, M. J., and Bateman, P. E. (1976). *Teratology*, **13**, 299.

Le Chat, M. F., Borlee, I., Bouckaert, A., and Mission, C. (1980). *Science*, **207**, 1296.

Lechtig, A., Delghado, H., Lasky, R. E., Klein, R. E., Engle, R. L., Yarburgh, C., and Habicht, J. P. (1975). *Am. J. Dis. Child.* **129**, 434.

Lenaers, A., Ansay, M., Nusgens, B. V., and La Piere, C. M. (1971). *Eur. J. Biochem.*, **23**, 533.

Lenz, W. (1961). Diskussionsbemerkung Tagung der Rhein Westfah Kinderartzeverinigung, Dusseldorf, 11th November.

Lenz, W. and Knapp, K. (1962). *Deutsche, Med. Wochenschr.*, **7**, 253.

Levin, D. L., Stanger, P., Kitterman, J. A., and Heyman, M. A. (1975). *Circulation*, **52**, 500.

Loevy, H. (1962). *Anat. Rec.*, **142**, 375.

McBride, W. (1961). *Lancet*, **2**, 1358.

McClain, R. M. and Becker, B. A. (1975). *Toxicol. Appl. Pharmacol.*, **31**, 72.

Marin-Padilla, M. (1966). *J. Embryol. Exp. Morphol.*, **15**, 261.

Mellin, G. N. (1964). *Am. J. Obstet.*, **90**, 1169.

Menkes, B. and Deleanu, M. (1964). *Rev. Roum. Embryol. Cytol. Ser. Embryol.*, **1**, 69.

Mims, C. A. (1968) *Prog. Med. Virol.*, **10**, 194.

Moore, M. R., Hughes, M. A., and Goldberg, D. J. (1979). *Int. Arch. Occup. Health*, **44**, 81.

Morriss, G. M. (1972). *J. Anat.*, **113**, 241.

Naeye, R. L. and Blanc, W. (1965). *J. Am. Med. Assoc.*, **194**, 1277.

Naeye, R. L., Blanc, W., and Paul, C. (1973). *Pediatrician*, **52**, 492.

Nicholl, C. S. (1980). *Fed. Proc.*, **39**, 2563.

Nishimura, H., and Miyamoto, S. (1969). *Acta Anat.*, **74**, 121.

Overall, J. C. (1972). *Am. Heart J.*, **84**, 823.

Overall, J. C. and Glasgow, L. A. (1970). *J. Pediat.*, **77**, 315.

Palm, P. E., Arnold, E. P., Rachwell, R. C., Leyczek, J. C., Teague, K. W., and Kensler, C. J. (1978). *Toxicol. Appl. Pharmacol.*, **44**, 1.

Palmer, A. K. (1974). *Teratology.*, **10**, 301.

Pinsky, L. and Di George, A. M. (1965). *Science*, **147**, 203.

Pope, W. D. B., Halsey, M. J., Lansdown, A. B. G., Simmons, A., and Bateman, P. E. (1978). *Anaesthesiology*, **48**, 11.

Pope, W. D. B., Halsey, M. J., Lansdown, A. B. G., and Bateman, P. E. (1975). *Acta Anaesthesiol. Belg.*, **26**, 169.

Poswillo, D. E. (1975), *Br. Med. Bull.*, **31**, 101.

Rabinowitz, M. B., Wetherill, G. W., and Kopple, J. D. (1976). *J. Clin. Invest.*, **58**, 260.

Rest, J. R., (1987). *Vet. Rec.*, **31**, 426.

Ross, L. M. and Walker, B. (1967). *Am. J. Anat.*, **121**, 509.

St. Hilaire, E. G. (1832). *In*: 'Histoire generale et particuliere des animales de l'organisation chez l'homme et les animaux' *Eds* J. B. Baliere et fils, Paris.

Saunders, J. W. and Fallon, J. F., (1966). *In*: 'Major Problems in Developmental Biology', *Ed.* M. Locke, Academic Press, New York, p. 289.

Saxen, L. (1970). *Int. J. Gynec. Obstet.*, **8**, 784.

Saxen, L. (1976). *J. Embryol. Exp. Morphol.*, **36**, 1.

Saxen, L. and Kaitilia, I. (1972). *Adv. Exp. Med. Biol.*, **27**, 205.

Scandinavian Council for Environmental Information (1977). 'Workshop on the utility of an evaluated data bank in teratology and its use for a state-of-the-art report on adverse effects of lead on reproduction and developments', Nyashamn, Sweden.

Sever, J. L., Meier, G. W., Windle, W. F., Schiff, G. M., Monif, G. R., and Fabiyi, A. (1966). *J. Infect. Dis.*, **116**, 21.

Smith, R. L., and Caldwell, J. (1977). *In*: 'Drug Metabolism from Microbe to Man', *Eds* D. Parke and R. L. Smith, Taylor and Francis, London, p. 331.

Stadler, J., Kessidjan, M.-J., and Perraud, J. (1983). *Food Cosmet. Toxicol.*, **21**, 631.

Sternberg, S. S. (1979). *Pharmacol. Therap.*, **6**, 147.

Stockard, C. R. (1921). *Am. J. Anat.*, **28**, 115.

Swan, C. P., Tostevin, A. L., Moore, B., Mayo, H., and Black, G. H. B., (1943). *Med. J. Austr.*, **2**, 201.

Takeuchi, T. and Matsumoto, H. (1969). *In*: 'Methods for Teratological Studies in Experimental Animals and Man', *Eds* H. Nishimura and J. R. Miller, Igaku Shoin, Tokyo, p. 280.

Tuchmann-Duplessis, H. (1984). *Pharmacol. Therap.*, **26**, 273.

Ullberg, S. (1958). Proc. 2nd Int. Conf. on Peaceful Use of Atomic Energy, Sweden, **24**, 248.

Valdheim, C. M. (1985). *Diabetes*, **34**, 21A.

Valdheim, C. M. (1986). *N. Engl. J. Med.*, **315**, 1314.

Wildschutt, H. I. J., Tutein Nolthenius-Puyaert, M. C. B. J. E., Wiedijk, V., Treffers, P. E., and Huber, J. (1987). *Br. Med. J.*, **295**, 894.

Wilson, J. G. (1959). *J. Chronic Dis.*, **10**, 11.

Wilson, J. G. (1964a). *J. Cell. Comp. Physiol.*, **43**, 11.

Wilson, J. G., (1964b). *J. Pharmacol. Exp. Therap.*, **144**, 429.

Wilson, J. G. (1972). *Am. J. Anat.*, **136**, 129.

Wilson, J. G. (1977). *In*: 'Handbook of Teratology'., *Eds* J. G. Wilson and F. C. Fraser, Plenum, New York, 1, p. 47.

Wolpert, L. (1976). *Br. Med. Bull.*, **32**, 65.

Yamazaki, J., Wright, S., and Wright, P. (1954). *Am. J. Dis. Child.*, **87**, 448.

CHAPTER 13

Genetic Toxicology

DIANA ANDERSON

1 Introduction

Genetic toxicology is a branch of general toxicology which addresses the problems of toxicity to the DNA. It is a discipline which has arisen from the early studies of Müller (1927) and Auerbach (1947) in which irradiation and chemicals were shown to induce mutation in *Drosophila*, the fruit fly. When it was shown that mutations could be induced in mammals, the possibility arose that some of the hereditary diseases in man might be induced by environmental agents. Such materials include not only man-made chemicals, but natural carcinogens and mutagens found in fungi and plants and produced by cooking processes (Anderson and Purchase, 1983; Felton and Knize, 1991; IPCS, 1992; Rowland *et al.*, 1984; Sugimura, 1978; Wakabayashi *et al.*, 1993). The impact of these agents on human health of present and future generations could be of consequence if the extent of exposure is sufficient to permit expression of genotoxic properites in somatic or germ cells. Over the last generation many substances shown to be mutagenic were also shown to be carcinogenic, and early correlations as high as 90% between mutagenicity and carcinogenicity were claimed (McCann *et al.*, 1975a; 1976; Purchase, 1978; Sugimura, 1977). Such correlations stimulated a new interest in the somatic theory of cancer of the 1950s and has resulted in intensive study of the genetic effects of chemical substances.

It has been estimated that there is a genetic element in at least 10% of all human pathological conditions; thus genetic changes, if produced in man, could be of serious consequence. Abnormalities could arise as a result of either gene or chromosome mutation (Carter, 1977). McKusick (1975) has segregated 'traits' inherited at the level of the gene as dominant or recessive, and as autosomal or sex-linked. A dominant gene is immediately expressed in the next generation (*e.g.* Huntingdon's chorea and achondroplasia). A recessive mutation may take many generations to be expressed (*e.g.* phenylketonuria) except for sex-linked recessives (*e.g.* haemophilia). Many constitutional and degenerative

diseases such as epilepsy or schizophrenia could be caused by irregularities of gene expression or arise from multiple genes.

Chromosome abnormalities may arise either by errors in the distribution of chromosomes leading to abnormalities of chromosome numbers,
such as non-disjunction [Down's syndrome (mongolism) and Klinefelter's
and Turner's syndromes], where the effect is seen in the next generation;
or chromosome breakage (ataxia telangiectasia and Fanconi's anaemia)
which arise from recessive autosomal gene changes. Some chromosome
changes are thought to give rise to early embryonic loss and thus have
little genetic impact on society. Deleterious mutations that are compatible with survival may be a heavy burden to society if the affected
person requires medical or institutional care (mongolism, for example,
which occurs in 1 in 1000 births).

It is not possible to give a quantitative estimate of the contribution of
chemical agents to the incidence of genetic disease, but that they could
constitute an aetiological factor in such disease must be accepted. Hence,
in toxicological assessments, there is a need to define by the appropriate
tests the mutagenic activity of chemical substances.

This is a complex and difficult task, and no single method gives
conclusive information about genetic risk. A variety of test systems has
been put forward, and some governmental agencies have produced
guidelines for recommended methods (Department of Health and Social
Security (DHSS), UK, 1981; Department of Health (DoH), 1989; Official
Journal of the EEC, 1979; and the OECD, 1983).

Structurally DNA, the target of mutagenic action, is identical in all
living organisms but there are differences in the organisation of the
genetic material in different species so that some organisms may be
considered to constitute better predictive models for man. There are
many interactions between chemicals and organisms which may determine whether genetic damage is expressed—unlike ionising irradiation,
which is immediately active in terms of mutagenic potential. Irradiation
generates extremely short-lived free radicals, some of which are close to
or within the genetic target molecules, whereas there are many factors
that influence whether a chemical compound may reach or react with
such targets. Among such factors are the chemical structure of the
compound, the duration of treatment, the route of administration,
absorption, distribution, and excretion, the macromolecular binding, the
metabolic transformation (which may all be dependent on species, strain,
and sex), the membrane barriers, and the presence of pertinent defence
mechanisms.

Other factors determine whether the damage is expressed as genetic
damage. Such factors include the innate susceptibility of the cell (*i.e.*
genetic repair capacity), the numbers of susceptible cells, the type of
genetic target, the selection processes involved, and the mode of
inheritance (Matter, 1976).

Many compounds that require testing may not bind covalently with biologically important molecules unless they are first metabolised to highly reactive forms. Others disturb other functions and cause a genetic effect indirectly. Thus bases and nucleosides can themselves be mutagenic, an effect thought to arise as a result of unbalanced DNA precursor pools (Anderson *et al.*, 1981; Meuth, 1984; Clode *et al.*, 1986; Clode and Anderson *et al.*, 1988a and b). Clearly the opportunities that theoretically exist for a given compound to attack genetic material are widespread, as are the obstacles and defences against such an attack being effective. As a consequence, the methods available for assessing such effects are numerous and in many respects very specialised.

2   Mutagenicity Tests

The mutagenic activity of chemical substances has been studied in a number of test systems including micro-organisms, plants, insects, mammalian, and human cells using *in vitro* and *in vivo* techniques. The mutagenic effect detectable in one system may not be found in another or even in different tissues of the same system.

The main sub-mammalian test systems currently available are those which reflect the present state of knowledge and may be improved with the progress of experimental work in this field. Not all of the available tests are described here. Three categories are used: (a) tests for gene mutations; (b) tests for chromosome damage including aneuploidy/non-disjunction; (c) tests for DNA repair. The tests and cells and organisms used in them are listed in Table 1.

A   Gene Mutation Tests

In molecular terms, gene mutations consist of the substitution, deletion or insertion of one or more nucleotide base-pairs in DNA. Mutations due to substitution are known as base-pair substitutions and those due to deletions or insertions as frameshifts. When a base-pair substitution mutation occurs, a wrong base is inserted, which then pairs with its natural partner during replication (adenine with thymine, cytosine with guanine) so that a new pair of incorrect bases is inserted into the DNA. When a frameshift mutation occurs, if there is base-pair loss, the messenger DNA, which reads the DNA in triplet codons, incorporates the first base-pair from the next triplet codon. Thus the subsequent code of the DNA becomes scrambled until a 'nonsense' or terminating codon is reached. A similar process happens with a base-pair gain.

Gene (or point) mutations cannot be detected by cytological methods but can be distinguished as variants of characteristics controlled by specific gene loci or as recessive lethal conditions (generally linked to sex chromosomes). Mutations of specific sites can arise in either a 'backward' (from mutant to normal 'wild' type) or a 'forward' direction (from a wild type to mutant).

Table 1   *Tests, cells, and organisms used in mutagenicity tests*

Mutagenicity Tests

Gene mutation tests	Chromosome damage tests	DNA repair tests
Microbial tests	Microbial tests	Microbial tests

Gene mutation tests

Microbial tests

Bacteria
Salmonella typhimurium
Histidine independence
(reverse mutation)
8-Azaguanine resistance
Arabinose resistance
(forward mutation)
Escherichia coli
Tryptophan independence
(reverse mutation)
5-Methyltryptophan
Streptomycin
(forward mutation)
Prophage reduction
Fluctation test

Fungi
Saccharomyces cerevisiae
Schizosaccharomyces pombe
Neurospora crassa
Aspergillus nidulans

Insect test
Drosophila

Plants
Tradescantia hair

Mammalian cells *in vitro*
HPRT ase locus
Chinese hamster V79
Chinese hamster ovary
Human lymphocytes
Human fibroblasts
TK ase locus
L5178Y
P388F

Mammalian cells *in vivo*
The spot test
The specific locus test
The dominant skeletal test
The cataract test

Others
Biochemically based tests
involving protein change,
histocompatability loci,
Pigment loci
Inversion techniques to
measure recessive lethals
Sperm morphology changes
Transgenic mice

Chromosome damage tests

Microbial tests

Nondisjunction in fungi
Chromosome loss
Saccharomyces cerevisiae
Aspergillus nidulans
Sordaria breviocollis

Plants
Vicia faba
Allium cepa
Tradescantia
Zea Mays
Tricitum vulgare
Hordeum vulgare
Lycopersium exulentum
Nicotiania tobacum
Pisum sativum

Insects

Drosophila melanogaster
Bombyx mori

Mammalian cells *in vitro*
Chinese hamster ovary cells
Human fibroblasts
Human lymphocytes

Mammalian cells *in vivo*
The rat bone marro metaphase
assay
The micronucleus test
Dominant lethal assay
The heritable translocation
test
Direct spermatocyte test
Sex chromosome losses
The fertilised oocyte test

DNA repair tests

Microbial tests

Spot tests for
inhibition zones
Salmonella tyhimurium
Bacillus subtilis
Escherichia coli
Saccaromyces cerevisiae
Mitotic recombination
Gene conversion
Saccharomyces cerevisiae
*Schizosaccharomyces
pombe*
Unscheduled DNA
synthesis
in mammalian cells
in vitro and *in vivo*
synthesis
Breakage of single chain
DNA in individual cells
using gradient or elution
techniques
Single strand DNA breaks
and alkali labile damage
individual cells using
single cell gel electro-
phoresis (COMET assay)

Backward (reverse) mutation is generally studied in mutant cells for which the base-pair substitutions or frameshift mutations are known. Normal functional activity is re-established following a new substitution, or a second deletion and insertion near the first one. Backward mutation is highly specific in the type of DNA interaction required, and a chemical compound that does not increase the frequency of this type of mutation may nevertheless still cause other genetic effects.

Forward mutations may arise from effects which range from the substitution, insertion, or deletion of the nucleotide bases of a gene, to the deletion of the entire gene and neighbouring genes. They are detected when there is a loss of an enzyme function (auxotrophy) or from acquired resistance to various toxic chemicals, substrate analogues and agents such as heat. In some instances where only one specific locus may be affected, the forward mutation can be just as specific as the backward mutation.

The specificity of some of these mutation systems, together with secondary factors such as the effect of the particular nucleotide sequence near the original mutation, makes a quantitative comparison or a between-species extrapolation of the mutagenicity of different chemicals very difficult. For a list of other factors that can influence effects *in vitro*, see Ashby and Styles (1978) and Brusick (1987).

Bacterial Tests

Point mutations can be detected in various bacterial species. Those most commonly used are *Salmonella typhimurium* and *Escherichia coli*.

In addition to its use in the reverse mutation assay of Ames, detecting independence to histidine as a medium nutrient (Ames *et al.*, 1973; 1975; McCann *et al.*, 1975a and b; Ashby and Tennant, 1988; McCann and Ames, 1976; Maron and Ames, 1983; Tennant *et al.*, 1987; Zeiger, 1987) *S. typhimurium* can be used to detect forward mutations to 8-azaguanine resistance (Skopek *et al.*, 1978) and arabinose resistance (Peuyo, 1978). The Ames test has been reviewed in the EPA Gene-Tox Programme (Kier *et al.*, 1986) and by Gatehouse *et al.*, 1990. Strains most commonly used are TA1535, TA1537, TA1538, TA97, TA98, TA100, and TA102.

E. coli has also been reviewed in the EPA Gene-Tox Programme (Brusick *et al.*, 1980) and can detect (a) the induction of mutations from arginine dependence to independence (base-pair substitution reverse mutation) (Mohn *et al.*, 1974) or from tryptophan dependence to independence (base-pair and frameshift reverse mutation) (Green and Muriel, 1976; Venitt and Crofton-Sleigh, 1981; Venitt *et al.*, 1983); (b) the induction of forward mutations from inability to ability to ferment galactose (Saedler *et al.*, 1968) and from sensitivity to resistance to 5-methyltryptophan (Mohn, 1973) or streptomycin (Wild, 1973). These forward systems allow the detection of base-pair substitutions, frameshift mutations, or small deletions; (c) the induction of prophage (inductest), which induces lysis of bacterial cells harbouring a latent bacteriophage, through changes that lead to the destruction or deactivation of the phage repressor. These effects can be observed in a single strain which carries the necessary genetic markers and the bacteriophage in the prophage form (Ho and Ho, 1981; Moreau *et al.*, 1976; Speck *et al.*, 1978); (d) reverse fluctuation test (Green *et al.*, 1976).

Tests using DNA-repair deficient bacterial strains have been reviewed by Leifer *et al.* (1981) (Gene-Tox).

Fungal Tests

The yeasts *Saccharomyces cerevisiae* and *Schizosaccharomyces pombe* and the Ascomycetes *Neurospora crassa* and *Aspergillus nidulans* are among the fungi most used for the detection of point mutations.

In the yeasts *Saccharomyces cerevisiae* and *Schizosaccharomyces pombe*, mutations at each of the two genetic loci that control adenine biosynthesis cause a red pigmentation of the colonies. In *Saccharomyces cerevisiae*, the effect has been used to develop forward mutation tests either in haploid or diploid strains (Mortimer and Manney, 1971). In this system it is possible to distinguish (by established morphological criteria) the effect of the mutation from that of mitotic recombination (Zimmerman, 1973; 1975; Zimmerman *et al.*, 1984, Gene-Tox).

Schizosaccharomyces pombe has been used to detect forward mutation in wild or adenine-dependent haploid strains where mutations have been introduced that increase sensitivity to chemical mutagens (Mortimer and Manney, 1971). The use of these two yeasts for mutagenicity assays has been extensively reviewed (Loprieno *et al.*, 1974; Loprieno *et al.*, 1983, Gene-Tox; Zimmerman, 1973; 1975). A system of forward mutation from canavanine sensitivity to resistance (Brusick, 1972) has also been developed in *Saccharomyces cerevisiae* and the organism can also be used in a fluctuation test, for example, the D-7 strain is used for detecting isoleucine dependence (Parry, 1977a and b; Parry *et al.*, 1984).

Strains of *Neurospora crassa* can also detect the induction of forward mutation in each of two genetic loci which control adenine biosynthesis (Ong and de Serres, 1972). In addition, strains have been developed that allow identification of recessive point mutations at each of the two genetic loci mentioned, dominant lethal mutations in the genetic region in which the two loci are situated and recessive mutations in the whole genome (Brockman *et al.*, 1984, Gene-Tox; de Serres and Malling, 1971; Ong and de Serres, 1972). Strains of aspergillus can be used for the induction of forward mutation to 8-azaguanine and *p*-fluorophenylalanine and back mutation to methionine independence (Roper, 1971). Aspergillus haploid and diploid strains, respectively, have been reviewed in the gene-tox programme (Kafer *et al.*, 1982; Scott *et al.*, 1982).

Insect Tests

The most widely used point mutation test in *Drosophila melanogaster* is that which detects the induction of sex-linked recessive lethals. These become evident in the second generation of treated individuals and can be observed at much lower doses than those needed to induce chromosome loss or dominant lethal mutations in this system. The presence of the X chromosome (which is about 20% of the genome) makes it possible to classify most of the sex-linked recessive lethals as gene mutations or multi-gene deletions. Autosomal recessive lethal mutations can be observed in flies of second generation. The induction of visible mutations is measured by crossing wild males (treated or control) with females that are homozygous for the visible mutations. The function of Drosophila in genetic toxicology testing has been described by Abrahamson and Lewis (1971), Vogel and Sobels (1976) and others (Auerbach, 1977; Baker *et al.*, 1976; Bootman and Kilbey, 1983; Lee *et al.*, 1983, Gene-Tox; Millet and Weilenmann, 1976; Mollet and Wurgler, 1974; Valencia *et al.*, 1984, Gene-Tox; Vogel, 1977; Wurgler *et al.*, 1977).

Mammalian Cell Tests In Vitro

There are various genetic systems for the detection of point mutations in cultures of mammalian cells (Cole *et al.*, 1990). They are based on the use of phenotypic markers generally resulting from resistance to antimetabolites or temperature sensitivity. In addition to gene mutation, variants of somatic cells in culture can be generated by other effects such as gene suppression. Hamster, mouse, and human diploid fibroblasts are among the cell lines most often used to detect genetic changes. For some markers many cell lines may be used. For example, the locus controlling the hypoxanthine guanine phosphoribosyl transferase (HPRT) enzyme occurs in Chinese hamster embryo lung cells [V-79 cells] (Arlett, 1977; Bradley *et al.*, 1982, Gene-Tox; Chu *et al.*, 1976; Cole *et al.*, 1983; Fox *et al.*, 1976; Huberman and Sachs, 1974; Krahn and Heidelberg, 1977), Chinese hamster ovary cells [CHO cells] (Hsie *et al.*, 1981, Gene-Tox; Neill *et al.*, 1977) and primary cultures of lung cells (Dean and Senner, 1977), human fibroblast (Jacob and Mars, 1977) and lymphoblast cells (Thilly *et al.*, 1976). The locus controlling adenine phosphoribosyl transferase (APRT) is present in human fibroblasts (de Mars, 1974). The thymidine kinase (TK) locus occurs in L5178Y (Clive, 1973; Clive *et al.*, 1972; Clive *et al.*, 1983, Gene-Tox; Clive and Spector, 1975; Cole *et al.*, 1983; Knaap and Simons, 1975; Nakamura *et al.*, 1977) and P388F mouse lymphoma cells (Anderson, 1975a; Anderson and Cross, 1982; Anderson and Cross, 1985; Anderson and Fox, 1974; Fox and Anderson, 1976).

A fluctuation test (Cole *et al.*, 1976) and a host-mediated assay (Fischer *et al.*, 1974) have been developed using L5178Y cells, and other cell lines from the mouse have been employed for mutagenesis assays (Anderson, 1975b).

Tests for measuring cell transformation are also available (Heidelberger *et al.*, 1983, Gene-Tox).

Mammalian Cell Tests In Vivo

The 'spot' test (Fahrig, 1978; Russel *et al.,* 1981b, Gene-Tox). This test detects somatic mutations. Mice that are heterozygous for certain coat colour genes are treated *in utero* by administration of the test compound to the mother. If a mutation occurs during the development of the hair follicles, this may be observed in the mouse when the coat is fully developed. Thus, the female is treated during pregnancy and the offspring grown until the coat colour has fully developed. Each mouse is then examined for the coat markings indicative of a clone arising from a single mutant cell. This system has the advantage of being one of the few assays for gene mutations in mammals. The main disadvantages are that the somatic mutations for coat colour may occur as a consequence, not of gene mutations, but of chromosomal mutations, and the coloured spots can be difficult to distinguish from the natural pigmentation zones of the coat.

The specific locus test (Russel, 1951; Russel *et al.,* 1981a, Gene-Tox; Searle, 1975). This test in mice is at present the only one available that can detect heritable mutations in mammalian germ cells. The test uses mouse strains, homozygous for certain dominant wild-type alleles and strains that are homozygous for recessive alleles at these loci. The dominant, wild-type strains are treated and then mated with the recessive strains. Changes induced at any of the loci cause visible changes in the offspring, in respect of eye colour, ear size, and coat colour. Although it is recognised that these events are gene mutations, their exact nature cannot be identified in most instances and they are assumed to be small multi-locus deletions. The main disadvantage of the specific locus test is that the spontaneous rate for these loci is low. Hence, for a negative result to be recorded, large numbers of offspring (up to 30,000) need to be scored, making the test extremely costly and laborious.

Biochemically-based tests. There are a number of biochemical tests for detecting gene mutations in somatic and germ cells in mammals. Enzyme activity can be used as a genetic marker to detect induced microlesions in mammals. They depend on variations in specific proteins caused by mutations in the genes controlling the synthesis of the enzymes or proteins (Mays *et al.,* 1978; Neel, 1979; Neel *et al.,* 1979; Valcovic and Malling, 1973).

Other approaches to detection of mutants in mice include changes in mandible shape (Festing and Wolfe, 1979), skeletal mutations (Ehling, 1970; Selby and Selby, 1977), histocompatibility loci (Kohn, 1973), recessive lethals measured by the inversion technique (Evans and Phillips, 1975; Roderick, 1979) and pigment loci (Searle, 1979).

Transgenic animals tests. Transgenic animals provide efficient systems for detecting gene mutations *in vivo.* Such systems rely on the use of a bacteriophage lambda shuttle vector which has been integrated into the genome of an inbred mouse *via* microinjection so that as division occurs every somatic and germ cell in the animal contains this gene which is quickly recovered using standard laboratory techniques. Two systems are commercially available—Mutamouse and the Big Blue mouse. Mutamouse was created by incorporating DNA from the bacteriophage lambda gt10 lacZ into the genome of the CD2 hybrid mouse. Each Mutamouse cell contains approximately 40 copies of the transgene arranged head-to-tail at a single insertion site. Mutamouse has become homozygous for the transgene and contains approximately 80 lac Z sites within every normal diploid cell and 40 per haploid gamete.

In the case of the Big Blue mouse the target organ is the Lac1^q gene in C57B16/6 mouse strain.

These Lac genes are susceptible to mutation like any other endogenous genes in the cell. After the transgenic mouse is exposed to a test compound, genomic DNA is isolated from the desired tissues such as bone marrow, germ tissue, and liver. The integrated lambda vectors are then easily and efficiently recovered by subjecting the genomic DNA to an *in vitro* packaging extract (proprietorial) which excises the vector DNA and packages it into a lambda phage head. Each packaged phage is then used to infect *E. coli* cells. The packing extract is used to give high target gene rescue efficiencies.

When *E. coli* infected with the lambda vectors are plated and incubated (18 hr) on indicator agar dishes containing the chromagenic substrate X-gal, mutations with the Lac1^q target gene cause the formation of blue plaques, while unmutated or intact Lac1^q targets result in clear plaques. In the case of the gt10 LacZ gene (Mutamouse) mutations are the opposite from those in the Big Blue mouse, *i.e.* they form clear plaques while the unmutated ones are blue.

Details of the transgenic systems have been described in several publications (Kohler *et al.,* 1990; Kohler *et al.,* 1991a and b).

It has been found, for example, that for male germ cells the spontaneous mutation frequency is lower in these cells than in somatic tissues (Kohler *et al.,* 1991a).

Transgenic rats containing the same lambda/Lac1 shuttle vector have been developed for inter-species comparison of mutagenesis testing results, which may offer a better understanding of the specific mechanisms involved in mutagenesis at the molecular level *in vivo* (Provost *et al.,* 1993).

The Use of Metabolic Activation Systems

Microbial and mammalian cell systems *in vitro* generally do not possess the metabolic capability of cells of the intact animal. Several supplementary systems have been developed for use with microbial and mammalian test systems.

The most widely used systems for metabolic activation is the rat liver S-9 fraction supplemented with the co-factors NADP and glucose-6-

phosphate (Ames *et al.*, 1973; Ames *et al.*, 1975; Maron and Ames, 1983; Kier *et al.*, 1986, Gene-Tox). The S-9 fraction is the supernatant fraction of the liver homogenate obtained by centrifugation at 9000 g for 10 min. It contains a preponderance of microsomes and the attendant oxidases. Before the fraction is prepared, the animals are commonly treated with inducing agents such as polychlorinated biphenyls (PCB), phenobarbital (PB), or 3-methylcholanthrene. PCB can induce microsomal enzymes capable of metabolising many different types of environmental mutagens and carcinogens, but a combination of β-naphthoflavone and PB is to be preferred (Matsushima *et al.*, 1976).

Rat liver is the most widely used source but the livers from other species, including man, have also been used. The use of human liver might be important in evaluating the risk of environmental mutagens and carcinogens to man (Bartsch *et al.*, 1975; Tang and Friedman, 1977).

Several experimental procedures exist for microbial systems: (a) *the liquid method*. The bacteria are incubated with the S-9 mix and the test substance, washed, and the revertants and surviving bacteria counted; (b) In *the plate method* (Ames *et al.*, 1975; Maron and Ames, 1973), the mixture of bacteria, S-9 mix and test substance is added directly to molten-soft agar and poured onto hard agar. This method is relatively simple and quick; (c) In a variant of this procedure (Nagao *et al.*, 1978; Sugimura *et al.*, 1976) the mixture of bacteria, test substance and S-9 mix is incubated for 20 min at 37°C before adding molten-soft agar. With this *pre-incubation method*, dimethylnitrosamine, a definite mutagen and carcinogen, gives positive results (Sugimura *et al.*, 1976) which are not readily detected by other methods.

An alternative approach, where mammalian cells in culture constitute the test system, utilises 'feeder' layers of mammalian cells (Huberman, 1975). These are usually of embryonic origin, with their enzyme system fully intact but with cell replication inactivated by radiation or a high dose of an antimitotic agent such as mitomycin C.

A method involving *in vivo* metabolic activation of a drug has been reported (Legator *et al.*, 1982, Gene-Tox; Legator and Malling, 1971; Legator *et al.*, 1977). The test substance is administered to animals by gastric tube or subcutaneous injection and the test bacteria (indicator organisms) are injected intraperitoneally, withdrawn after several hours, and examined for mutation using an *in vitro* assay. This 'host-mediated assay' has been used to investigate the mutagenicity of undefined or unsterilised material thought to contain mutagens and carcinogens. The test organism may also be given intravenously or into the testes. This assay is relatively insensitive by comparison with the plate incorporation assay. High doses of test compound have to be used to produce a response but it is conceivable that this method gives a realistic measure of the mutagenic impact *in vivo*. Clearly the antibacterial activity of the host may be a confounding factor. *Salmonella typhimurium* (Legator and Malling, 1971), *Escherichia coli* (Mohn and Ellenberger, 1973),

Saccharomyces cerevisiae (Fahrig, 1974; 1977), *Schizosaccharomyces pombe* (Legator and Malling, 1971), *Neurospora crassa* (Legator and Malling, 1971), and mouse lymphoma cells (Fischer *et al.*, 1974) have all been used as indicator organisms.

Chemical substances which have undergone metabolism are excreted in the urine as degradation or conjugated products and as such could be identified by suitable microbial indicators, after deconjugation (Commoner *et al.*, 1974; Durston and Ames, 1974). Although of little value in the detection of new mutagens, analysis of the mutagenic activity of the urine or body fluids (Legator *et al.*, 1982, Gene-Tox; Legator *et al.*, 1976; Siebert, 1973) or faeces (Bruce *et al.*, 1977; Combes *et al.*, 1984) might be of value to screen workers with known exposures.

Metabolic activation systems have been described with tests for gene mutation, but can also be applied with tests for chromosome damage and DNA repair.

B Chromosome Damage Tests

Chromosome damage is defined as a modification of the number or structure of chromosomes. It can be detected both by cytological and genetic methods.

Variations in the number of chromosomes (aneuploidy and polyploidy) may result from endoreduplication (continued chromosome division), metaphase arrest, anaphase retardation, and non-disjunction in mitosis and meiosis.

Structural changes (Figure 1) are mainly the result of breaks in the chromatid arms. Depending on the number of breaks and the way in which they may be rejoined a series of unstable structural modifications may arise that are not transmitted to successive cellular generations. These include 'gaps' (achromatic interruptions, but see Anderson and Richardson, 1981), breaks of one or both the chromatids, chromatid interchanges, acentric fragments, ring and dicentric chromosomes. Stable structural modifications that are transmissible may also arise, such as inversions, translocations, and deletions (Evans, 1976; Buckton and Evans, 1973; Preston *et al.*, 1981, Gene Tox; Scott *et al.*, 1983; Scott *et al.*, 1990).

Microbial Tests

It is possible to measure non-disjunction in micro-organisms. Tests for detecting mitotic non-disjunction have been developed using the fungus *Aspergillus nidulans* (Bignami *et al.*, 1974; Kafer *et al.*, 1976; Kafer *et al.*, 1982, Gene-Tox; Scott *et al.*, 1982, Gene-Tox) and the yeast *Saccharomyces cerevisiae* (Parry *et al.*, 1977b; Parry *et al.*, 1984; Parry and Zimmerman, 1976; Sora and Magni, 1988; Zimmerman, 1975); meiotic non-disjunction can be measured in the fungus *Sordaria brevicollis* (Bond, 1976).

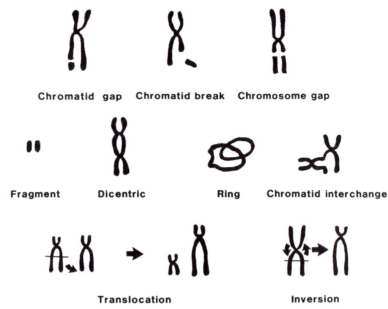

Figure 1   *Diagram of different categories of chromosome damage*

Mitotic chromosome loss has also been examined in *Saccharomyces cerevisiae* (D61M) (Zimmerman *et al.*, 1985). The test relies upon the recovery and expression of multiple recessive markers reflecting the presumptive loss of the chromosome VII homologue carrying the corresponding wild type alleles. Deviations from the conventional aneuploid or diploid state often causes serious perturbations in normal growth and development. Whittaker *et al.* (1989; 1990) have carried out an interlaboratory assessment of this assay. Albertini (1989) has examined chromosomal mal-segregation and aneuploidy and Albertini *et al.* (1993) have examined various chemicals in the EEC aneuploidy programme.

Plant Tests

The radical apices (root tips) of *Vicia faba,* broad bean, (Kihlman, 1971; Te-Hsiu Ma, 1982a, Gene-Tox) *Allium cepa,* common onion (Gran, 1982, Gene-Tox; Kihlman, 1971) and some species of the genus *Tradescantia* (Marimuthu, 1970; Te-Hsiu Ma, 1982b, Gene-Tox; Van't Hof and Schairer, 1982, Gene-Tox) can be used for the detection of chromosome aberrations, but such systems may not be suitable for extrapolation to mammalian systems because different degrees of chromosome damage have been recorded in onion root tips when compared with Chinese hamster cells, due possibly to variation in the efficiency of the repair system (Kihlman, 1971). Other plants, such as barley, offer systems for the analysis of chromosomal aberrations (Constantin and

Nilan, 1982a and b, Gene-Tox; Ehrenberg, 1971; Nilan and Vig, 1976; Plewa, 1982, Gene-Tox; Vig, 1982, Gene-Tox). The role of plants for genetic and cytogenic screening has been reviewed by Constantin and Owens (1982, Gene-Tox).

Insect Tests

Effects at the chromosome level such as non-disjunction, loss of the X chromosome, deletions, translocations, dominant lethal mutations, and mitotic and meiotic recombination have been observed in *Drosophila melanogaster* (see earlier references, also Valencia *et al.*, 1984, Gene-Tox; Wurgler *et al.*, 1977). Changes are generally shown by observing the phenotypes of the progeny from the appropriate crosses. For example, reciprocal translocations may be detected in the second generation of treated individuals carrying defined recessive markers in the autosomes. The absence of any of the expected phenotypes in the progeny indicates that the translocations occurred in the parent reproductive cells. The loss or acquisition of a chromosome, or the loss of part of a chromosome, can be detected by observing phenotypes in the progeny of crosses in which only one of the parents have been treated.

Mammalian Cell Tests In Vitro

Chromosome damage can easily be observed by cytological methods in mammalian cells in culture (Preston *et al.*, 1981, Gene-Tox; Scott *et al.*, 1983). Human fibroblasts, lymphocytes, and rodent cells (hamster fibroblasts and mouse lymphocytes) are used. Translocations, inversions, and other stable rearrangements indicate genetic damage that can be inherited. Gaps probably indicate early toxic effects unless accompanied by other types of damage (Anderson and Richardson, 1981) but have in fact persisted for three months when other categories of damage have been eliminated after vinyl chloride treatment.

Sister chromatid exchange. Sister chromatid exchange in somatic mammalian cells is a measure of the reciprocal exchange in segments of homologous loci between sister chromatids, *i.e.* the number of crossovers during replication that occur between paired chromatids following treatment with the substance (Latt *et al.*, 1981, Gene-Tox; Perry and Evans, 1975; Perry *et al.*, 1984; Stetka and Wolff, 1976). Cells are exposed to 5-bromodeoxyuridine for one round of replication. A second round without it follows, so that one arm of the chromosome is a substituted chromatid (these are recognised by differential staining techniques) and the frequency of such chromatid exchanges is increased by mutagens. It is considered to be a sensitive method for detecting chromosome damage but the significance of the test is not yet fully understood. It is considered as a test that indirectly measures DNA damage which involves repair.

Mammalian Cell Tests In Vivo

Methods exist for the assay of chromosomal aberrations in somatic cells after the administration of a test chemical to the intact animal (Albanese *et al.*, 1984; Preston *et al.*, 1981, Gene-Tox; Richold *et al.*, 1990; Topham *et al.*, 1983). Peripheral lymphocytes (Lilly *et al.*, 1975) and bone marrow cells (Cohen and Hirschorn, 1971; Legator *et al.*, 1973; Nichols *et al.*, 1973; Richold *et al.*, 1990; Schmid *et al.*, 1971; Tijo and Whang, 1962) are commonly used. Bone marrow cells are a naturally proliferating cell population but peripheral lymphocytes are a synchronised population of G_o cells that rarely proliferate and have to be stimulated by mitogens for analysis. Such systems have large numbers of cells available and show good potential for analysis of human cells, provided solid baseline data are available (Anderson *et al.*, 1988).

The micronucleus test. In anaphase chromosomes, abnormal chromosome or chromatid fragments lag behind the other migrating chromosomes, and deformed anaphase bridges occur. At telophase, these lagging elements form micronuclei (Heddle, 1973; Heddle *et al.*, 1983, Gene-Tox; Maier and Schmid, 1976; Salamone *et al.*, 1980; Schmid, 1976) because they do not become incorporated into the nuclei of the daughter cells. Micronuclei are most often observed in polychromatic erythrocytes but they can be observed in embryonic and other cells in culture by suitable staining techniques. The test is also of value in the detection of mitotic spindle poisons. There has been debate about the number of cells to be analysed, up to 2000 recommended for suitable sensitivity (Ashby and Mirkova, 1987).

Dominant lethal test. Basically, this assay depends on an increase in the foetal abortion rate due to defective sperm or ova, measured as a decrease in the numbers of uterine implants or an increase in early foetal death, *i.e.* pre- and post-implantation loss, respectively (Anderson *et al.*, 1983; Bateman, 1977; Bateman and Epstein, 1971; Ehling *et al.*, 1978; Epstein and Rohrborn, 1970; Green *et al.*, 1985, Gene-Tox). An effect on fertility may be measured in the same study. Pre-implantation losses can occur due to other than genetic reasons (including lack of fertility) but the early implantation death is thought to be due to chromosomal abnormalities produced in germ cells. In order to sample all stages of sensitivity of the mating germ cell, an 8-week mating period is required in the mouse and a 10-week period in the rat (Epstein and Rohrborn, 1970), these being the times taken to produce mature sperm from the germinal cells.

The assay may be extended by examining surviving offspring for congenital malformations (Anderson *et al.*, 1993; Frances *et al.*, 1990; Jenkinson *et al.*, 1987; Jenkinson *et al.*, 1990; Knudsen *et al.*, 1977).

The dominant lethal assay can also be carried out using Drosophila (Wurgler *et al.*, 1977) (see earlier references).

Sperm-head morphology. Some mutagens give rise to abnormally shaped spermatocytes (Topham, 1980b; Wyrobek and Bruce, 1975) and this can result in inherited sperm abnormalities after parental treatment in animals (Topham, 1980b; Wyrobek *et al.*, 1983a, Gene-Tox). An evaluation has been made of human sperm as indicators of chemically-induced alterations of spermatogenic function (Wyrobek *et al.*, 1983b, Gene-Tox; Wyrobek *et al.*, 1984).

Heritable translocation test. The presence of a heritable translocation (*i.e.* a balanced chromosome translocation) is thought to confer sterility or partial sterility on the F_1 offspring. Its presence can be confirmed either cytogenetically or by mating on several occasions suspect males who will continue to produce litters of reduced numbers (Cachiero *et al.*, 1974; Generoso *et al.*, 1977, Gene-Tox; Generoso *et al.*, 1980).

Direct spermatocyte test. Spermatogonia are examined at or near metaphase 1 for the abnormal mitotic figures that result from reciprocal translocation (Léonard, 1975; 1977).

Sex chromosome losses. The test also measures the heritability of chromosomal damage by scoring the progeny of exposed animals for infertility (Russel, 1979).

The fertilised oocyte test. The first cleavage embryos from treated male and female germ cells are scored for structural chromosome damage. As the male and female pronuclei condense at different times, both genomes can be examined. Chromosome aberrations in the male pronuclei have been shown to correlate with dominant lethality (Brewen *et al.* 1975; Albanese, 1987).

The various germ cell methods have been reviewed by Albenese (1987).

C DNA Damage and Repair Tests

When DNA is damaged, repair normally follows (Cleaver, 1975; 1977; Larsen *et al.*, 1982, Gene-Tox). The initial lesions in DNA may be lethal, may remain without being repaired, may be repaired correctly to restore a normal genome, or incorrectly to produce errors and an abnormal genome. There are several types of repair possibilities.

Excision repair. Cut and patch or pre-replication repair occurs when a lesion, recognised by an endonuclease, is excised by an exonuclease and the missing part re-synthesised by a polymerase to reconstitute the original strand. The new part is joined to existing DNA by a ligase. Excision repair is normally error-free and is known to occur in human cells. Hydroxyurea has been used to inhibit normal DNA replication, which then allows the detection of excision repair, but it is now suspected that this compound itself may have some effect on excision repair so that the results from experiments using hydroxyurea should be considered with caution (Pearson and Styles, 1984).

Post-replication (by-pass) repair. The lesion is by-passed in newly synthesised daughter DNA, thus leaving a gap that is sealed by insertion of a DNA segment by recombination into the new daughter DNA. This process is error-prone. The post-replication gap may be filled by DNA synthesised *de novo* and thus correct errors copied by replication.

Recombination repair is a post-replication repair process in bacteria which involves the recombination of daughter strands of DNA to reconstruct the correct genome. This process is error-free, but error can occur if the repair requires *de novo* synthesis. This has been established in bacteria but has not been demonstrated conclusively in mammalian cells.

Damage to the DNA molecule may be considered as a primary lesion that could be involved in the process of mutation and the extent of the repair is an indicator of the amount of damage that has occurred to DNA.

Several methods exist for detecting DNA repair phenomena. These include differential zones of inhibition or killing in bacterial strains with and without repair processes (Ichinotsubo *et al.*, 1977; Leifer *et al.*, 1981, Gene-Tox; Tanooka, 1977) gene conversion in strains of yeast (Zimmerman, 1973; Zimmerman *et al.*, 1984, Gene-Tox), sister chromatid exchange in mammalian cells (Latt *et al.*, 1981, Gene-Tox; Perry and Evans, 1975; Perry *et al.*, 1984; Steka and Wolff, 1976) and the direct measurement of DNA damage and repair (Cleaver, 1975; 1977; Larsen *et al.*, 1982, Gene-Tox).

Bacterial Tests

'Spot' tests measuring differential zones of inhibition (Tweats *et al.*, 1984). A Petri dish is seeded or streaked with the test organism (*Salmonella typhimurium*) (Ames *et al.*, 1975; Leifer *et al.*, 1981, Gene-Tox), *Bacillus subtilis* (Tanooka, 1977), *Escherichia coli* (Leifer *et al.*, 1981, Gene-Tox; Sugimura *et al.*, 1977), *Saccharomyces cerevisiae* (Zimmerman, 1975; Zimmerman *et al.*, 1984 Gene-Tox) or *Aspergillus nidulans* (Kafer *et al.*, 1976; Kafer *et al.*, 1982, Gene-Tox; Roper, 1971). The test compound is placed in the dish and the inhibition zone or lethal effect produced by the compound is evaluated in two different strains of the test organism, one being the wild-type strain and one being deficient in a DNA repair system (*e.g.* pol A$^-$ in *B. subtilis* and rec A$^-$ and uvr$^-$ in *E. coli*). When a greater zone of inhibition is produced in the repair-deficient strain than in the wild-type strain, the compound is considered to be capable of affecting DNA. The assay can be carried out with and without metabolic activation (S-9 mix) incorporated in the agar. If minimal medium is used, both mutation and inhibition zones can be detected.

Fungal Tests Measuring Mitotic Recombination or Gene Conversion

In eukaryotes, it is possible to measure an increase in the frequency of mitotic recombination or gene conversion when a recessive phenotype is expressed in the transition from a heterozygote to a homozygote situation. These tests are thought to be related to the exchange following breakage of the chromatids of two homologous chromosomes. Such changes may allow for the expression of recessive mutation, as it is known that meiotic recombination allows the expression of recessive gene mutations in man.

Such tests are carried out in the yeast *Saccharomyces cerevisiae* and *Schizosaccharomyces pombe*. A fluctuation test can also be used to detect mitotic-gene conversion (Zimmerman *et al.*, 1984). Other fungi such as *Aspergillus nidulans* have also been used (Kafer *et al.*, 1982, Gene-Tox).

Mammalian Cell Tests In Vivo

Tests to measure sister chromatid exchange in mammalian cells have already been described.

Tests to measure DNA repair by synthesis. DNA damage can be detected by determining unscheduled DNA synthesis which occurs as a result of DNA excision repair. One of the techniques reveals repair synthesis by determining radioactivity (tritiated thymidine) incorporated into DNA during the repair process (Mitchell *et al.*, 1983, Gene-Tox; Waters *et al.*, 1984). Tests to measure unscheduled synthesis in liver cells *in vivo* are now suggested if a negative result has been obtained in the bone marrow assay [see DoH 1989 Guidelines (Figure 2)].

The radioactivity can be measured either by autoradiography or by direct counting of the incorporated thymidine by liquid scintillation. Metabolic activation can also be included with such systems.

Another technique measures *the breakage of the single DNA chain in alkali,* using gradient or elution techniques. The gradient technique is the most sensitive but alkaline elution is simple and faster, measuring the rate of elution of DNA through a filter, a function of the relative molecular mass of the DNA (Cleaver, 1975; 1977).

The COMET or single cell gel electrophoresis assay has more recently become established as a useful technique for detecting DNA damage (*e.g.* Green and Lowe, 1992). It detects single strand breaks and alkali labile damage in individual cells. Cells are electrophoresed under alkaline conditions. Stained nuclei with increased DNA damage display increased migration of single stranded DNA towards the anode. The length of migration of the tail can be measured with a graticule. Density profiles can also be measured using image analysis. This method has been reviewed by McKelvey-Martin *et al.* (1993).

STAGE 1

Initial Screening
Two tests required (a + b) except where human exposure would be expected to be extensive and/or sustained, and, difficult to avoid, when all three tests are necessary

In Vitro *Tests*

(a) Bacterial assay for gene mutation

(b) Test for clastogenicity in mammalian cells, for example metaphase analysis

(c) Test for gene mutation in mammalian cells (for example the L5178Y TK+/− assay)

STAGE 2

Tests for:
Compounds positive in one or more tests in Stage 1,
and
All compounds where high, or moderate, prolonged levels of human exposure are anticipated

In Vivo *Tests*

(a) Bone marrow assay for chromosome damage (metaphase analysis or micronucleus test)

Plus, if above negative, and any *in vitro* test positive

(b) Test(s) to examine whether mutagenicity or evidence of DNA damage can be demonstrated in other organs (*e.g.* liver, gut, *etc.*)

STAGE 3

If risk assessent for germ cell effects is justified (on basis of properties including pharmacokinetics, use, and anticipated exposure).

In Vivo *Tests for Germ Cell Effects*

(a) Tests to show interaction with DNA

Dominant lethal assay (most useful)

Cytogenetics in spermatogonia

One cell embryo test

(b) Tests to show potential for inherited effects

Dominant lethal assay gives indication of likelihood of inherited effects

Cytogenetics in spermatocytes for reciprocal translocations

Non-disjunction in the mouse (10-day embryo)

(c) Test for quantitative assessment of heritable effects

Quantitative studies need strong justification in view of their complexity, long duration, costs, and use of large numbers of animals

Mouse heritable translocation test

Mouse specific locus test

*General guidance only is given in this flow diagram. Decisions regarding, for example, whether a specific compound is expected to produce high or moderate, but prolonged, exposure would normally be taken by Regulatory Authorities having regard to other relevant data, and on a case-by-case basis.

There may be instances where alternative tests to those specified might be more appropriate, and it is important that a flexible approach is adopted. Each compound should be considered on a case-by-case basis with regard to the selection of tests as well as the interpretation of results.

Figure 2   *Flow diagram for testing strategy for investigating mutagenic properties of substance*
(After Department of Health Mutagenicity Guidelines, HMSO, 1989)

Conclusion

Not all of these systems described have been equally well-studied and some are used more for specialised study than screening chemicals. The most widely used screening system today is the Ames test. Its use is widespread because of its extensive validation and the assertion of its potential to detect genotoxic carcinogens.

3 Usefulness of Some of the Systems for Screening Purposes

Microbial assays. The major advantage of the microbial methods is that they are rapid, inexpensive and relatively simple to carry out for the experienced scientist, though not for the novice.

A Bacterial Assays

The plate incorporation test of *Salmonella typhimurium* (the Ames test) is the test used routinely for screening purposes. It was shown in early blind trials to have correlations with known carcinogens and non-carcinogens as high as 90% and also to detect carcinogen and non-carcinogen pairs equally well (Purchase *et al.*, 1978). It is a test which can be carried out by many laboratories (de Serres and Ashby, 1981; Ashby *et al.*, 1985). It has a stable phenotype which is demonstrated by its lack of genetic drift (Anderson *et al.*, 1984).

Correlations with known carcinogens have been much lower in recent years (Zeiger, 1987; Tennant *et al.*, 1987) but this is primarily because non-genotoxic carcinogens have been included in the testing programme. The assay still has high predictivities for genotoxic carcinogens (Ashby and Tennant, 1988). This will be discussed in more detail later—see Prediction of Human Carcinogens.

The liquid incubation or the pre-incubation method has been used for detecting those mutagens difficult to determine in the plate incorporation assay. The fluctuation assay has been suggested as being more sensitive than the plate incorporation assay, in that it can detect compounds at lower dose levels of the test compound. A forward mutation assay is also available for *Salmonella* but these deviations from the standard Ames test method have not been validated. The repair-deficient/proficient microbial tests and spot tests, where a 'spot' of the test compound is placed in the centre of the bacterial plate, are really only suitable as pre-screens. The Ames test can be completed in about three days.

B Yeast Assays

Both *Saccharomyces cerevisiae* and *Schizosaccharomyces pombe* are suitable for use in routine screening assays. Strains can be cultivated in both the diploid and haploid phases, which allows for the detection of a wide range of mutation events. Yeasts can be used to detect both point mutations and DNA repair events in terms of mitotic recombination as evaluated by crossing-over and gene conversion. A disadvantage is that the chromosomes are too small for direct cytological observation but chromosome damage can be measured by tests for non-disjunction. Yeast systems take a few days longer than bacterial systems for colony growth but the overall time scale involved is not greatly different. They tend to be used as supplementary assays.

C Plant Assays

Plant systems can detect most types of damage. They have short generation times and the cost, handling, and space requirements are relatively small; genetics of seeds can be investigated under a wide range of environmental conditions such as pH, water content, and temperature, and chromosomal organisation is similar to the human system (Nilan and Vig, 1976). Difficulty is experienced, however, in extrapolating the results to mammalian systems, including man. Several agents, such as cytosine arabinoside, daunomycin, and adriamycin, are known to be ineffective on the plant genomes and yet cause severe genetic damage to mammalian cells. This may be because the cell wall inhibits absorption and because of greater ability to repair DNA lesions. Such systems are probably not satisfactory for predicting potential human mutagens. They are used as supplementary assays but may be useful for testing chemicals which are sprayed on plants such as pesticides.

D Insect Assays

Drosophila has a short generation time of 10–12 days and is cheap and easy to breed in large numbers with relatively simple facilities. Extensive studies on the metabolism of insecticides performed over 15 years (Würgler, 1977) have revealed that insect microsomes are capable of similar enzymic activity to those of the mammalian liver, but insects do not have any specific organ in which the enzymes are predominantly located.

In *Drosophila,* mutagenic activity can be tested at different germ cell stages which is important where mutagens have specificity of action. *Drosophila* permits the scoring for the whole spectrum of genetic effects. The observation that the lowest effective concentration (LEC) values (and therefore the highest mutagenic effectiveness) have been recorded for recessive lethals indicates the superior discriminating power of this test. (The X chromosome represents a fifth of the whole genome.) By comparison, the test for dominant lethality is of limited value. High doses are required, and dominant lethals sometimes fail to arise when agents cause the induction of recessive lethals. Changes in hatchability some-times produce false-positives. However, *Drosophila* is a good 'catch-all' system owing to the variety of genetic and end-points that can be detected.

Vogel (1987) points out that, based on Gene-Tox Report data and two international collaborative trials, the sex-linked recessive lethal test does not have a high ability to detect carcinogens when genotoxins other than direct acting and simple pro-mutagens are included. However, it has a high predictability for non-carcinogens. The tests detecting somatic mutation/mitotic recombination (SMART) have higher predictibilities than the sex-linked recessive lethal assays. *Drosophila* assays, although they were included in some regulatory guidelines, *e.g.* DHSS, 1981, are not currently included in those of DoH, 1989 (Figure 2).

E Mammalian Cell Assays *In Vitro*

Mammalian cell systems are generally considered more valid than non-mammalian systems in terms of extrapolation to man, because mammalian DNA is more similar to that of man. However, the use of mammalian cell mutation assays is currently in debate. Few chemicals are known which are detected exclusively in mammalian cell mutation assays, and are not detected by the similar (in terms of cost) mammalian cell *in vitro* cytogenetic assays. For this reason the use of mammalian cell mutation assays is questioned. In addition, the V79 and the CHO cell mutation systems, whilst suitable for detecting a positive response are basically inadequate in terms of cell numbers for detecting a negative response. By increasing cell numbers, whilst the mutant fraction remains constant, there would be sufficient cell numbers available for statistical purposes. Suitable cell numbers are only available at present in the cell suspension assay of mouse lymphoma—the L5178Y system. The L5178Y cell system of Clive (Clive *et al.*, 1983, Gene-Tox) detects both large and small colonies. The large are thought to arise from gene mutation and the small from chromosome damage. It has been suggested that this system may be too sensitive and lack discriminating power for the detection of carcinogens and non-carcinogens. The thymidine kinase (TK) and hypoxanthine guanine phosphoriboxyl transferase (HPRT) loci are commonly used in mammalian cell assays but the latter, which is used in V79 and CHO cells, is less sensitive than the TK locus (McGregor, personal communication). If mutation cell systems are to be used, however, cells grown in suspension are easier to handle. They do not require trypsinisation, are easily sub-cultured and are not subject to metabolic co-operation, so do not suffer from the reduced sensitivity that occurs through metabolic co-operation when mutated cells are in close contact with non-mutated cells.

[Current DoH guidelines (1989) suggest that chemicals giving negative responses in three *in vitro* systems (bacterial mutation, chromosomal and cellular point mutation assays) with and without metabolic activation do not require further testing in animal systems.]

F   Mammalian Assays *In Vivo*

The advantage to *in vivo* studies is that the test compound is metabolised in the animal. Both the bone marrow metaphase and micronucleus assays thus have this advantage. Positive results in such assays indicate that the test chemical is a mutagen to somatic mammalian cells *in vivo*. Results correlate well with carcinogenicity studies, particularly human carcinogens (Shelby, 1988).

The dominant lethal assay in rodents has been claimed to be insensitive to detecting chemicals, but the lack of sensitivity may reflect the real situation because of the so-called 'testes barrier' formed by the Sertoli cells surrounding the germ cells. In addition to the pharmacokinetic hurdles and organ specificity, *etc.*, and the short half-life of the chemical, the blood–testes barrier may be important. Even if the compound reaches the testes, it may not be metabolised. With dimethylnitrosamine, for example, there is less alkylation of the DNA in the testes than in any other organ (Swann and Magee, 1968) and the compound gives a negative result in the dominant lethal assay (Propping *et al.*, 1972).

These considerations, of course, also affect the other assays concerned with the germ cells, such as the specific locus and heritable translocation assays. The drawbacks to the latter two assays are that they require vast numbers of animals in order to detect a response and thus are very costly in terms of resources and time. However, the mammalian *in vivo* assays are required for determining genetic hazard and risk estimates. The dominant lethal assay although not useful for predicting carcinogens (Green *et al.*, 1985, Gene-Tox) could be useful for predicting heritable hazard and the specific locus and heritable translocation assays are useful for quantitative risk assessment.

Kirkland (1987) discusses the implications of germ cell cytogenetic tests in the regulatory process. One of the responses to a questionnaire sent to regulators was that germ cell tests are rarely requested as a matter of course. Where germ cell tests are indicated, a dominant lethal test (most often) or a heritable translocation test (sometimes) would be seen as helpful in elucidating germ cell effects. The mouse specific locus test is rarely requested due to the large number of animals involved and the small number of laboratories with the relevant experience.

The current DoH guidelines (1989) (Figure 2) recommend germ cell assays only for risk assessment as the last tier. In previous guidelines (DHSS, 1981) germ cell assays were included in a battery of assays.

G   DNA Repair Assays

Measurements of excision repair are determined as mean values for a population of cells, whereas mutation is a rare event in individual cells. The amount of excision repair after exposure to an agent will depend on several factors, such as the extent of reaction with the DNA (the total number of damaged sites), the number of sites that can be excision repaired, the size of the repaired regions, the kinetics of excision repair as functions of time and dose, and the extent to which chemical interactions modify other sites, and possibly inhibit excision repair (Cleaver, 1975; 1977). The amount of excision repair will therefore be greatest for mutagens that induce the greatest proportion of extensively damaged sites requiring repair by the large substitution. The number of mutations depends on the severity of pre- and post-replication damage.

Studies of the relationship between DNA damage, excision repair, post-replication repair, and mutagenesis must take account of the numbers and varieties of lesions involved in mutagenesis and the modes of repair. Exclusion reliance on any one measurement is useless. It is best to consider DNA repair, for example, only in conjunction with some other parameters before assessing the possible mutagenicity of an agent. However, the COMET assay is proving useful for the rapid measurement of genetic damage (McKelvey-Martin, 1993).

4 Cell Transformation Assays

These assays are thought to 'bridge the gap' between mutagenicity and carcinogenicity. Cell transformation has been defined as the induction in cultured cells of certain phenotypic alterations that are related to neoplasia (Barrett *et al.*, 1986). There are several types of endpoints for cell transformation, including loss of anchorage dependence and alterations in morphology, viral dependence, and altered growth in agar (McGregor and Ashby, 1985). Morphological transformation and altered growth have been used most extensively. These systems are the Syrian hamster (SHE) assay (Barrett and Lamb, 1985; Berewald and Sachs, 1963; DiPaolo *et al.*, 1969a and b; 1971; 1972; Huberman and Sachs, 1966) and the mouse C3H/IOT1/2 and mouse BALB/c3T3 assays (Heidelberger *et al.*, 1983; Kakunaga, 1973; Reznikoff *et al.*, 1973). Of these, the SHE cell transformation assay is unique in that it uses normal diploid cells (Berwald and Sachs, 1963; 1965). The other two systems are based on established cell lines which have undergone some adaptive changes in culture, resulting in an aneuploid karyotype and a potentially preneoplastic phenotype. A recommended protocol for all three assays based on a survey of current practice has been suggested (Dunkel *et al.*, 1991). SHE cells have a limited lifespan in culture and rarely become tumorigenic unless treated with carcinogens (Barrett *et al.*, 1977). Therefore the cellular events underlying the morphological transformation of SHE cells might be indicative of earlier neoplastic changes compared with those underlying cell transformation observed in the other cell lines. The acquisition of a fully neoplastic phenotype in these cells in a multistep process analogous to that *in vivo* and this system is therefore particularly useful for studying the cellular and molecular events involved in neoplastic development (Barrett *et al.*, 1986; Koi and Barrett, 1986). Fitzgerald and Yamasaki

(1990) have addressed the issue of tumour promotion describing models and assay systems.

In the SHE transformation assay morphologically transformed colonies are identified by a disorientated pattern of piled up cells (Berwald and Sachs, 1965). The cells have a considerable range of metabolic activities but chemicals requiring further metabolic activation can be studied with incorporation of appropriate sub-cellular fractions or a second cell type.

5 Test significance and Interpretation

The available sub-mammalian test systems, used without mammalian *in vivo* studies, would not be acceptable by any governmental authority for estimating risk to man. They do provide a useful tool for a preliminary screening of possible human mutagens. A positive result in a well constructed and validated system is generally regarded as a warning sign. However, when considering the many and diverse chemicals which are of unknown mutagenic potential which require metabolic activation and which may react differently with different cells and organs before producing genotoxic effects, it is not surprising that many give equivocal data that cannot be resolved from experience gained from classical studies.

By comparison, the handling of positive data is much more clear-cut, but it is desirable that dose–response curves should be established in routine testing and results should be reproducible. Difficulties may arise when mutagens have a strong killing effect so that a genetic effect is obviously not as readily detected. This can be exemplified by mammalian cell mutation assay systems where the activity of a chemical can produce an absolute increase in the number of mutants per treated cell or an absolute decrease if there is a strong killing effect. In the former case the rate of increase of mutants (over the spontaneous) is then greater than the rate of inactivation, per unit dose; in the latter, this is not so. Weak mutagens are more difficult to evaluate when there is an increase only in mutants per surviving cell (that is, after correction for survival) and not per treated cell. The apparently weak positive effect could be due to the induction of new mutants but may be due to a greater resistance of spontaneous mutants to inactivation by the agent used. Reconstruction experiments (where known numbers of mutant and wild-type cells are mixed together) can solve the former problem, but to the latter there is no good solution.

Dose–response curves may have linear or diphasic shape, or may be diphasic at high dosage with a linear function at low doses. Thresholds may exist at some chemical concentrations when the chemical is without effect below a certain concentration. At higher doses, dose–response curves tend to flatten or plateau or decline when the killing effect overrides the mutagenic effect. When a dose–response relationship is established it is easier to reach a conclusion regarding the mutagenicity of a chemical.

In *in vitro* mammalian cell systems, particular emphasis should be placed on using doses of chemicals which are not too high and so do not alter the pH or affect the osmolality of the test system. The problems arising in such instances, where false positives can be generated, have been highlighted in a special issue of Mutation Research (Brusick, 1987).

6 Strategies for the Protection of Man

Auerbach (1975) stated that the procedures for the estimation of possible hazards from genotoxins are full of uncertainties. This is still the case. It is, nevertheless, important to attempt to develop approaches that will allow risk estimation in men exposed to potentially genotoxic agents. The *first* approach is to use a group of short term tests to detect possible mutagenic activity while recognising that such an approach is subject to limitations in terms of test variability, species sensitivity, and interspecies extrapolation (but this is true for toxicological tests in general).

The *second* is to measure induced genetic damage directly in man.

The *third* is to use results from an evaluation of effects in the gonads which can be combined with the pattern of expected human exposure from which a judgement as to the amount of risk can be estimated. This can be attempted by combining human exposure data with (a) the known dose-response observed in animal studies, or (b) measured target/germ cell concentrations combined with mutagenic responses defined by the best understood *in vitro* assay systems. Concepts such as radiation-equivalents and doubling doses may be of value in this approach.

A   Approach 1—Use of Short Term Test Battery

Prediction of Human Mutagens

Developments in short term test procedures for genotoxicity have generally focused on their ability or inability to identify potential carcinogens. However, in regulatory practice in the UK and some other European countries, such tests are conducted to determine mutagenicity *per se* with a view to identifying potential human mutagens.

There are no examples of induced mutations in man with proven causality. Cigarette smoke, vinyl chloride, lead, and anaesthetic gases are the agents which are best documented but no effects in man are yet unequivocally determined. Thus a validated mutational assay for detecting potential human mutagens is not attainable, although attempts are being made to address this issue.

Before this can be achieved, there is a need to examine the data on the performance of the test systems and to establish how consistent these data are when derived from different sources.

Within test variation. An assessment of the variation that is obtained within a test is most easily achieved by comparing the performance of tests which independently assay the same chemical. Over a decade ago, three such studies were completed (de Serres and Ashby, 1981; Dunkel, 1979; Poirier and de Serres, 1979). All showed errors of about 10% in detecting positive effects and about the same in detecting negative effects. It is worth considering the consequences of a 10% discrepancy when 6 test systems are used. Thus if a single test is used, 90% of the mutagens will be correctly identified. If it is assumed that each one of the six tests provides an independent assessment of the mutagenicity of the chemical, and that the error rate of each test for both positive and negative results is 10%, six tests will identify 99.999% of the mutagens. At the same time, however, only 56% of the non-mutagens will be negative in all six test systems.

Between test variation. When considering test systems which, although they may assess the same genetic end-point, use different organisms, the range of variability is wide. One study (Poirier and de Serres, 1979) using three assay systems found agreement for fifty-five chemicals (twenty-five all negative and thirty all positive) and disagreement for forty-four. Similar findings occurred in an international study (de Serres and Ashby, 1981, as shown in Table 2).

The use of such schemes for regulatory purposes was much discussed (Brusick, 1982; Purchase, 1980) but as yet has still not been satisfactorily resolved. The reproducibility of the *Salmonella typhimurium* and *Escherichia coli* mutagenicity assays has been examined by Dunkel *et al.* (1985). It is generally recognised that a single positive is not sufficient to define mutagenicity, but it was proposed that a single response from some test systems might have a greater weighting than from others (Brusick, 1982). However, as more work has been done with these tests it is realised that some of these weightings may not apply. Assessment panels have addressed the evaluation of mutagenicity assays for genetic risk assessment (Brusick *et al.*, 1992; Russel *et al.*, 1984), whilst Ray *et al.* (1987) have examined the various assays for identifying classes of chemicals.

Evaluation of the performance of short term tests in identifying germ cell mutagens. Waters *et al.* (in press), evaluated the performance of various STTs in identifying germ cell mutagens. Using a combined data set derived from the US EPA/IARC Genetic Activity Profile (GAP) database and the US EPA Gene-Tox database, a total of 56 germ cell mutagens were identified. These chemicals had given positive results in one or more of the following assays: the mouse specific locus test; *in vivo* tests for chromosomal aberrations in the germ cells; the dominant lethal test in mice or rats; and the mouse heritable translocation test. The same two databases were used to provide information (where available) on the activity of these chemicals in bacterial mutagenicity assays, in two *in vitro* mammalian cell assays (one for chromosome aberrations, another for gene mutation), and in two *in vivo* tests on the bone marrow (chromosome aberrations or micronucleus formation). The performance of the various STTs is summarised in Figure 3. Although the sample size was only small, the data indicated that the two *in vitro* assays with mammalian cells were able to identify 86% (gene mutation) and 93% (chromosome aberration) of the germ cell mutagens, whilst bacterial mutagenicity assays were slightly less sensitive (75%). [Unfortunately, in the absence of any data on chemicals that are not germ cell mutagens, it is not possible to assess the specificity of the assays (*i.e.* their ability to correctly identify such chemicals as non-mutagenic to the germ cells)]. The sensitivity of the individual assays was increased when the results from two or three of the assays were combined. Of the 36 germ cell mutagens that had been tested in a bone marrow assay (for chromosomal aberrations or micronucleus formation), 33 gave positive results, and the evidence for germ cell mutagenicity of two of the three not identified was called into question. The problem of strain variability among rodents was raised as a possible source of discrepancy. This study indicates that STTs can provide valuable information on the potential of a chemical to induce germ cell mutations.

Ashby (1986) proposed a testing strategy to detect genotoxic agents *in vivo*, where after a positive response in *Salmonella* or *in vitro* cytogenetics, a chemical should be tested in a mouse micronucleus or a rat liver unscheduled DNA synthesis assay *in vivo*. This strategy was debated by Garner and Kirkland (1986) and Gatehouse and Tweats (1986). However, the guidelines for mutagenicity testing (DoH, 1989) use a similar scheme, but *in vivo* tests for germ cell effects may be justified on the basis of properties including pharmacokinetics, use, and anticipated exposure (Figure 2). Guidelines in most countries and internationally are constantly under scrutiny (*e.g.* OECD Guidelines Harmonisation Meeting at the 6th International Environmental Mutagen Meeting in Melbourne 1993). As a result of such moves, it is hoped that better models for evaluating human mutagenic risk may emerge.

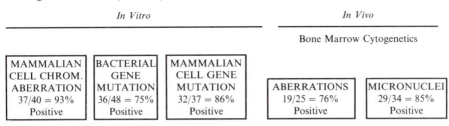

Figure 3   *Test performances are given for the germ cell mutagens from the combined EPA/IARC GAP and GENE-TOX databases. Performance is indicated by the fraction of agents with positive test results divided by the number of agents tested and is expressed also as the percentage positive*
[After Waters *et al.*, (in press)]

Prediction of Human Carcinogens

The use of genetic toxicology assays in predicting the carcinogenicity of chemicals to rodents or humans has come under a great deal of scrutiny. In a 1975 publication in which 300 chemicals were tested in the Ames test, McCann *et al.* found that 90% of the 174 carcinogens were mutagenic. Similar figures were provided by other workers (see Introduction), and it seemed only a matter of time before complementary short-term tests (STTs) would be developed to detect the remaining 10% of carcinogens that were 'missed' by the Ames test.

Through the late 1970s and early 1980s there was a period of activity as the matrix of STTs and chemicals tested increased, and various groups of workers endeavoured to show that their favoured assays could identify known mammalian carcinogens. A series of internationally coordinated (through WHO/IPCS, UNEP, and ILO) validation studies (funded partly by the US National Institute of Environmental Health Sciences and the UK Medical Research Council) were conducted (Ashby *et al.,* 1985; 1988). The various STTs were assessed on their *sensitivities* (the percentage of known carcinogens correctly detected), their specificities (the percentage of non-carcinogens correctly identified), and their *concordances* (the overall accuracy in their identifications). These studies revealed that the predictive values of the tests were no longer as high as they had been in earlier investigations. As more and more rodent carcinogens were identified, mainly under the National Toxicology Program (NTP), it was found that the predictive power of the Ames test, in particular, declined. While the activity of most of the long-standing and well-established rodent carcinogens could be rationalised in terms of their electrophilic properties (because this was the primary stimulus for their initial selection for carcinogenicity testing), an increasing proportion of the newly-identified carcinogens (not pre-selected in this way) were both without a supporting chemical rationale for their activity and were non-mutagenic to Salmonella. The fact that these chemicals were appearing positive in the animal carcinogenicity studies but were consistently negative in STTs was of great concern, and had regulatory implications.

This can now be explained in terms of the two types of carcinogen that are thought to exist, those acting by a genotoxic mechanism (that would have produced positive results in the STTs), and those acting by a non-genotoxic mechanism (generally negative in the STTs).

Analysis of concordance between STT results and carcinogenicity data. The inconsistency between STT results and carcinogenicity findings prompted an investigation by the Cellular and Genetic Toxicology Branch of the NTP, the aim of which was to assess the ability of prokaryotic and eukaryotic STTs to detect rodent carcinogens. The resultant publication (Tennant *et al.,* 1987) revealed that four of the most used STTs (the Ames test, the mouse lymphoma L5178Y mutagenicity assay, and tests for chromosomal aberrations and sister chromatid exchange in Chinese hamster ovary cells) were poor predictors of carcinogenic activity. The assessment of 73 chemicals that had recently been tested for carcinogenicity by the NTP revealed that the concordances between the carcinogenicity results and the genotoxicity findings were only about 60% for each STT.

The individual tests exhibited different data profiles, the mouse lymphoma and SCE assays giving more positive results than the Ames or chromosomal aberration assays. Although some chemicals always gave consistent results, there were eleven chemicals which showed only a single positive STT result. No combination of the tests improved the accuracy of the cancer prediction: for instance, defining a chemical as positive if it gave a positive result in any of the four tests increased the overall sensitivity but decreased the specificity. The most difficult problem from the genotoxicity viewpoint was the failure of six carcinogens to show any positive results at all, despite the fact that three of these (dioxin, reserpine, and a polybrominated biphenyl mixture) were the most potent carcinogens in the whole group, at least on the basis of the dose producing statistically significant increases in tumour incidences. In an update of this study, Zeiger *et al.* (1990) tested a further 41 NTP chemicals and found a wider variation in the level of concordance for the four STTs (ranging from 54% for the SCE assay to 73% for the Ames test), but this was not considered to represent any substantial improvement in the predictive power of the Ames test. Again, no combinations of STT improved upon the concordance and predictivity of the Ames test alone.

The 1987 publication by Tennant and colleagues provoked immediate responses and counter-responses in the genetic toxicology journals (*e.g.* Ashby, 1988a; Ashby and Purchase, 1988; Auletta and Ashby, 1988; Brockman and DeMarini, 1988; Haseman *et al.,* 1988; Kier, 1988; Trosko, 1988; Young, 1988). The 73 compounds tested were said to be a distorted sample, as they represented compounds that had only recently been tested by the NTP and had been selected for testing on the basis of production volumes, degrees of human exposure and suspicion of carcinogenic potency. They were certainly more representative of the more subtle 1980s type of carcinogen than the classic potent carcinogens of the 1950s and 1960s. Brockman and DeMarini (1988) criticised the approach of blind testing, because no account was taken of what might already be known about a chemical's properties or those of structurally related compounds. Individually 'customized' protocols, it was argued, might have given results that more closely matched the carcinogenicity findings. They also considered that the results of animal carcinogenicity bioassays (particularly negative ones) did not deserve the exalted position that the scientific community had generally assigned to them because they had low statistical power, were rarely replicated, were seldom done under different sets of experimental conditions, and had many limitations that were unlikely to be overcome in the near future. Other investigators have also questioned the scientific validity of lifetime animal feeding studies in which the chemical is administered at the maximum tolerated dose (see later for further discussion) (Ames and Gold, 1990a and b; Ashby and Morrod, 1991). The problem with the numerical approach of Tennant *et al.* (1987) is that the mathematical sophistication does not make up for the inherent limitations of the carcinogenicity and genotoxicity assays. The expression of carcinogenicity bioassay findings as a simple positive or negative result is, in many cases, an over-simplification or misrepresentation of a chemical's true carcinogenic activity. Similarly, the results of a number of STTs for a particular chemical often include conflicting or equivocal findings which do not allow a simple positive or negative categorisation. Mammalian cell assays are said to present a particular problem in producing so-called 'false positive' results (Adler *et al.*, 1989; Scott *et al.*, 1991). Such limitations can seriously affect the interpretation of the data obtained, although various attempts may subsequently be made to rationalise the discordant results in the different STTs, as occurred for the 1987 study by Tennant *et al.* (*e.g.* Ashby, 1988b; Myhr and Caspary, 1991; Prival and Dunkel, 1988).

Despite the limitations, the mathematical approach has been used by a number of other investigators (*e.g.* Auletta and Ashby, 1988; Benigni and Giuliani, 1988; Ennever and Rosenkranz, 1989; Klopman and Rosenkranz, 1991; Kuroki and Matsushima, 1987; Loprieno *et al.*, 1991; Parodi *et al.*, 1991). Kuroki and Matsushima (1987) evaluated the performance of a range of STTs in detecting 71 established, probable or possible human carcinogens (classified in IARC Groups 1, 2A and 2B respectively) and concluded that the chromosome aberration test in mammalian cells *in vitro* provided results that were complementary to those obtained in the Ames test. No figures could be derived to assess the specificity of the chromosome aberration test because of the lack of clearly identified human non-carcinogens. Sorsa *et al.* (1992) evaluated the available cytogenetic data on 27 proven, 10 probable, and 15 possible human carcinogens and found that 19/27, 6/10, and 5/15 induced chromosomal aberrations, sister chromatid exchanges, and/or micronuclei in humans. A large prospective cohort study suggested that chromosomal aberrations, but not sister chromatid exchanges, are significant for prospective cancer risk (Sorsa *et al.*, 1992).

Klopman and Rosenkranz (1991) selected 253 compounds that had been tested by the NTP for carcinogenicity to animals, and used probability calculations to assess the predictivity of the Ames test and of assays for sister chromatid exchange (SCE) and chromosomal aberrations (CA). The unscheduled DNA synthesis (UDS) assay was similarly evaluated for 130 chemicals, the corresponding cancer bioassay data being taken from Williams *et al.* (1989). The analysis revealed concordances of 62% (Ames), 61.3% (SCE), 56.5% (CA), and 56.2% (UDS). The probabilities of a carcinogen being positive in individual tests were 56% (Ames), 88% (SCE), 75% (CA), and 53% (UDS), while the probabilities of a non-carcinogen being positive in individual tests were 27% (Ames), 89% (SCE), 78% (CA), and 33% (UDS). Although the former might suggest that the SCE and CA assays were better than the Ames and UDS assays at detecting carcinogens, the latter indicate that they may in fact be too sensitive to be of use in predicting carcinogenicity, as a high proportion of non-carcinogens also gave positive results in these assays.

The concordance figures reflect the fact that none of the four assays is particularly good at predicting carcinogenicity.

As noted previously, Tennant *et al.* (1987) and Zeiger *et al.* (1990) found that various combinations of *in vitro* STTs (Ames, CA, SCE, and the mouse lymphoma assay) were no better at predicting carcinogenicity than was the Ames test alone. However, Jenssen and Ramel (1980) and Shelby (1988) proposed that a combination of two genotoxicity assays, the Ames test and the (*in vivo*) bone marrow micronucleus test, could be used for the detection of genotoxic chemicals that might be predicted to be carcinogens. In an analysis of 23 chemicals designated by the International Agency for Research on Cancer (IARC) as Group 1 compounds (carcinogenic to humans), Shelby (1988) found that 20 of the 23 carcinogens (87%) were active in one or both STTs. Seventeen of the 22 that were tested in the Ames assay gave positive results, and the untested chemical (treosulphan) was considered likely to be active on structural grounds (subsequently confirmed by Zeiger and Pagano, 1989); 12 of these 17 were also active in the *in vivo* bone marrow test, and four of the remaining five were considered as likely positives due to structural similarities to known bone marrow clastogens. Three other chemicals (benzene, diethylstilboestrol, and arsenic) were inactive in the Ames test but gave positive results in the bone marrow test, while of the remaining three chemicals not tested in the bone marrow test, treosulphan was again considered a likely positive (as was subsequently demonstrated by Gulati *et al.*, 1990 and Shelby *et al.*, 1989), and asbestos and conjugated oestrogens (both negative in the Ames test) were not expected to affect the bone marrow. Thus, the latter two carcinogens would not be detected by this combination of assays, nor would their carcinogenicity be anticipated on structural grounds (Shelby, 1988).

Since that time, numerous studies have been conducted to assess the various possible protocols for the *in vivo* bone marrow test, including single or multiple exposures and different sampling times (*e.g.* Adler and Kliesch, 1990; Čihák and Vontorková, 1990; George *et al.*, 1990; Gulati *et al.*, 1990; Mavournin *et al.*, 1990; Mirkova, 1990; Tice *et al.*, 1990). The various investigators reached different conclusions on which protocol was most effective in detecting carcinogens, but as there was some evidence that a three-exposure protocol might be more effective than a single-exposure protocol (*e.g.* Gulati *et al.*, 1990; Tice *et al.*, 1990), Shelby *et al.* (1993) went on to test 49 NTP chemicals (25 rodent carcinogens and 24 non-carcinogens) using a three-exposure intraperitoneal protocol with a single sampling time. Only five of the 25 rodent carcinogens gave positive results, although a further two were found to be positive in a single-exposure protocol. Two of the seven (benzene and monuron) would not have been suspected from Ames test data or from their chemical structure. Four of the 24 chemicals that had shown no evidence of carcinogenicity in rodent bioassays (ascorbic acid, phenol, titanium dioxide, and 2,6-toluenediamine) were found to be active in the micronucleus test; all but

2,6-toluenediamine were non-mutagenic in the Ames test. This study indicates that the three-exposure (ip) protocol with single sampling time is not satisfactory for distinguishing between carcinogens and non-carcinogens; further work is needed before the micronucleus test can provide additional meaningful information on a chemical's genotoxic activity.

In an evaluation of a mammalian gene mutation assay, the Chinese hamster ovary/hypoxanthine guanine phosphoribosyl transferase (CHO/HGPRT) assay, Li *et al.* (1988) found that 40 out of 43 reported animal carcinogens (93%) gave positive results, while the only definitive non-carcinogen, caprolactam, was negative. Since then, a further nine chemicals that gave negative results in NTP bioassays have been tested in this assay; seven gave negative results, one (2-chloroethanol) was positive, and one (benzoin) was equivocal (Li *et al.*, 1991). The investigators concluded that the CHO/HGPRT assay seemed to have high specificity as well as high sensitivity, and that it could be used to complement the Ames test and cytogenetic assays. Oshiro *et al.* (1991) tested ten compounds deemed non-carcinogenic in the literature, and concluded that the CHO/HGPRT assay produced more relevant results than other genotoxicity tests in mammalian cells: only two of the chemicals [dichlorvos and 2-(chloromethyl)pyridine], were positive in this assay, and the non-carcinogenic status of both was considered questionable.

In an analysis by Loprieno *et al.* (1991), a test battery of two *in vitro* assays (Ames and CA tests) and one *in vivo* assay (the rodent bone marrow micronucleus test) was reported to have greater predictivity than the Ames test alone. In an analysis of 716 chemicals that had been tested in cancer bioassays, the accuracy of the three STTs was 68.6% for the Ames test (number of test chemicals, $n = 544$), 64.3% for the CA test ($n = 445$), and 70.6% for the *in vivo* micronucleus test ($n = 163$). This was increased to 71.6% for Ames + CA ($n = 310$), 85.0% for Ames + micronucleus ($n = 113$), 87.9% for CA + micronucleus ($n = 107$), and 92.5% for a combination of all three STTs ($n = 93$). Thus the three STTs together correctly identified 43 out of 45 carcinogens and 43 out of 48 non-carcinogens. The 716 chemicals that had been tested for carcinogenicity were part of a much larger data set of 3389 chemicals, 2898 of which had been tested in the Ames test (85.5%), 1399 (41.3%) in the CA test, and 319 (9.4%) in the *in vivo* micronucleus test. For the 270 chemicals that had been tested in all three STTs, 107 were positive in both *in vitro* assays, of which just over half (56) were also positive *in vivo*; a smaller proportion (25% or less) of the 16 or 71 chemicals that were positive only in the Ames test or only in the CA test (respectively) also gave positive results *in vivo*; and eleven of the 76 chemicals (14.5%) that gave an indication of genotoxicity *in vitro* were found to be active in the micronucleus test. Five of the eleven (chlorobenzene, *ortho-* and *para-*dichlorobenzene, toluene and trichloroethylene) have also demonstrated carcinogenic activity in rodents, while the other six (isoxaben, 1,3,5-, 1,2,4-, and 1,2,3-trichlorobenzene, trimethoprim, and vincristine) have not been adequately tested. The eleven chemicals were thought by the investigators to represent a class of compound for which the mechanism of genotoxicity could not be fully applied, and more information was being collected on them (Loprieno *et al.*, 1991).

Various supplementary assays have been developed, including tests for recombination, gene conversion and aneuploidy (chromosome loss) in different strains of the yeast, *Saccharomyces cerevisiae*. These have not been well validated, and inconsistencies have been reported, for example, in an aneuploidy test system using strain D61.M (Albertini, 1991). Nevertheless, there is some indication that tests for recombination (Schiestl, 1989; Schiestl *et al.*, 1989) or for gene mutation (Morita *et al.*, 1989) in yeast cells might be able to detect carcinogens that are 'missed' by the Ames test.

Ashby and Leigibil (1992) authored a commentary on the use of transgenic mouse mutation assays in the context of genotoxic and non-genotoxic mutagens. Recommendations for dose levels and treatment protocols for these mice were presented and the authors asked for further discussion of dosing regimens. Mirsalis (1993) and Ashby and Leigibil (1993) debated the proposed dosing approach for the detection of genotoxic and non-genotoxic carcinogens. Gunz *et al.* (1993) further examined whether non-genotoxic carcinogens can be detected with the lacl^q transgenic mouse assays.

Use of the Ames test and/or structural alerts in predicting carcinogenicity. A non-computational chemistry based method for identifying carcinogens has been proposed (Ashby, 1985). [This has been progressed in parallel with developments in quantitative structure activity relationships (QSAR)—see review by Phillips and Anderson (1993)]. This chemistry based approach utilises the electrophilic theory of carcinogenesis propounded by the Millers in the 1970s and relies on personal skill to identify any electrophilic centre(s) within the molecule under consideration. Such a feature constitutes a 'structural alert' for reaction with a nucleophilic site in DNA.

Following the 1987 study by Tennant *et al.*, Ashby and Tennant (1988) and Ashby *et al.* (1989) examined the Ames data, carcinogenicity verdicts and chemical structures of a set of 264 compounds tested in the NTP's rodent cancer bioassays. Chemicals were classified by the breadth of their cancer activity. For instance, 66 were assigned to Group A because they caused cancer in both rats and mice at one or more sites. Group B chemicals (19 in all) produced tumours at multiple sites in one species. Group C (27 compounds) and Group D (26 compounds) resulted in cancer at a single site either in both sexes or a single sex (respectively) in a single species. The two remaining classes consisted of 94 chemicals where there was no indication of carcinogenicity in either species (Group F) and 32 where there was an equivocal or uncertain result (Group E). Aspects of the chemical structure that were suggestive of electrophilic activity either of the compound or its metabolites were then identified. The investigators assessed how well the results of the Ames test and the presence or absence of structural alerts predicted the NTP carcinogenicity results. The concordance between the four carcinogenic and non-carcinogenic classes (A, B, C, D, and F) and the Ames test was 64%, as low as Tennant's earlier results. The concordance with the structural alerts was also 64%. The number of mismatches between the structural alerts and the Ames test results was 31 (12%). Thus, although there was a high correlation between the presence or absence of structural alerts in the 264 chemicals and their mutagenic activity in the Ames test, neither of these was particularly effective in predicting carcinogenicity. The observed sensitivity of the Ames test (in correctly identifying carcinogens) was only 58%, but this increased to 72% when the 66 two-species carcinogens (Group A) were considered on their own (Ashby *et al.*, 1989). This compared with only about 50% (Groups B and D) or 32% (Group C) for the 72 single-species carcinogens. The evaluation of a further 39 chemicals had little effect on these figures for sensitivity (Tennant and Ashby, 1991).

These differences between the four classes of carcinogens provided the impetus for a further analysis of all the chemicals that had been included in the three earlier surveys (*i.e.* Ashby *et al.*, 1989; Ashby and Tennant, 1988; Tennant and Ashby, 1991). The combined data set consisted of 301 chemicals, which were split into broad classes based, this time, on their known or projected chemical reactivity (Ashby and Tennant, 1991). Roughly half of the chemicals (154) were structurally-alerting chemicals, and these were further subdivided into aromatic amino/nitro-type compounds (84 chemicals), natural electrophiles including reactive halogens (46), and miscellaneous structurally-alerting groups (24). The 147 non-alerting chemicals were broadly categorised as compounds devoid of actual or potential electrophilic centres (61), compounds containing a non-reactive halogen (50), or compounds previously classed as non-altering in structure but with minor concerns about a possible structural alert (36). When the six classes were each analysed separately, Ashby and Tennant reported a clear distinction between the three groups of structurally-alerting chemicals and the three groups of non-alerting chemicals. Thus 65–70% of the structurally-alerting chemicals were carcinogenic compared with 25–52% of the non-alerting chemicals, and 65–86% were mutagenic in the Ames test compared with only 0.02–9% of the non-alerting chemicals. The sensitivity of the Ames test in predicting carcinogenicity was 93% for the nitro/amino compounds, 83% for the electrophiles, and 56% for the miscellaneous structurally-alerting compounds; however, for all three groups there was a high false-positive rate, as 71%, 55% (6/11), and 75% (3/4) of the non-carcinogens in the respective groups gave positive results in the Ames test. For the non-alerting chemicals, the sensitivity was low (13%) or non-existent, while the high specificity (with only one non-carcinogen in each group showing mutagenic activity) was of little value.

Genotoxic and non-genotoxic carcinogens. Ashby and Tennant (1991) distinguished between structurally-alerting carcinogens, *i.e.* chemicals with one or more electrophilic groups that could potentially interact with and damage the cell's DNA in some fairly immediate way ('genotoxic carcinogens'), and non-alerting carcinogens that had no obvious electrophilic centre and were thought to induce tumours by some mechanism that did not involve early attack of the chemical or a direct metabolite on the cell's DNA (putative 'non-genotoxic carcinogens'). Some examples of non-genotoxic (or epigenetic) routes to tumour formation include enhanced cell proliferation resulting from cytotoxicity or mitogenesis (induced cell division), hormonal changes, and peroxisome proliferation. At a 1990 meeting on early indicators of non-genotoxic carcinogenesis, several possible mechanisms were described (Anon., 1991).

Since the various STTs are specifically designed to detect genotoxic activity, any chemical that induces tumours by a non-genotoxic mechanism will not (or at least in theory ought not to) be active in these STTs.

Jackson *et al.* (1993) indicated that the situation is not that simple. Many chemicals that are described as non-genotoxic carcinogens in the literature have in fact shown evidence of genotoxic activity in appropriate STTs. In an analysis of 39 putative non-genotoxic carcinogens which had been tested in five or more STTs for gene mutation, chromosomal aberrations and/or aneuploidy, 14 showed evidence of activity (seven *in vivo* and seven *in vitro*), and a further ten showed limited evidence (*in vivo* and/or *in vitro*). Only two were considered to have been sufficiently tested to warrant the description non-mutagenic (*i.e.* non-genotoxic). The remaining 13 chemicals had not been adequately tested *in vitro* and *in vivo* for each of the three broad categories of DNA alterations (*i.e.* gene mutation, chromosome aberration, and aneuploidy). Thus, although all but one of the 39 chemicals gave predominantly negative results in the Ames test, and most were also devoid of structural alerts, over half showed genotoxic potential in other STTs, and another third might also do so if they were to be adequately tested. Whether their genotoxic activity plays an important role in their carcinogenicity is questionable, but the investigators noted that four out of six compounds that induce peroxisome proliferation, two out of five cytotoxic carcinogens and one out of three mitogens all demonstrated some mutagenic activity.

Distinguishing between genotoxic and non-genotoxic carcinogens on the basis of their tumour profile. Ashby *et al.* (1989) and Ashby and Tennant (1988, 1991) distinguished between two types of carcinogen: multi-species (and usually multi-site) carcinogens which could generally be predicted by either a positive result in the Ames test or (equally well) by a structural alert (the so-called genotoxic carcinogens); and the single-species (often single-site, and/or single-sex) carcinogens which were less likely to be mutagenic or structurally-alerting (the so-called non-genotoxic carcinogens). Analysis of the patterns of carcinogenic response provided by these two classes of carcinogen had revealed that for many tissues (notably the mouse liver), equal sensitivity to genotoxic and non-genotoxic carcinogens was observed, but that certain tissues (notably the Zymbal's gland in the rat) appeared only to be sensitive to genotoxic carcinogens (Ashby *et al.*, 1989; Ashby and Tennant, 1988). When Ashby and Tennant (1991) evaluated the target sites for 59 carcinogenic nitro/amino compounds (93% of which were Ames-positive) and 30 carcinogenic electrophilic agents (83% Ames-positive), they were all said to have been previously associated with genotoxic carcinogens. In contrast, the target sites for the 57 carcinogenic non-alerting chemicals (only two of which were mutagenic in the Ames test) were said to be almost exclusively confined to tissues connected with non-genotoxic carcinogens. This in fact may be an over-simplification. Whilst the 1988 study had identified 16 tissues that were affected only by genotoxic carcinogens, the 1991 study revealed a marked reduction in this number, as some non-genotoxic carcinogens also induced tumours in these tissues. Only the Zymbal's gland (rat) and the lungs (both species) appeared to be exclusively targeted by genotoxic carcinogens. The authors' post-analysis rationalisation of certain putative non-genotoxic carcinogens which were active in tissues previously only associated with genotoxic carcinogens did not explain away all of the discrepancies.

In a further analysis of the pattern of the carcinogenic response among tissues and test groups, Ashby and Paton (1993) aimed to test the tentative conclusions of the 1991 paper on a larger data set, restricting their analysis of genotoxicity to a simple consideration of chemical structure. The data for the analysis were taken from the carcinogen database compiled by Gold *et al.* (1991), which detailed the site of carcinogenesis for 522 chemical carcinogens. Of the 511 rodent carcinogens, 251 had been tested in both rats and mice, 168 in rats only and 92 in mice only. One chemical, benzene, was eliminated from the analysis because of its 'exceptional' carcinogenic effects (it induces an extensive range of tumours in both sexes of rats and mice often at unusual sites, and yet no useful structural alert can be derived without condemning all benzenoid chemicals). Around 70% of the remaining 510 rodent carcinogens were structurally-alerting, *i.e.* 300 of the 418 chemicals tested in rats and 236 of the 342 tested in mice. When these were analysed in terms of the number of tissues affected, 27.7% (rat) and 18.6% (mouse) of the structurally-alerting chemicals were active in more than two tissues, compared with only 5.9% (rat) and 0% (mice) of the non-alerting chemicals. In contrast, a much higher percentage of the non-alerting chemicals affected only a single site (rat, 58.8%; mouse 65.9%), compared with the structurally-alerting chemicals (rat, 34.2%; mouse, 34.6%). Of the 250 chemicals tested in both species, roughly half of the 43 rat-specific carcinogens and of the 64 mouse-specific carcinogens were structurally-alerting chemicals, compared with 77% of the 143 two-species carcinogens. The investigators considered this a confirmation of their earlier hypothesis, that structurally-alerting (or genotoxic) carcinogens tend to be active in more than one species and at multiple sites, whereas the non-alerting (putatively non-genotoxic) carcinogens are more likely to affect only a single species and a single tissue.

In an earlier paper, Ashby and Purchase (1992) had selected five 'genotoxic carcinogens' (positive in the Ames test and with structural alerts) and five 'non-genotoxic carcinogens' (negative Ames and no structural alerts), and had shown that all five genotoxins were carcinogenic to both species and both sexes, and that four of the five were multi-site carcinogens; in contrast, the five non-genotoxins were active in only one species, three of the five affected only a single sex, and four were single-site carcinogens. These ten compounds well supported their earlier hypothesis. However, when Ashby and Paton (1993) tried to demonstrate the same effect amongst over 500 carcinogens, the findings were not so clear cut. Thus, around 34% of the structurally-alerting carcinogens were active in only a single tissue and 35–40% of the non-alerting chemicals were active in more than one tissue. Furthermore, although 110 of the 143 two-species carcinogens were structurally-alerting, this still left 33 which had no structural alerts. For the 107 carcinogens that were tested in both species and found to be active in only one, 53 were structurally-alerting and 54 were not. In a separate study by Gold *et al.* (1993), involving an analysis of 351 mutagenic (in the Ames test), or non-mutagenic rodent carcinogens from the Carcinogenic Potency Database, it was found that 42% of the single-site, single-species carcinogens were mutagenic while 31% of the two-species carcinogens were not. This does not greatly support the distinction between genotoxic carcinogens (multi-species, multi-site) and non-genotoxic carcinogens (single-species, single-site).

Gold *et al.* (1993) also compared the target organs for the mutagenic and non-mutagenic carcinogens, and concluded that both groups induced tumours in a wide variety of sites, that most organs were target sites for both, and that the same sites tended to be the most common targets for both. When the more unusual tumour sites had been excluded (because the number of mutagenic and/or non-mutagenic chemicals that affected them was too small to be meaningful), Gold *et al.* found that only the Zymbal's gland was targeted exclusively by mutagenic carcinogens. The one discordant chemical was benzene, which Ashby and Paton (1993) had eliminated from their own analysis because it was clastogenic but non-alerting and non-mutagenic. The latter investigators also reported the Zymbal's gland to be the sole tissue that was affected only by structurally-alerting chemicals (after benzene had been excluded). Ashby and Tennant (1988) attributed this 'property' to 15 specific sites in 1988, but ten of the sites were subsequently shown also to be susceptible to non-alerting chemicals (Ashby and Paton, 1993); it remains to be seen whether the other sites (including the Zymbal's gland, subcutaneous tissues, clitoral gland, spleen, and ovary) will also be shown, eventually, to be susceptible to non-genotoxic carcinogens. Gold *et al.* (1993) noted that out of 351 rodent carcinogens, 20 induced Zymbal's gland tumours in rats and two induced them in mice (these two being included in the 20 that were active in the rat). The 20 chemicals were also multi-site carcinogens. Of the 230 carcinogens in the overall Carcinogenic Potency Database that were tested in both rats and mice, only 42 (18%) affected multiple sites in both species; but of the 14 that were tested in both species and that induced Zymbal's gland tumours, ten (71%) affected multiple sites in both species. This indicates that chemicals that induce Zymbal's gland tumours are generally multi-site, multi-species carcinogens.

It can be seen that successive attempts to distinguish between genotoxic and non-genotoxic carcinogens on the basis of their tumour profile have become less convincing as the number of carcinogens analysed has increased.

Screening for non-genotoxic carcinogens. Whilst there are a number of STTs that can be used to detect genotoxic chemicals, the problem remains of how to screen for non-genotoxic carcinogens when the actual mechanisms involved are not known in any detail. Carcinogenesis is a complex, multi-stage process, and there are many ways in which its onset and progress may be affected (Green, 1991). The various stages of cancer development have been defined operationally as initiation, promotion, progression, and metastasis. Initiation appears to represent damage to key genes involved in the regulation of cell growth, and genotoxic chemicals are believed to contribute to tumour initiation as a result of their damaging effect on DNA. Other factors such as viruses, UV light, ionising radiation, and error-prone DNA replication may also lead to initiation. The clonal expansion of these initiated cells may result from the action of, for example, growth factors, hormones, or many non-genotoxic carcinogens.

1   Cell proliferation
The induction of cell proliferation appears to be a key factor in non-genotoxic carcinogenicity, and may involve either a direct mitogenic effect on the target without apparent cytolethality or a cytotoxic effect which produces cell death in the target tissue followed by regenerative cell proliferation (Butterworth, 1990). Cell proliferation is a key factor in genotoxic as well as non-genotoxic carcinogenicity, as mutagenic activity may occur as a secondary event in the carcinogenic process (Butterworth *et al.*, 1992). As noted previously, use of the maximum tolerated dose (MTD) in animal carcinogenicity bioassays has been the subject of much debate; such a high dose may cause cell death and subsequent cell proliferation, the development of tumours being secondary to this excessive organ-specific toxicity. Thus, tumours that are induced only at the MTD are of questionable relevance to humans who are generally exposed at much lower levels (Ames and Gold, 1990a and b; Ashby and Morrod, 1991). Information on a chemical's capacity to induce cell proliferation may therefore be very useful in setting bioassay doses or in

evaluating bioassay results. Two measures that have been used to assess the extent of cell proliferation are the mitotic index and the labelling index. These indicate, respectively, the fraction of cells that are in the process of mitosis and the percentage that have taken up a radiolabelled DNA precursor (Butterworth *et al.*, 1992). More complex studies may be conducted to identify the specific receptors that mediate the mitogenic action of some non-genotoxic carcinogens, or to understand how the various proto-oncogenes and tumour suppressor genes that regulate cell growth are influenced by non-genotoxic carcinogens (Green, 1991). These types of study are based on mechanistic considerations and have not been applied to any great extent in testing and validation programmes.

2 Cell transformation assays

The most well-established assays for detecting non-genotoxic (and also genotoxic) carcinogens are the various cell transformation assays, which are based on carcinogen-induced loss of contact inhibition of cultured cells resulting in a piling up of transformed cells in a criss-cross fashion. These assays have been critically evaluated by various investigators and expert groups, who considered them highly relevant for the process of carcinogenesis *in vivo* because they involve the same endpoint (*i.e.* the transformed cells are tumorigenic in appropriate hosts) and because they may well share many of the same cellular and molecular mechanisms (Dunkel *et al.*, 1981; Heidelberger *et al.*, 1983; IARC/NCI/EPA Working Group, 1985). Studies evaluating the performance of cell transformation tests in predicting animal or human carcinogens have generally given promising results (Barrett and Lamb, 1985; DiPaolo *et al.*, 1972; Fitzgerald *et al.*, 1989; Jones *et al.*, 1988; Pienta *et al.*, 1977; Swierenga and Yamasaki, 1992), although the lack of a dose response, the low transformation frequency and the difficulties in obtaining consistently reproducible results in repeat assays have impeded the development of this system for routine use (Jones *et al.*, 1988; Tu *et al.*, 1986).

In an assessment of the Syrian hamster embryo (SHE) cell transformation assay, Jones *et al.* (1988) tested 18 coded chemicals in three different laboratories using the same basic protocol. Rodent carcinogenicity data were available for 16 of the 18 chemicals. When the four chemicals that gave discordant transformation responses in two laboratories were counted as positive, the investigators found the results of the two systems (for carcinogenicity and cell transformation) to be in agreement for 14 of the 16 chemicals. Four rodent carcinogens that were inactive in the Ames test gave positive results in the SHE assay. However, two of the eight purported rodent non-carcinogens, caprolactam and geranyl acetate, were found to be active in the SHE assay in all three laboratories. A non-carcinogen might turn out to be carcinogenic, however, if tested in the appropriate species or strain and under appropriate conditions, but it is interesting to note that caprolactam is the only chemical classified by the International Agency for Research on Cancer (IARC, 1987) as a Group 4 compound (probably not carcinogenic to humans).

Swierenga and Yamasaki (1992) evaluated the performance of transformation tests in identifying IARC Group 1 compounds (carcinogenic to
humans) and Group 2A compounds (probably carcinogenic to humans).
The data from cell transformation tests were considered collectively, so
that a chemical was positive if it gave a positive response in any such assay.
Out of 28 Group 1 carcinogens for which data were available, 25 (89%)
gave a positive response in the cell transformation assay, while for the
three chemicals that were negative, only a single test result was identified.
For comparison, the Ames test and the chromosome aberration test
identified 25 out of 41 (61%) and 25 out of 30 (83%) Group 1 carcinogens,
respectively. For Group 2A compounds, the cell transformation assay
again showed a high level of sensitivity, with 21 of the 25 (84%) giving
positive results; the corresponding figures for the Ames test and the
chromosome aberration test were 34 out of 41 (83%) and 30 out of 31
(96%), respectively. The specificity of the cell transformation assays (in
identifying human non-carcinogens) is more difficult to assess because of
the lack of clearly identified non-carcinogens, but Ennever *et al.* (1987)
reported that only four out of twelve probable human non-carcinogens
(33%) gave negative results in at least one cell transformation assay.
According to Swierenga and Yamasaki (1992), some of these probable
human non-carcinogens have since shown evidence of carcinogenic
activity in animals, but the exclusion of these from the analysis apparently
does not increase the specificity of the cell transformation test. These
investigators nevertheless concluded that the importance of cell transformation tests may have been underestimated in previous IARC
evaluations while that of the Ames test may have been overestimated.
They considered the various assays for all transformation to have
generally shown good agreement, and suggested that further refinements
were needed to include the use of human cells, human xenobiotic
metabolism and appropriate tissue-specific target cells (Swierenga and
Yamasaki, 1992).

3   Gap-junction intercellular tests

It has been suggested that, since normal cells surrounding those that
contain transforming oncogenes are able to suppress the transformed
phenotype, the disruption of intercellular communication *via* gap junctions may play a role in carcinogenesis (Green, 1991; Yamasaki, 1990). In
an evaluation of the performance of gap junction intercellular tests,
Swierenga and Yamasaki (1992) reported that three out of four Group 1
compounds (established human carcinogens) and four out of five Group
2A compounds (probable human carcinogens) gave positive results in this
test. For the two compounds that were not detected, one, crystalline silica,
was thought to exert its carcinogenic effects without disturbing cell-to-cell
communication *via* the gap junction, and the other, diethylstilboestrol was
thought to be cell-type specific, as two steroidal oestrogens gave positive
results in this assay. Four out of five organochlorine pesticides also gave
positive results, and three of the four were classified by IARC as Group 2B

compounds (possible human carcinogens). Whilst these findings are encouraging, the data set is obviously far too small for any definitive conclusions on this type of test. Even so, Swierenga and Yamasaki (1992) considered that the results of such tests should be accorded greater weight than they currently receive.

4 Use of toxicity data

An alternative approach might be to use existing toxicological data on a chemical as an indication of its carcinogenic potential. A relationship between carcinogenicity and systemic toxicity might be expected, given that the highest dose used in carcinogenicity tests is selected on the basis of toxicity. The possible tautologous nature of this relationship has been discussed by Bernstein *et al.* (1985) and Crouch *et al.* (1987). In fact, only limited correlations have been shown between LD_{50} values and carcinogenic potencies of known animal carcinogens (McGregor, 1992; Metzger *et al.*, 1989; Zeise *et al.*, 1984). McGregor (1992) suggested that better correlations might be obtained if the maximum tolerated dose (MTD) were used as a measure of toxicity rather than the LD_{50} value, but that even this assumed that the mechanisms of death and of carcinogenicity were the same in animals exposed to a particular chemical. Haseman and Seilkop (1992) looked for such an association between the MTD and carcinogenicity in an analysis of 326 NTP studies conducted in rodents, and they found the overall concordance between toxicity and carcinogenicity to be only 56%. When 130 NTP carcinogenicity studies, involving around 1500 sex-species-exposure groups, were analysed to see if there was a direct causal relationship between organ toxicity and carcinogenicity, it was concluded that the available data did not support a correlation between chemically-induced toxicity and carcinogenicity (Hoel *et al.*, 1988; Huff, 1992; Tennant *et al.*, 1991). Some chemicals caused organ toxicity without cancer, others induced site-specific cancer with no associated toxicity, and a third group caused toxicity and cancer in the same organ; only seven of the 53 positive carcinogenicity studies were said to have exhibited the types of target organ toxicity that could have been the cause of all observed carcinogenic effects (Hoel *et al.*, 1988). Huff (1992) concluded that it would be scientifically premature to make any inference about the influence of toxicity or of cell replication on chemical carcinogenesis.

A wider approach suggested by Travis *et al.* (1990a and b, 1991) involves making use of all available information on a compound's biological activity; they found that a combination of acute and reproductive toxicity data, mutagenicity data and sub-chronic and chronic tumorigenicity data provided a far better prediction of carcinogenic potency than did mutagenicity data alone. Unfortunately the method cannot be used to distinguish between carcinogens and non-carcinogens, as it can only predict the carcinogenic potency of known mouse carcinogens.

It can be seen that, for non-genotoxic carcinogens, mechanistic considerations must play a major role in any assessment of risk. Until the various mechanisms of action are more fully understood, it is difficult to see how a suitable battery of short-term tests can be developed to screen against the carcinogenic effects of such chemicals. This area merits intensive study if it is to be systematised to the extent necessary to be of use in carcinogen regulation (Clayson and Arnold, 1991).

Despite all the problems identified, the short term tests do have a role in predicting carcinogens. IARC is currently using data from short term tests, alongside animal and epidemiological evidence to reclassify carcinogens. Until 1987, there were 23 human chemical carcinogens and 7 from industrial processes. Now as a result of reclassification, there are 55 recognised human carcinogens.

In addition to their use in predicting carcinogens, short term tests when used with due consideration, have a role to play in mechanistic studies, for examining complex mixtures and air pollutants and for the early identification of the genetic toxicity of new chemical products.

B   Approach 2—The Direct Measurement of Genetic Damage in Man

An approach to measuring genetic damage in man has been to investigate the incidence of chromosome damage in peripheral lymphocytes of subjects exposed to a given chemical (Anderson, 1988; Kilian *et al.*, 1975; Purchase *et al.*, 1978). It has been reported by several authors, for example, that workers occupationally exposed to vinyl chloride have an increase in the incidence of chromosome damage (Ducatman *et al.*, 1975; Fleig and Theiss, 1977; Kilian *et al.*, 1975; Natarajan *et al.*, 1978; Szentesi *et al.*, 1976). In the study of Purchase *et al.* (1978b), 81 workers were investigated (57 VC exposed and 24 controls) and effects found in the exposed group. Eighteen months later, the incidence of chromosome damage was still present, though it had decreased in those workers who had changed jobs. At 42 months the incidence of damage had returned to control values (Anderson *et al.*, 1980). Many other chemicals had been investigated in a similar way, *e.g.* ethylene oxide, epichlorohydrin, styrene, butadiene, acrylonitrile, asbestos, benzene, chloroprene, cyclophosphamide, chromium, lead, *etc.*

Such studies have to be well-controlled with age- and sex-matched individuals and confounding factors, such as drinking and smoking, taken into account (Anderson *et al.*, 1990; Brinkworth *et al.*, 1992; Dewdney *et al.*, 1986). As yet, the studies yield results which can only be interpreted on a group basis and not used for individual counselling. In these circumstances, chromosome analysis is useful for determining whether exposure levels of a chemical do or do not induce chromosome damage. At a meeting in Luxembourg in 1987 (convened by the EEC, IPCS, WHO, IARC, and Institute of Occupational Health, Helsinki) on Human

Monitoring, a concensus opinion was that increased levels of chromosome damage were indicative of an increased risk of cancer even though no causal relationship had yet been established between chromosome damage and cancer. However, most of the agents causing chromosome damage have been shown to be carcinogenic to animals. Sorsa *et al.* (1990) in a preliminary study, however, did suggest a causal relationship between chromosome damage and cancer, but not between SCE and cancer.

Other techniques for direct application to man are available, such as the use of urine or blood plasma from exposed workers in combination with a microbial assay (Legator *et al.*, 1982, Gene-Tox), electrophoretic monitoring of enzymatic markers in man (Neel, 1979b) detection of variants in haemoglobin molecules (Nute *et al.*, 1976; Popp *et al.*, 1979), determining increases in the formation of haemoglobin adducts (Farmer, 1982; Farmer *et al.*, 1986; Tornquist *et al.*, 1988) and DNA adducts (Pfeifer *et al.*, 1993; Phillips *et al.*, 1988; Weston, 1993), investigations of sperm morphology (Wyrobek *et al.*, 1983b, Gene-Tox), increases in YY bodies in sperm (Kapp *et al.*, 1979), thioguanine mutant frequency in human lymphocytes (Albertini *et al.*, 1988; Cole *et al.*, 1988), and oncoproteins in plasma (Brinkworth *et al.*, 1992).

A special issue of Mutation Research reviews Human Monitoring methods (Anderson, 1988) and there is another issue in preparation (Anderson, 1994).

C Approach 3—Measurement of Gonadal Effects in Mammalian Systems and Extrapolation to Man Based on Principles Determined from Radiation Genetics

The virtual universality of DNA as the genetic material furnishes a rationale for using various sub-mammalian and non-human test systems for such predicting mutagenic potential. Nevertheless, some organisms may be considered more similar to man than others and thus more suitable for the purpose of evaluating human genetic risk. In man, however, there are many pharmacokinetic factors which affect the mutagenic efficiency of a compound.

The assessment of heritable mutagenic risk to humans involves determining gonadal effects and acknowleding the influence of many variables. Results can be assessed on the basis of expected human exposure and known mutagenic potential from such such studies as dominant lethal, heritable translocation, or specific locus assays. Similarly, the concentration of the mutagen in germ cells can be assessed biochemically, extrapolated to human exposure levels, and combined with mutagenicity data from the most-used and well-understood *in vitro* test systems.

It has been suggested that the population effects of chemical mutagens should take radiation as an equivalent and equate the population dose of those mutagens to the radiation dose admissible for that population (Bridges, 1974; Crow, 1973); or should not exceed a dose which doubles the spontaneous mutation rate of that population. Both concepts, however, have been criticised (Auerbach, 1975; Schalet and Sankaranarayanan, 1976; Sobels, 1977). An approach currently regaining favour is the parallelogram approach of Sobels (1977) where with data from rodent germ cells and human and rodent somatic cells, human germ cell data can be predicted/estimated. An EEC/US EPA collaboration has been instigated (1993) to examine this approach.

7   General Conclusions

The present methods available for the testing for mutagenicity are not equally reliable or reproducible. Even the systems most frequently used (*e.g.* Ames test) give different results in different circumstances.

Since no one test system satisfactorily detects all genetic end-points, a combination of tests is required. Such a 'battery' should preferentially consist of a microbial test and a test detecting chromosome damage. This combination would be the minimum required. Currently in some countries an *in vitro* assay for gene mutation is also carried out (Arlett and Cole, 1988). If a larger 'battery' is to be used to identify those chemicals which are potentially hazardous to man, a mammalian *in vivo* system should be considered. By making the test battery too large a greater number of false results may be generated (Purchase *et al.,* 1971; 1978a; 1980; Tennant *et al.,* 1987). It is possible that the 'gap' that exists between mutagenicity and carcinogenicity may be partially bridged at an early stage of testing with a cell transformation assay (Heidelberger *et al.,* 1983; Meyer *et al.,* 1984; Dunkel, 1991; DoH COC Guidelines, HMSO, 1991; Fitzgerald and Yamasaki, 1990).

The need for safety evaluation in general toxicological testing is well recognised and this is certainly true in the field of genetic toxicology. However, it is not certain that positive or negative results in laboratory model test systems are relevant to man because of man's unique metabolism and because of the absence of any convincing 'no-effect' level data for animals or man. Epidemiological evidence for germ cell mutation after chemical exposure (or, in fact, any agent) is lacking. In the industrial situation, it is often difficult to identify the exact chemical or agent that may be causing a problem. Unbiased abortion rates are difficult to determine by comparison with control or unexposed populations. Not all abortions are recorded.

The limitations of the simpler short term tests, which are more concerned with the concept of somatic mutation than of heritable genetic damage, are now better understood (Purchase, 1980; Tennant *et al.,* 1987; Ashby and Tennant, 1988; Ashby and Paton, 1993; Waters *et al.,* 1993) but to extrapolate to man in terms of heritable damage, *in vivo* animal studies are required. The logistics of tests for this purpose (the specific locus test and heritable translocation test), however, are difficult to satisfy.

There is a school of opinion which holds that just as mutagenicity tests might detect carcinogens, so might carcinogenicity tests detect mutagens. This opinion is based on the concept that cancer may be an easily observable phenotype of DNA mutation. If this is so, then the carcinogenicity and mutagenicity data for a chemical should be considered together. This is probably true for genotoxic carcinogens and IARC is using this approach.

It is hoped that as research progresses, our understanding and techniques will improve so that the results generated in our model systems will become unequivocal in terms of hazard to man. To achieve this aim, attention will have to be given to studies aimed at assessing the significance to man of positive mutagenic responses produced by a test system for a given chemical, in addition to the search for better assay procedures and the understanding and interpretation of effects from chemicals producing mutation by indirect means.

References

Abrahamson, S. and Lewis S. B. (1971). The detection of mutations in *Drosophila*: *In*: 'Chemical Mutagens, Principles and Methods for their Detection', *Ed.* A. Hollaender, Plenum Press, New York, London, Vol. 2, pp. 461–489.

Adler, I. D., Ashby, J., and Würgler, F. E. (1989). Screening for possible human carcinogens and mutagens: a symposium report. *Mutat. Res.*, **213**, 27–39.

Adler, I.-D. and Kliesch, U. (1990). Comparison of single and multiple treatment regimens in the mouse bone marrow micronucleus assay for hydroquinone (HQ) and cyclophosphamide (CP). *Mutat. Res.*, **234**, 115–123.

Albanese, R. (1987). Mammalian male germ cell cytogenetics. *Mutagenesis*, **2**, 79–87.

Albanese, R., Topham, J. C., Evans, E., Clare, M. G., and Tease, C. (1984). Mammalian germ-cell cytogenesis. The report of the UKEMS Sub-Committee on Guidelines for Mutagenicity Testing. Part 2, pp. 145–172.

Albertini, S. (1989). Influence of different factors on the induction of chromosomal-segregation in *Saccharomyces cerevisiae* D61.M by baviston and assessment of its genotoxic property in the Ames test and in *Saccharomyces cerevisiae* D7. *Mutat. Res.*, **216**, 327–340.

Albertini, S. (1991). Re-evaluation of the 9 compounds reported conclusive positive in yeast *Saccharomyces cerevisiae* aneuploidy test systems by the Gene-Tox Program using strain D61.M of *Saccharomyces cerevisiae*. *Mutat. Res.*, **260**, 165–180.

Albertini, S., Brunner, M., and Würgler, F. E. (1993). Analysis of six additional chemicals for *in vitro* assays of the European Economic Communities (EEC) aneuploidy programme using *Saccharomyces cerevisiae* D61.M and the *in vitro* Porcine Brain Tubulin Assembly assay. *Environ. Mol. Mutagen.*, **21**, 180–192.

Albertini, R. J., Sullivan, L. M., Berman, J. K., Greene, C. J., Stewart, J. A., Silveira, J. M., and O'Neill, J. P. (1988). Mutagenicity monitoring in humans by autoradiographic assay for mutant T lymphocytes. *Mutat. Res.*, **204**, 481–492.

Ames, B. N., Durston, W. E., Yamasaki, E., and Lee, F. D. (1973a). Carcinogens are mutagens: a simple test system combining liver homogenates for activation and bacteria for detection. *Proc. Natl. Acad. Sci. (USA)*, **70**, 2281–2285.

Ames, B. N. and Gold, L. S. (1990a). Chemical carcinogenesis: too many rodent carcinogens. *Proc. Natl. Acad. Sci (USA)*, **87**, 7772–7776.

Ames, B. N. and Gold L. S. (1990b). Too many rodent carcinogens: mitogenesis increases mutagenesis. *Science,* **249**, 970–971.

Ames, B. N., Lee, F. D., and Durston, W. E. (1973b). An improved bacterial test system for the detection and classification of mutagens and carcinogens. *Proc. Natl. Acad. Sci. (USA)*, **70**, 782–786.

Ames, B. N., McCann, J., and Yamasaki, E. (1975). Methods for detecting carcinogens and mutagens with the Salmonella/mammalian microsome mutagenicity test. *Mutat. Res.,* **31**, 347–364.

Anderson, D. (1975a). The selection and induction of 5-iodo-2-deoxyuridine and thymidine variants of P388 mouse lymphoma cells with agents with are used for selection. *Mutat. Res.,* **33**, 399–406.

Anderson, D. (1975b). Attempts to produce systems for isolating spontaneous and induced variants in various mouse lymphoma cells using a variety of selective agents. *Mutat. Res.,* **33**, 407–416.

Anderson, D. (1988). Human Monitoring—Special Issue. *Mutat. Res.,* **204**, No. 3.

Anderson, D. and Cross, M. F. (1982). Studies with 4-chloromethylbiphenyl in P388 cells resistant to 5-iodo-2-deoxyuridine. *Mutat. Res.,* **100**, 257–261.

Anderson, D. and Cross, M. F. (1985). Suitability of the P388 mouse lymphoma system for detecting potential carcinogens and mutagens. *Food Chem. Toxicol.,* **23**, 115–118.

Anderson, D., Dewdney, R. S., Jenkinson, P. C, Lovell, D. P., Butterworth, K. R., and Conning, D. M. (1985). Sister chromatid exchange in 106 control individuals. *In:* 'Monitoring of Occupational Genotoxicity'. Proceedings of the Fourth ICEM Satellite Meeting, Helsinki, June 30–July 2. Eds M. Sorsa and H. Norppa, pp. 38–58. Alan R. Liss, and abstract in *Hum. Toxicol.,* **4**, 79–116.

Anderson, D. and Fox, M. (1974). The induction of thymidine and IUdR resistant variants in P388 mouse lymphoma cells by X-rays, UV, and mono- and bifunctional alkylating agents. *Mutat. Res.,* **25**, 107–122.

Anderson, D., Francis, A. J., Godbert, P., Jenkinson, P. C., and Butterworth, K. R. (1990). Chromosome aberrations (CA), sister chromatid exchanges (SCE) and mitogen induced blastogenesis in cultured peripheral lymphocytes from 48 control individuals sampled 8 times over 2 years. *Mutat. Res.,* **250**, 467–476.

Anderson, D., Green, M. H. L., Mattern, I. E., and Godley, M. J. (1984). An international collaborative study of 'Genetic drift' in *Salmonella typhimurium* strains used in the Ames test. *Mutat. Res.,* **130**, 1–10.

Anderson, D., Jenkinson, P. C., Dewdney, R. S., Francis, A. J., Godbert, P., and Butterworth, K. R. (1988). Chromosome aberrations, mitogen-induced blastogenesis and proliferative rate index in peripheral lymphocytes from 106 control individuals of the UK population. *Mutat. Res.,* **204**, 407–420.

Anderson, D., Jenkinson, P. C., Edwards, A. J., Evans, J. G., and Lovell, D. P. (1993). Male-mediated F_1 abnormalities. *In:* 'Reproductive Toxicology', *Ed.* M. Richardson. VCH Verlagsgesellschaft, Weinheim, pp. 101–116.

Anderson, D., McGregor, B. D., and Bateman, A. J. (1983). Dominant lethal mutation assays. Report of the UKEMS sub-committee on guidelines for mutagenicity testing, pp. 43–164.

Anderson, D. and Purchase, I. F. H. (1983). The mutagenicity of food. *In*: 'Toxic Hazards in Food', *Eds* D. M. Conning and A. B. G. Lansdown. Croom Helm Ltd., London, pp. 145–182.

Anderson, D. and Richardson, C. R. (1981a). Issues relevant to the assessment of chromosome damage *in vivo*. *Mutat. Res.*, **90**, 261–272.

Anderson, D., Richardson, C. R., and Davies, P. J. (1981b). The genotoxic potential of bases and nucleosides. *Mutat. Res.*, **91**, 265–272.

Anderson, D., Richardson, C. R., Purchase, I. F. H., Weight, T. M., and Adams, W. G. F. (1980). Chromosomal analyses in vinyl chloride exposed workers. *Mutat. Res.*, **70**, 151–162.

Anon. (1991). Early indicators of non-genotoxic carcinogenesis. *Mutat. Res.*, **248**, 211–376 (Special Issue).

Arlett, C. F. (1977). Mutagenicity testing with V79 Chinese hamster cells. 'Handbook of Mutagenicity Test Procedures', *Eds* B. J. Kilbey, M. Legator, W. Nichols, and C. Ramel. Elsevier Biomedical Press, North-Holland, Amsterdam, pp. 175–191.

Arlett, C. F. and Cole, J. (1988). The role of mammalian cell mutation assays in mutagenicity and carcinogenicity testing. *Mutagenesis*, **3**, 455–458.

Ashby, J. (1985). Fundamental structural alerts to potential carcinogenicity or non-carcinogenicity. *Environ. Mutagen.*, **7**, 919–921.

Ashby, J. (1986a). The prospects of simplified and internationally harmonized approach to the detection of possible human carcinogens and mutagens. *Mutagenesis*, **1**, 3–16.

Ashby, J. (1986b). Letter to the Editor, *Mutagenesis*, **1**, 309–317.

Ashby, J. (1988a). The separate identities of genotoxic and non-genotoxic carcinogens. *Mutagenesis*, **3**, 365–366.

Ashby, J. (1988b). An opinion on the significance of the 19 non-clastogenic gene-mutagens reported by Tennant *et al.* (1987). *Mutagenesis*, **3**, 463–465.

Ashby, J., de Serres, F. J., Draper, M., Ishidate, M., Jr., Margolin, B. H., Matter, B. E., and Shelby, M. D. (1985). Evaluation of short term tests for carcinogens. Report of the International Programme on chemical safety's collaborative study on *in vitro* assays. Vol. 5. Elsevier-Science Publishers, Amsterdam, Oxford, New York.

Ashby, J., de Serres, F. J., Shelby, M. D., Margolin, B. H., Ishidate, M., and Becking, G. (1988). Evaluation of short term tests for carcinogens. Report of the International Programme on Chemical Safety's Collaborative Study on *in vitro* assay, Vol. 1; and *in vivo* assays, Vol. 11. Cambridge University Press on behalf of the World Health Organisation.

Ashby, J. and Liegibel, U. (1992). Transgenic mouse mutation assays. Potential for confusion of genotoxic and non-genotoxic carcinogenesis. A proposed solution. *Environ. Mol. Mutagen.*, **20**, 145–147.

Ashby, J. and Liegibel, U. (1993). Dosing regimens for transgenic animals mutagenesis assays. *Environ. Mol. Mutagen.*, **21, 2**, 120–121.

Ashby, J. and Mirkova, E. (1987). The activity of MNNG in the mouse bone marrow micronucleus assay. *Mutagenesis*, **2**, 199–205.

Ashby, J. and Morrod, R. S. (1991). Detection of human carcinogens. *Nature (London)*, **352**, 185–186.

Ashby, J. and Paton, D. (1993). The influence of chemical structure on the extent and sites of cacinogenesis for 522 rodent carcinogens and 55 different human carcinogen exposures. *Mutat. Res.,* **286**, 3–74.

Ashby, J. and Purchase, I. F. H. (1988). Reflections on the declining ability of the *Salmonella* assay to detect rodent carcinogens as positive. *Mutat. Res.,* **205**, 51–58.

Ashby, J. and Purchase, I. F. H. (1992). Non-genotoxic carcinogens: an extension of the perspective provided by Perera. *Environ. Hlth. Perspect.,* **98**, 223–226.

Ashby, J. and Styles, J. A. (1978). Factors influencing mutagenic potency *in vitro*. *Nature (London),* **274**, 20–22.

Ashby, J. and Tennant, R. W. (1988). Chemical structure, *Salmonella* mutagenicity and extent of carcinogenicity as indicators of genotoxic carcinogenesis amoung 222 chemicals tested in rodents by the US NCI/NTP. *Mutat. Res.,* **204**, 17–115.

Ashby, J. and Tennant, R. W. (1991). Definitive relationships among chemical structure, carcinogenicity and mutagenicity for 301 chemicals tested by the US NTP. *Mutat. Res.,* **257**, 229–306.

Ashby, J., Tennant, R. W., Zeiger, E., and Stasiewicz, S. (1989). Classification according to chemical structure, mutagenicity to *Salmonella* and level of carcinogenicity of a further 42 chemicals tested for carcinogenicity by the US National Toxicology Program. *Mutat. Res.,* **223**, 73–103.

Auerbach, C. (1975). The effects of six years of mutagen testing on our attitude to the problems posed by it. *Mutat. Res.,* **33**, 3–10.

Auerbach, C. (1977). The role of *Drosophila* in mutagen testing. *In*: 'Topics in Toxicology, Mutagenesis in Sub-mammalian Systems. Status and Significance', *Ed.* G. Paget. MTP Press Ltd., Lancaster, pp. 13–20.

Auerbach, C., Robson, J. M., and Carr, J. G. (1947). The chemical production of mutations. *Science,* **105**, 243.

Auletta, A. and Ashby, J. (1988). Workshop on the relationship between short-term test information and carcinogenicity; Williamsburg, Virginia, January 20–23, 1987. *Environ. Mol. Mutagen.,* **11**, 135–145.

Baker, B. S., Boyd, J. B., Carpenter, A. T. C., Green, M. M., Nguyen, T. D., Tipoll, P., and Smith, P. D. (1976). Genetic controls of meiotic recombination and somatic DNA metabolism in *Drosophila melanogaster. Proc. Natl. Acad. Sci. (USA),* **73**, 4140–4144.

Barrett, J. C. (1985). Cell culture models of multistep carcinogens. *In*: 'Age-related factors in carcinogenesis', *Eds* A. Likhacker, V. Arisimov, and R. Montesano. IARC, Lyon, pp. 181–202.

Barrett, J. C., Crawford, B. D., Grady, D. L., Hester, L. D., Jones, P. A., Benedict, W. F., and Ts'o, P. O. P. (1977). Temporal acquisition of enhanced fibrinolytic activity by Syrian hamster embryo cells following treatment with benzo(*a*)pyrene. *Cancer Res.,* **37**, 3815–3823.

Barrett, J. C., Kakunaga, T., Kuroki, T., Neubert, D., Troske, J. E., Vasilieu, J. M., Williams, G. M., and Yamasaki, H. (1986). Mammalian Cell Transformation in Culture. *In*: 'Long and short term test assays for carcinogens. A critical appraisal', *Eds* R. Montesano, H. Bartsch, H. Vainio, J. Wilbourn, and H. Yamasaki. IARC Sci Publ. No. 83, Lyon, pp. 267–286.

Barret, J. C. and Lamb, P. W. (1985). Tests with the Syrian hamster embryo cell transformation assay. *In*: 'Progress in Mutation Research. Volume 5. Evaluation of Short-Term Tests for Carcinogens. Report of the International

Programme on Chemical Safety's Collaborative Study on *In Vitro* Assays', *Eds* J. Ashby, F. J. de Serres, M. Draper, M. Ishidate Jr., B. H. Margolin, B. E. Matter, and M. D. Shelby. Elsevier, Amsterdam, pp. 623–628.

Bateman, A. J. (1977). The dominant lethal assay in the mouse. *In*: 'Handbook of Mutagenicity Test Procedures', *Eds* B. J. Kilbey, M. Legator, W. Nichols, and C. Ramel. Elsevier Biochemical Press, North-Holland, Amsterdam, pp. 225–255.

Bateman, A. J. and Epstein, S. S. (1971). Dominant lethal mutations in mammals. *In*: 'Chemical Mutagens, Principles and Methods for their Detection'. Plenum Press, New York, Vol. 2, pp. 541–568.

Bartsch, H., Malaveille, C., and Montesano, R. (1975). Human rat and mouse liver-mediated mutagenicity of vinyl chloride in *S. typhimurium* strains. *Int. J. Cancer*, **15**, 429–437.

Benigni, R. and Giuliani, A. (1988). Statistical exploration of four major genotoxicity data bases: an overview. *Environ. Mol. Mutagen.*, **12**, 75–83.

Bernstein, L., Gold, L. S., Ames, B. N., Pike, M. C., and Hoel, D. G. (1985). Some tautologous aspects of the comparison of carcinogenic potency in rats and mice. *Fund. Appl. Toxicol.*, **5**, 79–86.

Berwald, Y. and Sachs, L. (1963). *In vitro* transformation with chemical carcinogens. *Nature (London)*, **200**, 1182–1184.

Berwald, Y. and Sachs, L. (1965). *In vitro* transformation of normal cells to tumour cells by carcinogenic hydrocarbons. *J. Natl. Cancer Inst.*, **35**, 641–661.

Bignami, M., Morpugo, C., Pagliani, R., Carere, A., Conte, G., and Di Guiseppe, G. (1974). Non-disjunction and crossing-over induced by pharmaceutical drugs in *Aspergillus nidulans*. *Mutat. Res.*, **26**, 159–170.

Bond, D. J. (1976). A system for the study of meiotic non-disjunction using *Sodaria brevicollis*. *Mutat. Res.*, **39**, 213–220.

Bootman, J. and Kilbey, B. J. (1983). Recessive lethal mutations in *Drosophila*. The report of the UKEMS sub-committe on guidelines for mutagenicity testing. Part 1.

Bradley, M. O., Bhuyan, B., Francis, M. C., Langenbach, R., Peterson, A., and Huberman, E. (1981). Mutagenesis by chemical agents in V79 Chinese hamster cells: A review and analysis of literature. A report of the US Environmental Protection Agency Gene-Tox program. *Mutat. Res.*, **87**, 81–142.

Brewen, J. C., Payne, H. S., Jones, K. P., and Preston, R. J. (1975). Studies on chemically-induced dominant lethality. 1. The cytogenetic basis of MMS-induced dominant lethality in post-meiotic male germ cells. *Mutat. Res.*, **33**, 239–250.

Bridges, B. A. (1974). The three-tier approach to mutagenicity screening and the concept of the radiation equivalent dose. *Mutat. Res.*, **26**, 335–340.

Brinkworth, M. H., Yardley-Jones, A., Edwards, A. J., Hughes, J. A., and Anderson, D. (1992). A comparison of smokers and non-smokers with respect to oncogene products and cytogenetic parameters. *J. Occup. Med.*, **34**, 1181–1188.

Brockman, H. E., de Serres, F. J., Ong, T.-M., DeMarini, D. M., Katz, A. J., Griffiths, A. J. F., and Stafford, R. S. (1984). Mutation tests in *Neurospora crassa*. A report of the US Environmental Protection Agency Gene-Tox Program. *Mutat. Res.*, **133**, 87–134.

Brockman, H. E. and DeMarini, D. M. (1988). Utility of short-term tests for genetic toxicity in the aftermath of the NTP's analysis of 73 chemicals. *Environ. Mol. Mutagen.*, **11**, 421–435.

Bruce, W. R., Varghese, A. J., Furrer, R., and Land, P. C. (1977). A mutagen in the faeces of normal human. *In*: 'Origins of Human Cancer', *Eds* H. H. Hiatt, J. D. Watson, and J. A. Winsten. Cold Spring Harbor Laboratory, New York, Vol. C, pp. 1641–1646.

Brusick, D. J. (1972). Induction of cyclohexamide resistant mutants in *Saccharomyces cerevisiae* with *N*-methyl-*N*-nitronitrosoguanidine and ICR-70. *J. Bacteriol.*, **109**, 1134.

Brusick, D. J. (1982). Genetic toxicology. *In*: 'Principles and Methods of Toxicology', *Ed*. A. W. Hayes. Raven Press, New York, pp. 223–272.

Brusick, D. J. (1987). Genotoxicity produced in mammalian cell assays by treatment conditions. Special issue. *Mutat. Res.*, **189**.

Brusick, D. J. and Auletta, A. (1985). Development status of bioassays in genetic toxicology. A report of Phase II of the US Environmental Protection Agency Gene-Tox Program. *Mutat. Res.*, **153**, 1–10.

Brusick, D. J., Gopalan, H. N. B., Heseltine, E., Huismans, J. W., and Lohman, P. H. M. (1992). Assessing the risk of genetic damage. *UNEP/ICPEMC*, Hodder and Stoughton, pp. 1–52.

Brusick, D. J., Simmons, V. F., Rosenkranz, H. S., Ray, V. A., and Stafford, R. S. (1980). An evaluation of the *Escherichia coli* WP2 and WP2 *uvrA* reverse mutation assay. *Mutat. Res.*, **76**, 169–190.

Buckton, K. E. and Evans, H. J. (1973). Methods for the analysis of human chromosome aberrations. WHO Publication, Geneva.

Butterworth, B. E. (1990). Consideration of both genotoxic and nongenotoxic mechanisms in predicting carcinogenic potential. *Mutat. Res.*, **239**, 117–132.

Butterworth, B. E., Popp, J. A., Conolly, R. B., and Goldsworthy, T. L. (1992). Chemically induced cell proliferation in carcinogenesis. *In*: 'Mechanisms of Carcinogenesis in Risk Identification', *Eds* H. Vainio, P. N. Magee, D. B. McGregor, and A. J. McMichael. IARC, Lyon, pp. 279–305.

Cachiero, N. L. A., Russel, L. B., and Swarthout, M. S. (1974). Translocation, the predominant cause of total sterility in sons of mice treated with mutagens. *Genetics*, **76**, 73–91.

Carter, C. O. (1977). The relative contribution of mutant genes and chromosome aberrations in genetic ill health. *In*: 'Progress in Genetic Toxicology', *Eds* D. Scott, B. A. Bridges, and F. H. Sobels. Elsevier Biomedical Press, North-Holland, Amsterdam, pp. 1–14.

Chu, E. G. Y., Brimer, P., Jacobson, K. B., and Merriam, E. V. (1976). Mammalian cell genetics. 1. Selection and characterisation of mutations auxotrophic for 1-glutamine or resistance to 8-azaguanine in Chinese hamster cells *in vitro*. *Genetics*, **62**, 359–377.

Čihák, R and Vontorková, M. (1990). Activity of acrylamide in single-, double-, and triple-dose mouse bone marrow micronucleus assays. *Mutat. Res.*, **234**, 125–127.

Clayson, D B. and Arnold, D. L. (1991). ICPEMC Publication No. 19. The classification of carcinogens identified in the rodent bioassay as potential risk to humans: What type of substance should be tested next? *Mutat. Res.*, **257**, 91–106.

Cleaver, J. E. (1975). Methods for studying repair of DNA damaged by physical and chemical carcinogens. *In*: 'Methods of Cancer Research', *Ed*. H. Busch. Academic Press, New York, Vol. 9, pp. 123–165.

Cleaver, J. E. (1977). Methods for studying excision repair of DNA damaged by physical and chemical mutagens. *In*: 'Handbook of Mutagenicity Test Procedures', *Eds* B. J. Kilbey, M. S. Legator, W. Nichols, and C. Ramel. Elsevier Biomedical Press, North-Holland, Amsterdam, pp. 19–47.

Clive, D. W. (1973). Recent developments with the L5178TK heterozygote mutagen assay system. *Environ. Hlth. Perspect.*, **6**, 119–125.

Clive, D. W., Flamm, W. G., Machesko, M. R., and Bernheim, N. H. (1972). Mutational assay system using the thymidine kinase locus in mouse lymphoma cells. *Mutat. Res.*, **16**, 77–87.

Clive, D. W., McCuen, R., Spector, J. F. S., Piper, C., and Mavournin, K. H. (1983). Specific gene mutations in L5178Y cells in culture. A report of the US Environmental Protection Agency Gene-Tox Program. *Mutat. Res.*, **155**, 225–251.

Clive, D. W. and Spector, J. F. S. (1975). Laboratory procedure for assessing specific locus mutations at the TK locus in cultured L5178Y mouse lymphoma cells. *Mutat. Res.*, **31**, 17–29.

Cohen, M. M. and Hirschhorn, K. (1971). Cytogenetic studies in animals. *In*: 'Chemical Mutagens, Principles and Methods for their Detection', *Ed*. A. Hollaender. Plenum Press, New York, Vol. 2, pp. 515–534.

Clode, S. A. and Anderson, D. (1988a). High performance liquid chromatographic analysis of bases and nucleosides in mouse testes following *in vivo* thymidine administration. *Mutat. Res.*, **200**, 63–66.

Clode, S. A. and Anderson, D. (1988b). Germ and somatic cell abnormalities following *in vivo* administration of thymidine and adenine. *Mutat. Res.*, **200**, 249–254.

Clode, S. A., Anderson, D., and Gangolli, S. D. (1986). Studies with unbalanced precursor pools *in vivo*. *In*: 'Genetic Toxicology of Environmental Chemicals, Part A, Basic principles and mechanisms of action', *Eds* C. Ramel, B. Lambert, and J. Magnusson, Liss, New York, pp. 551–561.

Cole, J. and Arlett, C. F. (1976). Ethyl methanesulphonate mutagenesis with L5178Y mouse lymphoma cells. A comparison of ouabain, thioguanine, and excess thymidine resistance. *Mutat. Res.*, **24**, 507–531.

Cole, J., Arlett, C. F., and Green, M. H. L. (1976). The fluctuation test as a more sensitive system for determining induced mutation in L5178Y mouse lymphoma cells. *Mutat. Res.*, **41**, 377–386.

Cole, J., Fox, M., Garner, R. C., McGregor, D. B., and Thacker, J. (1983). Gene mutation assay in cultured mammalian cells. The report of the UKEMS sub-committee on guidelines for mutagenicity testing. Part 1, pp. 65–102.

Cole, J., Green, M. H. L., James, S. E., Henderson, L., and Cole, H. (1988). A further assessment of factors influencing measurements of thioguanine-resistant mutant frequency in circulating T-lymphocytes. *Mutat. Res.*, **204**, 493–507.

Cole, J., McGregor, D. B., Fox, M., Thacker, J., Garner, R. C. (1990). Gene mutation assays in cultured mammalian cells. *In*: 'Basic Mutagenicity Tests. UKEMS Recommended Procedures', *Ed*. D. J. Kirkland. Cambridge University Press, Cambridge, pp. 87–114.

Combes, R. D., Anderson, D., Brooks, T. M., Neale, S., and Venitt, S. (1984). Mutagens in urine, faeces and body fluids. The report of the UKEMS sub-committee on guidelines for mutagenicity testing. Part 2, pp. 203–243.

Commoner, B., Vithayathal, A. J., and Henry, J. I. (1974). Detection of metabolic carcinogen intermediates in urine of carcinogen fed rats by means of bacterial mutagenesis. *Nature (London)*, **249**, 850.

Constantin, M. J. and Nilan, R. A. (1982a). Chromosome aberration assays in barley (*Hordeum vulgare*). A report of the US Environmental Protection Agency Gene-Tox Program. *Mutat. Res.*, **99**, 13–36.

Constantin, M. J. and Nilan, R. A. (1982b). The chlorophyll-deficient mutant assay in barley (*Hordeum vulgare*). A report of the US Environmental Protection Agency Gene-Tox Program. *Mutat. Res.*, **99**, 37–49.

Constantin, M. J. and Owens, E. T. (1982). Introduction and perspectives of plant genetic and cytogenetic assays. A report of the US Environmental Protection Agency Gene-Tox Program. *Mutat. Res.*, **99**, 1–12.

Crouch, E., Wilson, R., and Zeise, L. (1987). Tautology or not tautology? *J. Toxicol. Environ. Hlth.*, **20**, 1–10.

Crow, J. (1973). Impact of various types of genetic damage and risk assessment. *Environ. Hlth. Perspect.*, **6**, 1–5.

Dean, B. J. and Senner, K. R. (1977). Detection of chemically induced mutation in Chinese hamsters. *Mutat. Res.*, **46**, 403–405.

DeMars, R. (1974). Resistance of cultured human fibroblasts and other cells to purine and pyrimidine analogues in relation to mutagenesis detection. *Mutat. Res.*, **24**, 335–364.

de Serres, F. J. and Ashby, J. (*Eds*) (1981). Evaluation of short term tests for carcinogenesis. *In*: 'Chemical Mutagens. Principles and Methods for their Detection'. Elsevier Biomedical Press, North-Holland, Amsterdam.

de Serres, F. J. and Malling, H. V. (1971). Measurement of recessive lethal damage over the entire genome and at two specific loci in the Ad-3 region of a two component heterokaryon of *Neurospora crassa*. *In*: 'Chemical Mutagens, Principles and Methods for their Detection', *Ed.* A. Hollaender. Plenum Press, New York, London, Vol. 2, pp. 311–341.

Dewdney, R. S., Lovell, D. P., Jenkinson, P. C., and Anderson, D. (1986). Variation in sister chromatid exchange using 106 members of the general UK population. *Mutat. Res.*, **171**, 43–51.

DHSS (1981), No. 24. Guidelines for the testing of chemicals for mutagenicity, Report on Health and Social Subjects. Committee on mutagenicity of chemicals in food, consumer products and the environment. Department of Health and Social Security (8 July, 1981), London, HMSO.

DoH (1989), No. 35. Guidelines for the testing of chemicals for mutagenicity. Committee on mutagenicity of chemicals in food, consumer products and the environment. Department of Health and Social Security (April, 1989), London, HMSO.

DoH (1991), No. 42. Guidelines for the evaluation of chemicals for carcinogenicity. Committee on carcinogenicity of chemicals in food, consumer products and the environment. Department of Health and Social Security (July, 1991), London, HMSO.

DiPaolo, J. A., Donovan, P. J., and Nelson, R. L. (1969a). Qualitative studies of *in vitro* transformation by chemical carcinogens. *J. Natl. Cancer Inst.*, **42**, 867–876.

DiPaolo, J. A., Nelson, R. L., and Donovan, P. J. (1969b). Sarcoma producing cell clones derived from clones transformed *in vitro* by benzo(*a*)pyrene. *Science*, **167**, 917–918.

DiPaolo, J. A., Nelson, R. L., and Donovan, P. J. (1971). Morphological, oncogenic and karyological characteristics of Syrian hamster embryo cells transformed *in vitro* by carcinogenic polycyclic hydrocarbons. *Cancer Res.*, **31**, 1118–1127.

DiPaolo, J. A., Nelson, R. L., and Donovan, P. J. (1972). *In vitro* transformation of Syrian hamster embryo cells by diverse chemical carcinogens. *Nature (London)*, **235**, 278–280.

Ducatman, A., Hirshhorn, K., and Selikoff, I. J. (1975). Vinyl chloride exposure and human chromosome aberrations. *Mutat. Res.*, **31**, 163–168.

Dunkel, V. (1979). Collaborative studies on the *Salmonella* microsome mutagenicity assay. *J Assoc. Anal. Chem.*, **62**, 874–882.

Dunkel, V. C., Pienta, R. J., Sivak, A., and Traul, K. A. (1981). Comparative neoplastic transformation responses of Balb/3T3 cells, Syrian hamster embryo cells, and Rauscher murine leukemia virus-infected Fischer 344 rat embryo cells to chemical carcinogens. *J. Natl. Cancer Inst.*, **67**, 1303–1315.

Dunkel, V. C., Rogers, C., Swierenga, S. H. H., Brillinger, R. L., Gilman, J. P. W., and Nestmann, E. R. (1991). Recommended protocols based on a survey of current practice in genotoxicity testing laboratories. III. Cell transformation in C3H/10T$\frac{1}{2}$ mouse embryo cell, BALB/c3T3 mouse fibroblast and Syrian hamster embryo cell cultures. *Mutat. Res.*, **246**, 285–300.

Dunkel, V., Zeiger, E., Brunswick, D., McCoy, E., McGregor, D., Mortelmans, K., Rosenkranz, H. S., and Simmons, V. F. (1985). Reproducibility of microbial mutagenicity assays. II. Testing of carcinogens and non-carcinogens in *Salmonella typhimurium* and *Escherichia coli*. *Environ. Mutagenesis*, **77**, (Suppl. 5), 1–248.

Durston, W. and Ames, B. N. (1974). A simple method for detection of mutagens in urine, studies with the carcinogen 2-acetylaminofluorene. *Proc. Natl. Acad. Sci. (USA)*, **71**, 737–741.

Ehling, U. H. (1970). Evaluation of presumed dominant skeletal mutations. *In*: 'Chemical Mutagenesis in Mammals and Man', *Eds* F. Vogel and F. Rohrborn. Springer-Verlag, Berlin, Heidelberg, New York, pp. 162–166.

Ehling, U. H., Machemer, L., Buselmaier, W., Dycka, J., Frohbert, H., Kratochvilova, J., Lang, R., Lorke, D., Muller, K. D., Pehn, J., Rohrborn, G., Rollm, R., Schulze-Schencking, M., and Wiseman, H. (1978). Standard protocol for the dominant lethal test on male mice set up by the work group 'Dominant Lethal Mutations of the *ad hoc* committee chemogenetics'. *Arch. Toxicol.*, **39**, 173–185.

Ehrenberg, L. (1971). Higher plants. *In*: 'Chemical Mutagens, Principles and Methods for their Detection', *Ed.* A. Hollaender. Plenum Press, New, York, London, Vol. 2, pp. 365–386.

Ennever, F. K., Noonan, T. J., and Rosenkranz, H. S. (1987). The predictivity of animal bioassays and short term genotoxicity tests for carcinogenicity and non-carcinogenicity to humans. *Mutagenesis*, **2**, 73–78.

Ennever, F. K. and Rosenkranz, H. S. (1989). Application of the carcinogenicity prediction and battery selection method to recent National Toxicology Program short-term test data. *Environ. Mol. Mutagen.*, **13**, 332–338.

Ennever, F. K. and Rosenkranz, H. S. (1987). Prediction of carcinogenic potency by short term genotoxicity tests. *Mutagenesis*, **2**, 39–44.

Epstein, S. S. and Rohrborn, G. (1970). Recommended procedures for testing genetic hazard from chemicals based on the induction of dominant lethal mutation in mammals. *Nature (London)*, **230**, 264.

Evans, E. P. and Phillips, R. J. S. (1975). Inversion heterozygosity and the origin of XO daughters of Bpa/+ female mice. *Nature (London)*, **256**, 40–41.

Evans, H. J. (1976). Cytological methods for detecting chemical mutagens. *In*: 'Chemical Mutagens, Principles and Methods for their Detection', *Ed.* A. Hollaender. Plenum Press, New York, London, Vol. 4, p. 1.

Fahrig, R. (1974). Development of host-mediated mutagenicity tests. 1. Differential response of yeast cells injected into testes of rats and peritoneum of mice and rats to mutagens. *Mutat. Res.*, **26**, 29–36.

Fahrig, R. (1977). Host-mediated mutagenicity tests—yeast systems. Recovery of yeast cells out of testes, liver, lung and peritoneum of rats. *In*: 'Handbook of Mutagenicity Test Procedures', *Eds* B. J. Kilbey, M. Legator, W. Nichols, and C. Ramel. Elsevier Biomedical Press, North-Holland, Amsterdam, pp. 135–147.

Fahrig, R. (1978). The mammalian spot test: A sensitive *in vivo* method for the detection of genetic alterations in somatic cells of mice. *In*: 'Chemical Mutagens, Principles and Methods for their Detection', *Eds* A. Hollaender and F. de Serres. Plenum Press, New York, Vol. 5, pp. 152–176.

Farmer, P. B. (1982). Monitoring of human exposure to carcinogens. *Chem. Br.*, **18**, 790–794.

Farmer, P. B., Bailey, S. M., Gorf, M., Tornqvist, S., Osterman-Golkar, A., Kautianen, and Lewis-Enright, D. P. (1986). Monitoring human exposure to ethylene oxide by the determination of haemoglobin adducts using gas chromatography–mass spectrometry. *Carcinogenesis*, **7**, 637–640.

Festing, M. F. W. and Wolff, G. L. (1976). Quantitative characteristics of potential value in studying mutagenesis. *Genetics*, **92**, No. 1, Part 1, S.173.

Felton, S. J. and Knize, M. G. (1991). Occurrence, identification, and bacterial mutagenicity of heterocyclic amines in cooked food. *Mutat. Res.*, **259**, 205–217.

Fielding, R. (1987). Letter to the Editor. Mutagenicity Testing. *Mutagenesis*, **2**, 315–316.

Fischer, G. A., Lee, S. Y., and Calabresi, P. (1974). Detection of chemical mutagens using a host-mediated assay L5178Y mutagenesis system. *Mutat. Res.*, **16**, 501–511.

Fitzgerald, D. J., Piccoli, C., and Yamasaki, H. (1989). Detection of non-genotoxic carcinogens in the BALB/c3T3 cell transformation/mutation assay system. *Mutagenesis*, **4(4)**, 286–291.

Fitzgerald, D. J. and Yamasaki, H. (1990). Tumour promotion: models and assay systems. *Teratogen. Carcinogen. Mutagen.*, **10**, 89.

Fleig, I. and Theiss, A. M. (1977). External chromosome studies undertaken on persons and animals with VC Illness (Abstract 2nd Int. Conf. Environ., p. 219). *Mutat. Res.*, **52**, 187.

Fox, M. and Anderson, D. (1974). Induced thymidine and 5-iodo-2-deoxyuridine resistant clones of mouse lymphoma cells. *Mutat. Res.*, **25**, 89–105.

Fox, M., Boyle, J. M., and Fox, B. W. (1976). Biological and biochemical characterisation of purine analogue resistant clones of V79 Chinese hamster cells. *Mutat. Res.*, **35**, 289–310.

Francis, A. J., Anderson, D., Evans, J. G., Jenkinson, P. C., and Godbert, P. (1990). Tumours and malformations in the adult offspring of cyclophosphamide-treated and control male rats. *Preliminary Communication*, **229**, 239–246.

Funes-Cravioto, F. B., Lambert, B., Linsten, L., Ehrenberg, L., Natarajan, A. T., and Osterman-Golkar, S. (1975). Chromosome aberrations in workers exposed to vinyl chloride. *Lancet,* **i,** 459.

Garner, R. C. and Kirkland, D. J. (1986). Reply—Letter to the Editor. *Mutagenesis,* **1,** 233–235.

Gatehouse, D. G., Rowland, I. R., Wilcox, P., Callander, R. D., and Forster, R. (1990). Bacterial mutation assays in basic mutagenicity tests, UKEMS Recommended Procedures. *Ed.* D. K. Kirkland, Cambridge University Press, Cambridge, pp. 13–61.

Gatehouse, D. G. and Tweats, D. J. (1986). Letter to the Editor. *Mutagenesis,* **1,** 307–309.

Generoso, W. M., Bishop, J. B., Gosslee, D. G., Newell, G. W., Sheu, C.-J., and von Hallé, E. (1980). Heritable translocation test in mice. A report of the US Environmental Protection Agency Gene-Tox Program. *Mutat. Res.,* **76,** 191–215.

Generoso, W. M., Cain, K. T., Huff, S. W., and Gosslee, D. G. (1977). Heritable translocation in mice. *In*: 'Chemical Mutagens, Principles and Methods for their Detection', *Ed.* A. Hollaender, Plenum Press, New York, Vol. 5, p. 55.

George, E., Wootton, A. K., and Gatehouse, D. G. (1990). Micronucleus induction by azobenzene and 1,2-dibromo-3-chloropropane in the rat: evaluation of a triple-dose protocol. *Mutat. Res.,* **234,** 129–134.

Gold, L. S., Slone, T. H., Manley, N. B., and Bernstein, L. (1993). Comparison of target organs of carcinogenicity for mutagenic and non-mutagenic chemicals. *Mutat. Res.,* **286,** 75–100.

Gold, L. S., Slone, T. H., Stern, B. R., and Bernstein, L. (1991). Target organs in chronic bioassays of 533 chemical carcinogens. *Environ. Hlth. Perspect.,* **93,** 233–246.

Grant, W. F. (1982). Chromosome aberration assays in *Allium*. A report of the US Environmental Protection Agency Gene-Tox Program. *Mutat. Res.,* **99,** 273–291.

Green, M. H. L. and Lowe, J. E. (1992). Dietary vitamin C modulates the response of human lymphocytes to ionising radiation: Studies with the COMET Assay. '6th International Conference of Environmental Mutagens', Melbourne. Abstract. 27–1.

Green, M. H. L. and Muriel, W. J. (1976). Mutagen testing using tryp + reversion in *Escherichia coli*. *Mutat. Res.,* **38,** 3–32.

Green, M. H. L., Muriel, W. J., and Bridges, B. A. (1976). Use of a simplified fluctuation test to detect low levels of mutagens. *Mutat. Res.,* **38,** 33–42.

Green, S. (1991). The search for molecular mechanisms of non-genotoxic carcinogens. *Mutat. Res.,* **248,** 371–374.

Green, S., Auletta, A., Fabricant, J., Kapp, R., Manandhar, M., Sheu, C.-J., Springer, J., and Whitfield, B. (1985). Current status of bioassays in genetic toxicology—the dominant lethal assay. A report of the US Environmental Protection Agency Gene-Tox Program. *Mutat. Res.,* **154,** 49–67.

Gulati, D. K., Wojciechowski, J. P., and Kaur, P. (1990). Comparison of single-, double-, or triple-exposure protocols for the rodent bone marrow/peripheral blood micronucleus assay using 4-aminobiphenyl and treosulphan. *Mutat. Res.,* **234,** 135–139.

Gunz, D., Shephard, S. E., and Lutz, W. K. (1993). Can non-genotoxic carcinogens be detected with the lacl transgenic mutation assay? *Environ. Mol. Mutagen.,* **21**.

Hansteen, I., Hillestad, L., Thiis-Evensen, E., and Heldaas, S. S. (1978). Effect of vinyl chloride in man. A cytogenic follow-up study. *Mutat. Res.*, **51**, 172–178.

Haseman, J. K., Margolin, B. H., Shelby, M. D., Zeiger, E., and Tennant, R. W. (1988). Do short-term tests predict rodent carcinogenicity? Response. *Science*, **241**, 1233.

Haseman, J. K. and Seilkop, S. K. (1992). An examination of the association betwen maximum-tolerated dose and carcinogenicity in 326 long-term studies in rats and mice. *Fund. Appl. Toxicol.*, **19**, 207–213.

Heddle, J. A. (1973). A rapid *in vivo* test for chromosomal damage. *Mutat. Res.*, **18**, 187–190.

Heddle, J. A., Hite, M., Kirkhart, B., Mavournin, K., MacGregor, J. T., Newell, G. W., and Salamone, M. F. (1983). The induction of micronuclei as a measure of genotoxicity. A report of the US Environmental Protection Agency Gene-Tox Program. *Mutat. Res.*, **123**, 61–118.

Heidelberger, C., Freeman, A. E., Pienta, R. J., Sivak, A., Bertram, J. S., Casto, B. C., Dunkel, V. C., Francis, M. W., Kakunaga, T., Little, J. B., and Schechtman, L. M. (1983). Cell transformation by chemical agents—a review and analysis of the literature. A report of the US Environmental Protection Agency Gene-Tox Program. *Mutat. Res.*, **114**, 283–385.

Ho, Y. L. and Ho, S. K. (1981). Screening of carcinogens with the prophage clts 857 induction test. *Cancer Res.*, **41**, 532–536.

Hoel, D. G., Haseman, J. K., Hogan, M. D., Huff, J., and McConnell, E. E. (1988). The impact of toxicity on carcinogenicity studies: implications for risk assessment. *Carcinogenesis*, **9**, 2045–2052.

Hsie, A. W., Casciano, D. A., Couch, D. B., Krahn, D. F., O'Neill, J. P., and Whitfield, B. L. (1991). The use of Chinese hamster ovary cells to quantify specific locus mutation and to determine mutagenicity of chemicals. A report of the US Environmental Protection Agency Gene-Tox Program. *Mutat. Res.*, **86**, 193–214.

Huberman, E. (1975). Mammalian cell transformation and cell-mediated mutagenesis by carcinogenic polycyclic hydrocarbons. *Mutat. Res.*, **29**, 285–291.

Huberman, E. and Sachs, L. (1966). Cell susceptibility to transformation and cytotoxicity by the carcinogenic hydrocarbon benzo(*a*)pyrene. *Proc. Natl. Acad. Sci. USA*, **56**, 1123–1129.

Huberman, E. and Sachs, L. (1974). Cell-mediated mutagenesis of mammalian cells with chemical carcinogens. *Int. J. Cancer*, **13**, 326–333.

Huff, J. E. (1992). Chemical toxicity and chemical carcinogenesis. Is there a causal connection? A comparative morphological evaluation of 1500 experiments. *In*: 'Mechanisms of Carcinogenesis in Risk Identification', *Eds* H. Vainio, P. N. Magee, D. B. McGregor, and A. J. McMichael. IARC, Lyon, pp. 437–475.

IARC (1987). IARC Monographs on the Evaluation of Carcinogenic Risks to Humans. Overall Evaluations of Carcinogenicity: An updating of IARC Monographs Volumes 1 to 42. Supplement 7. p. 55. International Agency for Research on Cancer, Lyon.

IARC/NCI/EPA Working Group (1985). Cellular and molecular mechanisms of cell transformation and standardization of transformation assays of established cell lines for the prediction of carcinogenic chemicals: overview and recommended protocols. *Cancer Res.*, **45**, 2395–2399.

IPCS/WHO (1992). Final Report of the Steering Groups on Naturally Occurring toxins in plant antigen.

Ishinotsubo, D., Mower, H. F., Setliff, J., and Mandel, M. (1977). The use of REC bacteria for testing carcinogenic substances. *Mutat. Res.*, **46**, 53–62.

Jackson, M. A., Stack, H. F., and Waters, M. D. (1993). The genetic toxicology of putative non-genotoxic carcinogens. *Mutat. Res.*, **296**, 241–277.

Jacob, L. and de Mars, R. (1977). Chemical mutagenesis with diploid human fibroblasts. *In*: 'Handbook of Mutagenicity Test Procedures', *Eds.* B. J. Kilbey, M. Legator, W. Nichols, and C. Ramel. Elsevier Biomedical Press, North-Holland, Amsterdam, p. 193–220.

Jenkinson, P. and Anderson, D. (1990). Malformed foetuses and karyotypic abnormalities in the offspring of cyclophosphamide and allyl alcohol-treated male rats. *Mutat. Res.*, **229**, 173–184.

Jenkinson, P., Anderson, D., and Gangolli, S. D. (1987). Increased incidence of abnormal foetuses in the offspring of cyclophosphamide treated mice. *Mutat. Res.*, **188**, 57–62.

Jenssen, D. and Ramel, C. (1980). The micronucleus test as part of a short-term mutagenicity test program for the prediction of carcinogenicity evaluated by 143 agents tested. *Mutat. Res.*, **75**, 191–202.

Jones, C. A., Huberman, E., Callahan, M. F., Tu, A., Halloween, W., Pallota, S., Sivak, A., Lubet, R. A., Avery, M. D., Kouri, R. E., Spalding, J., and Tennant, R. W. (1988). An interlaboratory evaluation of the Syrian hamster embryo cell transformation assay using eighteen coded chemicals. *Toxic. In Vitro*, **2**, 103–116.

Kafer, E., Marshall, P., and Cohen, G. (1976). Well marked strains of *Aspergillus* for tests of environmental mutagens. Identification of induced recombination and mutation. *Mutat. Res.*, **38**, 141–148.

Kafer, E., Scott, B. R., Dorn, G. L., and Stafford, R. (1982). *Aspergillus nidulans*: Systems and results of tests for chemical induction of mitotic segregation and mutation. 1. Diploid and duplication assay systems. A report of the US Environmental Protection Agency Gene-Tox Program. *Mutat. Res.*, **98**, 1–48.

Kakunaga, T. (1973). A quantitative system for assay of a malignant transformation by chemical carcinogens using a clone derived from Balb/3T3. *Int. J. Cancer*, **12**, 463–473.

Kapp, R. W., Picciano, D. J., and Jacobson, C. B. (1979). Y-chromosomal non-disjunction in dibromochloropropane exposed workmen. *Mutat. Res.*, **64**, 47–51.

Kier, L. D. (1988). Comments and perspective on the EPA Workshop on 'The relationship between short-term test information and carcinogenicity'. *Environ. Mol. Mutagen.*, **11**, 147–157.

Kier, L. D., Brusick, D. J., Auletta, A. E., von Hallé, E. S., Simmons, V. F., Brown, M. M., Dunkel, V. C., McCann, J., Mortelmans, K., Prival, M. J., Rao, T. K., and Ray, V. A. (1986). The *Salmonella typhimurium*/mammalian microsome mutagenicity assay. A report of the US Environmental Protection Agency Gene-Tox Program. *Mutat. Res.*, **168**, 69–240.

Kihlman, B. A. (1971). Root tips for studying the effects of chemicals on chromosomes. *In*: 'Chemical Mutagens, Principles and Methods for their Detection', *Ed.* A. Hollaender. Plenum Press, New York, Vol. 2, pp. 365–514.

Kilian, D. J., Picciano, D. J., and Jacobson, C. B. (1975). Industrial monitoring: a cytogenetic approach. *Ann. N.Y. Acad. Sci.*, **249**, 411.

Kirkland, D. (1987). Regulatory aspects of genotoxicity testing. Implications of germ cell cytogenetic tests in the regulatory process. *Mutagenesis*, **2**, 61–67.

Klopman, G. and Rosenkranz, H. S. (1991). Quantification of the predictivity of some short-term assays for carcinogenicity in rodents. *Mutat. Res.*, **253**, 237–240.

Knaap, A. G. A. C. and Simmons, J. W. I. M. (1975). A mutational assay system for L5178Y mouse lymphoma cells using hypoxanthine guanine phosphoribosyl transferase deficiency as a marker. The occurrence of a long expression time for mutations induced by X-rays and EMS. *Mutat. Res.*, **30**, 97–109.

Knudsen, I., Hansen, E. V., Meyer, O. A., and Poulsen, E. (1977). A proposed method for the simultaneous detection of germ cell mutations leading to foetal death/dominant lethality and malformations (Male Teratogenicity) in mammals. *Mutat. Res.*, **48**, 267–270.

Kohler, S. W., Provost, G. S., Fieck, A., Kretz, P. L., Bullock, W. O., Sorge, J. A., Putnam, D. L., and Short, J. M. (1991a). Spectra of spontaneous and mutagen induced mutations in the lacI gene in transgenic mice. *Proc. Natl. Acad. Sci. USA*, **17**, 4707–4716.

Kohler, S. W., Provost, G. S., Fieck, A., Kretz, P. L., Bullock, W. O., Putnam, D. L., Sorge, J. A., and Short, J. M. (1991b). Analysis of spontaneous and induced mutations in transgenic mice using a Lambda ZAP/lacI shuttle vector. *Environ. Mol. Mutagen.*, **18**, 316–321.

Kohler, S. W., Provost, G. S., Kretz, P. L., Dycaico, M. J., Sorge, J. A., and Short, J. M. (1990). Development of a short term *in vivo* mutagenesis assay. The effects of methylation on recovery of a Lambda phage shuttle vector from transgenic mice. *Nucleic Acids Res.*, **18**, 3007–3013.

Kohn, H. I. (1973). H-gene (Histocompatibility) mutations induced by triethylene–melamine in the mouse. *Mutat. Res.*, **20**, 235–247.

Koi, M. and Barrett, J. C. (1986). Loss of tumour-suppressive function during chemically induced neoplastic progression of Syrian hamster embryo cells. *Proc. Natl. Acad. Sci. USA*, **83**, 5992–5996.

Krahn, D. F. and Heidelberger, C. (1977). Liver homogenate-mediated mutagenesis in Chinese hamster V79 cells by polycyclic 'Aromatic' hydrocarbons and aflatoxins. *Mutat. Res.*, **46**, 27–44.

Kuroki, T. and Matsushima, T. (1987). Performance of short term tests for detection of human carcinogens. *Mutagenesis*, **2**, 33–37.

Larsen, K. H., Mavournin, K. H., Brash, D., Cleaver, J. E., Hart, R. W., Maher, V. M., Painter, R. B., and Sega, G. A. (1982). DNA repair assays as tests for environmental mutagens. A report of the US Environmental Protection Agency Gene-Tox Program. *Mutat. Res.*, **98**, 287–318.

Latt, S. A., Allen, J., Bloom, S. E., Carrano, A., Falke, E., Kram, D., Schneider, E., Schreck, R., Tice, R., Whitfield, B., and Solff, S. (1981). Sister-chromatid exchanges: A report of the US Environmental Protection Agency Gene-Tox Program. *Mutat. Res.*, **87**, 17–62.

Lee, W. R., Abrahamson, S., Valencia, R., von Hallé, E. S., Wurgler, F. E., and Zimmering, S. (1983). The sex-linked recessive lethal test for mutagenesis in *Drosophila melanogaster*. A report of the US Environmental Protection Agency Gene-Tox Program. *Mutat. Res.*, **123**, 183–279.

Legator, M S., Bueding, E., Batzinger, R., Connor, T. H., Eisenstadt, E., Farrow,

M. G., Fiscor, G., Hsie, A., Seed, J., and Stafford, R. S. (1982). An evaluation of the host-mediated assay and body fluid analysis. A report of the US Environmental Protection Agency Gene-Tox Program. *Mutat. Res.*, **98**, 319–374.

Legator, M. S. and Malling, H. V. (1971). The host-mediated assay, a practical procedure for evaluating potential mutagenic agents in mammals. *In*: 'Chemical Mutagens, Principles and Methods for their Detection', *Ed.* A. Hollaender. Plenum Press, New York, Vol. 2, pp. 569–589.

Legator, M. S., Palmer, K. A., and Adler, I. A. (1973). A collaborative study of *in vivo* cytogenetic analysis. 1. Interpretation of slide preparations. *Toxicol. Appl. Pharmacol.*, **24**, 337.

Legator, M. S., Pullin, T. G., and Connor, T. H. (1977). The isolation and detection of mutagenic substances in body fluid and tissues of animals and body fluid of human subjects. *In*: 'Handbook of Mutagenicity Test Procedures', *Eds* B. J. Kilbey, M. Legator, W. Nichols, and C. Ramel. Elsevier Biomedical Press, North-Holland, Amsterdam, pp. 149–159.

Legator, M. S., Zimmering, S., and Connor, T. H. (1976). The use of indicator systems to detect mutagenic activity in human subjects and experimental animals. *In*: 'Chemical Mutagens, Principles and Methods for their Detection'. *Ed.* A. Hollaender. Plenum Press, New York, Vol. 4, pp. 171–192.

Leifer, Z., Kada, T., Mandel, M., Zeiger, E., Stafford, R., and Rosenkranz, H. S. (1981). An evaluation of tests using DNA repair-deficient bacteria for predicting genotoxicity and carcinogenicity. A report of the US Environmental Protection Agency Gene-Tox Program. *Mutat. Res.*, **87**, 211–297.

Léonard, A. (1975). Tests for heritable translocation in male mammals. *Mutat. Res.*, **31**, 291–298.

Léonard, A. (1977). Observations on meiotic chromosomes of the male mouse as a test of the potential mutagenicity of chemicals. *In*: 'Chemical Mutagens, Principles and Methods for their Detection', *Ed.* A. Hollaender, Plenum Press, New York, Vol. 3, pp. 21–56.

Li, A. P., Aaron, C. S., Auletta, A. E., Dearfield, K. L., Riddle, J. C., Slesinski, R. S., and Stankowski, Jr., L. F. (1991). An evaluation of the roles of mammalian cell mutation assays in the testing of genotoxicity. *Regulat. Toxicol. Pharmacol.*, **14**, 24–40.

Li, A. P., Gupta, R. S., Heflich, R. H., and Wassom, J. S. (1988). A review and analysis of the Chinese hamster ovary/hypoxanthine guanine phosphoribosyl transferase assay to determine the mutagenicity of chemical agents. A report of Phase III of the US Environmental Protection Agency Gene-Tox Program. *Mutat. Res.*, **196**, 17–36.

Lilly, L. J., Bahner, B., and Magee, P. N. (1975). Chromosome aberrations induced in rat lymphocytes by *N*-nitroso compounds as a possible basis for carcinogen screen. *Nature (London)*, **258**, 611–612.

Loprieno, N., Barale, R., Bauer, C., Baroncelli, S., Bronzetti, G., Cammellini, A., Cini, A., Corsi, G., Leporini, C., Nieri, R., Mozzolini, M., and Serra, C. (1974). The use of different systems with yeasts for the evaluation of chemically-induced gene conversion and gene mutations. *Mutat. Res.*, **25**, 197–217.

Loprieno, N., Barale, R., von Hallé, E. S., and von Borstel, R. C. (1983). Testing of chemicals for mutagenic activity with *Schizosaccharomyces pombe*. A report of the US Environmental Protection Agency Gene-Tox Program.

Mutat. Res., **115**, 215–223.

Loprieno, N., Boncristiani, G., Loprieno, G., and Tesoro, M. (1991). Data selection and treatment of chemicals tested for genotoxicity and carcinogenicity. *Environ. Hlth. Perspect.,* **96**, 121–126.

Maier, P. and Schmid, W. (1976). Ten model mutagens evaluated. *Mutat. Res.,* **50**, 325–338.

Marimuthu, K. M., Sparrow, A H., and Schairer, L. A. (1970). The cytological effects of space flight factors, vibration, clinostat and radiation on root tip cells of *Tradescantia. Radiat. Res.,* **42**, 105.

Maron, D. and Ames, B. N. (1983). Revised methods for the *Salmonella* mutagenicity test. *Mutat. Res.,* **113**, 173–215.

Matter, B. E. (1976). Problems of testing drugs for potential mutagenicity. *Mutat. Res.,* **38**, 243–258.

Matsushima, T., Sawamura, M., Hara, K., and Sugimura, T. (1976). A safe substitute for polychlorinated biphenyls as an inducer of metabolic activation systems. *In*: '*In vitro* Metabolic Activation of Mutagenesis Testing', *Eds* F. J. de Serres, J. R. Fout, J. R. Bend, and R. M. Philpot. Elsevier Biochemical Press, North-Holland, Amsterdam, pp. 85–88.

Mavournin, K. H., Blakey, D. H., Cimino, M. C., Salamone, M. F., and Heddle, J. A. (1990). The *in vivo* micronucleus assay in mammalian bone marrow and peripheral blood. A report of the US Environmental Protection Agency Gene-Tox Program. *Mutat. Res.,* **239**, 29–80.

Mays, J., McAninch, J., Feurs, R. J., Burkhart, J., Mohrenweiser, H., and Casciano, D. A. (1978). Enzyme activity as a genetic marker to detect induced microlesions in mammals. *Mutat. Res.,* **58**, No. 1, Abstract 58.

McCann, J. and Ames, B. N. (1976). Detection of carcinogens as mutagens in the *Salmonella*/microsome test: assay of 300 chemicals: Dicussion. *Proc. Natl. Acad. Sci. (USA),* **73**, 950–954.

McCann, J., Choi, E., Yamasaki, E., and Ames, B. N. (1975a). Detection of carcinogens as mutagens in the *Salmonella*/microsome test: assay of 300 chemicals. *Proc. Natl. Acad. Sci. (USA),* **72**, 5135–5139.

McCann, J., Spingarn, N. E., Kobori, J., and Ames, B. N. (1975b). Detection of carcinogens as mutagens: bacterial tester strains with R factor plasmids. *Proc. Natl. Acad. Sci. (USA),* **72**, 979–983.

McGregor, D. B. (1992). Chemicals classified by IARC: their potency in tests for carcinogenicity in rodents and their genotoxicity and acute toxicity. *In*: 'Mechanisms of Carcinogenesis in Risk Identification', *Eds* H. Vainio, P. N. Magee, D. B. McGregor, and A. J. McMichael, IARC, Lyon, 323–352.

McGregor, D. and Ashby, J. (1985). Summary report on the performance of the cell transformation assays. *In*: 'Evaluation of short-term tests for carcinogens', *Eds* J. Ashby, F. de Serres, M. Draper, M. Ishidate, B. H. Margolin, B. E. Matter, and M. D. Shelby. Elsevier, New York, pp. 103–115.

McKelvey-Martin, V. J., Green, M. H. L., Schmezer, P., Pool-Zobel, B. L., De Méo, M. P., and Collins, A. (1993). The single cell gel electrophoresis assay (COMET assay). *Mutat. Res.,* **288**, 47–63.

McKusick, V. (1975). 'Mendelian inheritance in Man', 4th Edn, John Hopkins Press, Baltimore.

Metzger, B., Crouch, E., and Wilson, R. (1989). On the relationship between carcinogenicity and acute toxicity. *Risk Analysis,* **9**, 169–177 (cited in Travis *et al.,* 1990a).

Meuth, M. (1984). The genetic consequences of nucleotide precursor pool imbalance in mammalian cells. *Mutat. Res., 126,* 107–112.

Meyer, A. L., McGregor, D. B., and Styles, J. A. (1984). *In vitro* cell transformation assays. The report of the UKEMS Sub-Committee on Guidelines for Mutagenicity Testing. Part 2, pp. 123–145.

Mirkova, E. (1990). Activity of the human carcinogens benzidine and 2-naphthylamine in triple- and single-dose mouse bone marrow micronucleus assays: results for a combined test protocol. *Mutat. Res., 234,* 161–163.

Mirsalis, J. (1993). Dosing regimes for transgenic animal mutagenesis assays. *Environ. Mol. Mutagen., 21,* 118–119.

Mitchell, A. D., Casciano, D. A., Meltz, M. L., Robinson, D. E., San, R. H. C., Williams, G. M., and von Hallé, E. S. (1983). Unscheduled DNA synthesis tests. A report of the US Environmental Protection Agency Gene-Tox Program. *Mutat. Res., 123,* 363–410.

Mohn, G. (1973). 5-Methyl-tryptophan resistance mutations in *Escherichia coli* K-12. Mutagenic activity of monofunctional alkylating agents, including organophosphorus insecticides. *Mutat. Res., 20,* 7–15.

Mohn, G. and Ellenberger, J. (1973). Mammalian blood-mediated mutagenicity tests using a multipurpose strain of *Escherichia coli* K-12. *Mutat. Res., 19,* 187–196.

Mohn, G., Kerklaan, R., and Ellenberger, J. (1984). Methodologies for the direct and animal mediated determination of various genetic effects in derivatives of strain 343/113 of *E. coli* K-12. *In*: 'Handbook of Mutagenicity Test Procedures', 2nd Edition, *Eds* B. J. Kilbey, M. Legator, W. Nichols, and C. Ramel. Elsevier, Amsterdam, pp. 189–214.

Mollet, P. and Wurgler, F. E. (1974). Detection of somatic recombination and mutations in *Drosophila*. A method for testing genetic activity of chemical compounds. *Mutat. Res., 25,* 421–424.

Moreau, P., Bailone, A., and Devoret, R. (1976). Prophage induction in *Escherichia coli* K-12 enVA uvrB: A highly sensitive test for potential carcinogens. *Proc. Natl. Acad. Sci. (USA), 73,* 3700–3704.

Morita, T., Iwamoto, Y., Shimizu, T., Masuzawa, T., and Yanagihara, Y. (1989). Mutagenicity tests with a permeable mutant of yeast on carcinogens showing false-negative in Salmonella assay. *Chem. Pharm. Bull., 37,* 407–409.

Mortimer, R. K. and Manney, T. R. (1971). Mutation induction in yeast. *In*: 'Chemical Mutagens, Principles and Methods for their Detection', *Ed.* A. Hollaender. Plenum Press, New York, Vol. 1, pp. 209–237.

Müller, H. J. (1927). Artificial transmutation of the gene. *Science, 66,* 84–87.

Myhr, B. C. and Caspary, W. J. (1991). Chemical mutagenesis at the thymidine kinase locus in L5178Y mouse lymphoma cells: results for 31 coded compounds in the National Toxicology Program. *Environ. Mol. Mutagen., 18,* 51–83.

Nagao, M., Yahagi, T., and Sugimura, T. (1978). Differences in effects of Norharman with various classes of chemical mutagens and amounts of S-9. *Biochem. Biophys. Res. Commun., 83,* 373–378.

Nakamura, J., Suzuki, N., and Okada, S. (1977). Mutagenicity of furylfuramide, a good preservative tested by using alanine-requiring mouse L5178Y cells *in vitro* and *in vivo*. *Mutat. Res., 46,* 355–364.

Natarajan, A. T., van Buul, P. W., and Raposa, T. (1978). An evaluation of the use of peripheral blood lymphocyte systems for assessing cytological effects induced *in vivo* by chemical mutagens. *In*: 'Mutagen Induced Chromosome Damage in Man', *Eds* H. C. Evans and D. C. Lloyd. Edinburgh University Press, Edinburgh, pp. 268–276.

Neel, J. V. (1978). Some trends in the study of spontaneous and induced mutation in man. *Genetics*, **92**, No. 1, Part 1, S25.

Neel, J. V., Kohrenweiser, H., Satoh, C., and Hamilton, B. H. (1979). A consideration of two biochemical approaches to monitoring human populations for a change in germ cell mutation rate. *In*: 'Genetic Damage in Man Caused by Environmental Agents', *Ed*. K. Berg. Academic Press, New York, London, pp. 29–47.

Neill, J. P., Brimer, P. A., Machanoff, R., Hirsch, G. P., and Hsie, A. W. (1977). A quantitative assay of mutation induction at HGPRT locus in Chinese hamster ovary cells: development and definition of the system. *Mutat. Res.*, **45**, 91–101.

Nesnow, S., Argus, M., Bergman, H., Ohu, K., Firth, C., Helmes, T., McGaughey, R., Ray, V., Staga, T. J., Tennant, R., and Weisburger, E. (1983). Chemical carcinogens: A review and analysis of the literature of selected chemicals and the establishment of the Gene-Tox carcinogen data-base. A report of the US Environmental Protection Agency Gene-Tox Program. *Mutat. Res.*, **185**, 1–195.

Nichols, W. W., Moorhead, P., and Brewen, G. (1973). Chromosome methodologies in mutation testing. *Toxicol. Appl. Pharmacol.*, **24**, 337.

Nikaido, O. and Fox, M. (1976). The relative effectiveness of 6-thioguanine and 8-azaguanine in selecting resistant mutants from two V79 Chinese hamster clones *in vitro*. *Mutat. Res.*, **35**, 279–288.

Nilan, R. A. and Vig, B. K. (1976). Plant test systems for detection of chemical mutagens. *In*: 'Chemical Mutagens, Principles and Methods for their Detection', *Ed*. A. Hollaender. Plenum Press, New York, Vol. 4, pp. 143–170.

Nute, P. E., Wood, N. E., Stammatoyannopoulos, G., Olivery, C., and Failkaw, P. J. (1976). The Kenya form of hereditary persistence of foetal haemoglobin: Structural studies and evidence for homogeneous distribution of haemoglobin F using fluorescent F antibodies. *Br. J. Haematol.*, **32**, 55.

OECD (1983). Organisation for Economic and Cultural Development. Guidelines for testing of chemicals. Genetic Toxicology Guidelines.

Official Journal of the European Communities. Council Directive 18 September 1979 Amending for the Sixth Time Directive 67/548/EEC on the approximation of the Laws, Regulations, and Administrative Provisions relating to the classification, packaging, and labelling of dangerous substances.

Ong, T. M. and de Serres, F. J. (1972). Mutagenicity of chemical carcinogens in *Neurospora crassa*. *Cancer Res.*, **32**, 1890–1893.

Oshiro, Y., Piper, C. E., Balwierz, P. S., and Soelter, S. G. (1991). Chinese hamster ovary cell assays for mutation and chromosome damage: data from non-carcinogens. *J. Appl. Toxicol.*, **11**, 167–177.

Parodi, S., Malacarne, D., and Taningher, M. (1991). Examples of uses of databases for quantitative and qualitative correlation studies between genotoxicity and carcinogenicity. *Environ. Hlth. Perspect.*, **96**, 61–66.

Parry, J. M. (1977a). The use of yeast cultures for the detection of environmental mutagens using a fluctuation test. *Mutat. Res.*, **46**, 165–176.

Parry, J. M. (1977b). The detection of chromosome non-disjunction in the yeast *Saccharomyces cerevisiae*. *In*: 'Progress in Genetic Toxicology', *Eds* D. Scott,

B. A. Bridges, and F. H. Sobels. Elsevier Biomedical Press, North-Holland, Amsterdam, pp. 223–29.

Parry, J. M., Brooks, T. M., Mitchell, I. de G., and Wilson, P. (1984). Gentoxicity studies using yeast cultures. Report of the UKEMS Sub-Committee on Guidelines of Mutagenicity Testing. Part 2, pp. 27–62.

Parry, J. M. and Zimmerman, F. K. (1976). The detection of monosomic colonies produced by mitotic chromosome non-disjunction in the yeast *Saccharomyces cerevisiae*. *Mutat. Res., 36,* 49–66.

Pearson, C. M. and Styles, J. A. (1984). Effects of hydroxyurea consistent with the inhibition of repair synthesis in UV irradiated HeLa cells. *Cancer Lett., 21,* 247–252.

Perry, P. and Evans, H. J. (1975). Cytological detection of mutagen-carcinogen exposure by sister chromatid exchange. *Nature (London), 258,* 121–125.

Perry, P. E., Henderson, L., and Kirkland, D. J. (1984). Sister chromatid exchange in cultured cells. *In*: 'Report of the UKEMS Sub-Committee on Guidelines of Mutagenicity Testing'. Part 2, pp. 89–123.

Pfeifer, G. P., Drouin, R., and Holmqvist, G. P. (1993). Detection of DNA adducts at the DNA sequence level of ligation-mediated PCR. *Mutat. Res., 288,* 39–46.

Phillips, D. H., Hemminki, K., Alhonen, A., Hewer, A., and Grover, P. L. (1988). Monitoring occupational exposure to carcinogens. Detection by ^{32}P-post-labelling of aromatic DNA adducts in white blood cells from iron foundry workers. *Mutat. Res., 204(3),* pp. 531–541.

Phillips, J. C. and Anderson, D. (1993). Predictive toxicology Structure Activity relationships and carcinogens. *Occup. Hlth. Rev., 41,* 27–30.

Pienta, R. J., Poiley, J. A., and Lebherz, W. B., III (1977). Morphological transformation of early passage golden Syrian hamster embryo cells derived from cryopreserved primary cultures as a reliable *in vitro* bioassay for identifying diverse carcinogens. *Int. J. Cancer, 19,* 642–655.

Plewa, M. J. (1982). Specific-locus mutation assays in *Zea mays*. A report of the US Environmental Protection Agency Gene-Tox Program. *Mutat. Res., 99,* 317–337.

Poirier, L. A. and de Serres, F. J. (1979). Initial National Cancer Institute studies on mutagenesis as a pre-screen for chemical carcinogens: an appraisal. *J. Natl. Cancer Inst., 62,* 919–926.

Popp. R. A., Hrisch, G. P., and Bradshaw, B. S. (1979). Amino acid substitution: Its use in detection and analysis of genetic variants. *Genetics, 92,* No. 1, Part 1, S.39.

Preston, R. J., Au, W., Bender, M. A., Brewen, J. G., Carrano, A. V., Heddle, J. A., McFee, A. F., Wolff, S., and Wassom, J. S. (1981). Mammalian *in vivo* and *in vitro* cytogenetic assays. A report of the US Environmental Protection Agency Gene-Tox Program. *Mutat. Res., 87,* 143–188.

Prival, M. J. and Dunkel, V. C. (1989). Reevaluation of the mutagenicity and carcinogenicity of chemicals previously identified as 'false positives' in the *Salmonella typhimurium* mutagenicity assay. *Environ. Mol. Mutagen., 13,* 1–24.

Propping, P., Rohrborn, G., and Buselmaier, W. (1972). Comparative investigations on the chemical induction of point mutations and dominant lethal mutations in mice. *Mol. Gen. Genet., 117,* 197.

Provost, G., Kretz, P. L., Hammer, R. T., Matthews, C. D., Rogers, B. J.,

Lundberg, K. S., Dycaico, M. J., and Short, J. M. (1993). Transgenic systems for *in vivo* mutation analysis. *Mutat. Res.,* **288,** 123–131.

Pueyo, C. A. (1978). A forward mutation assay using 1-Arabinose sensitive strain. Abstracts (Ab^{-2}). American Environmental Mutagen Society.

Purchase, I. F. H. (1980). Appraisal of the merits and shortcomings of tests of mutagenic potental. *In*: 'Mechanisms of Toxicity and Hazard Evaluation', *Eds* B. Holmstedt, R. Lauwerys, M. Mercier, and M. Roberfroid. Elsevier Biochemical Press, North-Holland, Amsterdam, pp. 105–119.

Purchase, I. F. H., Longstaff, E., Ashby, J., Styles, J. A., Anderson, D., Lefevre, P. A., and Westwood, F. R. (1978a). Evaluation of six short-term tests for detecting organic chemical carcinogens and recommendations for their use. *Nature (London),* **263,** 624–627; *Br. J. Cancer,* **37,** 873–903.

Purchase, I. F. H., Richardson, C. F., Anderson, D., Paddle, G. M., and Adams, W. G. F. (1978b). Chromosomal analysis in vinyl-chloride exposed workers. *Mutat. Res.,* **57,** 325–334.

Ray, V. A., Kier, L. D., Kannan, K. L., Haas, R. T., Auletta, A. E., Wassom, J. S., Nesnow, S., and Waters, M. D. (1987). An approach to identifying specialised batteries of bioassays for specific causes of chemicals and analysis using mutagenicity and carcinogenicity relationships and phylogenetic concordance and discordance patterns. 1. Comparison and analysis of the overall data base. A report of the US Environmental Protection Agency Gene-Tox Program. *Mutat. Res.,* **185,** 197–241.

Reznikoff, C. A., Bertram, J. S., Brankow, D. W., and Heidelberger, C. (1973). Quantitative and qualitative studies of chemical transformation of cloned C34 mouse embryo cells sensitive to post-confluence inhibition of cell division. *Cancer Res.,* **33,** 3239–3249.

Richold, M., Chandley, A., Ashby, J., Gatehouse, D. G., Bootman, J., and Henderson, L. (1990). *In vivo* cytogenetics assays, UKEMS Recommended Procedures, *Ed.* D. J. Kirkland. Cambridge University Press, Cambridge, pp. 115–141.

Roderick, J. H. (1979). Chromosomal-inversions—Studies of mammalian mutagenesis. *Genetics,* **92,** No. 1, Part 1, S.121.

Roper, J. A. (1971). *Aspergillus. In*: 'Chemical Mutagens, Principles and Methods for their Detection', *Ed.* A. Hollaender. Plenum Press, New York, Vol. 2, pp. 343–365.

Rowland, I. R., Rubery, E. D., and Walker R. (1984). Bacterial assays for mutagens in food. *In*: 'Report of the UKEMS Sub-Committee on Guidelines of Mutagenicity Testing', Part 2, pp. 173–202.

Russell, L. B. (1979). *In vivo* somatic mutations in the mouse. *Genetics,* **92,** No. 1, Part 1, S.153.

Russell, L. B., Aaron, C. S., de Serres, F. J., Generoso, W. M., Kannan, K. L., Shelby, M. P., Spring, J., and Voytek, P. (1984). Evaluation of mutagenicity assays for purposes of genetic risk assessment. A report of the US Environmental Protection Agency Gene-Tox Program. *Mutat. Res.,* **134,** 143–157.

Russell, L. B., Selby, P. B., von Hallé, E., Sheridan, W., and Valcovic, L. (1981a). The mouse specific-locus test with agents other than radiation: interpretation of data and recommendations for future work. *Mutat. Res.,* **86,** 329–354.

Russell, L. B., Selby, P. B., von Hallé, E., Sheridan, W., and Valcovic, L. (1981b). Use of the mouse spot test in chemical mutagenesis: interpretation of past data and recommendations for future work. *Mutat. Res., 86,* 355–359.

Russell, W. L. (1951). Specific locus mutations in mice. Cold Spring Harbor Laboratory. *Quant. Biol., 16,* 237.

Saedler, H., Gullon, A., Fiethen, L., and Starlinger, P. (1968). Negative control of galactose operon in *Escherichia coli. Mol. Gen. Genet., 102,* 79.

Salamone, M., Heddle, J., Stuart, E., and Katz, M. (1980). Towards an improved micronucleus test. Studies on 3 agents, Mitomycin C, cyclophosphamide and dimethylbenzanthracene. *Mutat. Res., 35,* 347–370.

Schantel, A. P. and Sankaranarayanan, K. (1976). Evaluation and re-evaluation of genetic radiation hazards in man. *Mutat. Res., 35,* 341–370.

Schiestl, R. H. (1989). Non-mutagenic carcinogens induce intrachromosomal recombination in yeast. *Nature (London), 337,* 285–288.

Schiestl, R. H., Gietz, R. D., Mehta, R. D., and Hastings, P. J. (1989). Carcinogens induce intrachromosomal recombination in yeast. *Carcinogenesis, 10,* 1445–1455.

Schmid, W. (1973). Chemical mutagen testing on *in vivo* somatic cells. *Agents Actions, 3,* 77–85.

Schmid, W. (1976). The micronucleus test for cytogenetic analysis. *In*: 'Chemical Mutagens, Principles and Methods for their Detection', *Ed.* A. Hollaender. Plenum Press, New York, Vol. 4, pp. 31–53.

Schmid, W., Arakaki, D. T., Breslau, N. A., and Culbertson, J. C. (1971). Chemical mutagenesis. The Chinese hamster bone marrow as an *in vivo* test system. *Humangenetik, 11,* 103–118.

Scott, D., Danford, N. D., Dean, B. J., Kirkland, D. J., and Richardson, C. R. (1983). *In vitro* chromosomal aberration assays. *In*: 'Report of the UKEMS sub-committee on Guidelines of Mutagenicity Testing,' Part 1, pp. 41–64.

Scott, D., Dean, B. J., Danford, N. D., and Kirkland, D. J. (1990). Metaphase chromosome aberration assays *in vitro. In*: 'Basic Mutagenicity Tests UKEMS Recommend Procedures', Cambridge University Press, Cambridge, pp. 62–86.

Scott, B. R., Dorn, G. L., Kafer, E., and Stafford, R. (1982). *Aspergillus nidulans*: Systems and results of tests for induction of mitotic segregation and mutation. II. Haploid assay systems and overall response of all systems. A report of the US Environmental Protection Agency Gene-Tox Program. *Mutat. Res., 98,* 49–94.

Scott, D., Galloway, S. M., Marshall, R. R., Ishidate, M., Brusick, D., Ashby, J., and Myhr, B. C. (1991). Genotoxicity under extreme culture conditions. *Mutat. Res., 257,* 147–206.

Searle, W. G. (1975). The specific locus test in the mouse. *Mutat. Res., 31,* 277–290.

Searle, W. G. (1979). The use of pigment loci for detecting reverse mutations in somatic cells of mice. *Arch. Toxicol., 38,* 105–108.

Selby, P. B. and Selby, P. R. (1977). Gamma-ray induced dominant mutations that cause skeletal abnormalities in mice. *Mutat. Res., 43,* 357–376.

Shelby, M. D. (1988). The genetic toxicity of human carcinogens and its implications. *Mutat. Res., 204,* 3–15.

Shelby, M, D., Erexson, G. L., Hook, G. J., and Tice, R. R. (1993). Evaluation of a three-exposure mouse bone marrow micronucleus protocol: results with 49 chemicals. *Environ. Mol. Mutagen., 21,* 160–179.

Shelby, M. D., Gulati, D. K., Tice, R. R., and Wojciechowski, J. P. (1989). Results of tests for micronuclei and chromosomal aberrations in mouse bone marrow cells with the human carcinogens 4-aminobiphenyl, treosulphan, and melphalan. *Environ. Mol. Mutagen.,* **13,** 339–342.

Siebert, D. (1973). Induction of mitotic conversions of *Saccharomyces cerevisiae* by lymph fluid and urine of cyclophosphamide-treated rats and human patients. *Mutat. Res.,* **21,** 202.

Skopek, T. R., Liber, J. L., Krowleski, J. J., and Thilly, W. G. (1978). Quantitative forward mutation assays in *Salmonella typhimurium* using 8-azaguanine resistance as a genetic marker. *Proc. Natl. Acad. Sci. (USA),* **75,** 410–411.

Sobels, F. H. (1977). Some thoughts on the evaluation of environmental mutagens. From the Address presented to the 7th Annual EMS Meeting in Atlanta, Georgia.

Sora, S. and Magni, G. E. (1988). Induction of meiotic chromosomal malsegregation in yeast. *Mutat. Res.,* **201,** 375–384.

Sorsa, M., Ojajarui, A., and Salomaa, S. (1990). Cytogenetic surveillance of workers exposed to genotoxic chemicals: preliminary experience from a prospective cancer study in a cytogenetic cohort. *Teratogen. Carcinogen. Mutagen.,* **10,** 215–221.

Sorsa, M., Wilbourn, J., and Vainio, H. (1992). Human cytogenetic damage as a predictor of cancer risk. *In*: 'Mechanisms of Carcinogenesis in Risk Identification', *Eds* H. Vainio, P. N. Magee, D. B. McGregor, and A. J. McMichael. IARC, Lyon, pp. 543–554.

Speck, W. T., Santella, R. M., and Rosenkranz, H. S. (1978). An evaluation of the prophage induction (inductest) for the detection of potential carcinogens. *Mutat. Res.,* **54,** 101–104.

Stetka, E. G. and Wolff, G. (1976). Sister chromatid exchange as an assay for genetic damage by mutagen-carcinogens: *In vivo* test for compounds requiring metabolic activation. *Mutat. Res.,* **41,** 333–342.

Stich, H. F., San, R. H. C., Lam, P., Koropatnick, J., and Lo, L. (1977). Unscheduled DNA synthesis of human cells as a short term assay for chemical carcinogens. *In*: 'Origins of Human Cancer', *Eds* H. H. Hiatt, J. D. Watson, and J. A. Winsten. Cold Spring Harbor Laboratory, New York, Vol. C, pp. 1499–1512.

Sugimura, T. (1978). Let's be scientific about the problems of mutagens in cooked foods. *Mutat. Res.,* **55,** 149–152.

Sugimura, T., Kawachi, T., Matsushima, T., Nagao, M., Sato, S., and Yahai, T. (1977). A critical review of sub-mammalian systems for mutagen detection. *In*: 'Progress in Genetic Toxicology', *Eds* D. Scott, B. A. Bridges, and F. H. Sobels. Elsevier Biochemical Press, North-Holland, Amsterdam, pp. 25–140.

Sugimura, T., Yahagi, T., Nagao, M., Takuichi, M., Kawacki, T., Hara, K., Yamasaki, E., Matsushima, T., Hashimoto, Y., and Okada, M. (1976). Validity of mutagenicity testing using microbes as a rapid screening method for environmental carcinogens. *In*: 'Screening Tests in Chemical Carcinogens', *Eds* R. Montesano, H. Bartsch, and T. Tomatis. IARC, Lyon, **12,** 81–101.

Swann, P. F. and Magee, P. N. (1968). Nitrosamine-induced carcinogenesis. The alkylation of nucleic acids of the rat by *N*-methyl-*N*-nitrosourea, dimethylnitrosamine dimethyl sulphate, and methyl methanesulphonate. *Biochem. J.,* **110,** 39–47.

Swierenga, S. H. H. and Yamasaki, H. (1992). Performance of tests for cell transformation and gap-junction intercellular communication for detecting nongenotoxic carcinogenic activity. *In*: 'Mechanisms of Carcinogenesis in Risk Identification', *Eds* H. Vainio, P. N. Magee, D. B. McGregor, and A. J. McMichael. IARC, Lyon, pp. 165–193.

Szentesi, I., Hornyak, E., Unvary, G., Czeizel, A., Bognar, Z., and Timor, M. (1976). High rate of chromosomal aberrations in PVC workers. *Mutat. Res.,* **37**, 313–316.

Tang, T. and Friedman, M. A. (1977). Carcinogen activation of human liver enzymes in the Ames mutagenicity test. *Mutat. Res.,* **46**, 387–394.

Tanooka, J. (1977). Development and applications of *Bacillus subtilis* test system for mutagens involving DNA repair deficiency and suppressible auxotrophic mutations. *Mutat. Res.,* **48**, 367–370.

Te-Hsui, Ma. (1982a). *Vicia* cytogenetic test for environmental mutagens. A report of the US Environmental Protection Agency Gene-Tox Program. *Mutat. Res.,* **99**, 293–302.

Te-Hsui, Ma. (1982b). Tradescantia cytogenetic tests (root-top mitosis, pollen mitosis, pollen mother-cell mitosis). A Report of the US Environmental Protection Agency Gene-Tox Program. *Mutat. Res.,* **99**, 293–302.

Tennant, R. W. and Ashby, J. (1991). Classification according to chemical structure, mutagenicity to Salmonella and level of carcinogenicity of a further 39 chemicals tested for carcinogenicity by the US National Toxicology Program. *Mutat. Res.,* **257**, 209–227.

Tennant, R. W., Elwell, M. R., Spalding, J. W., and Griesemer, R. A. (1991). Evidence that toxic injury is not always associated with induction of chemical carcinogenesis. *Mol. Carcinogen.,* **4**, 420–440.

Tennant, R. W., Margolin, B. H., Shelby, M. D., Zeiger, E., Haseman, J. K., Spalding, J., Caspary, W., Resnick, M., Stasiewicz, S., Anderson B., and Minor R. (1987). Prediction of chemical carcinogenicity in rodents from *in vitro* genetic toxicity assays. *Science,* **236**, 935–941.

Thilly, W. G., de Luca, J. G., Hoppe, I. V. H., and Penmann, B. W. (1976). Mutation of human lymphoblasts by methylnitrosourea. *Chem. Biol. Interact.,* **99**, 293–302.

Tice, R. R., Erexson, G. L., Hilliard, C. J., Huston, J. L., Boehm, R. M., Gulati, D., and Shelby, M. D. (1990). Effect of treatment protocol and sample time on the frequencies of micronucleated polychromatic erythrocytes in mouse bone marrow and peripheral blood. *Mutagenesis,* **5**, 313–321.

Tijo, J. W. and Whang, J. (1962). Chromosome preparation of bone marrow cells without prior *in vitro* culture of *in vivo* colchicin administrations. *Stain Technol.,* **37**, 17–20.

Topham, J. C. (1980). Chemically-induced transmissible abnormalities in sperm head shape. *Mutat. Res.,* **70**, 109–114.

Topham, J. C., Albanese, R., Bootman, J., Scott, D., and Tweats, D. (1983). *In*: 'Report of the UKEMS Sub-Committee on Guidelines of Mutagenicity Testing', Part 1, pp. 119–142.

Tornqvist, M., Osterman-Golkar, S., Kautianen, A., Naslund, M., Calleman, C. J., and Ehrenberg, L. (1988). Methylations in human haemoglobin. *Mutat. Res.,* **204**, 521–529.

Travis, C. C., Richter Pack, S. A., Saulsbury, A. W., and Yambert, M. W. (1990a). Prediction of carcinogenic potency from toxicological data. *Mutat. Res.,* **241**, 21–36.

Travis, C. C., Saulsbury, A. W., and Richter Pack, S. A. (1990b). Prediction of cancer potency using a battery of mutation and toxicity data. *Mutagenesis*, **5**, 213–219.

Travis, C. C., Wang, L. A., and Waehner, M. J. (1991). Quantitative correlation of carcinogenic potency with four different classes of short-term test data. *Mutagenesis*, **6**, 353–360.

Trosko, J. E. (1988). A failed paradigm: carcinogenesis is more than mutagenesis. *Mutagenesis*, **3**, 363–364.

Tu, A., Hallowell, W., Pallotta, S., Sivak, A., Lubet, R. A., Curren, R. D., Avery, M. D., Jones, C., Sedita, B. A., Huberman, E., Tennant, R., Spalding, J., and Kouri, R. E. (1986). An interlaboratory comparison of transformation in Syrian hamster embryo cells with model and coded chemicals. *Environ. Mutagen.*, **8**, 77–98.

Tweats, D. J., Bootman, J., Combes, R. D., Green, M. H. L., and Watkins, P. (1984). Assays for DNA repair in bacteria. *In*: 'Report of the UKEMS sub-committee on Guidelines of Mutagenicity Testing', Part II, pp. 5–27.

Valcovic, L. R. and Malling, H. V. (1973). An approach to measuring germinal mutations in the mouse. *Environ. Hlth. Perspect.*, **6**, 201–206.

Valencia, R., Abrahamson, S., Lee, W. R., von Hallé, E. S., Woodruff, R. C., Wurgler, F. E., and Zimmering, S. (1984). Chromosome mutation tests for mutagenesis in *Drosophila melanogaster*. A report of the US Environmental Protection Agency Gene-Tox Program. *Mutat. Res.*, **134**, 61–88.

Van't Hof. J. and Schairer, L. A. (1982). *Tradescantia* assay system for gaseous mutagens. A report of the US Environmental Protection Agency Gene-Tox Program. *Mutat. Res.*, **99**, 303–315.

Venitt, S. and Crofton-Sleigh (1981). Mutagenicity of 42 coded compounds in a bacterial assay using *Escherichia coli* and *Salmonella typhimurium*. *In*: 'Evaluation of Short Term Test for Carcinogens, Progress in Mutation Research', Vol. II, *Eds* F. J. de Serres and J. Ashby. Elsevier Biochemical Press, North-Holland, Amsterdam, pp. 351–360.

Venitt, S., Forster, R. C., and Longstaff, E. (1983). Bacterial mutation assays. *In*: 'Report of the UKEMS Sub-Committee on Guidelines of Mutagenicity Testing', Part I, pp. 5–40.

Vig, B. K. (1982). Soybean [*Glycine max* (L.) merrill] as a short term assay for the study of environmental mutagens. A report of the US Environmental Protection Agency Gene-Tox Program. *Mutat. Res.*, **99**, 339–347.

Vogel, E. (1976). The function of *Drosophila* in genetic toxicology testing. *In*: 'Chemical Mutagens, Principles and Methods for their Detection', *Ed*. A. Hollaender. Plenum Press, New York, Vol. 4, pp. 93–132.

Vogel, E. (1977). Identification of carcinogens by mutagen testing in *Drosophila*. The relative reliability for the kinds of genetic damage measured. *In*: 'Origins of Human Cancer', *Eds* H. H. Hiatt, J. D. Watson, and J. A. Winsten. Cold Spring Harbor Laboratory, New York, Vol. C, pp. 1483–1497.

Vogel, E. (1987). Evaluation of potential mammalian genotoxins using *Drosophila*: the need for a change in a test strategy. *Mutagenesis*, **2**, 161–171.

Wakabayashi, K., Ushiyama, H., Takahashi, M., Nukaya, H., Kim, S. B., Horse, M., Ochiai, M., Sugimura, T., and Nagao, M. (1993). Exposure to heterocyclic amines. *Environ. Hlth. Perspect.*, **99**, 129–133.

Waters, M. D., Stack, H. F., Jackson, M. A., and Bridges, B. A. (in press). Hazard identification-efficiency of short-term tests in identifying germ cell mutagens and putative nongenotoxic carcinogens. *Environ. Hlth. Perspect.*

Waters, R., Ashby, J., Barrett, R. H., Coombes, R. D., Green, M. H. L., and Watkins, P. (1984). Unscheduled DNA synthesis. *In*: 'Report of the UKEMS Sub-Committee on Guidelines of Mutagenicity Testing', Part 2, p. 63–89.

Weston, A. (1993). Physical methods for the detection of carcinogen-DNA adducts in humans. *Mutat. Res., 288*, 19–29.

Whittaker, S. C., Zimmermann, F. K., Dicus, B., Piegorsch, W. W., Resnick, M. A., and Fogel, S. (1990). Detection of induced mitotic chromosome loss in *Saccharomyces cerevisiae*—interlaboratory assessment of 12 chemicals. *Mutat. Res., 241*, 225–242.

Wild, D. (1973). Chemical induction of streptomycin-resistant mutations in *Escherichia coli*. Dose and mutagenic effect of dichlorvos and methyl methanesulphate. *Mutat. Res., 19*, 33–41.

Williams, G. M., Mori, H., and McQueen, C. A. (1989). Structure-activity relationships in the rat hepatocyte DNA-repair test for 30 chemicals. *Mutat. Res., 221*, 263–286.

Wurgler, F. E., Sobels, F. H., and Vogel, E. (1977). *Drosophila* as an assay system for detecting genetic changes. *In*: 'Handbook of Mutagenicity Test Procedures', *Eds* B. J. Kilbey, M. Legator, W. Nichols, and C. Ramel. Elsevier Biochemical Press, North-Holland, Amsterdam, pp. 335–373.

Wyrobek, A. J. and Bruce, W. R. (1975). Chemical induction of sperm abnormalities in mice. *Proc. Natl. Acad. Sci. (USA), 72*, 4425–4429.

Wyrobek, A. J., Gordon, L. A., Burkhart, J. G., Francis, M. W., Kapp, R. W. Jr., Letz, G., Malling, H. V., Topham, J. C., and Whorton, M. D. (1983a). An evaluation of the mouse sperm morphology test and other sperm tests in non-human mammals. A report of the US Environmental Protection Agency Gene-Tox Program. *Mutat. Res., 115*, 1–72.

Wyrobek, A. J., Gordon, L. A., Burkhart, J. G., Francis, M. W., Kapp, R. W. Jr., Letz, G., Malling, H. V., Topham, J. C., and Whorton, M. D. (1983b). An evaluation of human sperm as indicators of chemically induced alterations of spermatogenic function. A report of the US Environmental Protection Agency Gene-Tox Program. *Mutat. Res., 115*, 73–148.

Wyrobek, A. J., Watchmaker, G., and Gordon, L. (1984). An evaluation of sperm tests and indicators of germ cell damage in men exposed to chemical or physical agents. *Teratogen, Carcinogen. Mutagen., 4*, 83–107.

Young, S. S. (1988). Do short-term tests predict rodent carcinogenicity? *Science, 241*, 1232–1233.

Zeiger, E. (1987). Carcinogenicity of mutagens: Predictive capability of the *Salmonella* mutagenesis assay for rodent carcinogenicity. *Cancer Res., 47*, 1287–1296.

Zeiger, E., Haseman, J. K., Shelby, M. D., Margolin, B. H., and Tennant, R. W. (1990). Evaluation of four *in vitro* genetic toxicity tests for predicting rodent carcinogenicity: confirmation of earlier results with 41 additional chemicals. *Environ. Mol. Mutagen, 16*, (Suppl. 18), 1–4.

Zeiger, E. and Pagano, D. A. (1989). Mutagenicity of the human carcinogen treosulphan in Salmonella. *Environ. Mol. Mutagen., 13*, 343–346.

Zeise, L., Wilson, R., and Crouch, E. (1984). Use of acute toxicity to estimate carcinogenic risk. *Risk Analysis, 4*, 187–199 (cited in Travis *et al.*, 1990a).

Zimmermann, F. K. (1975). Procedures used in the induction of mitotic recombination and mutation in the yeast *Saccharomyces cerevisiae*. *Mutat. Res.*, **31**, 71–86.

Zimmermann, F. K., Mayer, V. W., Scheel, I., and Resnick, M. A. (1985). Acetone, methyl ethyl ketone, ethyl acetate, acetonitrile and other polar aprotic solvents are strong inducers of aneuploidy in *Saccharomyces cerevisiae*. *Mutat. Res.*, **149**, 339–351.

Zimmermann, F. K., von Borstel, R. C., von Hallé, E. S., Parry, J. M., Siebert, D., Zetterberg, G., Barale, R., and Loprieno, N. (1984). Testing of chemicals for genetic activity with *Saccharomyces cerevisiae*: a report of the US Environmental Protection Agency Gene-Tox Program. *Mutat. Res.*, **133**, 199–214.

CHAPTER 14

Molecular Toxicology

PAUL RUMSBY

1 Introduction

The techniques of gene cloning, nucleic acid manipulation, and detection known as molecular biology have infiltrated most branches of the biological and medical sciences. Toxicology has in general been slow in utilising this technology but the last 5 years have increasingly seen the application of basic knowledge of gene structure and regulation to answer questions in this field. Much of this technology is now common knowledge among undergraduates in the biological sciences so rather than explain it in depth, the basic 'tools' will be outlined and examples given of their use in toxicology. For those readers wishing for greater depth of knowledge on specific techniques various good text books and reviews have been suggested.

Due to tremendous advances in defining the genetic events in cancer, many of the best examples for toxicologists involve the assessment of chemically-induced carcinogenesis.

2 Techniques of Molecular Biology

A Gene Cloning

Gene cloning is the insertion of a DNA sequence into a vector, which can then be introduced into a host such as a bacterium for propagation (Figure 1). The host–vector system used depends on the size of DNA to be cloned and whether there is a need for expression of the gene. A list of the many and varied vector systems now available is beyond the scope of this chapter but a perusal of the catalogues of biotechnology companies such as Clontech, Stratagene, and Promega will give a flavour of the technology involved.

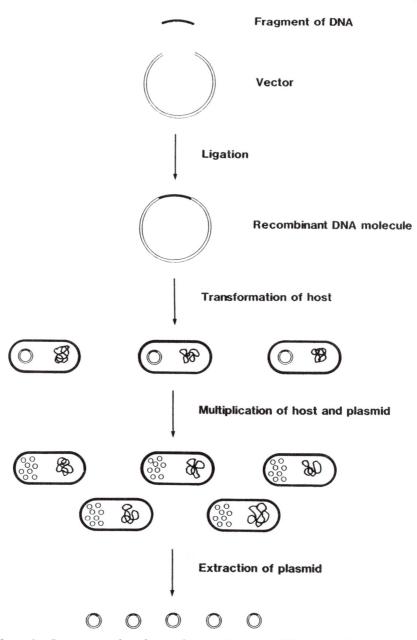

Figure 1   *Basic principles of gene cloning. Fragment of foreign DNA and plasmid vector are cut with an appropriate restriction enzyme and then ligated. The recombinant DNA is transported into the host cell, in this case a bacterium, where there is multiplication of the recombinant molecule and division of the host cells*

The vectors are usually based on plasmids for smaller fragments (up to 5–10 kb), bacteriophage lambda, and a combination of the two, called a cosmid (a plasmid with cohesive ends for efficient bacteriophage-like infection) for larger fragments (50 kb). A newer system, based on the Yeast Artificial Chromosome (YAC) for the isolation of DNA of up to 1 Mb, has proved vital to the Human Genome Mapping Project. The vector with its DNA insert will replicate autonomously within the host to produce multiple copies of the original DNA sequence.

The donor DNA for cloning can be isolated from the genome of any source or from complementary DNA (cDNA) synthesised from RNA by the action of the enzyme, reverse transcriptase.

A collection of clones representative of a population of cDNA or genomic DNA is called a library from which specific sequences can be isolated.

B Restriction Enzymes

The cloning of DNA was made possible by the isolation of a number of DNA modifying enzymes mainly from bacteria which enabled the cutting and joining of DNA. Restriction endonucleases, more than 100 of which are commercially available, cut DNA molecules at specific nucleotide sequences (typically 4 or 6 base-pairs). For example, the restriction enzyme, EcoR1 will only cut DNA at the sequence GAATTC which occurs in the mammalian genome, on average, once every 4096 base-pairs.

$$5'\ GAATTC\ 3' \rightarrow \begin{matrix} G \\ CTTAA \end{matrix} + \begin{matrix} AATTC \\ G \end{matrix}$$

This enzyme gives 'overhanging' ends which are the same in all EcoR1-treated DNA and can be easily joined in the presence of the enzyme, DNA ligase to construct recombinant DNA molecules. Most vectors now incorporate a number of common restriction enzyme recognition sites. Thus the vector and donor DNA are cut by a specific restriction enzyme, ligated by DNA ligase, and this mix is used to transform a recipient bacterium. The vector carries an antibiotic resistance gene so growth of the bacteria in the antibiotic confirms the presence of plasmid (Figure 1).

This is a simple outline of gene cloning. Further guides to its intricacies can be found elsewhere (*e.g.* Berger and Kimmel, 1987) including the molecular biologist's 'bible', 'Molecular Cloning—A Laboratory Manual' (Sambrook *et al.,* 1989). The expansion of this manual from its single volume first edition in 1982 (Maniatis *et al.,* 1982) to its present 3-volume form is, if anything, an understatement of the growth of techniques in the field.

Restriction enzymes are also useful for the detection of changes in DNA sequence (polymorphism) leading to gain or loss of restriction recognition sites (restriction fragment length polymorphism, RFLP). Examples of the use of RFLPs will be outlined later in the chapter.

C Detection of DNA and RNA

Over the last 20 years the chief means of analysing nucleic acids have been the Southern blot (Southern *et al.*, 1975) for DNA and the Northern blot for RNA (Alwine *et al.*, 1977). The basis for these two techniques is very similar and is outlined in Figure 2.

The probes used for the detection of DNA or RNA can be cloned DNA, RNA, or short synthesised sequences specific for the gene of interest. Typically these are radiolabelled using ^{32}P (a β-emitter of short half-life, 14 days). The labelling of the probe can be by a number of methods including random priming (Feinberg and Vogelstein, 1983) or endlabelling. Recently, there has been much work on non-radioactive labelling techniques using chemiluminescence or colour changes in an attempt to circumvent the short half-life and potential radiation hazard of ^{32}P.

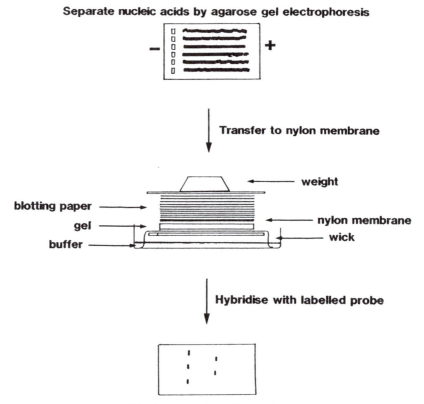

Figure 2 *Basic principles of Southern and Northern blotting. DNA can be cut with restriction enzymes before separation while RNA is separated by electrophoresis through a denaturing gel. DNA/RNA is then transferred to a nylon membrane which is incubated with a labelled probe and washed under controlled conditions. Autoradiography indicates where the labelled probe has bound to a complementary sequence on the blot.*

The conditions under which DNA sequences will anneal to a complementary strand is a science in itself (Young and Anderson, 1985). In practice hybridisation of two strands of DNA takes place at one set of conditions and separate conditions govern the separation of this duplex so that only the 'best fit' remains annealed.

D   The Polymerase Chain Reaction (PCR)

The mid-1980s saw the development of PCR as a method for the amplification of a specific region of DNA (Saiki *et al.,* 1985). This basic technique has proved to be the most important contribution to molecular biology since gene cloning was first described. Briefly, PCR consists of a cycle of events (Figure 3). Firstly, the double-strands of DNA are separated by heating. Next, primers, usually oligonucleotides of 20–30 base-pairs, designed to complement sequences either side of the region of

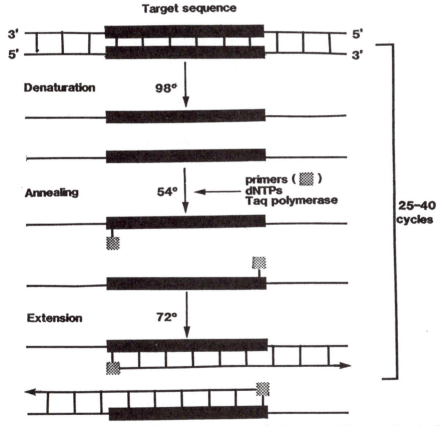

Figure 3   *The polymerase chain reaction as described in the test. The second cycle of primer annealing defines the 3' end so the PCR product is the length of the target sequence*

interest are annealed to the single-stranded DNA at an appropriate temperature. Finally, the action of *Taq* polymerase in the presence of excess deoxynucleotides leads to the formation of a new complementary strand of DNA commencing at double-stranded DNA where the primers have annealed to the template. The DNA is then heated again to separate the new and old strands and a further cycle of amplification commences. 25–40 cycles of strand separation, primer annealing, and elongation lead to an amplification of up to a million-fold of the region of interest.

These techniques can be used in very many ways to facilitate various aspects of gene cloning and the study of gene regulation. Of potential importance to toxicologists is the detection of very low levels of gene expression. This technique, usually called reverse transcriptase–PCR (RT–PCR), involves the production of complementary DNA (cDNA) from cellular RNA samples and the subsequent amplification of cDNA by PCR, thus increasing the sensitivity of the assay. A further method for the detection of DNA mutations of potential importance in toxicology is the use of small samples of tissues such as scrapes from archival paraffin-fixed sections (Shibata *et al.,* 1988). Examples of the use of PCR will be given later in the chapter. A number of books describing the method and its applications have been produced (Ehrlich, 1989; Innis *et al.,* 1990) although new uses are continually being described.

3   Applications of Molecular Biology to Toxicology

Although the techniques of molecular biology have now been utilised in all aspects of toxicology, it is in the field of chemical carcinogenesis that its weight has been most felt.

A   Metabolism of Chemicals

Many chemicals require enzymatic activation to become carcinogenic. The microsomal P450 oxygenases often metabolise chemicals to DNA damaging (genotoxic) electrophilic intermediates. These enzymes are part of a seemingly ever-expanding gene superfamily, divided into 13 families with 65 different protein sequences (for a review see Gonzalez, 1989). The genes encoding many of these members have been cloned. The gene for aryl hydrocarbon hydroxylase (AHH now called CYP1A1) is a good example of how molecular biology has elucidated gene structure and regulation.

Figure 4 shows the structure of the CYP1A1 gene in humans. In prokaryotes such as bacteria, an individual gene consists of a single coding sequence whereas in eukaryotes such as mammals, the DNA of each gene usually consists of exon (coding) sequences interspersed with intron (non-coding) sequences. These non-coding sequences are spliced from the primary RNA transcript leaving an mRNA which is ultimately translated to form the functional protein. The CYP1A1 gene consists of seven exons

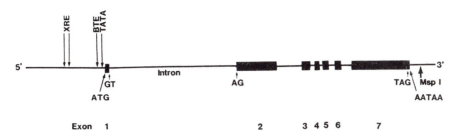

Figure 4 *Map of the human CYP1A1 gene showing exons as black boxes with introns in between. The binding sites for regulatory products include BTE—basic transcriptase element and XRE—xenobiotic response element. ATG and TAG mark the start and finish of the coding region; GT and AG, the boundaries of introns, and AATAA, the signal for the poly A tail.* Msp 1 *indicates the polymorphic site described in the text*

and spans 6311 base-pairs (Jaiswal *et al.*, 1985) while the mRNA is 2596 base-pairs long and encodes a protein of 512 amino acids (MW58151). The gene sequence contains a series of signals, for instance, to mark the start (ATG) and finish of the gene (TAG) and a sequence to mark the addition of a poly (A) tail to the mRNA. All six of the introns begin with GT and end with AG.

Sequencing of the 5′ end of the gene (known as upstream: mRNA is synthesised 5′ → 3′ so the 5′ end retains a phosphate) reveals a series of sequences which are conserved in a number of species. Some of these are common to the majority of genes such as the TATA box promoter lying 20–30 base-pairs from the site of transcriptional initiation. However, a number of other sequences are binding sites for proteins which control the transcription of the gene. How the presence of toxic chemicals affects the proteins which regulate genes such as CYP1A1 is currently the subject of much research and one to which I will return later in the Chapter.

About 20 years ago it was suggested that high inducibility of the CYP1A1 enzyme was an important risk factor in lung cancer. Induction was more frequently observed in lung cancer patients than those with benign disease implying that these patients could more readily activate the carcinogens in cigarette smoke (McLemore *et al.*, 1990).

The human CYP1A1 gene shows a sequence change (RFLP) revealed by the restriction enzyme, *Msp* 1 (Figure 5). Japanese patients homozygous for the rarer *Msp* 1 form of the gene (*i.e.* both copies of the gene have this uncommon sequence) have been shown to be susceptible to the type of lung cancer associated with smoking (Kawajiri *et al.*, 1990; Nakachi *et al.*, 1991). This homozygous form appears to be rare in Caucasian populations (Tefre *et al.*, 1991).

Another member of the P450 family, debrisoquine hydroxylase (CYP2D subfamily) also has mutant forms, detectable by RFLP analysis, which are associated with altered ability of humans to metabolise certain

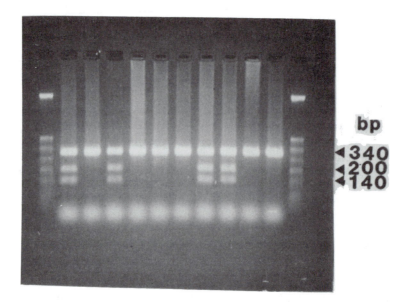

Figure 5 *Agarose gel electrophoresis of products from PCR of the 3' flanking region of the human CYP1A1 gene. The band of 340 base-pairs (bp) represents the common sequence of the gene while the bands at 200 bp and 140 bp represent an uncommon sequence recognised by the presence of an* Msp 1 *restriction site (RFLP). The gel shows PCR analysis of DNA from the white blood cells of 10 volunteers. Four have one copy (allele) of the common sequence and one copy of the uncommon sequence. There are no volunteers homozygous for the uncommon sequence. These would have no band at 340 bp.*
     The ladder of bands at each site of the gel are molecular weight markers

drugs and chemicals. There is now some evidence that certain forms of the gene product which lead to poor metabolism of the model substrate, debrisoquine, may have some protective effects against cigarette smoking-induced lung cancer (Gough *et al.,* 1990).

It is now clear that from the study of the structure and regulation of genes such as the P450 family, we can develop fairly simple procedures (PCR of the relevant regions of the gene and restriction enzyme digestion) to look at the genetic susceptibility of individuals to environmental chemicals such as cigarette smoke. Such differences in individual suscepti-bility to hazardous chemicals may be a major variable in the assessment of risk (Idle, 1991).

B   Risk and Hazard Assessment

The detection of potential chemical hazards and the assessment of their risk to an exposed workforce and the general population, particularly in the field of carcinogenesis, has been a preoccupation of toxicologists for many years. The use of molecular biology to uncover fundamental genetic events during the development of human cancer will lead to improved model systems and tests for toxicologists. At present, carcinogenic risk is assessed by *in vitro* cell mutation assays and rodent bioassay. Both of these components will be improved by recent discoveries using molecular biology. A continuing problem lies in the interpretation of tumours found in rodents after treatment with chemicals which are found to be non-mutagenic (non-genotoxic) in bacterial and/or mammalian cell mutation assays. Discovery of the mechanisms involved in the development of tumours induced by these non-genotoxic carcinogens should enable a better interpretation of their importance in risk assessment.

C Genetic Events in Human Cancer

Oncogenes

It was first shown by Rous in 1911 that sarcomas of chicken could be transmitted by the inoculation of tumour extracts (Rous, 1911). The first real breakthrough in defining genetic events in cancer came in the late 1970s from the study of transforming retroviruses (RNA tumour viruses). A large number of retroviruses have been identified and consist of a few genes which encoded the protein coat of the virus, a polymerase for synthesising a DNA copy of the virus for incorporation into the host genome and a transforming gene. The first of these genes to be described was called *src* (Hanafusa and Hanafusa, 1966) but subsequently many others were discovered and given 3 letter names derived from the parent retrovirus. It was then found that all these transforming genes had cellular counterparts, cellular or proto-oncogenes (Stehelin *et al.*, 1976).

Proto-oncogene products are involved in the whole range of mechanisms controlling growth and proliferation of the cell: (a) Growth factors and receptors such as c-*sis* (platelet-derived growth factor) and *erb* B (EGF receptor); (b) transduction of signals from the cell surface to the nucleus including the *ras* family and tyrosine kinases such as *src*; and (c) nuclear proteins such as transcriptional activators, *myc* and *fos*.

These genes are highly conserved, for example, the *ras* family is found in all species from yeast to man. The connection between the proto-oncogenes and human cancer was not made however until the development of DNA transfection assays (Shih *et al.*, 1979) in which DNA isolated from human tumour tissue is applied to a cell line. When treated in one of a number of ways (originally as a calcium phosphate precipitate) DNA is taken into the cells usually NIH3T3 mouse fibroblasts. Cells then showing a transformed phenotype, *e.g.* 'piling up' of cells to form foci or tumourigenicity in nude mice, are analysed for the presence of foreign DNA. DNA sequences consistently present in these transformed cells were considered to code for transforming factors. Most of these factors from tumours proved to be members of the *ras* family, Ha-*ras*, Ki-*ras*, and one not found in a retrovirus, N-*ras*. These *ras* genes appear to be activated in many cases by point mutation at one of 3 'hotspots', codons 12, 13, or 61 of the coding sequence. Mutations such as these are present in 10–20% of all human cancers (Weinberg, 1989). The gene product of *ras*, the p21 protein is located on the inside of the cell membrane and its function appears to be the transmission of signals from receptors at the cell surface *via* 'cascade' systems to the nucleus. In its active form, the p21 protein is bound to GTP and mutations in the gene yield a constantly activated protein.

Oncogenes may also be activated by inappropriate changes in the level of gene expression, either by gene amplification (increasing the number of gene copies) or by transcriptional activation. A number of oncogenes including *ras* and *myc* are amplified in human tumours (Weinberg, 1989), although the significance to tumour development is unknown in most cases. However, in childhood neuroblastoma, Stage I of the disease is localised, treatable, and N-*myc* amplification never seen while Stages II and III are widespread, usually fatal with N-*myc* amplification present in 50% of the cases. Stage IIIs is a variant with widespread disease but reversible with the tumour redifferentiating to nerve cells. N-*myc* amplification is never seen in this variant.

Transcriptional activation has been reported in a number of cases due to the translocation of oncogenes. The best known example is the translocation of c-*myc* in Burkitt's Lymphoma from its normal position on chromosome 8 to a region of chromosome 14 where an immunoglobulin is situated and under strong regulation (Rabbitts and Rabbitts, 1989). Other rarer variants of the disease involve chromosomes 2 and 22 where further immunoglobulin genes are located thus suggesting that this mechanism is important in Burkitt's Lymphoma.

A further case of chromosomal translocation is seen in the Philadelphia chromosome characteristic of chronic myelogenous leukaemia. In this case, the c-*abl* proto-oncogene is translocated from chromosome 9 to the breakpoint cluster region (*bcr*) of chromosome 22. A fusion protein, *bcr-abl*, is formed of 210 kd (as compared to the normal 145 kd) which has abnormal tyrosine kinase activity (De Klein, 1987).

From these examples it can be seen that oncogenes are present in normal cells as proto-oncogenes and when expressed aberrantly either as the result of mutation or transcriptional activation, are involved in the malignant transformation of the cell.

Tumour Suppressor Genes

Theodore Boveri first postulated in 1914 (Boveri, 1929) that chromosomal abnormalities might be a cause of cancer. Evidence for this theory accumulated over several decades and includes that described above for Burkitt's Lymphoma and chronic myelogenous leukaemia. Further evidence has recently come from research into two cancers, retinoblastoma and Wilms' Tumour, characterised by chromosomal deletions, 13q and 11p respectively, which can be inherited and predisposes the carrier to cancer. These deletions gave rise to the hypothesis of tumour suppressor genes which when *lost* give rise to cancer. For instance, cancer would occur after the loss of the second copy of the gene somatically in patients with a germline deletion. A rarer sporadic form of the disease consists of a somatic mutation/loss of both copies of the gene. The genes involved in retinoblastoma, Rb-1 and Wilms' Tumour, WT-1 have both been isolated by detailed mapping of the deleted chromosomal region, a process known as reverse genetics. A further tumour suppressor gene, p53 was already isolated but its true nature not revealed until its location on chromosome 17 was found to coincide with a region often lost in human neoplasm, particularly colorectal cancer. Alterations in the p53 gene have proved to be the most common genetic event yet found in human cancer, occurring in up to 70% of tumours (Hollstein *et al.*, 1991). Many of the mutations are very specific to certain cancer types suggesting organ specificity perhaps due to DNA repair sensitivity or the effects of environmental mutagens. A specific mutation (GC to TA) in the p53 gene found in human hepatocellular carcinoma in Southern Africa and China may reflect the exposure of the population to aflatoxins (Hsu *et al.*, 1991; Bressac *et al.*, 1991).

The best example of a sequence of genetic events in the development of human cancer is shown in Figure 6. Seven histological stages and six genetic or epigenetic events have been described in colon cancer (Fearon and Vogelstein, 1990) involving both the oncogene, Ki-*ras* and a number of tumour suppressor genes on chromosome 5 (mutated in colon cancer, MCC, and adenomatous polyposis coli, APC), 17 (p53), 18 (deleted in colon cancer, DDC), and recently changes to a gene on chromosome 2 (familial colon cancer, FCC; Marx, 1993). How these multistages can be interpreted in terms of models for human risk assessment remains a challenge to toxicologists. It would seem that simple models of initiation, promotion, and progression may be no longer helpful in analysis of carcinogenesis; for example, does each of the genetic events in colorectal cancer require an initiation and promotion stage?

D Animal Studies

Do the animal models used for the assessment of chemically-induced cancer reflect the genetic events seen in the human disease? Many of these model systems have been analysed for oncogene mutation particularly the *ras* gene. The formation of mouse skin papillomas and carcinomas is the best documented example of multistage carcinogenesis in animals. This system can be divided into 3 distinct stages; initiation, promotion, and progression. Treatment with one of a number of genotoxic carcinogens such as dimethyl-benzanthracene (DMBA) or methyl-*N*-nitroso guanidine (MNNG) leads to formation of genetically altered ('initiated') cells. Treatment with a further agent, frequently a phorbol ester such as

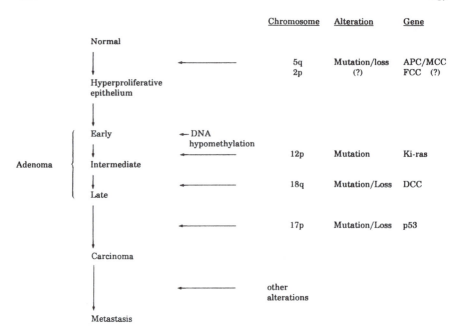

Figure 6   *The genetic and histological events in the development of human colon cancer*

12-*O*-tetradecanoyl-phorbol-13-acetate (TPA) leads to the promotion of these initiated cells to form a precancerous lesion, a papilloma. Promoters applied alone or before the initiator do not lead to papillomas. Prolonged application of the promoter may lead to the spontaneous development of malignant carcinomas. This event can be accelerated by treatment with a further mutagen. Therefore, 2 genetic events, one early and one late may be the minimum required plus promotion of the original initiated cells.

Almost all of the tumours induced by DMBA possess a specific mutation (AT to TA) in codon 61 of the Ha-*ras* oncogene, whereas those induced by MNNG have a different mutation (GC to AT, Balmain *et al.*, 1984; Quintanilla *et al.*, 1986). These mutations are consistent with the specific action of the chemical in forming adducts with DNA bases and so activation of the oncogene appears to be the initiating event in this cancer (Quintanilla *et al.*, 1986).

In rat mammary tumours, induced by methylnitrosourea (MNU), there is a 100% incidence of GC to AT transition in codon 12 of Ha-*ras*. Using DNA analysis this mutation can be found in mammary cells with no overt signs of pathological abnormality within 2 weeks of treatment (Sukumar *et al.*, 1983; Kumar *et al.*, 1991).

One of the main targets for chemically-induced tumours in the rodent bioassay is the mouse liver. A number of non-genotoxic carcinogens only give tumours in this organ. For this reason the mechanism of tumour development induced by genotoxins and non-genotoxins in the mouse liver has been the subject of much study. A number of groups have found mutations at codon 61 of the Ha-*ras* oncogene in both spontaneous and genotoxic carcinogen-induced tumours (Reynolds *et al.*, 1986, 1987; Wiseman *et al.*, 1986) although the frequency appears low in mouse strains resistant to chemically-induced hepatocarcinogenesis (Buchmann *et al.*, 1991; Dragani *et al.*, 1991; Rumsby *et al.*, 1992). With a number of chemicals the spectrum of mutations appears specific to the compound suggesting that the Ha-*ras* mutation is an initiating event (Wiseman *et al.*, 1986; Rumsby *et al.*, 1991). However, non-genotoxic agents, such as chloroform, ciprofibrate and phenobarbitone, all have a frequency of mutations lower than that found in spontaneous tumours (Fox *et al.*, 1990; Rumsby *et al.*, 1991). It has been suggested that non-genotoxic carcinogens act by promoting spontaneous lesions. If this is true in the mouse liver, these lesions must be Ha-*ras* mutation free or there is a separate mechanism of tumour development. It does appear that muta-tions in the *ras* oncogene are initiating events in a number of animal systems and can be matched to the action of the chemical carcinogen. So the measurement of this genetic event, *e.g.* Ha-*ras* codon 61 mutations in the mouse liver, may give some indication of the mechanisms involved which is of value in assessing carcinogenic risk to man. Some of the molecular methods used for the detection of oncogene mutations are described in Figure 7.

It appears that oncogene activation is often not the initiating event in human cancer. For example, late adenomas in human colorectal cancer have frequent Ki-*ras* mutations (58%) whereas it is present in only 9% of the early adenomas.

The tumour suppressor genes have not as yet been so extensively studied in animal models. Mutations in p53 have been shown to be involved in the progression of papilloma to carcinoma in the mouse skin (Bremner and Balmain, 1990) but appear not to be involved in the development of mouse liver tumours (Kress *et al.*, 1992; Goodrow *et al.*, 1992). However, p53 mutations are found in rat liver tumours induced chemically where *ras* mutations are rare. Toxicologists must try to understand such specific differences and use them to analyse the value of animal models in assessing human toxicity.

Although mutations in the p53 tumour suppressor genes do appear to cluster at 'hot spots' in exons 5, 7, and 8, these regions are too large to enable the use of allele-specific hybridisation to screen for mutation. A recent technical advance called single-strand conformation polymorphism (SSCP, Orita *et al.*, 1989) together with PCR has proved useful for such screening. This method takes advantage of the observations that single-stranded DNA's (usually PCR products) with a single base change have an altered conformation configuration which can be detected as a change in electrophoretic mobility in a polyacrylamide gel. Samples showing altered mobility in the SSCP screen can then be sequenced to define the mutation.

Detection of Ha-*ras* codon 61 mutations

1)   Oligonucleotide hybridisation

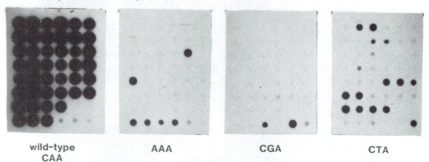

wild-type   AAA   CGA   CTA
   CAA

2)   Direct sequencing of PCR products

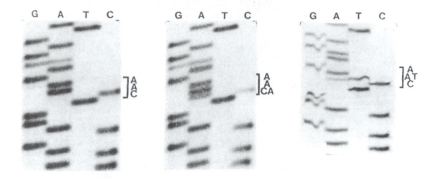

3)   Detection of CAA→CTA transversion by creation of *Xba*I

      restriction site (lanes 1-4, 9-11)

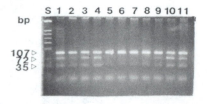

Figure 7 *Detection of Ha-ras mutations. This figure indicates the methods for the analysis of the Ha-ras codon 61 mutations. PCR is used to amplify the region of exon 2 around codon 61.*

1. *PCR products are blotted onto a nylon membrane which is hybridised to a panel of radiolabelled oligonucleotides (20 bases long) complementary to single base mutations at codon 61. Stringent washes mean that only a perfect match anneals the radiolabelled probe to the dot while a 1 base mismatch washes off.*

    In this example, each dot represents DNA from a tumour having at least one copy of the gene with the normal sequence, CAA at codon 61. Hybridisation with

E *In Vitro* Studies

A number of *in vitro* cell culture assays have been developed for the measurement of mutagenicity. However, both the Ames Salmonella assay and the mammalian short-term tests have been criticised for poor correlation with the result of long-term carcinogenicity bioassays due mainly to non-genotoxic carcinogens. These assays detect mutation in target genes leading to phenotypic changes detected by survival in a selective medium.

Molecular analysis of the mutation in mammalian cell assay such as the Chinese hamster *hprt* (Fuscoe *et al.*, 1983; Thacker *et al.*, 1985; Vrieling *et al.*, 1985; Davies *et al.*, 1993) and mouse lymphoma *tk* (Applegate *et al.*, 1990; Clive *et al.*, 1990; Yandell *et al.*, 1990; Davies *et al.*, 1993a) has indicated that certain chemicals act mainly as point mutagens while others cause large-scale chromosomal damage. However, the assays differ in their ability to detect a range of damage; for example, large-scale damage to regions around the *hprt* gene essential for growth which are present as a sole copy on the X chromosome may lead to cell death and no detection of the damage. The *tk* gene and its environs are present in two copies (one *tk* allele is disabled) and so large scale damage to one copy of an essential region is not invariably fatal and can be detected as a mutant. Such a system may also detect changes involving two alleles such as recombination. Further molecular analysis of these assays would enable the assessment of their ability to detect events we know to be important in the development of human cancer and to devise more accurate *in vitro* assays.

F Transgenic Animals

The techniques of molecular biology have led to the widespread use of transgenic animals to study expression of individual genes. Genes can be introduced into the germline of rodents and thus into every cell of the offspring. The use of promoters leads to the expression of the gene in specific sites, and it is now also possible to selectively delete genes. This technology has also been used to develop *in vivo* mutation assays (Stratagene, 'Big Blue'; Hazleton, 'Mutamouse'). A mouse with a 'target'

other oligonucleotides shows some tumours with the sequences *AAA*, *CGA* or *CTA* in the other copy of the gene.
2. The above screen can be verified by direct sequencing of the PCR product. Each track represents a base and by reading each track in turn from the bottom, a normal CAA sequence can be detected (as in the left-hand picture) or the presence of two bases in the same position indicates one copy of the gene with CAA and the other copy with either AAA (as in the middle picture) or CTA (right hand picture).
3. The detection of RFLPs can prove a sensitive method if a new restriction site is produced by the mutation. In this example, the normal CAA sequence gives a 107 bp PCR product. Mutation at codon 61 to CTA in one copy of the gene gives the Xba 1 recognition site, TCTAGA and so bands of 72 pb and 35 bp are seen in lanes 1–4 and 9–11.

bacterial transgene incorporated is treated with a test chemical. The 'target' gene can then be recovered from any organ, transformed into a bacterium, and mutations detected by a colour reaction. These assays have the advantage of measuring organ-specific mutations in a system using *in vivo* metabolism of the chemical.

G Non-genotoxic Carcinogenesis

There has recently been marked progress in describing events which may be important in non-genotoxic carcinogenesis. Much work has centred on measurement of cellular proliferation mediated by non-genotoxic chemicals. Increased proliferation may increase the likelihood of spontaneous mutational events, the importance of which has been previously outlined. Receptors have been described (and their genes isolated) which mediate the action of groups of non-genotoxic chemicals; the aromatic hydrocarbons such as 2,3,7,8-tetra-chlorodibenzo-*p*-dioxin (TCDD) *via* the A_h receptor (Burbach *et al.*, 1992), and peroxisome proliferators *via* PPAR (peroxisome proliferator activated receptor, Isseman and Green *et al.*, 1990). The latter, a member of the steroid superfamily (oestrogen also has mitogenic effects), apears to bind to a ligand, the peroxisome proliferator, and also to specific sites on DNA (Isseman and Green, 1990). The A_h receptor action is more complicated as there seems to be two proteins, the A_h receptor which binds ligand and specific DNA sites (Burbach *et al.*, 1992) and ARNT (A_h-receptor nuclear translocator) which may form a dimer with the A_h receptor and aid DNA binding (Hoffman *et al.*, 1991). There is now evidence that such receptors bind to the regulatory sites of the genes for proteins such as the P450s leading to increased transcription. The role of such receptors in the normal regulation of cellular control (for instance, the identification of an endogenous ligand) and how this is disrupted in chemically-induced cancer is the next step for molecular biology and toxicology.

4 Future Developments

Although the use of molecular biology in toxicology is most advanced in the field of chemically-induced cancer, the techniques are now being used in all branches of toxicology.

In immunology, molecular techniques are being used to dissect the network of cytokines which regulate the cells of the immune system. Toxicologists are using these sensitive techniques to examine changes in this regulation mediated by toxic chemicals in, for example, allergic reactions and sensitisation. This new field has been reviewed by Meredith (1992).

Developmental biology has undergone a revolution in understanding using the techniques of molecular biology to study development of the embryo and foetus. Regions of spatial and temporal development have been defined by the expression of highly conserved genes called homeobox genes. Expression of some of these genes in transgenic animals has led to foetal abnormalities similar to those seen with teratogenic chemicals such as the retinoids (Balling *et al.*, 1989). The effect of toxic compounds on the expression of homeobox genes may open the way for mechanistic studies in teratology and possibly *in vitro* screens using altered gene expression.

A number of different processes, such as apoptosis and stress response, are currently of intense research interest and may help to explain some fundamental problems in toxicology. Apoptosis is the way in which cell populations control their normal development by programmed cell death. A number of genes have been implicated in this process including the oncogenes *myc*, *bcl*-2, and the p53 tumour suppressor gene, thus suggesting a link with carcinogenesis. It has been suggested that p53 may be a 'molecular policeman' (Lane, 1992), extending the G_1 stage of the cell cycle in damaged cells to allow for DNA repair and in some way leading cells to apoptosis, if the damage is too severe. Mutations leading to malfunction and no p53 response would mean no extended repair time or apoptosis in time of DNA damage with mutations present and the inherent risk of the clonal expansion of initiated cells.

The aim of toxicologists must be the development of a series of universal principles to explain the response of a cell to toxic insult whether chemical, including metals, or physical agents such as radiation and heat. A number of stress response proteins have been identified which may fulfil these criteria and in the future be the 'target' for sensitive analysis as a screen for toxic insult (Welch, 1993). The heat shock protein (hsp) family appears to act as 'molecular chaperones' in binding protein which becomes denatured as a result of toxins. The transcription factors, *myc* and particularly *fos* are also very early responses to modulatory stimuli. The analysis of these stress response genes is at an early stage and their manipulation in toxicology is for the future, as one of the many challenges which have been set to toxicologists by the advances brought about by molecular biology.

References

Alwine, J. C., Kemp, D. J., and Stark, G. R. (1977). Method for detection of specific RNAs in agarose gels by transfer to diazobenzyloxy-methyl paper and hybridization with DNA probes. *Proc. Natl. Acad. Sci. (USA)*, **74**, 5350–5354.

Applegate, M. L., Moore, M. M., Broder, C. B., Burrell, A., Juhn, G., Kasweck, K. L., Lin, P.-F., Wadhams, A., and Hozier, J. C. (1990). Molecular dissection of mutations at the heterozygous thymidine kinase locus in mouse lymphoma cells. *Proc. Natl. Acad. Sci. (USA)*, **87**, 51–55.

Balling, R., Mutter, G., Gruss, P., and Kessel, M. (1989). Craniofacial abnorm-alities induced by ectopic expression of the homeobox gene Hox-1.1 in transgenic mice. *Cell*, **58**, 337–347.

Balmain, A., Ramsden, M., Bowden, G. T., and Smith, J. (1984). Activation of the mouse cellular Harvey-*ras* gene in chemically-induced benign skin papillomas. *Nature (London)*, **307**, 658–660.

Berger, S. L. and Kimmel, A. R. (*Eds*), (1987). Guide to molecular cloning techniques. *Methods Enzymol.*, **152**. Academic Press, San Diego, CA.

Boveri, T. (1929). 'The origin of malignant tumors'. Williams and Watkins, Boston.

Bremner, R. and Balmain, A. (1990). Genetic changes in skin tumor progression: correlation between presence of a mutant *ras* gene and loss of heterozygosity on mouse chromosome 7. *Cell*, **61**, 407–417.

Bressac, B., Key, M., Wands, J., and Ozturk, M. (1991). Selective G to T mutations of p53 in hepatocellular carcinoma from Southern Africa. *Nature (London)*, **350**, 429–431.

Buchman, A., Bauer-Hofmann, R., Mahr, J., Drinkwater, N. R., Luz, A., and Schwarz, M. (1991). Mutational activation of the c-Ha-*ras* gene in liver tumors of different rodent strains: correlation with susceptibility to hepatocarcinogenesis. *Proc. Natl. Acad. Sci. (USA)*, **88**, 911–915.

Burbach, K. M., Poland, A., and Bradfield, C. A. (1992). Cloning of the A_h-receptor cDNA reveals a distinctive ligand-activated transcription factor. *Proc. Natl. Acad. Sci. (USA)*, **89**, 8185–8189.

Clive, D., Glover, P., Applegate, M., and Hozier, J. (1990). Molecular aspects of chemical mutagenesis in L5178Y/TK$^{+/-}$ mouse lymphoma cells. *Mutagenesis*, **5**, 191–197.

Davies, M. J., Phillips, B. J., Anderson, D., and Rumsby, P. C. (1993). Molecular analysis of mutation at the *hprt* locus of Chinese hamster V79 cells induced by ethyl methanesulphonate and mitomycin C. *Mutat. Res.*, **291**, 117–124.

Davies, M. J., Phillips, B. J., and Rumsby, P. C. (1993a). Molecular analysis of mutations at the *tk* locus of L5178Y mouse lymphoma cells induced by ethyl methanesulphonate and mitomycin C. *Mutat. Res.* In press.

De Klein, A. (1987). Oncogene activation by chromosomal rearrangement in chronic myelocytic leukaemia. *Mutat. Res.*, **186**, 161–172.

Dragani, T. A., Manenti, G., Colombo, B. M., Falvella, F. S., Garibondi, M., Pierotti, M. A., and Della Porta, G. (1991). Incidence of mutations at codon 61 of the Ha-*ras* gene in liver tumors of mice genetically susceptable and resistant to hepatocarcinogenesis. *Oncogene*, **6**, 333–338.

Ehrlich, H. A. (*Ed.*), (1989). 'PCR Technology', Stockton Press, New York.

Fearon, E. R. and Vogelstein, B. (1990). A genetic model for colorectal tumorigenesis. *Cell*, **61**, 759–767.

Feinberg, A. P. and Vogelstein, B. (1983). A technique for radiolabeling DNA restriction endonuclease fragments to high specific activity. *Anal. Biochem.*, **132**, 6–13.

Fox, T. R., Schumann, A. M., Watanabe, P. G., Yano, B. L., Maher, V. M., and McCormick, J. J. (1990). Mutational analysis of the Ha-*ras* oncogene in spontaneous C57BL/6 × C3H/He mouse liver tumors and tumors induced with genotoxic and non-genotoxic hepatocarcinogens. *Cancer Res.*, **50**, 4014–4019.

Fuscoe, J. C., Fenwick, R. G., Ledbetter, D. H., and Caskey, C. T. (1983). Deletion and amplification of the *hgprt* locus in Chinese hamster cells. *Mol. Cell Biol.*, **3**, 1086–1096.

Gonzalez, F. J. (1989). The molecular biology of cytochrome P450s. *Pharmacol. Rev.*, **40**, 243–287.

Goodrow, T. L., Storer, R. D., Leander, K. R., Prahalada, S. R., van Zwieten, M. J., and Bradley, M. O. (1992). Murine p53 intron sequences 5-8 and their use in polymerase chain reaction/direct sequencing analysis of p53 mutations in CD-1 mouse liver and lung tumors. *Mol. Carcinog.*, **5**, 9–15.

Gough, A. C., Miles, J. S., Spurr, N. K., Moss, J. E., Gaedigk, A., Eichelbaum, M., and Wolf, C. R. (1990). Identification of the primary gene detect at the cytochrome P450 CYP2D locus. *Nature (London)*, **347**, 773–776.

Hanafusa, H. and Hanafusa, T. (1966). Determining factor in the capacity of Rous sarcoma virus to induce tumors in animals. *Proc. Natl. Acad. Sci. (USA)*, **55**, 531–535.

Hoffman, E. C., Reyes, H., Chu, F.-F., Sander, F., Conley, L. H., Brooks, B. A., and Hankinson, O. (1991). Cloning of a factor required for activity of the A_h (Dioxin) receptor. *Science*, **252**, 954–958.

Hollstein, M., Sidransky, D., Vogelstein, B., and Harris, C. C. (1991). p53 mutations in human cancers. *Science*, **253**, 49–53.

Hsu, I. C., Metcalf, R. A., Sun, T., Welsh, J. A., Wang, N. J., and Harris, C. C. (1991). Mutational hotspot in the p53 gene in human hepatocellular carcinomas. *Nature (London)*, **350**, 427–428.

Idle, J. R. (1991). Is environmental carcinogenesis modulated by host polymorphism? *Mutat. Res.*, **247**, 259–266.

Innis, M. A., Gelfand, D. H., Sninsky, J. J., and White, T. J. (*Eds*), (1990). 'PCR Protocols: a guide to methods and applications', Academic Press, San Diego.

Issemann, I. and Green, S. (1990). Activation of a member of the steroid hormone receptor superfamily by peroxisome proliferators. *Nature (London)*, **347**, 645–650.

Jaiswal, A. K., Gonzalez, F. J., and Nebert, D. W. (1985). Human P_1-450 gene sequence and correlation of mRNA with genetic differences in benzo[*a*]pyrene metabolism. *Nucl. Acids Res.*, **13**, 4503–4520.

Kawajiri, K., Nakachi, K., Imai, K., Yoshi, A., Shimoda, N., and Watanabe, J. (1990). Identification of genetically high risk individuals to lung cancer by DNA polymorphism of the cytochrome P450 IAI gene. *FEBS Lett.*, **263**, 131–133.

Kress, S., König, J., Schweizer, J., Löhrke, H., Bauer-Hofmann, R., and Schwarz, M. (1992). p53 mutations are absent from carcinogen-induced mouse liver tumors but occur in cell lines established from these tumors. *Mol. Carcinogen.*, **6**, 148–158.

Kumar, R., Sukumar, S., and Barbacid, M. (1990). Activation of *ras* oncogenes preceding the onset of neoplasia. *Science*, **248**, 1101–1104.

Lane, D. P. (1992). p53, guardian of the genome. *Nature (London)*, **358**, 15–16.

Maniatis, T., Fritsch, E. F., and Sambrook, J. (1982). 'Molecular cloning—a laboratory manual', Cold Spring Harbor Laboratory, Cold Spring Harbor, New York.

Marx, J. (1993). New colon cancer gene discovered. *Science*, **260**, 7512–752 and others in the same edition of *Science*.

McLemore, T. L., Adelberg, S., Liu, M. C., McMahon, N. A., Yu, S. J., Hubbard, W. C., Czerwiniski, M., Wood, T. G., Storeng, R., Lubet, R. A., Egglestone, J. C., Boyd, M. R., and Hines, R. N. (1990). Expression of CYP1A1 gene in patients with lung cancer: evidence for cigarette smoke-induced gene expression in normal lung tissue and for altered gene regulation in primary pulmonary carcinomas. *J. Natl. Cancer Inst.*, **82**, 1333–1339.

Meredith, C. (1992). Molecular Immunotoxicology. *In*: 'Principles and practice of immunotoxicology'. *Eds.* K. Miller, J. L. Turk, and S. Nicklin. Blackwell Scientific Publications, Oxford, pp. 344–356.

Nakachi, K., Imai, K., Hayashi, S.-I., Watanabe, J., and Kawajiri, K. (1991). Genetic susceptibility to squamous cell carcinoma of the lung in relation to cigarette dose. *Cancer Res., 51*, 5177–5180.

Orita, M., Suzuki, Y., Sekiya, T., and Hayashi, K. (1989). Rapid and sensitive detection of point-mutation and DNA polymorphisms using the polymerase chain reaction. *Genomics, 5*, 874–879.

Quintanilla, M., Brown, K., Ramsden, M., and Balmain, A. (1986). Carcinogen-specific mutations in mouse skin tumours: *ras* gene involvement in both initiation and progression. *Nature (London), 322*, 78–80.

Rabbitts, T. H. and Rabbitts, P. H. (1989). Molecular pathology of chromosomal abnormalities and cancer genes in human tumors. *In*: 'Oncogenes'. *Eds.* D. M. Glover and B. D. Hames. IRL Press, Oxford, pp. 67–111.

Reynolds, S. H., Stowers, S. J., Maronpot, R. R., Anderson, M. W., and Aaronson, S. A. (1986). Detection and identification of activated oncogenes in spontaneously occurring benign and malignant hepatocellular tumours of B6C3F1 mouse. *Proc. Natl. Acad. Sci. (USA), 83*, 33–37.

Reynolds, S. H., Stowers, S. J., Patterson, R., Maronpot, R. R., Aaronson, S. A., and Anderson, M. W. (1987). Activated oncogenes in B6C3F1 mouse liver tumors: implications for risk assessment. *Science, 237*, 1309–1316.

Rous, P. (1911). Transmission of a malignant new growth by means of a cell-free filtrate. *J. Am. Med. Assoc., 56*, 198.

Rumsby, P. C., Barrass, N. C., Phillimore, H. E., and Evans, J. G. (1991). Analysis of the Ha-*ras* oncogene in C3H/He mouse liver tumours derived spontaneously or induced with diethylnitrosamine or phenobarbitone. *Carcinogenesis, 12*, 2331–2336.

Rumsby, P. C., Evans, J. G., Phillimore, H. E., Carthew, P., and Smith, A. G. (1992). Search for Ha-*ras* codon 61 mutations in liver tumours caused by hexachlorobenzene and Aroclor 1254 in C57BL/10 ScSn mice with iron overload. *Carcinogenesis, 13*, 1917–1920.

Saiki, R. K., Scharf, S., Faloona, F., Mullis, K. B., Horn, G. T., Erlich, H. A., and Arnheim, N. (1985). Enzymatic analysis of β-globin sequences and restriction site analysis of sickle-cell anemia. *Science, 230*, 1350–1352.

Sambrook, J., Fritsch, E. F., and Maniatis, T. (1989). 'Molecular cloning—a laboratory manual', Vols. 1, 2, and 3, 2nd Edn. Cold Spring Harbor Laboratory, Cold Spring Harbor, New York.

Shibata, D., Martin, W. J., and Arnheim, N. (1988). Analysis of DNA sequences in forty-year-old thin-tissue sections: a bridge between molecular biology and classical histology. *Cancer Res., 48*, 4564–4566.

Shih, C., Shilo, B.-Z., Goldfarb, M. P., Dannenberg, A., and Weinberg, R. A. (1979). Passage of phenotypes of chemically transformed cells *via* transfection of DNA and chromatin. *Proc. Natl. Acad. Sci. (USA), 76*, 5714–5718.

Southern, E. M. (1975). Detection of specific sequences among DNA fragments separated by gel electrophoresis. *J. Mol. Biol., 98*, 503–507.

Stehelin, D., Varmus, H. E., Bishop, J. M., and Vogt, P. K. (1976). DNA related to the transforming gene(s) of avian sarcoma viruses is present in normal avian DNA. *Nature (London), 260*, 170–172.

Sukumar, S., Notario, V., Martin-Zanca, D., and Barbacid, M. (1983). Induction of mammary carcinomas in rats by nitroso-methylurea involves malignant activation of Ha-*ras*-1 locus by single point mutation. *Nature (London)*, **306**, 658–661.

Tefre, T., Ryberg, D., Hangen, A., Nebert, D. W., Skaug, V., Brogger, A., and Borressen, A. L. (1991). Human CYPIAI (cytochrome P450) gene: lack of association between the *Msp* 1 restriction fragment length polymorphism and incidence of lung cancer in a Norwegian population. *Pharmacogenetics*, **1**, 20–25.

Thacker, J. (1985). The molecular nature of mutations in cultured mammalian cells, a review. *Mutat. Res.*, **150**, 431–442.

Vrieling, H., Simons, J. W., Arwent, F., Natarajan, A. T., and van Zeeland, A. A. (1985). Mutations induced by *X*-rays at the *hprt* locus in Chinese hamster cells are mostly large deletions. *Mutat. Res.*, **144**, 281–286.

Weinberg, R. A. (*Ed.*), (1989). 'Oncogenes and the molecular origins of cancer', Cold Spring Harbor Laboratory, Cold Spring Harbor, New York.

Welch, W. J. (1993). How cells respond to stress. *Sci. Am.*, **268**, 34–41.

Wiseman, R. W., Stowers, S. J., Miller, E. C., Anderson, M. W., and Miller, J. W. (1986). Activating mutations of the *c*-Ha-*ras* proto-oncogene in chemically-induced hepatomas of the male B6C3F1 mouse. *Proc. Natl. Acad. Sci. (USA)*, **83**, 5825–5829.

Yandell, D. W., Dryja, T. P., and Little, J. B. (1990). Molecular genetic analysis of recessive mutations at a heterozygous autosomal locus in human cells. *Mutat. Res.*, **229**, 89–102.

Young, B. D. and Anderson, M. L. (1985). Quantitative analysis of solution hybridisation. *In*: 'Nucleic acid hybridisation—a practical approach', *Eds* B. D. Haines and S. J. Higgins, IRL Press, Oxford, pp. 73–111.

CHAPTER 15

Testing for Carcinogenicity

P. GRASSO

1 Introduction

Over the years, experience has shown that animals, like man, are afflicted by the development of tumours as they approach the end of their lifespan. The natural history of the animal tumours has many features in common with human tumours; some grow rapidly, metastasise widely, are readily invasive and rapidly fatal, whereas others grow slowly, do not invade or metastasise, and do not threaten life. There is also some resemblance in the histological structure between the animal and human tumours that arise from the same type of tissue; but there may be striking differences—structures that in man are strongly suggestive of a malignant potential may not have the same meaning in animals. The reverse may also be true.

Certain chemicals have caused cancer to develop in both animals and man. The susceptibility of animals to cancer development by those chemicals which are known to be human carcinogens has been the main justification for using laboratory animals to investigate the carcinogenic potential of chemicals. At the moment there are some forty chemicals and processes which are known to cause cancer in man and all of these, except benzene and arsenic, have caused cancer in animals (see Table 1 for examples).

2 Methods Employed in Carcinogenicity Studies

The routes employed in carcinogenicity studies are summarised in Table 2. There is, at the moment, some general agreement among scientists that the route most suitable for testing the carcinogenicity of a substance is that by which the chemical in question is likely to reach man. Although this principle is sound, it could give misleading results. For example, the inhalation of cigarette smoke appears to be carcinogenic to man, yet studies in animals have failed to produce lung tumours (other than pulmonary adenomata in some strains of mice) by the inhalation of cigarette smoke. On the other hand, cigarette smoke condensate has

Table 1 *Chemical carcinogens of established activity in man and laboratory rodents*

Chemical	Man		Lab. rodent		
	Target organ	Route of exposure	Target organ	Lab. rodent	Route of exposure
Aflatoxin	liver	oral	liver	rats, mice (benign tumours only)	oral oral
4-Aminobiphenyl	bladder	inhalation skin oral	bladder		
Asbestos	lung pleura	inhalation	lung pleura	mice, rats hamsters	inhalation and intra- pleural injection oral
Auramine (manufacture of)	bladder	inhalation skin oral	liver	mice	oral
Benzidine	bladder	inhalation skin oral	liver	rats, hamsters	oral
N,N-bis(2-chloroethyl)-2-naphthylamine	bladder	oral	lung	mice	intraperitoneal
Bis(chloromethyl)ether	lung	inhalation	lung skin	mice, rats mice	inhalation topical appl.

Chromium compounds (unspecified in man, chromium, strontium and zinc chromate in animals)	lung	inhalation	lung	rats	intratracheal installation
Cyclophosphamide	bladder	oral or parenteral injection	lung lymphomas	mice, rats	intraperitoneal intravenous (rats only)
Diethylstilbestrol	vagina (of mother treated in first 3 months)	oral to mother	mammary glands uterus	mice, rats hamsters	oral
Melphalan	bone-marrow (leukemia)	parenteral injection	lymphosarcomas	mice, rats	intraperitoneal injection
Mustard gas	lung	inhalation	lung	mice	inhalation or intraperitoneal injection
2-Naphthylamine	bladder	inhalation	bladder liver	hamsters mice	oral
Nickel (unspecified compound in man, nickel subsulfide and nickel carbonyl in rodents)	lung	inhalation	lung	rats	inhalation
Phenacetin	renal pelvis	oral (in analgesic abuse)	urinary tract	rats	oral
Soots, tars and mineral oils	skin (? other sites)	topical (inhalation)	skin	mice	topical application
Vinyl chloride	liver (other sites)	inhalation	liver	rats, mice hamsters	inhalation

Table 2 *Route employed in carcinogenicity testing*

(1) **Oral**	High dose administration by gavage or mixed in diet for 'lifespan' at 2 or more levels.
(2) **Parenteral**	(i) *Subcutaneous* or (ii) *Intraperitoneal*—oil or water are used as solvents for test substance. Administration at the same site once or twice weekly for a limited period or 'lifespan'. (iii) *Intravenous*—used for a limited period only. (iv) *Intrapleural*—used mainly for dusts.
(3) **Inhalation**	Route employed for testing dusts, aerosols, and vapour. Maximum exposure 5 hr day^{-1}, 5–6 days a week for 'lifespan'. Special expertise required to establish, maintain, and monitor dose of test substance inhaled.
(4) **Topical application**	Test substance usually dissolved in an aqueous or organic solvent and applied topically to skin of mice. Mineral oils are usually applied neat. Application usually once or twice weekly for a limited period or for 'lifespan'.
(5) **Intravaginal application**	Test substance applied onto epithelium of vagina or cervix in female mice once or twice weekly or for 'lifespan'.

been shown to be powerfully carcinogenic when painted on the skin of mice, a test system which is radically different from inhalation (Roe, 1981). The type of information sought should be clear before choosing the route of exposure. In exploring the carcinogenic potential of a chemical, the route of application should be the one likely to yield the optimum response for that particular type of chemical. On the other hand, in assessing the risk to man, the carcinogenic potential of the chemical by a route of exposure similar to that of man has been found of some value.

A number of points need to be considered when planning carcinogenicity studies, and many have been the subject of much debate in recent years. Although no final answer has been found, there would appear to be a general consensus among scientists on the best way of resolving the uncertainties.

The important points that need to be paid particular attention are:
(a) The test chemical
    (1) Chemical class
    (2) Physical properties
    (3) Methods for analysis in animal feed
    (4) Suitability for a particular route
    (5) Choice of vehicle
(b) The test animal
    (1) Species and strain
    (2) Background incidence of tumours
    (3) Longevity and resistance to disease
    (4) Diet and caging
    (5) Number of animals per group
(c) The test procedure
    (1) Route of administration

(2) Dose-levels and frequency of administration. Volume (by gavage or injection)

(3) The meaning of the maximum tolerated dose

(4) Criteria for terminating the study

(5) Clinical observations—*e.g.* bodyweight, general behaviour, *etc.*

(6) Termination of study

(7) *In utero* exposure

(8) *Post-mortem* procedure—record keeping, fixation for histology, which organs to be sampled if no tumour is present

(9) Pathological expertise and method of reporting

(10) Statistical considerations: randomisation at beginning of the study, statistical analysis of data at termination.

A Choice of Species and Strain

It is often recommended that a species should be chosen which closely resembles man in its physiology and in the metabolism of the compound to be tested. While this recommendation is ideal, it is rarely achieved. For purely practical reasons studies are restricted to three species, the mouse, hamster, and rat. Other species present serious difficulties, principally because of their size and long lifespan. The larger animals require much more floor space for accommodation than the laboratory rodents and their lifespan is usually so long that the test may embrace the period of employment of two or more toxicologists. These two factors immeasurably add to the cost of the test and to the complexity of administration since 'hand overs' are seldom completely effective. The restriction of choice to the three rodent species is not necessarily a drawback in a screening test for carcinogenicity, since experience has shown that they readily develop tumours with a large number of chemicals including those which have produced tumours in man (IARC, 1979).

There has been much debate on the use of outbred or inbred strains for testing carcinogenic activity. In the case of hamsters and rats, inbred strains are not as easy to obtain as are such strains in the mouse, so that the choice is limited to one of a number of outbred strains. In the case of the mouse, the number of inbred strains available is very large indeed and the choice is usually wide.

In theory, inbred strains are preferable because the tumour incidence is generally known with some degree of precision, whereas in outbred mice the incidence usually varies markedly from one generation to the next. In practice, outbred mice are more frequently used than the inbred variety because it has been the traditional practice for several years, and some influential governmental agencies have advocated their use (FDA, 1971; Ministry of Health and Welfare Canada, 1973), and it may happen that the inbred mouse may metabolise the substance in a way which avoids the production of the ultimate carcinogenic species (Page, 1977). It is possible to circumvent this pitfall by employing several inbred strains, on the lines advocated by Festing, to test pharmacological or biochemical responses to a chemical in mice (Festing, 1975). The advice is quite sound since using several strains reduces the likelihood of missing some important response. It does imply, however, that a particular test has to be done in several strains and although this may be possible in the sort of short-term tests that are usually employed in pharmacology or biochemistry, it is rarely practicable in a long term carcinogenicity test where funds usually put strict limits on the design, scope and size of the test.

B Diet

It is usual to allow animals free access (*ad libitum*) to food and water throughout the carcinogenicity studies. The food currently made available to the test animals usually contains an adequate amount of vitamins and minerals, but is usually rich in fat and protein. It is a common experience that rodents, particularly those in the control group, gain rapidly in weight in their first weeks on treatment and become obese as the experiment progresses. The obese mice and rats lose their sleek clean appearance as well as their agility and general spontaneous activity. The rapid onset of obesity indicates that the animals 'overeat and oversleep'. There is now clear and fully documented evidence that some benign tumours in both these species may be increased in incidence by excessive food intake (chiefly calorie intake) (Roe, 1979). It is clearly undesirable to have a high background incidence of tumours since it may affect the validity of the results in carcinogenicity studies. It is not possible however at this stage to advocate a reduction of calorie intake in long term tests since there is no knowledge of whether it might alter the sensitivity of the animals to carcinogenic agents. Obviously further research is urgently needed in this important field.

C Caging

Conventionally, rodents are housed 5 to a cage: the animals in a single cage belong to the same sex. This type of communal caging has resulted in problems, the principal ones being loss of animals from fighting (particularly in certain mice strains) (I. F. Gaunt, personal communication, 1982) and cannibalism of dead animals (IARC, 1980). Some authorities advise caging singly to overcome difficulties of this sort but this may cause problems of a different nature. For example, the incidence of liver tumours has been known to increase in single-caged (C_3H) mice compared with conventionally caged ones (Peraino, Fry, and Staffeldt, 1973).

D The Test Substance

Carcinogenicity studies in rodents were launched by the demonstration that a complex mixture of substances (coal tar) was carcinogenic when painted repeatedly on the skin of rabbits (Yamigawa and Ichikawa, 1918). It soon became obvious that such complex mixtures contributed very little to the identification of specific chemicals possessing carcinogenic properties so that the tendency in scientific circles has been to test only single chemicals, with particular attention to purity. Although this practice has been found of considerable value in the pharmaceutical, food additive, and chemical industries, other sections such as the oil, tobacco, and cosmetics industries, are compelled to test complex mixtures.

In selecting chemicals or products for carcinogenicity testing it is essential to have some basic information on:

(a) Chemical structure, physical properties, and impurities
(b) Stability, particularly in the medium in which it is intended for administration
(c) Probable daily exposure level for man and the route by which man is exposed
(d) If man is exposed to a mixture of substances (*e.g.* mineral oil or tobacco smoke) then a representative specimen should be selected. This advice may be difficult to follow since it is not always possible to reach agreement on what would constitute a representative specimen of, for example, a mineral oil.

The degree of purity of the chemical to be tested very often raises considerable difficulties in the selection of the source of the chemical for carcinogenicity testing. While pharmaceutical chemicals are, in general, of a high degree of purity, most chemicals in industrial use contain a variable amount of impurities. Such chemicals are known as 'technical grade' chemicals to distinguish them from the much more purified 'laboratory grade' of chemical. Man is exposed to the 'technical grade' of a chemical, and there has been some debate whether this grade or the 'pure' form should be tested. Ideally, of course, both grades should be tested simultaneously (IARC, 1980) but in a limited programme of testing, it would appear preferable to test the 'technical grade' in the first instance. In testing 'technical grade' materials there is always the chance of missing the carcinogenic activity of some component which is present in amounts too small to exert a clear effect in the sort of rodent study that is normally carried out. If such a situation is suspected, then the impurities may have to be tested separately.

E   Route of Administration

There is some agreement among oncologists that the route of administration should be closely similar to the one by which human exposure occurs (IARC, 1980). This rule, however, need not be inflexible if, for any reason, there is a strong suspicion that by following such a rule there is the risk of obtaining false results. In such circumstances, some other more promising route should be attempted. For example, it is not advisable to test for carcinogenicity the aerosol from a new type of cigarette or from an oil obtained from a new refining process by inhalation in rodents when topical application of the condensate may yield much more acceptable results.

Oral Exposure

This route is appropriate for most substances ingested by man. The substance may be administered mixed in the diet, dissolved in the drinking water, or by gavage, dissolved or suspended in a suitable medium—most substances given by gavage are suspended or dissolved in water or some edible oil. Administration by gavage has the advantage of a fairly accurate control of dosage. On the other hand, it results in a high concentration of the test chemical in the stomach and intestines with the possibility of causing serious mucosal damage. Such high concentrations in the gut enhance absorption and result in high blood levels, although these are of limited duration. Incorporation of the test substance in food or water avoids the possibility of these undesirable local pathological effects and of the large fluctuations of blood concentrations. It allows a much more uniform level of exposure and, in general, a larger daily dose can be administered by dietary incorporation of the test substance than by gavage.

Inhalation Exposure

This route is usually advocated to test the carcinogenicity of dusts, mists, aerosols, and vapours. The technology required to generate an atmosphere containing the required concentration of any of these test substances is complicated and requires considerable expertise and special equipment (WHO, 1978b). The design of the experiment, the type of equipment necessary, and the precautions to ensure an acceptable result must be carried out in collaboration with an inhalation toxicologist and a respiratory physiologist (IARC, 1980). In conducting such a test it is important to bear in mind:

(1) The respiratory rate is, on average, 18 per minute for man, 120 for rat and 150 for the mouse (WHO, 1987a)
(2) The respiratory exchange of the rat is approximately 150–200 ml/min, that of the mouse is 25 ml/min and that of man is 8.51 l/min (Mauderly and Kritchevsky, 1979; Sanockij, 1978)
(3) Rodents respire only through the nose, while man may respire through the mouth as well
(4) Airway resistance to flow may develop in experimental animals if the inhaled material is too irritant (Mead, 1960)
(5) 'Head only' exposure is recommended to avoid oral intake of the test substance from licking of one another's fur or to prevent the animals from burying the nose in their neighbour's fur thus filtering the inhaled air. Both of these undesirable complications could

occur in total body exposure of animals grouped conventionally (for example, 5 to a cage) (IARC, 1980)

(6) 'Head only' exposures usually involve a considerable degree of restraint and are not normally conducted for periods longer than 6 hours a day, five days per week. Longer periods of exposure may lead to serious damage to the animal's health from 'stress' (Powell and Hosey, 1965).

Topical Application

Skin painting is usually conducted in mice to assess the carcinogenicity of substances that normally would come into contact with human skin, for example mineral oils (Grimmer, Dettbarn, Brune, Deutsch-Wenzel, and Misfeld, 1982). It is also a useful method for detecting the carcinogenic activity of polycyclic hydrocarbons (Clayson, 1962), alkylating agents, and other proximate carcinogens (Grasso and Crampton, 1972). Other rodent species are less sensitive.

The test substance is usually applied in a vehicle which favours its passage across the keratin layer of the skin so that it comes into contact with the epidermal layer of cells. It is not usually recommended to apply the material daily; 2 or 3 times weekly would normally suffice. A mild degree of irritation produced locally by the applied material is permissible but ulceration of the skin must be avoided as this may destroy the tissue destined to develop into tumours.

A preliminary experiment to 'gauge' the epidermal reaction to the test substance is usually recommended.

Topical application is suitable for experimental designs intended to detect 'initiating' or 'promoting' potency of a particular chemical (Berenblum, 1955).

Parenteral Injection

This involves subcutaneous, intraperitoneal, or intravenous injection. Subcutaneous injection was very popular at one time because of the advantages it offers: accurate control of dosage, prevention of aerial oxidation, and the small amounts that may be used—an important consideration for a newly synthesised and very expensive chemical. The test substance is usually administered for 2 to 5 days a week. It is usually dissolved in a vehicle which may be water or a vegetable oil. Other vehicles are to be avoided as they are generally very irritant. Substances in liquid form have on occasion been administered undiluted but this practice is rarely commendable unless the test material is non-irritant (Grasso and Golberg, 1966).

Solids have often been implanted into the subcutaneous tissues of the rat for carcinogenicity testing. This practice is now no longer pursued (Bischoff and Bryson, 1964).

Repeated intraperitoneal injection has on occasion been used in place of repeated subcutaneous injection. There is always a serious risk of inducing a 'chemical peritonitis' if the substance is too irritant or of puncturing the gut with the consequence of inducing an infective type of peritonitis. The use of this route is to be discouraged even though it offers the advantages of rapid diffusion of the substances throughout the tissues.

Occasionally carcinogenicity testing has been attempted by the intravenous route. As in the intraperitoneal route, substances administered intravenously diffuse rapidly throughout the body. Unfortunately, only a limited number of intravenous injections can be given to rodents because the accessible veins soon become thrombosed. The total dose of a chemical which can be given by this route is necessarily much smaller than that which can be given by other routes, so that intravenous administration has a very limited scope in testing substances for carcinogenicity.

Other Routes of Exposure

Chemicals have been tested by other routes of exposure for some specific reason. For example, some anti-fertility compounds intended for local application have been applied intravaginally to mice (Boyland, Roe, and Mitchley, 1966), dusts have been applied intrapleurally (Wagner, 1962) and many compounds have been applied intratracheally in order to maximise the reaction of the lung to the test substance or to bypass the nasal passages of the experimental animal (Grimmer, Dettbarn, Brune, Deutsch-Wenzel, and Misfeld, 1982). Although the results from these routes are insufficient to determine the carcinogenic activity of a chemical, they often yield valuable information and are particularly suited to study some specific aspect of the carcinogenic process by a particular substance.

F   Selection of Dose

The dose selected for lifetime administration during carcinogenicity testing has been the subject of considerable scientific debate. It has been stated that 'high test dose selection is the most controversial and perhaps the most important element' in designing the protocol for a carcinogenicity test (Food Safety Council, 1978).

It has been the practice for the last two decades to select the highest dose compatible with long term survival, (Maximum Tolerated Dose). Doses of this sort are essential in order to ensure that the tissues of the test animal are exposed to the highest possible dose of the chemical. Such a precaution is essential in these experiments because the numbers of animals employed are, for logistic reasons, strictly limited. The doses applied must be large enough, therefore, to ensure that optimum conditions exist for tumour development to take place in statistically significant numbers. The alternative is to use very large numbers of animals (milli- or mega-mouse/rat) experiments. In theory, such large numbers would ensure that a sufficient number of sensitive animals, which would develop tumours with relatively low doses of chemicals, have been included in the test groups. For many reasons, large experiments of this sort are not practicable. For instance, in an experiment carried out on a total of 24,000 mice (the so-called ED_{01} experiment), there was no clear evidence that tumours of the bladder or liver were induced by the low doses (35 and 30 p.p.m.) of AAF administered, despite the fact that at each of these dose levels over a thousand mice had survived for 24 months.

The definition of a 'high dose' for any particular substance is, however, a difficult task. This topic has been discussed in detail in several publications and there seems to be some agreement on what is to be achieved and what is to be avoided (Food Safety Council, 1978; Munro, 1977).

In general, the toxic potential of the chemical should be established by conventional toxicity tests and some knowledge of the pharmacokinetics and metabolism should be available. These should form the basis for establishing the dose levels to be administered. Effects to be avoided include: depression of growth rate greater than 10% of untreated controls, recognisable injury of the target tissue, excessive stimulation of glandular activity through normal or pathological mechanisms, and abnormal pharmacological effects of excessive dosage (Grasso, 1979).

For substances which are non-toxic, *e.g.* some food colours, it is usual to limit the administration to 5% of the diet (Grasso, 1979).

There is at the moment no established procedure for estimating the dose at which novel foods (*e.g.* from single cell protein) could be administered to rodents for carcinogenicity testing. It may, perhaps, be more accurate to talk of 'levels' since they usually replace a substantial proportion (up to 50%) of the test animal's diet (de Groot, Til, and Feron, 1970). Some of the major factors that may influence the choice of the level at which such substances are to be administered are: the relative contents of phospholipid and protein, the nature of the protein and the amount of nucleic acid residues present (de Groot, Til, and Feron, 1970).

G   Duration of Experiment

Exposure to the test chemical usually commences when the animals are a few weeks old (4–6 weeks) and have been weaned for about 2 weeks. It is then continued for the lifespan of the animal (IARC, 1980). Strictly speaking, 'lifespan' means until the animal dies either from cancer induced by the chemical or from some naturally occurring tumour or other disease. Experience has shown that this type of 'lifespan' study is impracticable. In every experiment, there are usually some animals which substantially exceed the average lifespan and a 'lifetime' experiment in rats could easily go into the fourth year, adding considerably to the cost.

There is a more cogent reason against this sort of practice. Normally by 27 months, most strains of rats do not exhibit an incidence of tumours greater than 25% (approximately the same as that of man). After this date, the incidence rises exponentially and in one 'lifetime' study conducted at the British Industrial Biological Research Association (BIBRA) on nitrosamines, the tumour rate in untreated controls reached a level of 80% in animals that survived beyond 27 months and up to $3\frac{1}{2}$ years (P. Grasso, personal observation). This would mean that the longer the animals are allowed to live, the greater will be the 'background noise' and the less 'sensitive' in statistical terms will the experiment become. It is likely that similar problems occur in experiments with mice and hamsters.

Aware of this difficulty, some investigators advocate the termination of the experiment at 2 years in rats and at 18 months in mice and hamsters. Such a regime is, however, too inflexible and might mean that the animals are killed at the time of development of tumours induced by the test substance.

The International Agency for Research on Cancer (IARC), in their Supplement 2 (1980) (IARC, 1980) draw attention to the difficulties involved in deciding when to terminate a long term test and suggest a sensible compromise which is worth citing: 'the survivors in all groups are killed and the whole experiment is terminated if mortality in the control or low dose group ever reaches 75%. However, ... the study is not continued beyond week 130 of age of rats, 120 for mice and 100 weeks for hamsters, irrespective of mortality'. They also add this important rider: 'An experiment is not really satisfactory . . . . if the mortality in the control or low dose group is higher than 50% before the end of week 104 of age for rats, week 96 for mice and week 80 for hamsters'. This may perhaps be a counsel of perfection but it is worth bearing in mind.

H   The Numbers per Group

It is essential to have sufficient animals at the end of a carcinogenicity study to carry out a valid statistical analysis. Experience has shown that, in general, this can normally be achieved by having group sizes of 50 males and 50 females per dose level at the beginning of the experiment, provided that no excessive deaths from natural causes occur. Smaller numbers may make it impossible to evaluate differences in tumour incidences between test and control even if natural wastage is average, while increasing group sizes beyond the numbers given will not improve matters to an extent commensurate with the increase in effort (and cost).

In spite of these considerations, it is not always prudent to adhere rigidly to these numbers, particularly in the case of mice and hamsters, where, on occasion, an unexplained higher than expected mortality may occur. Ideally, groups sizes of about 80 to 100 of each sex per dose group in both rats, mice, and hamsters at the start of the experiment would be recommended.

An important issue in the design of carcinogenicity testing is the number of animals to be allocated to control groups and the type of controls that should be 'carried' in an experiment. Obviously, untreated control groups of the same number per sex as the test groups are essential to act as a reference point (*i.e.* to give some idea about the natural incidence of tumours) but it is unrealistic to have a single control group if 3 or more dose levels are contemplated, since this would increase the chance of obtaining a false 'negative' or 'positive' result. In fact, 2 control groups (*i.e.* double the test group size) should be included if the number of dose levels in the experiment is 3 or more.

The term 'untreated' does not mean that the animals are left entirely alone. To do this would mean that 'non-specific stress' factors which usually accompany certain test procedures, such as the restraint essential in 'head only' exposure in inhalation tests, repeated injection in subcutaneous tests, or the handling necessary in gavage experiment, will affect only the test animals. Since such procedures could affect the incidence of tumours it is essential that the controls be subject to procedures of this sort as well. In some instances it may be necessary to have also one control group which is essentially left untreated and undisturbed.

I Positive Controls

It has sometimes been the practice to include another group or groups of animals treated with a known carcinogen. This group then served as a 'positive control' designed to demonstrate that the strain of animals used can respond to carcinogens. This practice is less favoured today and some authorities go as far as to recommend against it (Committee on Carcinogenicity of Chemicals in Food, Consumer Products and the Environment, 1978; Page, 1977). Nevertheless, if the chemicals under test are close in chemical structure to that of known carcinogens, it may be advisable to include a positive control of an additional small group of animals (30 of each sex).

J Historical Controls

It is often recommended that as much information as possible is obtained on the 'background' incidence of tumours in the particular strain of animals employed in a specific test (IARC, 1980). This background incidence can be obtained from the breeder or from other long term tests using the same strain. It provides useful information on tumour incidence and can be of value in the interpretation of a borderline or doubtful 'negative' or 'positive' result, particularly if there is an unusual incidence of tumours (high or low) in concurrent controls. The information obtained from sources of this sort are not as valuable as those obtained from the concurrent control because the diet of animals or the 'non-specific stress' factors may have been different. Both of these, but particularly the diet, may influence profoundly the incidence of certain tumours.

K Analysis of Diet

The basic composition of the diet needs to be known with the aim of ensuring that an adequate amount of essential minerals and vitamins is present. Analysis for the main components can often reveal the presence of excess of fat or proteins (IARC, 1980). Knowledge of the composition of the feed is of particular importance when the test material itself is a nutrient (*e.g.* industrially treated protein or starch, single-cell protein, or irradiated food). This is essential to ensure that an appropriately balanced diet is provided, since these materials may be incorporated at levels as high as 20–60% (de Groot, Til, and Feron, 1970).

Analysis of diet should include a search for common dietary constituents which may influence carcinogenesis. Some of the constituents to be looked for are antioxidants, chlorinated hydrocarbons, substances with oestrogen-like activity, nitrites and nitrates, nitrosamines, heavy metals, polycyclic aromatic hydrocarbons, and mycotoxins.

Analysis of diet is also important to ensure that the concentration of the test substance is as close to the desired level as possible. It may reveal that some degradation of the test compound has occurred with the production of substances that may possibly interfere with the conduct of the test.

L Two-generation Studies

These have been advocated in the belief that a carcinogen stands a better chance of being detected if exposure is commenced *in utero* (IARC, 1980). With this in view, it is usual to mate animals dosed with the required dose of the test substance and then to commence treatment of the offspring soon after weaning. The offspring are then treated with the carcinogen in the conventional way.

Despite the theoretical attraction of this model, based on evidence that foetal tissues are more sensitive to carcinogens than those of adults of the same strain (Toth, 1968), it may not be possible to obtain meaningful results in a specific test. First because the transplacental passage is known to vary considerably for the same compound within the same litter and it is likely to differ even more so between different litters (Ruddick, Ashanullah, Craig, and Stavric, 1978). Secondly, the rate and site of metabolism of the compound may influence considerably the availability of the compound itself or of its reactive metabolites for transplacental passage. Finally, there is the possibility that the compound may be inactivated by the maternal tissues, by the placenta, or by the mammary glands (before weaning) (IARC, 1980).

The imponderables in such studies outweigh the theoretical sensitivity of the foetal tissues to carcinogens. Considering the substantial increase in the duration of the experiment and in cost, this type of approach is not recommended.

M   Observations During the Test

A careful check needs to be kept of the food and water consumption, of the body weights of the animals, and of their state of health and general behaviour. These are standard procedures in any well kept animal house so that no further details need to be given here. Data from urinalysis, if required, could be obtained without undue discomfort to the animals. However, the infliction of some injury is unavoidable in obtaining blood for haematological data and may cause some degree of stress.

If other types of investigations are required, for example the use of radiolabelled tracers, it is advisable to have separate or 'satellite' groups. The same advice applies if it is necessary to carry out interim kills. These extra groups should be kept apart from the main experiment and reported on separately.

N   Conducting the Necropsy

The necropsy is probably one of the most important events in the course of carcinogenicity testing. Unless it is properly conducted, valuable information may be lost and this may not only reduce the value of the test but may make it uninterpretable. Every effort should be made to describe carefully, preferably with diagrams or pictures (the Polaroid-type camera is a useful gadget in this respect) of every 'lump and bump', and a tentative diagnosis made. This sort of information can be of great help later on when examining the tissues microscopically. In fact some authors (Roe, 1981) think (and for many a good reason) that thoroughness at the necropsy table is just as important as proficiency at histological examination, and probably more so. In order to minimise errors at necropsy the following points are worth considering:

  (a) A careful watch should be kept on the state of health of animals so that those which show signs of being '*in extremis*' are killed before they die naturally. This will diminish the risk of autolysis.
  (b) The necropsy team for a particular study should consist as far as possible of the same individuals (even at weekends) in order to ensure uniformity.
  (c) A standard number of specimens should be taken from each organ and as far as possible from the same region (*e.g.* thryoid gland to include parathyroids, adrenals to include the medualla, *etc.*).
  (d) If, for any reason, more than the standard number of samples are taken from any organ, additional samples should be taken from controls.
  (e) It is desirable to give an opinion whether the tumour found was the cause of death or not.

O   The Histological Examination

It is usual to take tissues from a pre-determined set of organs and to have a standard number of sections from each tissue sample. Where tumours are large it is prudent to take a number of sections from various sites to ensure that the sections read are truly representative. The histologist must make every effort to be consistent in the diagnostic nomenclature. This may prove more difficult with unfamiliar tumours at the 'first time round' and may necessitate re-reading of the slides. The histologist must also grade the tumours into benign or malignant and, if possible, give the grade of malignancy. Any evidence of invasion or metastasis is of crucial importance and should be reported.

The pathologist must confirm or revise the opinion given at necropsy about the cause of death. If it is uncertain whether the tumour is the cause of death, some form of words may be used to express this uncertainty. A proposed form of words would be:

(a) cause of death
(b) probable cause of death
(c) probably not the cause of death
(d) incidental finding – not cause of death.

The principles involved have been described in detail by Glaister (1986).

3   Evaluation of Results

The objective of a carcinogenicity study is to determine the ability of a substance to enhance tumours in animals. An increase in tumour incidence in the test animals, as compared with controls, throws some degree of suspicion on the compound's ability to cause cancer but is not sufficient to attribute causally the increased tumour incidence to the test compound.

A number of factors need to be considered before such a causal link can be established. An important first step is to establish whether any difference in tumour incidence between test and controls could have occurred by chance. There are a number of statistical methods that could be used to achieve this end. [This subject is discussed in detail in IARC Supplement 2 (IARC, 1980)]. But even if the difference is statistically significant, it may not be sufficient to establish a cause-and-effect relationship. An important consideration is the dose–response relationship.

A   Dose-Response Relationship

Over the last 3 or 4 decades, several experiments have shown that the larger the dose of a carcinogen, the greater the number of tumours it will induce. In this respect, carcinogens resemble most other toxic and pharmacologic substances which elicit a stronger response the higher the dose. The basis of the dose–response relationship of carcinogens was established by Druckrey in the late 1950s and has been confirmed amply in subsequent experiments (Druckrey and Schmähl, 1962). This link between dose and response is one of the most important pieces of evidence causally linking a test compound with an increased incidence of tumours. At least three dose levels are essential in the experimental design to make it possible to establish such a link, although the greater the number of dose levels (within reason) the better can such a link be established.

Data from dose–response relationships are sometimes employed to estimate the expected tumour incidence at dose levels very much lower than those which could possibly be employed in conventional experiments. A number of mathematical models have been employed in such estimates (FDA, 1971; IARC, 1980). They include (1) the tolerance distribution models—the probit and logit method are the best known (Cornfield, Carlborg, and Van Ryzin, 1978; Hill, 1963); (2) simple linear extrapolation (Gross, Fitzhugh, and Mantel, 1970); (3) various hit models—for example the 'one hit' (Hoel, Gaylor, Kirschtein, Saffiotii, and Schneiderman, 1975), the 'multi-hit' model (Cornfield, Carlborg, and Van Ryzin, 1978); (4) the Armitage-Doll multi-stage model (Armitage and Doll, 1961); (5) the Weibull model (Peto and Lee, 1973); (6) a simplified statisticopharmacokinetic model (Cornfield, 1977); (7) time to tumour (Hogan and Hoel, 1982).

The value of these models in yielding reasonably reliable estimates of risk of cancer from low levels of carcinogens were critically assessed by the Scientific Committee of the Food Safety Council (*Food Cosmet. Toxicol.*, 1980, **18,** 711). The Committee expressed the view that the choice of the mathematical model for low-dose extrapolations has no firm biological basis and must, to some extent, be arbitrary. However, they recommend the use of the more flexible models such as the Armitage-Doll, Weibull, or multi-hit. In particular, such models are recommended where a linear response at low doses would be difficult to imagine, for example in those instances where a carcinogenic response is thought to result from prolonged tissue damage. On the other hand, the Committee is of the opinion that in the case of classical electrophilic carcinogens (*e.g.* mutagenic agents), then the models which assume a linear response at very low doses may well have an important role to play. Nevertheless, data from rodent carcinogenicity could itself be misleading due to the high levels given to animals when compared to the low levels of human exposure (Clayson, 1987).

4   Nature and Type of Tumour Induced

A   Malignant and Benign Tumours

In order to establish the carcinogenic property of a test compound, some of the tumours induced must be malignant. Although there is no general agreement with this view it would seem illogical to label a compound as a 'carcinogen' unless it has been shown to induce 'cancer' in experimental animals and 'cancer' is widely held to be the occurrence of a fatal malignant tumour.

This statement does not mean that benign tumours are to be ignored. In the experience of most workers in this field, chemical carcinogens induce both benign and malignant tumours. Many hold that benign tumours may progress into malignant tumours (Foulds, 1975), while others hold the view that benign tumours represent an 'end-point' and do not progress further. At the moment it is doubtful whether this controversy could be resolved, since the evidence for either view is inconclusive, but the fact that both types of tumours of the same histogenetic origin occur in animals treated with carcinogens indicates that both must be taken into account when assaying chemical carcinogenesis. Benign tumours do not have the same significance as malignant tumours, however, and could be viewed with less concern. The scheme recently devised by Squires supports this view and attempts to give some numerical rating to distinguish the gravity between these two types of tumours (Squires, 1981).

Hormones, compounds with hormone-like actions, and a gross imbalance in normal hormonal homeostasis, however it is brought about, can enhance both benign and malignant tumours (Foulds, 1975). The proportion of each type of tumour induced varies with the organ involved. For example, tumours of the pituitary and of the islet cells of the pancreas tend to be principally benign, but mammary tumours tend to have a fairly high proportion of malignancies, at least in some strains of rats and mice (Foulds, 1975).

5 Factors Affecting Tumour Incidence

Certain types of tumour are less reliable as an index of carcinogenicity than other types, irrespective of their benign or malignant nature. Such tumours often have a high and variable natural incidence. Their incidence often can be changed by clearly definable factors unconnected with the test substance, or else it can be determined by factors which are clearly connected with local pathological changes in the target organ.

Some of the more important factors which can influence the incidence of tumours in a carcinogenicity experiment will be outlined and some indication will be given of the way in which they could act as 'confounding' factors in the evaluation of results:

A Genetic

The development of three commonly occurring tumours in mice (pulmonary adenoma, mammary adenocarcinoma, and lymphoma) appears to be under some form of genetic control. The pulmonary adenoma appears to be the best studied in this respect. This tumour originates from the Type II pneumocyte and is, generally speaking, benign in nature. The high natural incidence of this tumour (*e.g.* 7% in C57BL and 71% in Strain A) has led several investigators to study the genetic factors involved in its development. According to Bentvelzen and Szalay (1966) a single gene is reported to have a major effect in determining the high incidence of this tumour in some strains, while others seem to have found evidence that more than one pair of genes are involved (Shimkin and Stoner, 1975). Earlier experiments had shown that hybrids of a high-tumour and low-tumour strain resulted in F_1 generations that resembled the high-tumour strain. The back-cross generations resembled the strain to which the F_1 mice were mated. This was interpreted as indicating that susceptibility to the development of pulmonary tumours was inherited in a dominant manner (Falconer and Bloom, 1964; Heston and Ulahakis, 1961).

Genetic factors play a less dominant, but none the less important, role in murine mammary neoplasia. This type of tumour was found as long ago as 1911 (Murray, 1911), to occur in a high incidence in the female progeny of mice whose mothers also displayed a high incidence of this tumour. Investigations by several authors have now established that a genetic basis for mammary tumour development does exist but that other factors also play an important role (Nandi and McGrath, 1973).

The gene operations involved in the development of lymphoreticular tumours in mice were for a long time overshadowed by the active role of tumour viruses in the development of this tumour. It would now appear that at least three genes are involved, namely Fv-1 which governs the likelihood that leukaemia virus is successful in producing the disease, the Fv-2 which determines the capacity of cell transformation, and H-2 which determines the host response to the tumour antigens (Lilly and Pincus, 1973).

Although genetic factors in relation to tumour development have not been as clearly defined in the rat as they are for some of the mouse tumours, there is some reason for suspecting that a genetic basis exists for the development of some of the rat tumours, particularly the leukaemias in F344 rats and the mammary gland tumours of Sprague-Dawley rat because of their high incidence (Gala and Loginsky, 1973; Moloney, Boschetti, and King, 1970).

B Hormonal

Under naturally occurring conditions, hormones appear to play a part in determining the incidence of mammary tumours in mice. It has long been known that mammary gland tumours occur with much greater frequency in female than in male mice of any strain (Bittner, 1957). Furthermore, there is considerable experimental evidence to show that oestrogens (both synthetic and naturally occurring) and prolactin can enhance the incidence of this type of tumour (Gardner, Pfeifter, and Trentin, 1959). It is now acknowledged, however, that hormones are only one of the factors that determine the incidence of this tumour, and that genetic and viral factors operate as well (Nandi and McGrath, 1973).

Natural or synthetic oestrogens are known to enhance the development of mammary tumours in certain strains of rat, *e.g.* Sprague-Dawley and Charles River, under experimental conditions (Gibson, Newberne, Kuhn, and Elsea, 1967). It would seem reasonable to assume that they may also have some role in the development of these tumours under natural conditions.

C Viruses

Although several different types of viruses are known to be capable of inducing tumours in a variety of organs in the mouse, when experimentally introduced in the host, it would seem that under natural conditions only the lymphoma virus causes tumours in any substantial numbers (Grasso, Crampton, and Hooson, 1977). Lymphoma viruses in mice have been known to exist since their discovery by Gross in 1951 (Gross, 1951). It is now known that such viruses are widespread in mice, although they produce the disease in only a relatively small proportion of infected animals (Grasso, Crampton, and Hooson, 1977). The 'stress' imposed by the experimental procedure (*e.g.* restraint during an inhalation experiment) may cause a latent infection to become manifest (Lemonde, 1959). A test chemical may well have the same effect through some mechanism other than 'stress' (such as immune suppression) which would tip the balance in favour of virus activity.

It is not known whether a similar situation occurs in rats. When oncogenic viruses are introduced experimentally in rats, tumours develop in the tissue sensitive to the particular virus (Gross, Schidlovsky, Feldman, Dreyfus, and Moore, 1975) but there is no substantial evidence as yet to indicate that an analogous situation exists under natural conditions. There is some suspicion that the high incidence of mononuclear cell leukemia in F344 rats may have some viral etiology but no virus has been isolated as yet (Moloney, Boschetti, and King, 1970). The high incidence of mammary tumours in Sprague-Dawley rats could also conceivably have a viral origin since C-type virus readily induces mammary tumours when injected into rats but there is no experimental evidence to support this suggestion (Gross, Schidlovsky, Feldman, Dreyfus, and Moore, 1975).

D Diet

The incidence of tumours of the pituitary, the liver, and mammary glands in mice can be influenced by both the calorie and protein content of the diet (Roe, 1979). It has been shown by a number of authors that reduction of food intake results in a marked reduction of these tumours and in a better survival of the animals (Conybeare, 1980). The beneficial effect apparently results from a restriction of the calorie content rather than from the reduction of any particular type of food (Tannenbaum, 1959). Quite modest reductions in food intake (6%) are sufficient to achieve a substantial reduction of tumours and to increase longevity (Tucker, 1979).

Increase in protein and fat intake, however, has the reverse effect, particularly in the case of liver tumours: substantial increases in tumour incidence have been shown to result from an increase in these two dietary components (Gellatly, 1975). A high fat intake can also result in a high incidence of mammary tumours (Tannenbaum, 1959).

Similar results occur in rats in the case of pituitary and mammary tumours. Other tumours, including the liver, do not appear to be affected in this species (Conybeare, 1980).

E   Other Factors

Apart from diet, hormones, viruses, and genetic constitution of the animals, there are a number of other factors which may make the induction of certain tumours uninterpretable in terms of risk to man. Such factors are usually physico-chemical in nature and do not involve damage to the DNA, but they cause injury to the tissue in which cancer will ultimately develop. Factors of this type operate in:

(a) The induction of subcutaneous sarcoma in rodents by the surgical implantation of solids of diverse chemical composition in the subcutaneous tissue. It has been shown by many investigators that induction of these tumours occurs when large pieces of solid, biologically inert material (*e.g.* gold, glass) (approximately $2\,cm^2$) are implanted, but the same material implanted in shredded or powder form does not produce any local tumours (Bischoff and Bryson, 1964).

(b) The induction of subcutaneous sarcoma by repeated injection of solutions or suspensions of diverse materials at the same site for several weeks. Glucose and table salt in hypertonic solution, hydrochloric acid at pH 1, and surface active agents have been shown to induce local sarcomas under these conditions (Grasso, 1976).

(c) Transitional cell carcinoma of the urothelium from stones formed endogenously in the bladder or from foreign bodies implanted surgically. Glass beads and cholesterol crystals produce tumours readily when implanted in this viscus (Ball, Field, Roe, and Walters, 1964).

(d) Carcinoma of thyroid by compounds which interfere with the synthesis of secretion of thyroxine (Innes, Ulland, Valerio, *et al.*, 1969).

(e) Lympho-reticular tumours from immunosuppression by anti-lymphocyte serum (Gleichmann and Gleichmann, 1973).
(f) In cancer of the liver when chronic liver injury is caused by non-genotoxic agents. Injury of this sort may involve repeated cycles of necrosis and proliferation such as may occur when large doses of hepatotoxic agents (*e.g.* CCl_4) are administered or it may involve some biochemical change in the liver which, in some way which is not quite understood, leads to nodular hyperplasia and then to cancer (*e.g.* hypolipidemic agents and peroxisome proliferation) (Cohen and Grasso, 1981).

It is not clear whether physico-chemical factors operate as well in other tissues, *e.g.* skin, but the evidence available from subcutaneous tissue, urothelium, and the other tissues mentioned in the preceding paragraph indicate that some caution needs to be exercised in the interpretation to be given to tumour induction in rodents in conventional long term tests.

In particular, one should seek to exclude the operation of the 'confounding' factors outlined in this section before concluding that a compound is carcinogenic. In order to do this, one has to apply common sense and knowledge gained from experience. Only then will the biological significance of any excess tumour incidence in test over control become apparent.

6 Conclusions

Carcinogenicity studies require careful planning and execution and an even greater care in interpretation. A false positive result may cause unnecessary worry among those who are exposed to a widely used chemical, which is quite harmless but which has given rise to tumours because of the factors discussed above. It may also lead to the loss of a valuable chemical. A false negative result, on the other hand, may give rise to a false 'sense of security' and may result in the tragic development of tumours among those exposed.

References

Armitage, P. and Doll, R. (1961). Stochastic models for carcinogenesis. *In*: 'Proceedings of the 4th Berkeley Symposium on Mathematical Statistics and Probability', No. 4. *Eds* Lecam and Neyman.

Ball, J. K., Field, W. E. H., Roe, F. J. C., and Walters, M. (1964). The carcinogenic and co-carcinogenic effects of paraffin wax and glass beads in the mouse bladder. *Br. J. Urol.*, **36**, 225.

Bentvelzen, P. A. J. and Szalay, G. (1966). Some genetic aspects of difference in susceptibility to the development of lung tumours between inbred strains of mice. *In*: 'Lung Tumours in Animals'. Proceedings of the 3rd Quadrennial Conference on Cancer, University of Perugia, 1965. *Ed.* L. Severi. p. 835.

Berenblum, I. (1955). The significance of the sequence of irritating and promoting actions on the process of skin carcinogenesis in the mouse. *Br. J. Cancer*, **9**, 268.

Bischoff, F. and Bryson, G. (1964). Carcinogenesis through solid state surfaces. *In*: 'Progress in Experimental Tumour Research'. Vol. 5. *Ed.* F. Homberg. p. 85. S. Karger, Basle.

Bittner, I. I. (1957). Recent studies on the mouse mammary tumour agent CMTA. *Ann. N.Y. Acad. Sci.,* **68,** 636.

Boyland, E., Roe, F. J. C., and Mitchley, B. C. V. (1966). Tests of certain constituents of spermicide for carcinogenicity in genital tract of female mice. *Br. J. Cancer.,* **20,** 18.

Cairns, T. (1979). The ED_{01} Study: Introduction, objectives and experimental design. *J. Exp. Pathol. Toxicol.,* **3,** 1.

Clayson, D. B. (1962). *In*: 'Chemical Carcinogenesis'. p. 135. J. and A. Churchill Ltd., London.

Clayson, D. B. (1987). ICPEMC Publ. No. 13. The need for biological risk assessment in reaching decisions about carcinogens. *Mutat. Res.,* **185,** 243–269.

Cohen, A. J. and Grasso, P. (1981). Review of the hepatic response to hypolipidaemic drugs in rodents and assessment of its toxicological significance to man. *Food Cosmet. Toxicol.,* **19,** 585–605.

Committee on Carcinogenicity of Chemicals in Food, Consumer Products and the Environment (1978). Guidelines on carcinogenicity testing. London.

Conybeare, G. (1980). Effect of quality and quantity of diet on survival and tumour incidence in outbred Swiss mice. *Food Cosmet. Toxicol.,* **18,** 65.

Cornfield, J. (1977). Carcinogenic risk assessment. *Science,* **198,** 693.

Cornfield, J., Carlborg, F. W., and Van Ryzin, J. (1978). Setting tolerance on the basis of mathematical treatment of dose–response data extrapolated to low doses. *Proc. First Int. Toxicology Congress.*

de Groot, A. P., Til, H. P., and Feron, V. J. (1970). Safety evaluation of yeast grown on hydrocarbons. I. One year feeding study in rats with yeast grown on gas-oil. *Food Cosmet. Toxicol.,* **8,** 267–276.

Druckrey, H. and Schmähl, D. (1962). Quantitative analyse der experimentellentellen Knebserzeugung. *Naturwissenschaften,* **49,** 19.

Falconer, D. S. and Bloom, J. L. (1964). Changes in susceptibility to urethane induced lung tumours produced by selective breeding in mice. *Br. J. Cancer,* **18,** 322.

Festing, F. W. (1975). A case for using inbred strains of laboratory animals in evaluating the safety of drugs. *Food Cosmet. Toxicol.* **13,** 369.

Food and Drug Administration Advisory Committee on Protocols for Safety Evaluation (1971). Panel on Carcinogenesis Report on Cancer Testing in the Safety of Food Additives and Pesticides.

Food Safety Council (1978). Chronic toxicity testing. Proposed system for food safety assessment. *Food Cosmet. Toxicol.,* **16,** Suppl. 2, 97–108.

Foulds, L. (1975). 'Neoplastic Development'. Vol. 2. p. 6. Academic Press, London.

Foulds, L. (1975). 'Neoplastic Development'. Vol. 2. p. 345. Academic Press, London.

Gala, R. R. and Loginsky, S. J. (1974). Correlation between serum prolactin levels and incidence of mammary tumours induced by 7,12-dimethylbenz[a]-anthracene in the rat. *J. Natl. Cancer Inst.,* **51,** 593–597.

Gardner, W. U., Pfeifter, C. A., and Trentin, I. I. (1959). Hormone factors in experimental carcinogenesis. *In*: 'The Physiopathology of Cancer'. 2nd Edn. p. 152. *Ed.* F. Homburger. Hoeber Harper, New York.

Gellatly, I. B. M. (1975). The natural history of hepatic parenchymal nodule formation in a colony of C57BL mice with reference to the effect of diet. *In*: 'Mouse Hepatic Neoplasia'. *Ed.* W. H. Butler and P. M. Newberne. Elsevier Scientific Publishing Company, Amsterdam.

Gibson, J. P., Newberne, J. W., Kuhn, W. L., and Elsea, J. R. (1967). Comparative chronic toxicity of three oral oestrogens in rats. *Toxic. Appl. Pharmacol.*, **11**, 489.

Glaister, J. (1986). 'Principles of Toxicological Pathology'. Taylor and Francis, London and Philadelphia. pp. 1–223.

Gleichmann, H. and Gleichmann, E. (1973). Immunosuppression and neoplasia. I. A critical review of experimental carcinogenesis and the immuno-surveillance theory. *Klin. Wochensch.*, **51**, 255.

Grimmer, G., Dettbarn, G., Brune, H., Deutsch-Wenzel, R., and Misfeld, J. (1982). Quantification of the carcinogenic effect of polycyclic aromatic hydrocarbons in used engine oil by topical application onto the skin of mice. *Int. Arch. Occup. Environ. Health*, **50**, 95.

Grasso, P. (1976). Review of tests for carcinogenicity and their significance to man. *Clin. Toxicol.*, **9**, 745–760.

Grasso, P. (1979). Carcinogenic risk from food—real or imaginary? *Chem. Ind. (London)*, 3rd February.

Grasso, P. and Crampton, R. F. (1972). The value of the mouse in carcinogenicity testing. *Food Cosmet. Toxicol.*, **10**, 418.

Grasso, P., Crampton, R. F., and Hooson, J. (1977). The mouse and carcinogenicity testing. The British Industrial Biological Research Association, Woodmansterne Road, Carshalton, Surrey, UK.

Grasso, P. and Golberg, L. (1966). Subcutaneous sarcoma as an index of carcinogenic potency. *Food Cosmet. Toxicol.*, **4**, 297.

Gross, L. (1951). Spontaneous leukaemia developing in C3H mice following inoculation, in infancy, with AK-leukaemic extracts, or AK-embryos. *Proc. Soc. Exp. Biol. Med.*, **76**, 27.

Gross, L., Schidlovsky, G., Feldman, D. M., Dreyfus, T., and Moore, L. A. (1975). *Proc. Natl. Acad. Sci. (USA)*, **72**, 3240–3244.

Gross, M. A., Fitzhugh, O. G., and Mantel, N. (1970). Evaluation of safety for food additives: an illustration involving the influence of methyl salicylate on rat reproduction. *Biometrics*, **26**, 101.

Heston, W. E. and Ulahakis, G. (1961). Elimination of the effect of the A^y gene on pulmonary tumours in mice by alteration of its effects on normal growth. *J. Natl. Cancer Inst.*, **27**, 1189.

Hill, B. M. (1963). The three parameter lognormal distribution and Bayesian analysis of a point-source epidemic. *J. Am. Stat. Assoc.*, **58**, 72.

Hoel, D. G., Gaylor, D. W., Kirschstein, R. L., Saffiotii, V., and Schneiderman, M. A. (1975). Estimation of risks of irreversible delayed toxicity. *J. Toxicol. Environ. Health*, **1**, 133.

Hogan, M. D. & Hoel, D. G. (1982). Extrapolation to man. *In*: 'Principles and Methods of Toxicology'. *Ed.* A. W. Hayes. Raven Press, New York.

IARC Working Group (1979). 'Monographs on the Evaluation of the Carcinogenic Risk of Chemicals to Humans'. Suppl. 1. International Agency for Research on Cancer, Lyon.

IARC Working Group (1980). 'Monographs on the Evaluation of the Carcinogenic Risk of Chemicals to Humans'. Suppl. 2. Long term and short term screening Assays for carcinogens: a critical appraisal. International Agency for Research on Cancer, Lyon.

Innes, J. R. M., Ulland, B. M., Valerio, M. G., Petrucelli, L., Fishbein, L., Hart, E. R., Pallotta, A. J., Bates, R. R., Falk, H. L., Gart, J. J., Klein, M., Mitchell, I. and Peters, J. (1969). Bioassay of pesticides and industrial chemicals for tumorigenicity in mice. A preliminary note. *J. Natl. Cancer Inst.*, **42**, 1101.

Kipling, M. D. (1976). Scots, tars and oils are causes of occupational cancer. *In*: 'Chemical Carcinogens'. *Ed.* C. E. Searle. ACS Monograph 173. American Chemical Society, Washington, DC.

Lemonde, P. (1959). Influence of fighting on leukaemia in mice. *Proc. Soc. Exp. Biol. Med.*, **102**, 292.

Lilly, F. and Pincus, T. (1973). Genetic control of murine viral leukaemogenesis. *Adv. Cancer Res.*, **17**, 231.

Mauderly, J. L. and Kritchevsky, J. (1979). Respiration of unsedated F-344 rats and the effect of confinement in exposure tubes. In 'ITRI' Annual Report 1978–1979. LF-69 pp. 475–478. Available from the National Technical Information Service, Springfield, Va.

Mead, J. (1960). Control of respiratory frequency. *J. Appl. Physiol.*, **15**, 325.

Ministry of Health and Welfare Canada (1973). The testing of chemicals for carcinogenicity, mutagenicity and teratogenicity.

Moloney, W. C., Boschetti, A. E. and King, V. P. (1970). Spontaneous leukaemia in Fischer rats. *Cancer Res.*, **30**, 41–43.

Munro, I. C. (1977). Considerations in chronic toxicity testing: the chemical, the dose, the design. *J. Envir. Path. Toxicol.*, **1**, 183.

Murray, I. A. (1911). Cancerous ancestry and the incidence of cancer in mice. *Sci. Rep. Invest. Imp. Cancer Res. Fund*, **4**, 114.

Nandi, S. and McGrath, C. M. (1973). Mammary neoplasia in mice. *Adv. Cancer Res.*, **17**, 353.

Page, N. P. (1977). Current concepts in a bioassay programme in environmental carcinogenesis. *In*: 'Environmental Cancer'. p. 114. *Eds* H. F. Kraybill and M. A. Mehlman. John Wiley and Sons, New York, London.

Page, N. P. (1977). Concepts of a bioassay programme in environmental carcinogenesis. *In*: 'Environmental Carcinogenesis'. *Eds* H. Kraybill and M. Mehlman. p. 7–171. Hemisphere Publishers, Washington.

Peraino, C., Fry, R. D. M., and Staffeldt, E. (1973). Enhancement of spontaneous hepatic tumourigenesis in C3H mice by dietary phenobarbital. *J. Natl. Cancer Inst.*, **51**, 1349.

Peto, R. and Lee, P. N. (1973). Weibull distributions for continuous-carcinogenesis experiments. *Biometrics*, **29**, 457–470.

Powell, C. H. and Hosey, A. D. (1965). The industrial environment, its evaluation and control (US DHS Publ. 614).

Roe, F. J. C. (1979). Food and Cancer. *J. Hum. Nutr.*, **33**, 405–415.

Roe, F. J. C. (1981). Testing *in vivo* for general chronic toxicity and carcinogenicity. *In*: 'Testing for Toxicity'. *Ed.* J. W. Gorrod. Taylor and Francis Ltd., London.

Ruddick, J. A., Ashanullah, M., Craig, J., and Stavric, B. (1978). Uptake and distribution of ^{14}C-saccharin in the rat foetus. *In*: Proceedings, Canadian Federation of Biological Societies, 21st Annual Meeting', London, Ontario. Abstract 635, p. 159.

Sanockij, I. V. (*Ed.*) (1978). Methods for determining toxicity and hazards of chemicals. Moscow. Medicina. pp. 62–63 (in Russian); cited in WHO Environmental Health Criteria..

Shimkin, M. B. and Stoner, G. D. (1975). Lung tumours in mice: application to carcinogenesis bioassay. *Adv. Cancer Res.*, **21**, 1.

Squires, R. A. (1981). Ranking Animal Carcinogens. A proper regulatory approach. *Science*, **214**, 877–880.

Tannenbaum, A. (1959). Nutrition and cancer. *In*: 'Physiopathology of Cancer'. 2nd Edn. *Ed*. F. Homberger. pp. 517–62. Hoeber-Harper, New York.

Toth, B. (1968). A critical review of experiments in chemical carcinogenesis using new born animals. *Cancer Res.*, **28**, 727–738.

Tucker, M. I. (1979). The effects of long-term food restriction on tumours in rodents. *Int. J. Cancer*, **23**, 803–807.

Wagner, J. C. 1962). Experimental production of mesothelial tumours of the pleura by implantation of dusts in laboratory animals. *Nature, London*, **196**, 180.

WHO (1978). Environmental Health Criteria 6. 'Principles and Methods for Evaluating the Toxicity of Chemicals'. Part I. World Health Organisation, Geneva.

WHO (1978a). Factors influencing the design of toxicity studies. *In*: 'Principles and Methods for Evaluating the Toxicity of Chemicals'. Part 1. World Health Organisation, Geneva. pp. 62–94.

WHO (1978b). Inhalation exposure. *In*: 'Principles and Methods for Evaluating the Toxicity of Chemicals'. Part 1. World Health Organisation, Geneva. pp. 199–235.

Yamigawa, K. and Ichikawa, K. (1918). Experimental study of the pathogenesis of carcinoma. *J. Cancer Res.*, **3**, 1.

CHAPTER 16

In Vitro *Methods for Teratology Testing*

DIANA ANDERSON

1 Introduction

Concern with congenital disease and malformations is ancient (Mark 9, 17–19) but prior to 1960 the only recommended protocol for testing chemicals during the reproductive cycle of animals was the 6 weeks' toxicity test in male and female rodents, which was evaluated over two pregnancies with a subsequent assessment of foetal survival. Following the thalidomide episode more stringent guidelines were adopted for reproductive studies for safety evaluation. Testing models for teratologic effects of foreign chemicals continue to rely chiefly upon mice, rats, and rabbits but they are, of necessity, time consuming and expensive to perform. Extrapolating findings from mammalian teratology testing models to those agents with the potential to cause human birth defects can never be absolute, since negative results can give no guarantee that a chemical will lack teratogenic effects in man, due to species variation in teratogenic response (Wilson, 1977). Effects at high doses will not necessarily be the same as at low doses. This, of course, is true for most branches of toxicology.

Because of inherent problems in models for mammalian teratology, the search continues for identifying predictive tests and various *in vitro* models have been established. For simplicity, *in vitro* here refers to the use of test subjects other than the pregnant mammal. The purpose of *in vitro* testing would be to identify compounds which require further whole animal testing rather than to define the pre-natal toxicity of the compounds. Smith *et al.* (1983) have suggested a selection of candidate compounds which might be used for *in vitro* test validation although Jensen *et al.* (1989) have criticised the list of compounds.

Selection of a representative tissue is important in the development of

Table 1 In Vitro *Teratogenicity Test Systems*

Sub-mammalian systems	Mammalian systems
Invertebrate Embryos Nematodes Coelenterates Echinoderms Insects	*Modified Whole Mammal System* The Chernoff Test
	Isolated Whole Embryos Pre-implantation Post-implantation
Vertebrate Embryos Pisces Amphibia Aves	*Isolated Embryonic Organs* Limb buds Palate
	Isolated Cell Systems Rat embryo limb and mid-brain cells Neural crest cells Chinese hamster ovary V79 cells and human embryonic and palatal mesenchyme Ascitic mouse ovarian tumour cells Other cell lines

an *in vitro* assay. The test systems vary as to the type of target employed and can be categorised according to whether DNA, single cells, organs, or whole embryos are used to assess pre-natal toxicity. The main problem is that little is known of the initial insults which lead to abnormal development. The insult may result in a change of one or more of the normal cellular functions of embryogenesis, such as proliferation, determination, aggregation, organisation, migration, and death. This may result from intracellular damage, perhaps involving genetic damage or inhibited enzymes, or extracellular damage such as altered cell membranes. *In vitro* systems are essentially too simple in biological terms to provide a complete model for a teratological event. Their value lies in isolating and investigating the constituent elements of teratological activity. However, some test systems have correctly predicted *in vivo* teratogenicity (ECETOC, 1989).

A test system should optimally incorporate all of the processes of growth and differentiation and detect retardation, malformation, and death. Two types of systems mainly satisfy these criteria. Those using developing embryos or embryo portions and those using regenerating tissue. Examples are sub-mammalian systems using invertebrate embryos [nematodes, coelenterates (Hydra), echinoderms (sea urchin), and insects (*Drosophila*)] and vertebrate embryos [pisces (zebra fish), amphibia (frog, salamander) and aves (chick)]; and mammalian systems utilising isolated whole embryos (pre- and post-implantation) and isolated embryonic organs (palate and limb-buds) and isolated cells in culture (Table 1).

2   Sub-mammalian Systems

A   Invertebrates

Large numbers of adult and embryonic and developmental stages can be maintained in the laboratory. The cell lineage of the nematode *Caenorhabditis elegans* from the fertilised egg to 900 cell adult has been elucidated (Sulston and Horvitz, 1977; Deppe *et al.*, 1978) and may prove a useful model. Planarians can also be used. Planarians can regenerate damaged fragments; absent, abnormal, or retarded regeneration, or loss of normal morphological appearances of intact animals are used as indicators of toxicity. Planarians appear to respond well to mammalian teratogens (Best and Monita, 1982). Similarly, coelenterates have proved useful. The adult Hydra (Attenuata) can be disrupted into individual cells, which on centrifugation form a pellet which may reform into a new organism. These artificial 'embryos' (Johnson, 1980) together with the adult have been used in an investigation of known teratogens and the data correlate well with mammalian data (Johnson *et al.*, 1982). Certain echinoderms, particularly the sea urchin (Gross *et al.*, 1972) can be cultured in large numbers and have been used in experimental embryology, and antimetabolite effects have been determined for embryogenesis (Lallier, 1965). The fruit fly *Drosophila* has been used extensively for short term tests for mutagenicity (Würgler *et al.*, 1977; de Serres and Ashby, 1981) but there are a great deal of genetic and developmental data for this organism and it may be suitable for teratogenicity studies (Schuler *et al.*, 1982). The mated female deposits eggs in a nutritive medium containing the test substance. It takes 6 days, after passing through larval and pupal stages, for the adults to emerge. These are anaesthetised and examined for abnormal development with a binocular microscope. In a study of 100 chemicals with single cell suspensions of *Drosophilia* eggs, 43 out of 47 teratogens and 2 non-teratogens were positive (Bournias-Vardiabasis *et al.*, 1982; 1983).

Despite the large phylogenetic difference between these non-mammalian species and man, some of these *in vitro* tests have performed well in predicting *in vivo* teratogenic activity (Collins, 1987; Daston and D'Amato, 1989; ECETOC, 1989).

B   Vertebrates

Fish, Amphibia

An integral part of toxicology studies on fish is the evaluation of the effects of chemicals on the development of eggs, embryos, and larvae (*e.g.* Birge *et al.*, 1979) and early experimental teratology studies often used fish as models (see Wilson, 1978). Fish such as zebra fish or minnows are easy to maintain in the laboratory, as are amphibia such as newts, salamander, and frogs. Dumart and co-workers (1982) have developed a teratogenicity screen known as FETAX (Frog Embryo

Teratogenesis Assay) using Xenopus embryos. Mid- to late-blastula stage embryos are exposed to test agents in water for up to 4 days and are assessed for mortality, pigmentation, growth, extent of development, and abnormalities. Preliminary validation of FETAX with various compounds including known teratogens and non-teratogens suggests it may be comparable to other screens (Dumont and Epler, 1984; Dawson *et al.,* 1989; Fort *et al.,* 1988; 1989).

Birds

The avian embryo, particularly that of the chick, has been used to investigate the direct effect of substances on developing morphogenetic systems since maternal influences are excluded. The chick embryo possesses its own intact metabolic system and comparative studies have claimed a predictive value of the chick embryo system comparable to other *in vitro* systems, as well as to whole animal systems (Jelinek, 1982).

WHO stated in 1967 that the use of the chick embryo for screening for teratogenicity is not to be recommended due to the absence of maternal–foetal relations, pharmacokinetic dissimilarities in the closed character of the avian egg, and high non-specific sensitivity. Jelinek (1982) argues that these three objections have arisen mainly from neglecting teratological principles, poor standardisation of test subjects, and inadequate administration techniques. Of late, the chick embryo has received more favourable reviews (Wilson, 1978) and this model would seem to require definitive validation.

The method is relatively simple. Eggs (generally White Leghorn) are incubated at 30°C. The test agent is administered through a hole bored in the shell, which may be resealed later with parafilm or wax. The chemical may be placed in the yolk sac, sub-germinal cavity, allantois, amnion, or air chamber (Wilson, 1978). Opinions differ as to appropriate treatment time, varying from 0 to 96 hours, but a detailed summary of these schedules and sacrifice times is available. The chick may be examined for abnormalities during incubation, hatching, or at full maturity to evaluate functional normality.

A more rapid 'chick embryotoxicity screening test' (CHEST) was proposed (Jelinek and Rychter, 1980) based upon the caudal morphogenetic system. Here, agents are administered directly below the caudal region of the embryo. According to Jelinek (1982) the chick embryo carries a complete set of morphogenetic control systems and has an advantage over other *in vitro* systems that employ isolated embryonic tissues which may have limited survival or lack a developing vascular bed, a frequent target for teratogens. Jelinek *et al.* (1984; 1985) tested 130 compounds, of these 119 exerted dose-related embryotoxic effects and the method permitted discrimination between substances toxic in low and high doses, suggesting a discrimination for teratogenic potency. Kučera and Burnand (1987) described an assay using Warren hen eggs.

3 Mammalian Systems

The various alternative test methods have been discussed by ECETOC (1989) and Daston and D'Amato (1989).

A A Modification of a Whole Mammal System—The Chernoff Test

This is a simplified *in vivo* teratogenicity assay where pregnant mice are treated at a single dose level (LDIO) with the test compound on days 8–12 of gestation. The new born are examined on days 1 and 3 post-partum for weight, litter size, and general appearance, and maternal weight during treatment is also recorded (Chernoff and Kavlock, 1980). Initially, 10 known teratogens affected one of the parameters measured whilst 11 out of 12 non-teratogens did not, and an independent validation with 55 compounds claimed a very good correspondence between teratogens and non-teratogens (Brown *et al.,* 1983). This test was independently validated under contract to NIOSH and has been reviewed (Hardin, 1987). Nevertheless, many mammals are still required to perform the assay even though they are half the number required for the usual animal study. It is at a different level than the other assays described here and may be useful to confirm data from a pre-screen.

B Whole Embryo Cultures

The two major test systems involving whole embryo culture use pre- or post-implantation embryos. The methodology of the two systems are well reviewed by Flynn (1987).

Pre-implantation

Culture techniques for the early mammalian conceptus from the one cell to blastocyst stages are well established (Daniel, 1971). The conceptus can be removed from the dam, maintained *in vitro* for several days, and re-implanted into a surrogate dam with development in term (Staples, 1971). Some authors have reported that the pre-implantation embryos are refractory to treatment with teratogenic agents (Staples, 1975) and high doses can result in embryo death, but instances of congenital malformation have resulted from insult to the blastocyst (Wilson, 1978). Hsu (1973; 1980) has perfected techniques to allow mouse blastocysts to be grown *in vitro* to the early somite stage, but this test is as yet too complicated for routine use and suffers from poor residual viability. Various biochemical and cytogenetic parameters can be assessed fairly readily (ECETOC, 1989). The DNA repair and metabolising capacity of the system after treatment with the test compound should be assessed (Spielmann and Vogel, 1989).

Post-implantation

Most studies of this kind have used the rat. Day 1 is defined as the time when a post-vaginal smear or a copulatory plug is detected. According to Brown and Fabro (1982), implantation is completed by 7.5–8 days of age. The embryo is then at the egg cylinder stage. Subsequent development proceeds through the following stages: age 8.5 days, primitive streak; 9 days, pre-somite neurula; 9.5 days, first somites and brainfolds; 10.5 days, 10–14 somites, yolk sac circulation; 11.5 days, tail bud embryo, 26–30 somites, forelimb bud; 12.5 days, complete embryo, 40–42 somites; 13.5 days, early foetus 48–50 somites. Techniques are available which support embryonic growth for 1–4 days. The dam is killed, the uterus removed and implantation sites dissected from the uterine wall with the aid of a dissecting microscope and microforceps. The maternal decidual tissue is removed, leaving the conceptus enclosed with the trophoblastic membrane. This is then removed and the conceptus transferred to culture medium using static or circulating methods. The latter enables the medium to be circulated around the immobilised conceptus and allows greater dilution of homologous rat or human serum. The growth of embryos explanted at 9.5 days and cultured over 48 hours is indistinguishable from that *in utero* over the equivalent period. Growth of older embryos is retarded due to a lack of placental flow of nutrients (the yolk sac placentation is sufficient for earlier embryos). The survival of embryos explanted earlier than 9.5 days is reduced compared with those explanted at 9.5 days, but they require lower oxygen concentrations (5% as opposed to 95% after 12.5 days). Tesh (1988) showed that the rat embryo has only limited metabolic competency at this stage of development.

A metabolising system may be incorporated into the system by using an hepatic microsomal fraction [the S-9 mix system of Ames *et al.* (1975) and Maron and Ames (1983)]. Thus Greenaway *et al.* (1982) found that 10 day embryos *in vitro* were unaffected by cyclophosphamide concentrations as high as 250 μg ml^{-1} alone, but 6.25 μg ml^{-1} of the drug produced abnormal embryos if S-9 mix was present. Another method uses serum from exposed animals or humans which is assumed to contain metabolites. For example, Chatot *et al.* (1980) showed that the serum of patients undergoing therapy with chemotherapeutic agents exerted significant embryotoxic effects on cultured rat conceptuses. Clearly such experiments require careful control and are subject to great problems of species specificity, dose determination, and uncertainty concerning the presence of other active factors that may be present in the serum. Anderson *et al.* (1986) and Jenkinson *et al.* (1986) have shown that not only chemicals, direct-acting or requiring metabolic activation, but also oxygen radical species can effect the development of rat embryos in culture.

Embryonic growth is relatively easy to measure and various parameters are suitable such as crown–rump length, head length, and embryonic protein or DNA content (Brown *et al.*, 1979). Yolk sac effects must be taken into account and data are best presented on those conceptuses which have a functional yolk sac circulation at the end of the culture period. When presenting incidences of malformation, some indication must be given of the proportion coming from abnormal yolk sacs. The exact relationship of yolk sac damage *in vitro* to actions of the test compound *in vivo* has yet to be clarified.

Cicurel and Schmid (1988a and b) concluded that this system had high predictive value for teratogens using 25 chemicals from 46 reference compounds (Smith *et al.*, 1983; Jensen *et al.*, 1989), however, criticised the use of these reference compounds. Van Maele-Fabry and Picard (1987) investigated the post-implantation embryo culture system in the mouse.

C Organ Culture

Organs from Embryos and Foetuses

Various organs have been maintained in culture (see Brown and Fabro, 1982; Lyng, 1989; 1990; Whitby, 1987) such as palatal shelves (mouse), palate (whole-embryo mouse), bone (mouse, rat), müllerian ducts (rat), testis, ovaries (rabbit, rat), and teeth (mice). The vast majority of these systems would not be appropriate as testing techniques since organs are explanted at an advanced stage of differentiation. They may be useful in studies of biochemical and histological development.

Limb buds. Limb buds are the most amenable organ to *in vitro* culture (Aydelotte and Kochlar, 1972; Kochlar *et al.*, 1982). Mouse is the main source of material but rat and rabbit limb buds have also been used. The best development is achieved with fore-limb buds explanted from embryos at the 40–43 somite stage. The limb buds, dissected from explanted embryos, are placed on membrane filters supported by metal grids in petri dishes. Sufficient culture medium is added to wet the underside of the filter and the culture is incubated in a humidified 5% CO_2-air atmosphere at 30°C. Maximum limb bud growth is obtained on day six of culture when cartilaginous rudiments of phalanges, ulna, radius, humerus, and scapula are formed. Neubert and Barrach (1977) maintained limb buds explanted in a roller-bottle system with an improvement in phalangeal development. Kocklar *et al.*, (1982) incorporated a metabolic system in the form of sera from drug-treated mice and exposed humans. In a novel approach, Agnish and Kochlar (1976) exposed the whole embryos in culture to bromodeoxyuridine and then organ cultured the fore-limbs for an additional nine days when structural development could be accurately assessed. Combining whole embryo and organ culture satisfies the conditions for exposure during the organogenesis period and allows assessment of differentiated structure, but cannot be considered rapid and inexpensive relative to whole animal testing.

The suitability of limb bud culture has been doubted (Wilson, 1978; Friedman, 1987) because the development of the organ does not closely parallel *in vivo* development, in that 6 days of culture is equivalent to approximately 3 days *in utero*, and there is organ distortion. The *in vitro* differentiation is reproducible and consistent in itself, however, and a scoring system has been devised to quantitate the extent of limb bud differentiation. The effect of exposure to test chemicals can be quantitatively assessed (ECETOC, 1989; Hales, 1992).

D   Cell Culture

Various assays have been established using isolated cells in culture.

Rat Embryo Limb Cells and Mid-brain Cells in Micromass Culture

Cultures are started by plating cells at high density in a single small drop. They adhere together to the dish within about two hours. The cells are prepared from a thirteen day embryo after trypsinisation before pre-cartilage condensation (Flint, 1979; 1982; 1983). After cell adhesion, medium containing the test substance floods the culture for 5 or 6 days. In the case of limb bud cells, untreated cultures develop foci of chondrogenic cells which secrete proteoglycan and this can be stained with alcian blue. Mid-brain (mesencephalon) cells develop into neurons. In a blind trial, where test chemicals with or without S-9 mix were added directly to the culture medium, Flint and Orton (1984) measured foci or neurons and found a concentration-dependent inhibition in the number of differentiating cell foci following exposure to 25 out of 27 teratogens and 2 out of 19 non-teratogens. This complementary assay is the best characterised of the primary cell culture methods. By treating the dams (intraperitoneally on day 12 of pregnancy) before moving embryonic tissue, Flint *et al.* (1984) have shown that of 31 chemicals tested, 17 out of 18 teratogens and one out of 13 non-teratogens were positive when a number of end-points were examined. The micromass test has been used in the early stages of pharmacological evaluation to screen out potentially teratogenic compounds (Flint, 1986). Criteria for a positive *in vitro* result have been suggested by Flint (1987). There have been several independent trials of the test (Bacon *et al.*, 1990; Parsons *et al.*, 1990; Uphill *et al.*, 1990; Newall and Beedles, 1993).

Neural Crest and Neuroblastoma Cells

Primary neural crest cells are explanted from chick embryos after $1\frac{1}{2}$ days incubation, into culture segments of neural tube (Wilk *et al.*, 1979). The tube is removed after the crest cells have migrated from the tube to form a monolayer. Cells are replated as small drops as for micromass cultures.

Depending on the type of serum present in the medium, such cells can develop into melanocytes or neurons and can be assayed for melanin or acetyl choline transferase, respectively. Only 14 agents have been tested; 10 out of 12 teratogens and none of the 4 non-teratogens were positive (Greenberg, 1980; Greenberg *et al.*, 1982).

Mummery *et al.* (1984a and b) using a clone (NIE 115) of the murine neuroblastoma cell line examined induction and inhibition of cell differentiation. Of 39 teratogens and 18 non-teratogens, the system detected as positive 35 teratogens and 4 non-teratogens.

Chinese Hamster Ovary V7 Cells and Human Embryonic Palatal Mesenchyme (HEPM) Cells

Welsch and Stedman (1984a and b) have used the disruption of cell to cell communication of these cell lines as a screen for teratogens. For V79 cells a toxic metabolite of 6-thioguanine is transferred from wild type to mutant cells by metabolic co-operation; for HEPM cells tritiated uridine is transferred through gap junctions. Treatment with a test compound reduces the transfer of the toxic metabolite causing an increase in mutant survival in V79 cells or an increase in incorporation of label as measured by autoradiography in HEPM cells. Eleven chemicals have been tested for increased mutant survival and 3 for gap junction transfer. Swierenga and Yamasaki (1992) have reviewed the disruption of gap-junction intercellular communication for screening chemicals.

Ascitic Mouse Ovarian Tumour Cells (MOT)

Braun and colleagues (1979 and 1982) have measured the inhibition of tumour cell attachment to Concanavalin-A-coated surfaces as an assay for teratogenic agents, since it is believed that cell adhesions may be critical to embryonic morphogenesis. MOT cells are grown as an ascites tumour in the peritoneum of mice and labelled with tritiated thymidine the day before harvest. After harvest they are treated with the test compound and then allowed to attach to the Con-A coated discs. In a study of 102 chemicals, 60 of 74 teratogens and 7 of the 28 non-teratogens were inhibitory. Cell growth and division are not involved in the cell attachment assay, so teratogens which affect DNA replication are not affected.

Thirteen of the teratogenic agents which were false negatives in this tumour cell attachment assay were tested with the HEPM cell line (above). Cell proliferation was used as an end-point for teratogenicity and this was decreased for 12 out of 13 of these agents (Pratt *et al.*, 1982). A system combining these two assays, known as BRAT, is being evaluated by the US National Toxicology Programme.

Other Cell Lines

The measurement of cell death or growth inhibition in cell cultures has been suggested by Freese (1982) as a means of identifying teratogens, since differentiating cells may be more vulnerable to treatment. The dose of compound required to inhibit 50% of cell growth was examined in various cells (*e.g.* Hela cell line and rat glial cell line C) but there was no obvious difference between teratogens and non-teratogens.

BSC 40 cells, a primate line of monkey origin, have been infected with pox virus and then cultured with the test chemical, after which cells are frozen. They are subsequently assayed for the number of virus colonies using a plaque-forming assay, on the assumption that viral replication is only possible if the host cell is actively proliferating. In a screen of 51 chemicals (Keller and Smith, 1982), measuring the inhibition of plaque number by 50%, 36 of 42 teratogens and 1 of 9 non-teratogens were positive.

It is not yet clear whether the cell culture systems described in 1–5 are able to determine teratogens as opposed to generally toxic chemicals.

4 Usefulness of the *In Vitro* Methodology

The tests described offer advantages of speed and ease of operation, and are relatively inexpensive to perform (ECETOC, 1989; Daston and D'Amato, 1989). *In vitro* tests permit precise regulation of exposure, and maternal factors such as the placental 'barrier' are excluded. A well-controlled *in vitro* test could give a precise estimate of the inherent teratogenic potential of a chemical under defined conditions. Fabro *et al.* (1982), Brown and Freeman (1984), and Brown (1987) have discussed the use of these systems in terms of developmental toxicity risk assessment, where concepts of teratogenic potential, potency, and hazard are described: potential is the ability of an agent under any circumstances to induce terata, potency is the dose required to induce abnormalities, and hazard is any estimate of the relationship between teratogenic doses and general or adult toxic doses. A chemical with high hazard is one which is teratogenic at doses which are well below maternally toxic doses. Both potency and hazard are important components of the estimation of risk to humans. According to Fabro *et al.* (1982), hazard is the more important for an initial screen which should be able to detect those chemicals which have specific action on development. Comparison of hazard *in vitro* and *in vivo* could determine whether the *in vitro* assay predicts strong and weak teratogens as determined by animal studies. (A strong teratogen is one which will induce a high incidence of malformation at doses well below maternally toxic doses and a non-teratogen is one which will not induce fatal abnormalities even up to maternally lethal doses.)

Obviously, the absence of the maternal–conceptus–placental-relationship reduces the practical value of *in vitro* screens in predicting possible mammalian teratogenicity and these tests may never replace the pregnant mammal as the primary system for teratogenicity testing. Assessment of an agent's developmental toxicity is often confounded by the close relationsip between the maternal and developing systems (Kimmel *et al.*, 1987). The relative performances of the *in vitro* tests will only become apparent when they have been tested with the same group of chemicals such as those suggested by Smith *et al.* (1983). To date, groups of chemicals used for validation purposes vary widely.

The tests do offer promise, however, as pre-screening procedures and provided such procedures are assessed as to their ability to distinguish mammalian teratogens from non-teratogens before the outcome of mammalian studies is known, they may find a role in reducing the burden of animal testing required (Faustman-Watts, 1988).

Once again, it is unlikely that the various models will be fully accepted until an understanding of the mechanisms of control of embryonic and foetal development is achieved and the methods directed towards the examination of the effect of compounds on these mechanisms.

References

Anderson, D., Jenkinson, P. C., and Gangolli, S. D. (1986). The effect of chemicals and radical species on rat embryos in culture. *Food Chem. Toxicol.*, **24**, (6.7) 637–638.

Agnish, N. D. and Kochlar, D. M. (1976). Direct exposure of post-implantation mouse embryos to 5-bromodeoxyuridine *in vitro* and its effect on subsequent chondrogenesis in the limbs. *J. Embryol. Exp. Morphol.*, **36**, 623–638.

Ames, B. N., McCann, J., and Yamasaki, E. (1975). Methods for detecting carcinogens and mutagens with the *Salmonella*/mammalian microsome mutagenicity test. *Mutat. Res.*, **31**, 347–364.

Aydelotte, M. B. and Kochlar, D. M. (1972). Development of mouse limb buds in organ culture: chondrogenesis in the presence of a proline analogue, L-azetidine-2-carboxylic acid. *Dev. Biol.*, **28**, 191–201.

Bacon, W. J., Duffy, P. A., and Jones K. (1990). Studies on variability of the micromass teratogen test. *Toxicol. In Vitro*, **4**, 577–581.

Best, J. B. and Morita, M. (1982). Planarians as a model system for *in vitro* teratogenesis studies. *Teratogen. Carcinogen. Mutagen.*, **2**, 277–291.

Birge, W. J., Black, J. A., Hudson, J. E., and Bruser, D. M. (1979). *In*: 'Aquatic Toxicology ASTM STP 667', *Eds* L. L. Marking and R. A. Kimerle. American Society for Testing Materials, pp. 131–147.

Bournias-Vardiabasis, N. and Teplitz, R. L. (1982). Use of *Drosophila* embryo cell cultures as an *in vitro* teratogen assay. *Teratogen. Carcinogen. Mutagen.*, **2**, No. 3/4, pp. 333–343.

Bournias-Vardiabasis, N., Teplitz, R. L., Chernoff, G. F., and Seccof, R. L. (1983). Detection of teratogens in the *Drosophila* embryonic cell culture test assay of 100 chemicals. *Teratology*, **28**, 109–122.

Braun, A. G., Emerson, D. J., and Nichinson, B. B. (1979). Teratogen drugs inhibit tumour cell attachment to lectin-coated surfaces. *Nature (London)*, **282**, 507–509.

Braun, A. G., Nichinson, B. B., and Horowicz, P. B. (1982). Inhibition of tumour cell attachment to Conconavalin-A coated surfaces as an assay for teratogenic agents, approaches to validation. *Teratogen. Carcinogen. Mutagen.*, **2**, No. 3/4, pp. 343–355.

Brown, N. A. (1987). *Arch. Toxicol.*, Suppl. 11, 105–114.

Brown, N. A. and Fabro, S. E. (1982). The *in vitro* approach to teratogenicity testing. *In*: 'Developmental Toxicology,' *Ed.* Keith Snell. Croom-Helm, London, pp. 33–54.

Brown, N. A., and Freeman, S. J. (1984). Alternative tests for teratogenicity. 'Alternatives to Laboratory Animals', **12**, 7–23.

Brown, N. A., Goulding, E. H., and Fabro, S. E. (1979). Ethanol embryotoxicity: direct effects on mammalian embryos *in vitro*. *Science*, **206**, 573–575.

Brown, J., Jorgenson, T. A., and Anderson, D. G. (1983). Validation of an *in vivo* pre-screen for the determination of embryo/foetal toxicity in mammals. *Teratology*, **27**, 34A.

Chatot, D. L., Klein, N. W., Piatek, J., and Pierre, L. J. (1980). Successful culture of rat embryos on human serum: use in the detection of teratogens. *Science*, **207**, 1471–1473.

Cicurel, L. and Schmid, B. P. (1988a). Post-implantation embryo culture: validation with selected compounds for teratogenicity testing. *Xenobiotica*, **18**, 617–624.

Cicurel, L. and Schmid, B. P. (1988b). Post-implantation embryo culture for the assessment of the teratogenic potential and potency of compounds. *Experientia*, **44**, 833–840.

Collins, T. F. X. (1987). Teratological research using *in vitro* systems. V. Nonmammalian model systems. *Environ. Health Perspect.*, **72**, 237–249.

Daniel, J. C. (*Ed.*) (1971). 'Methods in mammalian embryology'. Freeman and Company, San Francisco.

Daston, G. P. and D'Amato, R. A. (1989). *In vitro* techniques in teratology. *Toxicol. Ind. Health*, **5**, 555–585.

Dawson, D. A., Fort, D. J., Newell, D. L., and Bantle, J. A. (1989). Developmental toxicity testing with FETAX: evaluation of five compounds. *Drug Chem. Toxicol.*, **12**, 67–75.

Deepe, U., Schierenberg, E., Cole, T., Krieg, C., Schmitt, D., Yoder, B., and von Ehrenstein, G. (1978). Cell lineages of the embryo of the nematode *Caenorhabolitis elegans*. *Proc. Natl. Acad. Sci.* (*USA*), **75**, 376–380.

de Serres, F. J. and Ashby, J. (*Eds*) (1981). Evaluation of short term tests for carcinogens. 'Report of the International Collaborative Programme'. Elsevier/North-Holland Biomedical Press, Amsterdam, Holland.

Dumont, J. N., Schultz, T. W., and Newman, S. M. (1982). A frog embryo teratogenesis assay. Xenopus (FETAX)—a model for teratogen screening. *Teratology*, **25**, 37A.

Dumont, J. N. and Epler, R. G. (1984). Validation studies of the FETAX teratogenesis assay (frog embryos). *Teratology*, **29**, 38A.

ECETOC (1989). 'Alternative Approaches for the Assessment of Reproductive Toxicity (with Emphasis on Embryotoxicity/Teratogenicity)', Monograph No. 12, 50 pp.

Fabro, S., Shull, G., and Brown, N. A. (1982). The relative teratogenic index and teratogenic potency: proposed components of developmental toxicity risk assessment. *Teratogen. Carcinogen. Mutagen.*, **2**, No. 1, pp. 61–77.

Fantel, A. G. (1982). Culture of whole rodent embryos in teratogen screening. *Teratogen. Carcinogen. Mutagen.*, **2**, No. 3/4, pp. 231–243.

Fantel, A. G., Greenaway, J. C., Juchau, M. R., and Shephard, T. H. (1979). Teratogenesis bioactivation of cyclophosamide *in vitro. Life Sci.*, **25**, 67–73.

Faustman-Watts, E. M. (1988). *Mutat. Res.*, **205**, 355–384.

Flint, O. P. (1980). The effects of sodium salicylate, cytosine, arabinoside, and eosine sulphate in rat limb buds in culture. *In*: 'Teratology of the Limbs', *Eds* H. J. Merker, H. Nau, and D. Neubert de Gruyter. Berlin, pp. 325–338.

Flint, O. P. (1982). An *in vitro* assay for teratogens using embryonic cells exposed to compound *in utero* or directly in culture. *Teratology*, **15**, 40A.

Flint, O. P. (1983). A micromass method for rat embryonic neural cell. *J. Cell Sci.*, **61**, 247–262.

Flint, O. P. and Orton, T. C. (1984). An *in vitro* assay for teratogens using cultures of rat embryo cells, brain, and limb bud cells. *Toxicol. Appl. Pharmacol.*, **76**, 383–395.

Flint, O. P., Orton, T. C., and Ferguson, R. A. (1984). Differentiation of rat embryo cells in culture. Response following acute maternal exposure to teratogens and non-teratogens. *J. Appl. Toxicol.*, **4**, No. 2, pp. 109–116.

Flint, O. P. (1986). An *in vitro* test for teratogens: its practical application. *Food Chem. Toxicol.*, **24**, 627–631.

Flint, O. P. (1987). An *in vitro* test for teratogens using cultures of rat embryo cells. *In*: '*In vitro* Methods in Toxicology'. *Eds* C. K. Atterwill and C. E. Steele. Cambridge University Press, Cambridge, pp. 339–363.

Flynn, T. J. (1987). Teratological research using *in vitro* systems. I. Mammalian whole embryo culture. *Environ. Health Perspect.*, **72**, 203–210.

Fort, D. J., Dawson, D. A., and Bantle, J. A. (1988). Development of a metabolic activation system for the Frog Embryo Teratogenesis Assay: Xenopus (FETAX). *Teratogen. Carcinogen. Mutagen.*, **8**, 251–263.

Fort, D. J., James, B. L., and Bantle, J. A. (1989). Evaluation of the developmental toxicity of five compounds with the Frog Embryo Teratogenesis Assay: Xenopus (FETAX) and a metabolic activation system. *J. Appl. Toxicol.*, **9**, 377–388.

Freese, E. (1982). Use of cultured cells in the identification of potential teratogens. *Teratogen. Carcinogen. Mutagen.*, **2**, 355–360.

Friedman, L. (1987). Teratological research using *in vitro* systems. II. Rodent limb bud culture system. *Environ. Health Perspect.*, **72**, 211–219.

Greenberg, J. H. (1980). Normal and Abnormal Differentiation of Neural Crest Cells in Culture. *In*: 'Current Research Trends in Prenatal Craniofacial Development'. *Eds* R. M. Pratt and A. Christiansen. Elsevier/New York, pp. 65–80.

Greenberg, J. H. (1982). Detection of teratogens by differentiating embryonic neural crest cells in culture-evaluation as a screening system. *Teratogen. Carcinogen. Mutagen.*, **2**, No. 3/4, pp. 319–325.

Greenaway, J. C., Fantel, A. G., Shephard, T. J., and Juchau, M. R. (1982). The *in vitro* teratogenicity of cyclophosphamide on rat embryos. *Teratology*, **25**, 335–343.

Gross, P., Humphreys, T., and Anderson, E. (1972). 'The Sea Urchin: developmental urchin'. MSS Information Corporation, New York.

Hales, B. F. (1992). Teratogenicity. *In*: '*In Vitro* Toxicity Testing. Applications to Safety Evaluation'. *Ed.* J. M. Frazier. Marcel Dekker, Inc., New York, Ch. 9, pp. 205–220.

Hardin, B. D. (1987). Special Issue *Teratogen. Carcinogen. Mutagen.*, **7**, No. 1.

Hsu, Y. C. (1973). Differentiation *in vitro* of mouse embryos in the stage of early somite. *Dev. Biol.*, **33**, 403–411.

Hsu, Y. C. (1980). Embryo growth and differentiation factors in embryonic sera of mammals. *Dev. Biol.*, **76**, 465–474.

Jelinek, R. (1982). Use of Chick embryo in screening for embryotoxicity. *Teratogen. Carcinogen. Mutagen.*, **2**, No. 3/4, pp. 225–263.

Jelinek, R. and Rychter, Z. (1980). Morphogenetic systems and screening for embrytoxicity. *Arch. Toxicol. Suppl.*, **4**, 267–273.

Jenkinson, P. C., Anderson, D., and Gangolli, S. D. (1986). Malformations induced in cultured rat embryos by enzymically generated active oxygen species. *Teratogen. Carcinogen. Mutagen.*, **6**, No. 6, 547–554.

Jensen, M., Newman, L. M., and Johnson, E. M. (1989). Re: Problems in validation of *in vitro* developmental toxicity assays. *Fundam. Appl. Toxicol.*, **13**, 863–865.

Johnson, E. M. (1980). Detection of teratogenic potential of drugs and chemicals by an artificial embryo. *Anat. Rec.*, **196**, 89A.

Johnson, E. M., Gorman, R. M., Bradley, E. G., and George, M. E. (1982). The hydra attenuata system for detection of teratogenic hazards. *Teratogen. Carcinogen. Mutagen.*, **2**, No. 3/4, 263–277.

Keller, S. J. and Smith, M. K. (1982). Animal virus screen for potential teratogens. I. Pox virus morphogenesis. *Teratogen. Carcinogen. Mutagen.*, **2**, 361–374.

Kimmel, E. L., Kimmel, C. A., and Francis, E. Z. (1987). Special Issue *Teratogen. Carcinogen. Mutagen.*, **7**, No. 3.

Kochlar, D. M. (1982). Embryonic limb bud culture in assessment of teratogenicity of environmental agents. *Teratogen. Carcinogen. Mutagen.*, **2**, No. 3/4, 303–313.

Kučera, P. and Burnand, M. B. (1987). *Teratogen. Carcinogen. Mutagen.*, **7**, No. 5, 427–449.

Lallier, R. (1965) *J. Embryol. Exp. Morphol.*, **31**, 721–734.

Lyng, R. D. (1989). Test of six chemicals for embryotoxicity using fetal mouse salivary glands in culture. *Teratology*, **39**, 591–599.

Lyng, R. D. (1990). Use of fetal mouse salivary glands in culture to detect embryotoxicity: evaluation of eight additional chemicals. *Toxicol. Lett.*, **54**, 245–251.

Maron, D. and Ames, B. N. (1983). Revised methods for the *Salmonella* mutagenicity tests. *Mutat. Res.*, **113**, 173–215.

Mummery, C., van den Brink, C., van der Saag, P., and de Laat, S. (1984). A short-term screening test for teratogens using neuroblastoma cells *in vitro*. *Teratology*, **29**, 271–279, and abstracts from European Teratology Society, Holland 5–7 Sept, 1984.

Neubert, D. and Barrach, H. J. (1977). Techniques applicable to study morphogenetic differentiation of limb buds in organ culture. *In*: 'Methods in Prenatal Toxicology', *Eds* D. Neubert, H. J. Merker, and T. E. Kwasigroch. George Thieme Publishers, Stuttgart, pp. 241–251.

Newall, D. R. and Beedles, K. (1993). The stem-cell test—a novel *in vitro* assay for teratogenic potential. *In*: Book of Abstracts, Practical *In Vitro* Toxicology III, 25–29 July. Abstract 3.5, p. 20.

Parsons, J. F., Rockley, J., and Richold, M. (1990). *In vitro* micromass teratogen test: interpretation of results from a blind trial of 25 compounds using three separate criteria. *Toxicol. In Vitro,* **4,** 609–611.

Pratt, R. M., Grove, R. I., and Willis, W. D. (1982). Prescreening for environmental teratogens using cultured mesenchymal cells from the human embryonic palate. *Teratogen. Carcinogen. Mutagen.,* **2,** No. 3/4, 313–318.

Schuler, R. L., Hardin, B. D., and Niemeier, R. W. (1982). *Drosophila* as a tool for the rapid assessment of chemicals for teratogenicity. *Teratogen. Carcinogen. Mutagen.,* **2,** No. 3/4, 293–303.

Smith, M. K., Kimmel, G. L., Kochlar, D. M., Shephard, T. H., Spielberg, S. P., and Wilson, J. G. (1983). A selection of candidate compounds for *in vitro* teratogenesis test validation. *Teratogen. Carcinogen. Mutagen.,* **3,** 461–480.

Spielmann, H. and Vogel, R. (1989). Unique role of studies on preimplantation embryos to understand mechanisms of embryotoxicity in early pregnancy. *CRC Crit. Rev. Toxicol.,* **20,** 51–64.

Staples, R. E. (1971). *In* 'Methods in Mammalian Embryology', *Ed.* J. C. Daniel. Freeman and Company, San Francisco, pp. 190–204.

Staples, R. E. (1975). *In*: 'New Approaches to the Evaluation of Abnormal Embryonic Development', *Eds* D. Neubert and H. J. Merker. Thieme Edition Publishing Sciences Group, Berlin, pp. 71–81.

Sulston, J. E. and Horvitz, H. R. (1977). Post-embryonic cell lineages of the nematode, *Caenorhabolitis elegans. Dev. Biol.,* **56,** 110–156.

Swierenga, S. H. H. and Yamasaki, H. (1992). Performance of tests for cell transformation and gap-junction intercellular communication for detecting non-genotoxic carcinogenic activity. *In*: 'Mechanisms of Carcinogenesis in Risk Identification'. *Eds* H. Vainio, P. N. Magee, D. B. McGregor, and A. J. McMichael. Lyon International Agency for Research on Cancer, pp. 165–193.

Tesh, J. M. (1988). The application of whole-embryo culture to new product development. *Toxicol. In Vitro,* **2,** 189–194.

Uphill, P. F., Wilkins, S. R., and Allen, J. A. (1990). *In vitro* micromass teratogen test: results from a blind trial of 25 compounds. *Toxicol. In Vitro,* **4,** 623–626.

Welsch, F. and Stedman, D. B. (1984a). Inhibition of metabolic co-operation between Chinese hamster V79 cells by structurally diverse teratogens. *Teratogen. Carcinogen. Mutagen.,* **4,** 285–302.

Welsch, F. and Stedman, D. B. (1984b). Inhibition of intercellular communication between normal human embryonic palatal mesenchyme cells by teratogenic glycol ethers. *Environ. Health Perspect.* **57,** 125–133.

Whitby, K. E. (1987). Teratological research using *in vitro* systems. III. Embryonic organs in culture. *Environ. Health Perspect.,* **72,** 221–223.

Wilk, A. L., Greenberg, J. H., Horigan, E. A., Pratt, R. M., and Martin, G. R. (1979). Detection of teratogenic compounds using differentiating embryonic cells in culture. *In Vitro,* **16,** 269–276.

Wilson, J. G. (1977). Current status of teratology. *In*: 'Handbook of Teratology', *Eds* J. G. Wilson and F. C. Fraser. Vol. 1, pp. 47–74. Plenum Press, New York.

Wilson, J. G. (1978). Survey of *in vitro* systems, their potential use in teratogenicity screening. *In*: 'Handbook of Teratology', *Eds* J. G. Wilson and F. R. Fraser. Vol. 4, pp. 135–153. Plenum Press, New York.

World Health Organisation (1967). Principles for the testing of drugs for teratogenicity. Technical Report Series 167, No. 364.

Würgler, F. E., Sobels, F. H., and Vogel, E. (1977). *Drosophila* as assay system for detecting genetic changes. *In*: 'Handbook of Mutagenicity Test Procedures', *Eds* B. J. Kilbey, M. Legator, W. Nichols, and C. Ramel. Elsevier/North-Holland Biomedical Press, pp. 335–373.

CHAPTER 17

Assessing Chemical Injury to the Reproductive System

S. D. GANGOLLI AND J. C. PHILLIPS

1 Introduction

The mammalian reproductive process is long and complex, and thus uniquely susceptible to disruption by exogenous agents. Foreign compounds can interfere with either the neurophysiological processes involved in the process or in gametogenesis. In the male, for example, spermatogenesis can be altered directly by an effect on the germ cells at the various stages in their maturation process or indirectly by disruption of the endocrine functions essential for regulating their production and delivery. In the female, oogenesis can be affected directly (oocyte destruction) or indirectly by interference with the ovarian and uterine cycles. As the integrity of the reproductive system is of crucial importance to the survival of the species, the identification of foreign chemicals which can affect human reproduction is of major importance. Despite a lack of understanding of the biochemical mechanisms underlying the toxicity of many agents, a wide variety of compounds including drugs, pesticides, organic solvents, and heavy metals (Table 1) have been found to be reproductive toxins in both the male and female (Bernstein, 1984; Lucier, Lee, and Dixon 1977; Schrag and Dixon, 1985).

In this chapter, the investigation of chemical injury to the mammalian reproductive system will be discussed from a mechanistic standpoint, first by describing the essential components of the male and female reproductive system; secondly by outlining the toxic consequences resulting from chemical injury to specific male (testicular and extra testicular) and female (ovarian and uterine) sites by some model toxins, and thirdly by discussing practical techniques for identifying and assessing toxicity.

To this end, the chapter is divided into the following main sections:
The anatomy and physiology of the male reproductive system.
Anatomy and physiology of the female reproductive system.
Mechanisms regulating testicular function.
Mechanisms regulating ovarian function.

Table 1 *Some environmental agents associated with reproductive dysfunction in humans.*

Male: *confirmed effect†*
  Carbon disulphide; dibromochloropropane (DBCP), lead; oral contraceptives
    : *probable*
  Anaesthetic gases; benzene; carbaryl; chlordecone; chloroprene;
  dinitrotoluene; ethylene dibromide; hexane; hexachlorobenzene; glycol ethers;
  metals (arsenic, boron, cadmium, manganese); pentachlorophenol; radiation;
  dioxins; vinyl chloride monomer.
Female:
  Anaesthetic gases; aniline; benzene; carbon disulphide; chloroprene; ethanol;
  ethylene oxide; glycol ethers; formaldehyde; metals (lead, organic mercury);
  phthalate esters; polychlorinated biphenyls; styrene; tobacco smoke; toluene;
  viny chloride monomer.

---

† Adapted from Dixon (1986) and Schrag and Dixon, (1985).

  Site specificity of chemical injury by model toxins.
  Methods for studying reproductive toxicity in the male.
  Methods for studying reproductive toxicity in the female.
  Methods for assessing reproductive function.

2 Anatomy and Physiology of the Male Reproductive System

The principal components of the male reproductive system are the
primary sex organs (the testes) and the accessory sex organs, which
include the epididymis, vas deferens, prostate, Cowper's gland, and the
penis (Figure 1). As a number of other (extra testicular) sites are
involved in regulating the reproductive system, they will also be
considered in this context. These sites include the pituitary, adrenals,
thymus, and pineal gland.

The testis consists of two main components, the vascular interstitial
tissue containing Leydig cells and the seminiferous tubules containing
Sertoli and germ cells. The Leydig cells, which occur in clusters in the
sinusoidal spaces between seminiferous tubules, are directly exposed to
the lymph and are often closely associated with blood vessels. They
contain abundant smooth endoplasmic reticulum, mitochondria, and a
prominent Golgi complex. The normal function of the Leydig cell is to
produce testosterone and other steroid hormones *de novo* from acetate
via cholesterol (Figure 2). In the rat, the major pathway is the
Δ^4-pathway (Bell, Vinson, Hopkin, and Lacy, 1968), whereas the
Δ^5-pathway is important in other species and in man (Yanahara and
Troen, 1972).

The seminiferous tubules, in which spermatozoa are produced, form
long coiled loops opening at both ends into the rete testis. The number of
tubules, arranged in lobules, vary from approximately 30 in the rat to up

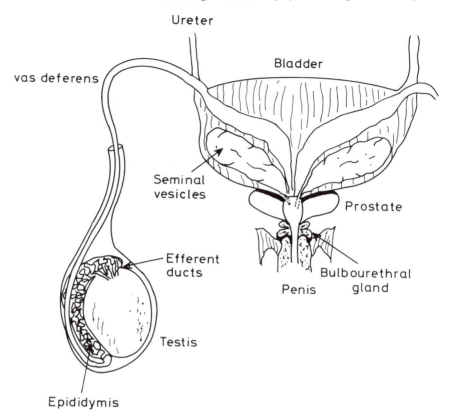

Figure 1   *Diagram of the human male reproductive system*

to 1000 in man. Within the tubule are the somatic Sertoli cells, and germ cells at various stages of maturity.

Spermatogenesis is the process of maturation of the germ cells, the spermatogonia (derived from the prenatal gonocyte) into sperm. During the first phase, the so-called type A spermatogonia undergo six mitotic divisions to form type B spermatogonia. Some spermatogonia remain undifferentiated to give rise to more stem cells. These may either be type A_4 cells or a separate long-cycling population (A_0). After the last mitotic stage, the type B spermatogonia divide mitotically to form preleptotene primary spermatocytes, which are able to pass through the blood–testis barrier into the more protected environment of the adluminal compartment of the seminiferous tubule. These spermatocytes replicate their DNA, initiating meiosis, to proceed to the leptotene step. This is followed by a zygotene step, during which the pairing of the homologous chromosomes occurs. Cells with completely paired chromosomes are termed pachytene primary spermatocytes. The progression from early pachytene through mid- and late-pachytene to form secondary spermatocytes is the longest in the meiotic prophase and also the stage at

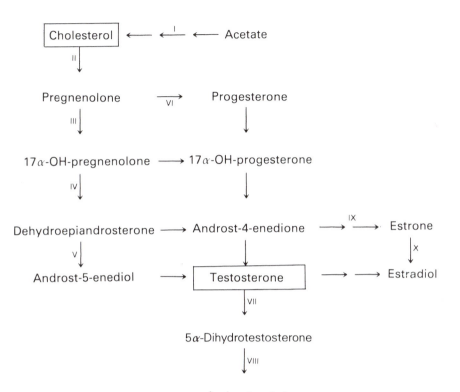

Figure 2 *Pathways of steroid synthesis in the testes. The enzyme(s) involved in the various pathways are as follows:*

I   *Multiple enzyme steps—HMG–CoA reductase is rate limiting*
II   *20α-hydroxylase, 22-hydroxylase, 20–22 lyase*
III   *17α-hydroxylase*
IV   *17–20 lyase*
V   *17β-hydroxysteroid dehydrogenase*
VI   *Isomerase, 3β-hydroxysteroid dehydrogenase*
VII   *5α-reductase*
VIII   *3α-reductase*
IX   *19α-hydroxylase, 19β-hydroxysteroid dehydrogenase, 'aromatase'*
X   *17β-hydroxysteroid dehydrogenase.*

The enzymes mediating the conversion of pregnenolone to androst-5-enediol (Δ^5-pathway) are the same as those mediating the conversion of progesterone to testosterone (Δ^4-pathway). Similarly, the conversion of androst-4-enedione to oestrone is mediated by the same enzymes that convert testosterone to oestradiol. Each intermediate in the Δ^5-pathway can be converted to its corresponding Δ^4-pathway intermediate by enzymes VI (isomerase/dehydrogenase) which the cells are particularly susceptible to damage. The secondary spermatocytes have a short lifespan, and without duplicating their DNA, they enter the second maturation division forming haploid spermatids. In a series of well-defined transformations (spermiogenesis), spermatids become spermatozoa. During the second phase of spermiogenesis, the cytoplasm of the Sertoli cell is gradually attenuated as the sperm are slowly extruded towards the lumen. The residual cytoplasm is subsequently phagocytosed.

The Sertoli cell, which extends from the base of the seminiferous epithelium into the lumen of the tubule, contains a large irregular-shaped nucleus, many spherical and rod shaped mitochondria, a prominent Golgi complex, and large amounts of smooth endoplasmic reticulum. The tight junction between Sertoli cells delimits two compartments in the seminiferous epithelium and is probably the morphological site of the blood–testis barrier. As well as participating in the maturation and release of spermatozoa from the germinal epithelium, Sertoli cells are the site of synthesis of some testis-specific proteins, such as androgen binding protein (ABP), transferrin, and ceruloplasmin.

Sperm deposited in the rete testis from the seminiferous tubules leave the testis through the efferent duct, which empties into the initial segment of the epididymis. During their passage through the epididymal duct, a long tortuous tube surrounded by connective tissue and smooth muscle, the sperm attain functional maturity, emerging with a strong unidirectional motility. The maturation process is androgen-dependent (Orgebin-Crist, Danzo, and Davies, 1975). The mature sperm travel along the vas deferens to the ejaculatory duct, which is made up of the proximal extremities of the seminal vesicles. The epithelium of these vesicles is packed with rough endoplasmic reticulum and dense excretory granules, and the secretions of this gland, which contain fructose, prostaglandins, and various reducing substances, make up a substantial proportion of the seminal plasma. The prostate, which surrounds the uretha below the seminal vesicles, also contributes to the ejaculate. The prostatic fluid is rich in proteolytic enzymes such as fibrinolysin, and also contains high concentrations of zinc, citric acid, and acid phosphatase. The functional activity of both the prostate and seminal vesicles is dependent on testosterone. A detailed description of the morphology of the male reproductive system can be found in Weiss and Greep (1977), and recent reviews of aspects of the physiology of the male reproductive system have been published (Lamb and Foster, 1988; Waller and Nikurs, 1986; Waller, Killinger, and Zaneveld, 1985).

3   Anatomy and Physiology of the Female Reproductive System

The female reproductive system consists of the ovaries, fallopian tubes (oviducts), uterus, and vagina (Figure 3). Hormonal control of the

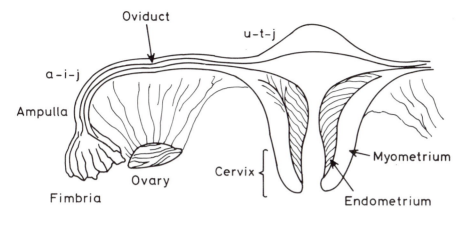

a-i-j: ampullary-isthmic junction

u-t-j: utero-tubal junction

Figure 3 *Reproductive organs of the human female*

uterine and ovarian cycles in sexually mature females is mediated by the hypothalamic–pituitary axis, in particular the pars distalis of the pituitary. The ovaries, which lie on either side of the uterus contain all of the eggs (oocytes) that the female will ever have. During the foetal period, the oogonia which are the stem cells for the oocytes and the female equivalent of spermatogonia, proliferate and shortly after birth are arrested in the primary oocytes stage, diplotene. The oocytes, and several granulosa cells (equivalent to testicular Sertoli cells), are surrounded by a basement membrane forming a follicle. The oocyte remains in this resting phase until just before ovulation. During each menstrual cycle a number of follicles increase in size, although only one releases its oocyte (*i.e.* ovulates). Following ovulation, the ovum is carried into the oviduct either by the action of cilia on the epithelial cells of the fimbria in sub-primates, or by muscular contraction in primates. The ovum travels down the ampulla to the ampullary–isthmic junction, where fertilisation takes place. During passage down the ampulla, the ovum progresses to the metaphase stage of the second meiotic division. If the ovum is fertilised, it implants in the endometrial lining of the uterus. In the rat, this occurs approximately 5 to 6 days after fertilisation.

Thus, the essential difference between gametogenesis in males and females is that in male, from puberty to death, there is a continuous production of mature germ cells from stem cells, whereas in females, germ cell maturation essentially ceases at birth, and thereafter there is a discontinuous release of these cells, which ceases at menopause. For a detailed description of the female reproductive tract see Weiss and Greep (1977) or Takizawa and Mattison (1983).

4   Mechanisms Regulating Testicular Function

A   Intra-testicular Mechanism

The most important intra-testicular sites of regulatory mechanisms are the Leydig and Sertoli cells. The function of these cells is mainly regulated by the principal testicular androgen, testosterone, and by the gonadotropic hormones, luteinising hormone (LH) and follicle stimulating hormone (FSH), produced by the anterior pituitary. However, recent studies have shown that refinements to this 'classical' picture are necessary (see below).

The Leydig cell is the major, if not the only source of *de novo* testosterone production, so that inhibition of Leydig cell activity can have important consequences for the regulation of testicular function. A reduction in testosterone synthesis, brought about by inhibition of the enzymes in the biosynthetic pathway (*e.g.* 17α-hydroxylase, 3β-hydroxysteroid dehydrogenase) may result in lowered plasma testosterone levels, reduced testicular and accessory sex organ weights, and decreased spermatogenesis (Dixon 1986; Willis, Anderson, Oswald, and Zaneveld, 1983).

The Sertoli cell provides a site for the proliferation and differentiation of germ cells, and it is thought that many of the effects of circulating hormones, such as the gonadotropin FSH, are mediated *via* the Sertoli cells. The major function of the Sertoli cell is the secretion of tubular fluid, which contains a number of important proteins, including plasminogen activator and androgen binding protein (ABP). Sertoli cells also secrete large amounts of lactate and pyruvate which appear to be important in maintainting ATP levels in spermatids and primary spermatocytes (LeGac, Attramadal, Horn, *et al.,* 1982). Other functions of the Sertoli cell include steroidogenesis, the production of intracellular proteins such as protein-kinase inhibitor (PKI) and the secretion of the peptide hormones, luteinising-hormone releasing hormone (LHRH)-like peptide (Hansson, Jégou, Attramadal, *et al.,* 1983).

Although Sertoli cells are not capable of *de novo* synthesis of androgens, their complement of steroidogenic enzymes can convert progesterone to testosterone, testosterone to androstenedione and 5α-diol, and testosterone, androstenedione and their 19-hydroxy derivatives to oestradiol-17β and oestrone (see Figure 2) (Tcholakian and Steinburger, 1978; Dorrington, Fritz, and Armstrong, 1976; Welsh and Wiebe, 1978).

The secretion of the testis- and epididymis-specific high affinity ABP (French and Ritzen, 1973) is regulated by both FSH and testosterone in the interstitial tissue, thereby facilitating its entry into the seminiferous tubules. The secretion rate of ABP *in vivo* can be used as an index of Sertoli cell function (Setchell, 1978; Gray, Beamand, and Gangolli, 1982). In *in vitro* studies with isolated rat seminiferous tubules, the rate of ABP secretion was found to vary throughout the spermatogenic cycle, being maximal at stages VIII to XI and minimal at stages IV to V (Ritzen, Boitani, Parvinen, *et al.*, 1982).

Other proteins known to be produced in the Sertoli cell include protein kinase inhibitor (Tash, Welsh, and Means, 1980) and inhibin (Van Thiel, Sherins, Myers, and DeVita, 1972; Steinberger and Steinberger, 1976). The function of the former is not clear; the latter is secreted and acts as a negative feedback control on the release of FSH (Figure 4). It is not yet known whether the effect of inhibin is a direct effect on the release mechanism or is a consequence of reduced FSH synthesis.

FSH binds to membrane receptors on Sertoli cells, thereby regulating the activity of adenylyl cyclase. The intracellular steady-state level of cyclic AMP, which controls the production of lactate and pyruvate and the activities of protein kinases, is dependent on the relative activities of

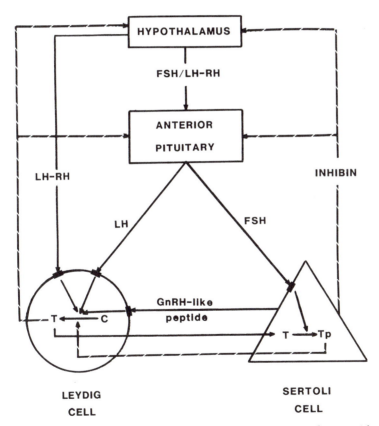

Figure 4 *Schematic diagram for hormonal regulation of testicular steroidogenesis showing classical 'long' feedback loops between testis and hypothalamus–pituitary axis and hypothetical 'short' feedback loops between Leydig and Sertoli cells. C = cholesterol, T = testosterone, Tp = polar testosterone metabolite, and GnRH = Gonadotropin releasing hormone. Inhibin is also known as Sertoli cell factor [Dashed line indicates down-regulation]*

adenylyl cyclase and phosphodiesterase. Germ cells may also produce cyclic AMP *via* a germ-cell specific adenylyl cyclase, located primarily in the spermatids. Whether this enzyme is regulated by Sertoli cell factor(s) is not known.

B   Extratesticular Mechanisms

The regulation of testicular function by the endocrine system has been reviewed by numerous authors (for example, see Steinberger and Steinberger, 1974; Setchell, 1978; DiZerega and Sherins, 1981; Sharpe, 1982a). The classical 'long' feedback loops between the hypothalamic–pituitary axis and the testis requires the regulation of Leydig cells by LH and the seminiferous epithelium by FSH (Steinberger, 1976). These glycoprotein hormones, which are released from the anterior pituitary by blood-borne releasing factors produced in the hypothalamus, bind to specific cell surface receptors to exert their effects. The immediate effect of LH is to stimulate testosterone synthesis (Rommerts, Bakker, and Van der Molen, 1983) but it is also believed to be involved in the early development of Leydig cell function (Chase, Dixon, and Payne, 1982) and in the maintenance of the integrity of the smooth endoplasmic reticulum (Ewing, Wing, Cochran, *et al.*, 1983). The major effect of FSH on the Sertoli cell is a general stimulation of protein synthesis, resulting in part in the stimulation of testosterone metabolism (Steinberger, Browning, and Grotjan, 1983).

Numerous other hormones have been shown to influence testicular function. Thus, for example, prolactin derived from the pituitary can increase the number of Leydig cell LH receptors leading to a stimulation of steroidogenesis and can stimulate cyclic AMP production (Baranao, Tesone, Calvo, *et al.*, 1983). Luteinising hormone-releasing hormone (LHRH) derived from the hypothalamus not only stimulates gonadotrophin production by the pituitary but has a direct effect on Leydig cells, initially stimulating steroidogenesis but inhibiting on prolonged exposure (Auclair, Kelly, Coy, *et al.*, 1977; Auclair, Kelly, Labrie, *et al.*, 1977; Sharpe, 1983). The effects of these and other hormones have been reviewed by Sharpe (1982a).

Although the regulatory mechanisms acting *via* the hypothalamic–pituitary axis are of major importance, recent studies have suggested that various 'short' loop regulatory mechanisms are involved in the fine tuning of testicular function. In particular it has been shown that there is intratesticular regulation of Leydig cell function by Sertoli cells, and *vice versa*. Thus, Leydig cell steroidogenesis may be regulated by as yet unidentified polar testosterone metabolites produced by the Sertoli cells and many of the processes of the Sertoli cell initiated by FSH can be maintained by testosterone secreted by the Leydig cells. In addition, gonadotropin releasing hormone (GnRH)-like peptide, secreted by the Sertoli cells, has been suggested as an inhibitor of Leydig cell function through an indirect effect on testosterone synthesis (Sharpe, Frazer, Cooper, and Rommerts, 1981; Steinberger, Browning, and Grotjan, 1983).

5 Mechanisms Regulating Ovarian Function

The details of the mechanisms underlying the process of follicle growth, selection, and ovulation (expulsion of the ovum as the gamete) are not well understood. It is clear, however, that the growth of the dominant follicle is influenced by FSH secreted by the pituitary, and that as it grows it produces oestrogen which acts initially as a feed-back inhibitor of FSH production. Oestrogen also stimulates proliferation of the endometrium and its appearance in the circulation coincides with an increase in plasma LH concentration. The continued secretion of oestrogen triggers a discharge of LH and FSH from the pituitary, the LH being involved in the preovulatory maturation of the oocyte. LH also stimulates the granulosa cells to secrete progesterone and the ovaries to synthesise prostaglandins (PGE_2 and PGF_2), resulting in the expulsion of the ovum. The follicle remaining after ovulation undergoes morphological changes

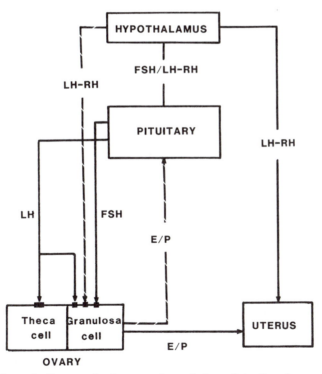

Figure 5 *Schematic diagram for hormonal regulation of the female reproductive system.* E/P = *Oestradiol-17β/Progesterone* [*Dashed line indicates a predominantly inhibitory effect*]

(luteinisation) forming the corpus luteum, which secretes both oestrogen and progesterone. These steroids are involved in preparing the endometrium for implantation (Figure 5). Luteinisation appears to be controlled by prolactin in both primate and sub-primate species and luteolysis, the breakdown of the corpus luteum, appears to be controlled by prostaglandins in sub-primates, but not in primates.

LHRH also has a role in the regulation of ovarian function, being a potent inhibitor of steroidogenesis in both the developing and post-ovulatory follicle. Because of its inhibitory effects on granulosa and luteal cell function, it has been postulated to be important in follicular atresia and luteal regression (Sharpe, 1982b).

6 Site Specificity of Chemical Injury by Model Toxins

A Testicular sites

Sertoli Cells

Many testicular toxins exert their effects on the Sertoli cell. Administration of the neurotoxins, n-hexane and methyl n-butyl ketone or their metabolite, 2,5-hexandione (2,5-HD), results in extensive atrophy of the germinal epithelium of the testes of experimental animals (Krasavage, O'Donoghue, DiVincenzo, and Terhaar, 1980). As the CNS controls the release of gonadotropins it was postulated that the effects of these compounds were due to changes in testicular homeostasis. However, it was found that 2,5-hexanedione treatment had no effect on serum testosterone concentration or on the circulating levels of FSH and LH (Chapin, Norton, Popp, and Bus, 1982), whereas a significant reduction in the activities of the Sertoli cell-localised enzymes, β-glucuronidase and γ-glutamyl transpeptidase, has been seen. Vacuolation of Sertoli cells after treatment with a subneurotoxic dose of 2,5-HD (Boekelheide, 1988) confirms a direct toxic effect on the Sertoli cell. The mechanism by which 2,5-hexanedione exerts its effects is not clear, although detailed morphological studies suggest this compound interferes with the microtubule network (Johnson, Hall, and Boekelheide, 1991).

Di-(2-ethylhexyl)phthalate (DEHP), a widely used industrial solvent, was first reported to cause testicular atrophy in the rat by Schaffer, Carpenter, and Smyth, (1945). Subsequent studies showed that the testicular injury produced by this and some other phthalate diesters was confined to a marked reduction in the diameter of seminiferous tubules and loss of germ cells (Gangolli, 1982). Sequential morphological studies of the effects of dipentyl phthalate on pre-pubertal rats showed a rapid development of vacuolation of the Sertoli cells, and disruption of the basal tight-junctions between Sertoli cells, which form the basis of the blood–testis barrier (Creasy, Beech, Gray, and Butler, 1987). Early degenerative changes were also apparent in germ cells (Creasy, Foster, and Foster, 1983). Other effects seen included a marked reduction in Sertoli cell mitochondrial succinate dehydrogenase activity and an inhibition of seminiferous tubule fluid and ABP secretion (Gray and Gangolli, 1986). Although the mechanisms underlying phthalate-induced testicular injury are not fully understood, it is clear that the proximate toxin is the monoester or a metabolite thereof (Sjöberg, Bondesson, Gray, and Plöen, 1986), and that the Sertoli cell is the principal target site.

Other chemicals targeting the Sertoli cell include nitrobenzenes and nitrotoluenes (Reader and Foster, 1990; Allenby, Sharpe, and Foster, 1990). The effects produced, such as cell vacuolation and germ cell sloughing, are similar to those of the toxic phthalate esters, although, as for phthalate esters, the biochemical mechanisms underlying the toxicity are not known.

Leydig Cells

Leydig cells are the target for many drugs and environmental chemicals including cadmium, ethane dimethylsulphonate (EDS), aminoglutethimide, and phthalate esters. EDS appears to be uniquely capable of depleting the testis of Leydig cells *in vivo* (Kerr, Donachie, and Rommerts, 1985) as a consequence of its cytotoxicity whereas other toxicants inhibit enzymic reactions in Leydig cells. Thus, for example, cannabinoids, particularly tetrahydrocannabinol, inhibit steroidogenesis by reducing cholesterol esterase activity (Burstein, Hunter, and Shoupe, 1979) and hexachlorocyclohexane inhibits the NAD-requiring Δ^5-3β-hydroxysteroid dehydrogenase (Δ^5-3β-HSD) and 17β-hydroxysteroid dehydrogenase activity (Shivanandappa and Krishnakumari, 1983). LHRH agonists ketoconazole and spironolactone inhibit steroidogenesis by reducing both 17α-hydroxylase and 17–20 lyase activities (Penhoat, Darbeida, Bernier, Saez, *et al.*, 1988; Sharpe, 1982b; Menard, Guenther, Kon, and Gillette, 1979; Brun, Leonard, Moronvaille, Caillard, *et al.*, 1991).

Ethanol and its principal metabolite, acetaldehyde, induce testicular atrophy in the rat and mouse (Van Thiel, Gavaler, Cobb, *et al.*, 1979; Cicero, Bell, Meyers, and Badger, 1980; Willis, Anderson, Oswald, and Zanveld, 1983). In most studies plasma testosterone levels are depressed by chronic ethanol treatment, suggesting an effect on the Leydig cells. Ultrastructural studies have shown intracellular changes in Leydig cells subsequent to chronic ethanol administration (Gavaler, Perez, and Van Thiel, 1982). The precise mechanism of ethanol-induced testicular dysfunction remains to be elucidated; however, large doses of ethanol have been shown to reduce the number of LH binding sites on Leydig cell membranes and several enzymatic steps in the biosynthesis of testosterone, including Δ^5-3β-HSD appear to be inhibited by ethanol and acetaldehyde (Boyden and Pamenter, 1983).

Germ Cells

A wide range of chemicals are known to affect spermatogenic cells directly, resulting in interference with the production of normal spermatozoa. The mechanisms of action vary as does the stage of spermatogenesis affected. Mono-functional alkylating agents, such as methyl methane sulphonate (MMS), react with germ cell DNA causing single-strand breaks. In both pre- and meiotic cells, including early spermatids (Lee, 1981), unscheduled DNA synthesis is induced; however, late spermatids and spermatozoa are unable to repair the DNA damage. Bifunctional alkylating agents, such as busulphan, stabilise DNA forming inter- or intrastrand links. These agents mainly affect spermatogonial cells, as do a number of other drugs such as the DNA-intercalators, adriamycin and daunomycin, and antimetabolites such as 6-mercaptopurine and 5-fluorouracil. The relatively prolonged spermatocyte development phase renders these cells particularly sensitive to injury by chemicals. A number of other compounds act during this phase, including oxygen-, sulphur- and nitrogen-containing heterocyclics, such as nitrofurans, thiophenes, and dinitropyrroles. Nitrofuran, for example, arrests primary spermatocyte development at the leptotene stage (Jackson, 1966).

Ethylene glycol monomethylether (EGME), a widely used industrial solvent, which has been shown to produce testicular atrophy in the rat, also exerts its effect at the primary spermatocyte stage. Morphological studies have shown that EGME and the corresponding ethyl ether (EGEE) are toxic specifically to pachytene spermatocytes, and that cells in meiotic division (Stage XIV) and in early pachytene (Stages I and II) are more susceptible than those in the late pachytene phase (Foster, Creasy, Foster, *et al.*, 1983; Creasy and Foster, 1984). Recently it has been shown that disruption of calcium homeostasis may be involved in the development of pachytene cell damage by EGME (Ghanayem and Chapin, 1990).

B   Extratesticular sites — Hypothalamus/Pituitary

Hypothalamus/Pituitary

It has been known for a long time that oestrogen can inhibit pituitary gonadotrophic activity (Moore and Price, 1932). Oestrogen treatment of male rats results in the inhibition of LH secretion, as gauged by the extent of androgenic stimulation of the sex accessory organs, but has no effect on FSH secretion. The testes of oestrogen-treated animals show atrophied Leydig cells, thickened tunica propria, and a reduction in the size of the seminiferous tubules. Androgens and progestogens, including testosterone and 17α-acetoxyprogestone derivatives, can also suppress the gonadotrophic activity of the pituitary–hypothalmic axis. A role for oestrogens in the pathophysiological effects of alcohol has also been suggested, although the evidence is not strong. However, there is evidence to suggest that ethanol may act directly at the hypothalamic level, in that chronic alcohol feeding impairs LH response to GnRH in the rat (Van Thiel, Gavaler, Cobb, *et al.*, 1979).

Sex Accessory Organs

As well as causing profound effects to the testes, phthalate esters also have marked effects on the prostrate and seminal vesicles. Treatment of rats with both mono- and di-(2-ethylhexyl) phthalate (DEHP) results in a decrease in prostate weight and a depletion of endogenous zinc levels (Thomas, Curto, and Thomas, 1982). Zinc has been shown to interfere with the binding of dihydrotestosterone to androgen receptors in this tissue in the mouse (Donovan, Schein, and Thomas, 1980). DEHP also reduces seminal vesicle weight in immature rats (Gray and Butterworth, 1980). The seminal vesicle/prostate complex in rats is also affected by chronic ethanol treatments. Willis, Anderson, Oswald, and Zanveld (1983) noted a diminution in the depth of the epithelial lining of the seminal vesicles and an associated reduction in the overall weight of the complex. These effects appeared to be a function of the duration and level of ethanol exposure.

C   Ovaries

Oocytes

Exposure to agents that damage oocytes can lead to reduced fertility in females. It has been long recognised that ionising radiation can cause destruction of female germ cells, and that the extent of sensitivity is both species- and age-dependent. Oocyte destruction has been demonstrated with X-rays, γ-radiation, neutron bombardment, and β-irradiation from ^3H_2O (see review by Dobson and Felton, 1983). The particular sensitivity of the immature mouse oocyte appears to be related to excessive vulnerability of the plasma membrane, rather than to nuclear damage. In most species, including man, the oocyte is not especially sensitive to radiation, so that the principal cause of cell death is likely to be damage to nuclear DNA.

A number of xenobiotics can also cause destruction of oocytes, although the mechanisms involved are not fully understood. Compounds known to be ovotoxins include polycyclic aromatic hydrocarbons (PAH), such as 7,12-dimethylbenzanthracene, 3-methylcholanthrene, and benzo[a]pyrene, and chemotherapeutic agents such as bleomycin, cyclophosphamide, and actinomycin D. As with radiation, the immature mouse oocyte is also particularly sensitive to chemical destruction, although not with respect to actinomycin D which does not attack the plasma membrane. Metabolic activation in the oocyte appears to be a necessary factor in the ovotoxicity of PAH (Mattison, Shiromizu, and Nightingale, 1983).

Uterus

As oestrogens act on the endometrium of the uterus to prepare it for implantation of the fertilised ovum, compounds that are either oestrogenic or anti-oestrogens are likely to affect fertility. Administration of technical grade DDT to the rat was found to be uterotropic (causing an increase in the weight of the uterus) and subsequent studies showed similar effects in both the rat and other species following administration of the o,p'-isomer. A typical oestrogenic response in the endometrium of uteri from ovariectomised rats was seen with o,p'-DDT, as well as a reduction in serum LH but not FSH levels. Studies on the mechanism of action of DDT showed that o,p'-DDT competitively inhibited the binding of oestradiol to uterine oestrogen receptor sites and translocated the oestrogen receptors from cystolic to nuclear compartments in a manner similar to oestradiol (see review by Bulger and Kupfer, 1983). A number of other pesticides, such as Methoxychlor (bis-p-methoxy DDT) and Kepone, have similar oestrogenic effects. The structural diversity of oestrogenic and antioestrogenic chemicals has been discussed by Katzenellenbogen, Katzenellenbogen, Tatee, *et al.* (1980).

7 Methods for Studying Reproductive Toxicity in the Male

In assessing chemically-induced reproductive toxicity in experimental animals, two aspects of the problem need to be investigated. These are, first, effects on the reproductive systems in male and female animals, and second, effects on their reproductive function (Section 9). A brief description of relevant methods employed for assessing effects on the male reproductive system follows. Some of the methods are also applicable in assessing toxicity in humans and these will be indicated in the text. For a more detailed discussion of methods for detecting alterations in the reproductive system of the male rat, see Zenick and Goeden, 1988, and quantitative data on inter-species differences in spermatogenesis, see Amann, 1986.

In selecting criteria for assessing injury to the male reproductive system, a number of considerations must be taken into account. The parameters measured must be objective and generate reproducible quantitative data amenable to statistical analysis. Furthermore, the tests must be sensitive and capable of indicating early signs of overt toxicity. In an evaluation of the usefulness of some potential indicators of reproductive toxicity in the rat, Blazak, Ernst, and Stewart (1985) examined sperm production rate, epididymal sperm numbers, transit time, and motility in groups of F334 strain animals of different ages. Their data suggested that testes weight and sperm numbers are unreliable indicators of sperm production rate, but that sperm production rate and sperm movement characteristics are sensitive indicators of adverse effects on male reproduction capacity. Recently, it has been suggested that the measurement of intracellular messenger protein levels in blood or semen may be a more sensitive method for detecting early effects on spermatogenesis and some encouraging results have been obtained with the Sertoli cell product, inhibin (Allenby *et al.*, 1991).

A   Species Selection

The use of more than one animal species is recommended for assessing the potential risk of reproductive toxicity in human males. Although a number of laboratory or domesticated animal species may be used, the rabbit and rat offer several advantages in comparison to the dog and sub-human primates, and are the animal models most commonly used. Their main advantage is that all the phases of the conception process, such as spermatogenesis, sperm maturation, ejaculation, sperm capacitation in the female, and fertilisation can be objectively evaluated, quantified, and manipulated throughout the year. Additionally, rabbits have a high predictable libido that may be useful in assessing effects on sexual behaviour. Interspecies extrapolation factors have been derived for a number of reproductive toxicants and have been used to develop a quantitative model for risk assessment in human males (Meistrich, 1992).

B   External Genitalia

Physical examination of the testes provides useful information in animal models and in man. Measurements of testis size and scrotal circumference can be taken easily in animals with a pendulous scrotum. The consistency of the testes, measured by using a tonometer, can provide objective information on intratesticular pressure and is useful in longitudinal surveillance studies to provide data on time- and treatment-related changes in the testes (Overstreet, 1984). The method has potential value in monitoring human populations.

C   Internal Morphology

The comparison of organ weights and volumes between treated and control animals can provide a valuable indication of treatment-related effects. Organs that should be examined are the testis, prostate, seminal vesicles, epididymis, coagulating glands, pituitary, and adrenals.

Histological examination by light and transmission electron microscopy of these organs and of the vas deferens is essential for detecting treatment-related morphological changes. Sequential studies enable both the identification of the initial target site and intracellular organelles affected, and the subsequent development of the lesion to be determined. These procedures have been found to be particularly useful in defining the stage specificity in toxicity to the germ cells and to the spermatogenic process (see Section 6). A description of methods commonly employed in morphological studies on the testes and accessory organs of the reproductive system is given by Christian (1983).

D Biochemical and Hormonal Parameters

The determination in testicular preparations of enzyme activities generally associated with intracellular metabolic functions or specifically associated with reproductive function have been widely used for identifying the initial biochemical lesion and the progression of testicular injury. Thus, for example, Shen and Lee (1977) measured eight enzyme activities including hyaluronidase, lactate dehydrogenase-isoenzyme X, and sorbitol and isocitrate dehydrogenase to distinguish normal from abnormal testicular tissue.

The levels of circulating androgens in experimental animals and in man can be used as an index of reproductive function, as can the activities of specific marker enzymes of steroidogenesis in testicular preparations. The enzymes mediating the conversion of cholesterol to testosterone *via* pregnenolone are indicated in Figure 2. Other biochemical markers of relevance include the circulating levels of androgen-binding protein as an indicator of Sertoli cell function, oxygen uptake, and carbon dioxide production as a measure of sperm respiration, and kinase activity as an index of phosphorylation. The hormones commonly measured as an index of testicular injury are FSH and LH. Blood levels of these hormones, measured by radioimmunoassay, are determined either directly or following the administration of gonadotropin-releasing hormone (the 'GnRH' test). This latter test is capable of detecting mild degrees of primary testicular dysfunction.

E Seminal Evaluation

The examination of seminal plasma can provide valuable information in assessing injury to the male reproductive tract. Human seminal plasma, approximately 30% of which is derived from the prostate and 60% from the seminal vesicles, is rich in acid phosphatase, lysozymes, fructose, citric acid, prostaglandins, zinc, and magnesium. However, the concentrations of these components would appear to be of limited diagnostic value at present and this indicates the need for further research in the interpetation of the results.

Sperm analysis provides data on sperm concentration in seminal fluid and information on spermatozoal motility, morphology, and *in vitro* function. Spermatozoal concentration, as an index of fertility in man, derives its relevance from the pioneering work of MacLeod in the 1940s, and the generally accepted lower limit of 'normal' sperm concentration at 20 million sperm per ml is based on a survey conducted on 2000 men (MacLeod and Gold, 1951). However, as a measure of human fertility, sperm concentration must be interpreted with considerable caution as wide intra-subject variation has been found (WHO, 1987). Thus, semen samples obtained from fertile donors over a period of one year contained between 7 and 170 million sperm per ejaculate. Furthermore, in a survey of infertile men, MacLeod and Wang (1979) found that less than 20% of the group had sperm concentrations below the norm of 20 million per ml and the rest of the patients in the group had well above normal sperm counts.

F   Sperm motility

Spermatozoa with adequate motility and fertilising capacity are essential for the successful outcome of the reproductive act, and it is considered that motility is an important determinant of fertility (Morrissey *et al.*, 1988). Whereas early methods for assessing sperm mobility were subjective, and therefore of limited value, methods have now been developed based on computer analysed videomicrography which generate objective data on sperm motility, including swimming speed and direction. At a recent workshop, standardisation of parameters for assessing sperm motility using computer-assisted sperm analysis (CASA) was discussed (Chapin *et al.*, 1992). Changes in motility may indicate abnormal spermatogenesis, abnormal functions of the epididymis, or the presence of antimotility factors in the semen. These latter factors may be due to foreign chemicals or their metabolites entering the semen *via* the prostate, bulbourethral glands, or vesicular glands. It should also be noted, however, that even a substantial reduction in the numbers of motile sperm may not have a marked effect on reproductive function. Thus, in a recent study, treatment of rats with nitrobenzene reduced the number of motile sperm by approximately 80%, but had only a marginal effect on numbers of implants (Blazak, Rushbrook, Ernst, *et al.*, 1985).

G   Spermatozoal Morphology

Although semen from normal humans can contain up to 40% of abnormal sperm cells, sperm morphology can provide valuable information on chemical toxicity to the male reproductive system. Unfortunately, there have been no methods available for generating objective data for assessing sperm morphology and there is no authoritative system of classification of overall sperm abnormalities, although the mouse sperm head morphology test (Wyrobek and Bruce, 1975) is widely used for screening for mutagenic chemicals. The abnormalities commonly observed include those related to sperm size and shape, either affecting the sperm head, midpiece, or tail. Videomicrography appears to be a potentially valuable tool for assessing sperm morphology (Overstreet, 1984) and the use of morphometric methods holds out promise in generating computer derived calculations of sperm head length, width, circumference, and other measurements (Katz, Overstreet, and Pelfrey, 1982). These procedures could be applied to spermatozoa from man and experimental animals.

Two additional criteria of sperm function have received increasing attention of late. These are, tests of sperm/cervical mucus interaction and tests of sperm/oocyte interaction. Sperm/cervical interaction occurs during the passage of spermatozoa through the mucus of the cervix to the site of fertilisation. Cervical mucus, a well defined fluid, may interact with and influence the flagellar activity of spermatozoa and the presence of anti-sperm antibodies in the cervical mucus has been shown to affect adversely the motility of sperm (Jager, Kremer, and Van Slochteren-Draaisma, 1979.) The tests are conducted either on a microscope slide or in a capillary tube. In the former test, the swimming behaviour of sperm placed in contact with cervical mucus on a microscope slide is investigated, and in the latter test, a capillary tube filled with cervical mucus is placed in contact with semen, and the distance travelled by sperm along the tube during a prescribed time measured (Katz, Overstreet, and Hanson, 1980). The disadvantage of the sperm/cervical mucus interaction test stems from the difficulty of obtaining an adequate supply of human cervical mucus. It is possible that bovine cervical mucus, available in large quantities, may prove a suitable substitute for the human material (Moghissi, Segal, Meinhold, and Agronow, 1982). Additionally, research is in progress to develop synthetic simulants for cervical mucus (Lorton, Kummerfield, and Foote, 1981).

The penetration of sperm into the ovum and the subsequent physiological changes, collectively known as capacitation, prior to the onset of morphological changes in the acrosomes form the basis of the sperm/oocyte interaction studies (Yanagimachi, 1981). Since human *in vitro* fertilisation studies are usually precluded for obvious ethical reasons, substitutes have been used for the human ovum. These have included zona pellucida-stored human follicular oocytes and the zona-free golden hamster egg. The removal of the zona pellucida of the hamster egg by hyaluronidase digestion provides a ready source of oocytes capable of fusing with sperms from many species including man.

Both capacitation and acrosome reaction are necessary to effect the fusion of human sperm with the zona-free hamster egg thus providing an appropriate system for assessing sperm function (Overstreet, 1983). The interpretation of the sperm/oocyte interaction findings as an index of chemical injury to the male reproductive system remains to be fully developed.

8 Methods for Studying Reproductive Toxicity in the Female

Many of the morphological and biochemical investigations carried out for assessing chemical injury to the male reproductive system are also employed in assessing toxicity in the female. A brief description of the relevant procedures used in the study of female reproductive toxicological follows.

A Morphological Studies

In addition to the visual examination of the external genitalia and the internal tissues of the reproductive tract, weights of the relevant organs, in particular the ovaries, adrenals, and pituitary, can provide useful information on the onset of chemical injury in female experimental animals. Histological examination by light microscopy and by electron microscopy of sections of the vagina, cervix, uterus, fallopian tubes, and ovaries may reveal early treatment-related morphological and ultra-structural changes in the female reproductive tract. An important indicator of reproductive toxicity is the number of oocytes present in the ovaries, measured in sections of ovaries fixed in Bouins's solution. The procedure provides a quantitative measure of effects on follicular development and degeneration.

B Biochemical and Hormonal Studies

Determinations of the circulating levels of oestradiol and its metabolites, oestrone and oestriol, and of the activities of enzymes mediating their biosynthesis provide useful information on the reproductive competence of experimental animals. The numbers of steroid receptors in the cytoplasm and nuclei of target tissues, particularly those for oestradiol and progesterone, are important indicators of chemical injury to the female reproductive system. Chemicals may adversely affect the steroid receptor function either by competing for, and occupying, the receptor sites in preference to the endogenous hormones, or by interfering with the biosynthesis of the steroid receptors. The former may lead to hormonal imbalance and reproductive abnormalities and the latter may affect the hormonally mediated function of the target tissues. Examples of chemicals in the former category are DDT, polychlorinated biphenyl, and dimethylbenzanthracene (DMBA); chemicals in the latter category include alkylating agents and inhibitors of protein synthesis.

The ovarian hormones, oestrogen and progesterone, exhibit an inverse relationship with FSH, LH, and LTH (luteotropic hormone), the three gonadotropins derived from the hypophysis. Additionally, as discussed earlier, the gonadotropins regulate the oestrous cycle and the preparation of the endometrium for the accommodation and nourishment of the fertilised ovum. Thus, the estimation of the three gonadotropic hormones, and of the two sex hormones provides a comprehensive picture of the development, maintenance, and function of the female reproductive system. Radioimmunoassay methods are extensively used for the determination of these hormones.

9 Methods for Assessing Reproductive Function

The procedures for the assessment of chemical injury to reproductive function in experimental animals have become relatively standardised and are designed primarily to meet various regulatory requirements. The guidelines set by the US Food and Drug Agency in 1966 have been the basis for official procedures prescribed by the US Environmental Protection Agency, by Japan, UK, OECD, and more than twenty other regulatory agencies. The US Food and Drug Agency Guidelines are currently being assessed with a view to revision (Collins *et al.*, 1991).

The approach commonly adopted in the evaluation of reproductive hazards is the use of a three segment scheme of studies. The 'Segment 1' study is designed to assess the toxic effect of the chemical on gonadal function and mating behaviour in both male and female (experimental) animals, usually rats. In addition, the effect on conception rates, the early and late stages of gestation, parturition, lactation, and the development of offspring is assessed.

The 'Segment 2' study is conducted to evaluate the teratogenic potential of a test compound. FDA guidelines prescribe both a rodent and non-rodent species, the rat and rabbit being commonly used. Female animals only are treated with the test compound post-mating and during the period of major organogenesis.

The 'Segment 3' study is intended to evaluate the effect of treatment on the peri- and post-natal development of the foetus. The study is usually performed in the rat. Sexually-mature female animals are mated and then treated with test compound beginning on day 15 of presumed gestation and continuing throughout delivery and until 21 days post-parturition. The effects of treatments on the duration of gestation, delivery, number, and viability of naturally delivered litters are evaluated. For further details of experimental methods, see Christian (1983).

In addition to the three-segment scheme for evaluating chemically induced reproductive toxicity of drugs, multigeneration reproduction studies are carried out on food additives and environmental contaminants. Treatment of the F_0 generation commences immediately following weaning and continues through mating, delivery, and nursing of the F_1 generation pups. These pups are individually treated after weaning, usually at 21 days of age, and the treatment continued through mating and delivery of F_2 generation pups. If required, selected F_2 pups are similarly treated through to the F_3 generation. In addition to the criteria previously mentioned, histopathological examination of all organ systems in the foetuses and post-natal 'behavioural' assessment are employed for evaluating reproductive toxicity in the experimental animals. A number of indices of reproductive function are used for comparing treated animals with controls. The indices commonly used are:

$$\text{Mating Index} = \frac{\text{Number of copulations} \times 100}{\text{Number of oestrus cycles required}}$$

$$\text{Fecundity Index} = \frac{\text{Number of pregnancies} \times 100}{\text{Number of copulations}}$$

$$\text{Male Fertility Index} = \frac{\text{Number of pregnant females} \times 100}{\text{Number of non-pregant females presented}}$$

$$\text{Female Fertility Index} = \frac{\text{Number of females that conceive} \times 100}{\text{Number of females exposed to fertile males}}$$

$$\text{Live Birth Index} = \frac{\text{Number of viable pups born} \times 100}{\text{Total number of pups born}}$$

For details of the procedures mentioned the reader is referred to official methods presented by the various national regulatory agencies (*e.g.* Food and Drugs Administration, 1966; Committee on the Safety of Medicines, 1974; Environmental Protection Agency, 1979).

Recently, the National Toxicology Program (NTP) in the United States has proposed a new test procedure, which is claimed to be a reliable and cost-effective alternative to the multigeneration study. This system, Reproductive Assessment by Continuous Breeding (RACB) has been evaluated with a wide range of chemicals and to date over 70 studies have been completed, almost exclusively with mice (Gulati *et al.*, 1991). The experimental design is based on four 'tasks', which in mice last over a period of 35 weeks, and include observations on F_0, F_1, and F_2 generations. Task 1 is a 14 day dose-selection study using conventional toxicity end points (body weight, clinical signs). Task 2 is a 14 week continuous breeding phase followed by a 3 week period in which the co-habiting pair are separated. During this task, data is collected on all newborn pups. If significant effects on fertility are seen, a cross-over mating trial (task 3) is carried out to determine which sex is more sensitive. The last litter of the continuous breeding phase (usually the 5th) is used for the second generation phase (task 4), the mothers (F_0) being dosed through weaning and the offspring (F_1) until mated (approximately 11 weeks). Finally, the litters (F_2) are examined. If task 3 is undertaken the offspring are also analysed as in task 4.

10 Conclusions

This chapter has attempted briefly to discuss chemical injury to the male and female reproductive systems from a mechanistic standpoint and to describe the various methods employed for investigating chemical injury, both in experimental animals and in man. No attempt has been made to address the question of toxicity expressed in the developing foetus or subsequently in the offspring (developmental and behavioural toxicology); many recent reviews have appeared on aspects of this area of toxicology (see, for example, Hemminki and Vineis, 1985; IPCS, 1984; Kimmel, 1988).

The limitations in the methodology currently employed have been indicated and the difficulties in interpreting experimental results pointed out. These difficulties arise principally as a result of the lack of detailed information on both the pathogenesis of chemically-induced testicular and ovarian injury and on the biochemical mechanisms underlying injury at these sites. Progress in understanding reproductive toxicity is likely to be enhanced by further research in a number of areas. These include the development of more objective methods for assessing toxic effects (*e.g.* the use of morphometric methods for examining sperm), investigations into the influence of diet, age, genetic, and other social factors (*e.g.* alcohol intake and drug interactions) on chemically-induced reproductive toxicity, and inter-species comparisons to allow the extrapolation of animal data to man.

An understanding of the biochemical mechanisms underlying reproduction is of central importance, and in this context the development and application of *in vitro* model systems may be valuable. Examples of *in vitro* systems examined recently include seminiferous tubule preparations, used in the studies on the toxicity of phthalate esters and glycol ethers (see Section 6) and the zona-free hamster egg for assessing sperm function (see Section 7). Alternative approaches for assessing embryotoxicity and teratogenicity have recently been reviewed by ECETOC (1989). Clearly, the realistic assessment of toxicity of chemicals to the human reproductive system requires further fundamental studies using both *in vivo* and *in vitro* techniques.

References

Allenby, G., Sharpe, R. M., and Foster, P. M. D. (1990). Changes in Sertoli cell function *in vitro* induced by nitrobenzene. *Fundam. Appl. Toxicol.*, **14,** 364–375.

Allenby, G., Foster, P. M. D., and Sharpe, R. M. (1991). Evaluation of changes in the secretion of immunoactive inhibin by adult rat seminiferous tubules *in vitro* as an indicator of early toxicant action on spermatogenesis. *Fundam. Appl. Toxicol.*, **16,** 710–724.

Amann, R. P. (1986). Detection of alterations in testicular and epididymal function in laboratory animals. *Environ. Health Persp.*, **70,** 149–158.

Auclair, C., Kelly, P. A., Coy, D. H., Schally, A. V., and Labrie, F. (1977). Potent inhibitory activity of [D-leu6, Des-gly-NH$_2$(10)]LHRH ethylamide on LH/hCG and PRL testicular receptor levels in the rat. *Endocrinology,* **101,** 1890–1893.

Auclair, C., Kelly, P. A., Labrie, F., Coy, D. H., and Schally, A. V. (1977). Inhibition of testicular luteinising hormone receptor levels by treatment with a potent luteinising hormone-releasing hormone agonist or human chorionic gonadotropin. *Biochem. Biophys. Res. Commun.*, **76,** 855–862.

Baranao, J. L. S., Tesone, M., Calvo, J. C., Luthy, I. A., Gonzalez, S. I., Charreau, E. H., and Calandra, R. S. (1983). Prolactin effects on the prepubertal male rat. *In*: 'Recent Advances in Male Reproduction: Molecular Basis and Clinical Implications'. *Eds* R. D'Agata, M. B. Lipsett, P. Polosa, and H. J. van der Molen. Raven Press, NY, pp. 305–319.

Bell, J. B. G., Vinson, G. P., Hopkin, D. J., and Lacy, D. (1968). Pathways for androgen biosynthesis from (7-³H)pregnenolone and (4-¹⁴C) progesterone by rat testis interstitium *in vitro*. *Biochem. Biophys. Acta*, **164**, 412–420.

Bernstein, M. E. (1984). Agents affecting the male reproductive system: effects of structure on activity. *Drug Metab. Rev.*, **15**, 941–996.

Blazak, W. F., Ernst, T. L., and Stewart, B. E. (1985). Potential indicators of reproductive toxicity: testicular sperm production and epididymal sperm numbers, transit time and motility in Fischer F344 rats. *Fundam. Appl. Toxicol.*, **5**, 1097–1103.

Blazak, W. F., Rushbrook, C. J., Ernst, T. L., Steward, B. E., Spak, D., DiBiasio-Erwin, D., and Black, V. (1985). Relationship between breeding performance and testicular/epididymal functioning in male Sprague-Dawley rats exposed to nitrobenzene. *Toxicologist*, **5**, 121.

Boekelheide, K. (1988). Rat testis during 2,5-hexanedione intoxication and recovery. I. Dose-response and reversibility of germ cell loss. *Toxicol. Appl. Pharmacol.*, **92**, 18–27.

Boyden, T. W. and Pamenter, R. W. (1983). Effect of ethanol on the male hypothalmic–pituitary–gonadal axis. *Endocrine Rev.*, **4**, 389–395.

Brun, H. P., Leonard, J. F., Moronvaille, V., Cailland, J. M., Melcion, C., and Cordier, A. (1991). Pig Leydig cell culture: a useful *in vitro* test for evaluating the testicular toxicity of compounds. *Toxicol. Appl. Pharmacol.*, **108**, 307–320.

Bulger, W. H. and Kupfer, D. (1983). Oestrogenic action of DDT analogs. *Am. J. Ind. Med.*, **4**, 163–173.

Burstein, S., Hunter, S. A., and Shoupe, T. S. (1979). Site of inhibition of Leydig cell testosterone synthesis by Δ'-tetrahydrocannabinol. *Mol. Pharmacol.*, **15**, 633–640.

Chapin, R. E., Norton, R. M., Popp, J. A., and Bus, J. S. (1982). The effects of 2,5-hexanedione on reproductive hormones and testicular enzyme activities in the F344 rat. *Toxicol. Appl. Pharmacol.*, **62**, 262–272.

Chapin, R. E., Filler, R. S., Gulati, D., *et al.* (1992). Methods for assessing rat sperm motility. *Reprod. Toxicol.*, **6**, 267–273.

Chase, D. J., Dixon, G. E. K., and Payne, A. H. (1982). Development of Leydig cell function. *Prog. Clin. Biol. Res.*, **112**, 209–219.

Christian, M. S. (1983). Assessment of reproductivity toxicity. State of the art. *In*: 'Assessment of Reproductive and Teratogenic Hazards. Advances in Modern Environmental Toxicology'. Vol. III, Chapter 8. *Eds* M. S. Christian, W. N. Galbraith, P. Voytek, and M. A. Mehlman. Princeton Scientific Publishers Inc., Princeton, USA, pp. 65–76.

Cicero, T. J., Bell, R. D., Meyer, E. R., and Badger, T. M. (1980). Ethanol and acetaldehyde directly inhibit testicular steroidogenesis. *J. Pharmacol. Exp. Ther.*, **213**, 228–233.

Collins, T. F. X., Black, T. N., Graham, S. L., Jackson, B. A., and Welsh, J. J. (1991). Updating developmental toxicity testing guidelines for the safety assessment of direct food additives and color additives used in food: results of a survey. *J. Am. Coll. Toxicol.*, **10**, 461–475.

Committee on the Safety of Medicines (1974). Notes for guidance on reproduction studies. Dept. of Health and Social Security, Great Britain.

Creasy, D. M., Beech, L. M., Gray, T. J. B., and Butler, W. H. (1987). The ultrastructural effects of di-n-pentyl phthalate on the testis of the mature rat. *Exp. Mol. Pathol.*, **46**, 357–371.

Creasy, D. M., Foster, J. R., and Foster, P. M. D. (1983). The morphological development of di-n-pentyl phthalate-induced testicular atrophy in the rat. *J. Pathol.*, **139**, 309–321.

Creasy, D. M. and Foster, P. M. D. (1984). The morphological development of glycol ether induced testicular atrophy. *Exp. Mol. Pathol.*, **40**, 169–176.

Dixon, R. L. (1986). Toxic responses of the reproductive system. *In*: 'Toxicology—the basic science of poisons', 3rd Edn. *Eds* C. D. Klaassen, M. O. Amdur, and J. Doull. Chapter 16. Macmillan, New York, pp. 432–477.

DiZerega, G. S. and Sherins, R. J. (1981). Endocrine control of adult testicular function. *In*: 'The Testis'. *Eds* H. Burger and D. Dekrester. New York Raven Press, pp. 127–140.

Dobson, R. L. and Felton, J. S. (1983). Female germ cell loss from radiation and chemical exposures. *Am. J. Ind. Med.*, **4**, 175–190.

Donovan, M. P., Schein, L. G., and Thomas, J. A. (1980). Inhibition of androgen-receptor interaction in mouse prostate gland cytosol by divalent metal ions. *Mol. Pharmacol.*, **17**, 156–162.

Dorrington, J. H., Fritz, I. B., and Armstrong, D. T. (1976). Site at which FSH regulates estradiol-17β biosynthesis in Sertoli cell preparations in culture. *Mol. Cell Endocrinol.*, **6**, 117–122.

ECETOC (1989). Alternative approaches for the asssessment of reproductive toxicity (with emphasis on embryotoxicity/teratogenicity). Monograph No. 12, Brussels, Belgium.

Environmental Protection Agency (1979). Proposed Health Effects Test Standards for Toxic Substances Control Act Test Rules and Proposed Good Laboratory Practice Standards for Health Effects. Fed. Reg. pt. IV **44** (145) Sub-part F. Teratogenic/Reproductive Health Effects, July 26th.

Ewing, L. L., Wing, T.-Y., Cochran, R. C., Kromann, N., and Zirkin, B. R. (1983). Effect of luteinising hormone on Leydig cell structure and testosterone secretion. *Endocrinology*, **112**, 1763–1769.

Food and Drugs Administration (1966). Guidelines for reproductive studies for safety evaluation of drugs for human use. Washington, DC.

Foster, P. M. D., Creasy, D. M., Foster, J. R., Thomas, L. V., Cook, W. M., and Gangolli, S. D. (1983). Testicular toxicity of ethylene glycol monomethyl and monoethyl ethers in the rat. *Toxicol. Appl. Pharmacol.*, **69**, 385–399.

French, F. S. and Ritzen, E. M. (1973). Androgen-binding protein in efferent duct fluid of rat testis. *J. Reprod. Fertil.*, **32**, 479–483.

Gangolli, S. D. (1982). Testicular effects of phthalate esters. *Environ. Health Perspect.*, **45**, 77–84.

Gavaler, J. S., Perez, H., and Van Thiel, D. H. (1982). Ethanol produced morphologic alterations of Leydig cells. *Alcoholism Clin. Exp. Res.*, **6**, 296.

Ghanayem, B. I. and Chapin, R. E. (1990). Calcium channel blockers protect against ethyleneglycol monomethylether (2-methoxyethanol)-induced testicular toxicity. *Exp. Mol. Pathol.*, **52**, 279–290.

Gray, T. J. B. and Butterworth, K. R. (1980). Testicular atrophy produced by phthalate esters. *Arch. Tox.* (Suppl 4), 452–455.

Gray, T. J. B., Beamand, J. A., and Gangolli, S. D. (1982). Effects of phthalate esters on rat testicular cell cultures and on Sertoli cell function in the intact testis. *Toxicologist*, **2**, 78, Abs 276.

Gray, T. J. B. and Gangolli, S. D. (1986). Aspects of the testicular toxicity of phthalate esters. *Environ. Health Perspect.*, **65**, 229–235.

Gulati, D. K., Hope, E., Teague, J. L., and Chapin, R. E. (1991). Reproductive toxicity assessment by continuous breeding in Sprague-Dawley rats: a comparison of two study designs. *Fundam. Appl. Toxicol.*, **17**, 270–279.

Hansson, V., Jégou, B., Attramadal, H., Jahnsen, T., Le Gac, F., Tvermyr, M., Frøysa, A., and Horn, R. (1983). Regulation of Sertoli cell function and response. *In*: 'Recent Advances in Male Reproduction: Molecular Basis and Clinical Implications', *Eds* R. D'Agata, M. B. Lipsett, P. Polosa, and H. J. van der Molen. Raven Press, New York.

Hemminki, K. and Vineis, P. (1985). Extrapolation of evidence on teratogenicity of chemicals between humans and experimental animals: Chemicals other than drugs. *Teratol. Carcinogen. Mutagen.*, **5**, 251–318.

IPCS: International Programme on Chemical Safety (1984). Principles for evaluating health risks to progeny associated with exposure to chemicals during pregnancy. *Environ. Health. Criteria*, **30**. WHO, Geneva.

Jackson, H. (1966). 'Antifertility compounds in the male and female', Thomas, Springfield, Ill.

Jager, S., Kremer, J., and van Slochteren-Draaisma, T. (1979). Presence of sperm agglutinating antibodies in infertile men and inhibition of *in vitro* sperm penetration into cervical mucus. *Int. J. Androl.*, **2**, 117.

Johnson, K. J., Hall, E. S., and Boekelheide, K. (1991). 2,5-Hexanedione exposure alters rat Sertoli cell cytoskeleton I. Microtubules and seminiferous tubule fluid secretion. *Toxicol. Appl. Pharmacol.*, **111**, 432–442.

Katz, D. F., Overstreet, J. W., and Hanson, F. W. (1980). A new quantitative test for sperm penetration into cervical mucus. *Fertil. Steril.*, **33**, 179–186.

Katz, D. F., Overstreet, J. W., and Pelfrey, R. J. (1982). Integrated assessment of the motility, morphology and morphometry of human spermatozoa. *In*: 'Human Fertility Factors', *Eds* P. Spira and P. Jouannet. Paris, Inserm, p. 97.

Katzenellenbogen, J. A., Katzenellenbogen, B. S., Tatee, I., Robertson, D. W., and Landvatter, S. W. (1980). The chemistry of oestrogens and anti-oestrogens: Relationships between structure, receptor binding and biological activity. In: 'Oestrogens in the Environment' *Ed.* J. A. McLachlan. Vol. 1, Elsevier/North-Holland, NY. pp. 33–52.

Kerr, J. B., Donachie, K., and Rommerts, F. F. G. (1985). Selective destruction and regeneration of rat Leydig cells *in vivo*. A new method for the study of seminiferous tubular-interstitial tissue interactions. *Cell Tissue Res.*, **242**, 145–156.

Kimmel, C. A. (1988). Current status of behavioural teratology. Science and Regulation. *CRC Crit. Rev. Toxicol.*, **19**, 1–10.

Krasavage, W. J., O'Donoghue, J. L., DiVincenzo, G. D., and Terhaar, C. J. (1980). The relative neurotoxicity of methyl n-butyl ketone, n-hexane, and their metabolites. *Toxicol. Appl. Pharmacol.*, **52**, 433–441.

Lamb, J. C. and Foster, P. M. D. (1988). 'Physiology and Toxicology of Male Reproduction', Academic Press Inc., San Diego, CA, USA.

Lee, I. P. (1981). Effect of drugs and chemicals on male reproduction. *INSERM*, **103**, 311–331.

Le Gac, F., Attramadal, H., Horn, R., Tvermyr, M., Frøysa, A., and Hansson, V. (1982). Hormone stimulation of lactate/pyruvate secretion by cultured Sertoli cells and maintenance of ATP levels in primary spermatocytes and round spermatids. Program 2nd Eur. Workshop Mol. Cell. Endocrin. of the Testis, Rotterdam, 11–14 May, Abstract C-16.

Lorton, S. P., Kummerfield, H. L., and Foote, R. H. (1981). Polyacrylamide as a substitute for cervical mucus in sperm migration tests. *Fertil. Steril.,* **35,** 222–225.

Lucier, G. W., Lee, I. P., and Dixon, R. L. (1977). Effects of environmental agents on male reproduction. *In*: 'The Testis' Vol. IV. *Ed.* R. A. Gomes, Academic Press, NY, pp. 578–604.

MacLeod, J. and Gold, R. Z. (1951). The male factor in fertility and infertility. II. Spermatozoon counts in 1000 men of known fertility and in 1000 cases of infertile marriage. *J. Urol.,* **66,** 436–449.

MacLeod, J. and Wang, Y. (1979). Male fertility potential in terms of semen quality: a review of the past, a study of the present. *Fertil. Steril.,* **31,** 103–116.

Mattison, D. R., Shiromizu, K., and Nightingale, M. S. (1983). Oocyte destruction by polycyclic aromatic hydrocarbons. *Am. J. Ind. Med.,* **4,** 191–202.

Meistrich, M. L. (1992). A method for quantitative assessment of reproductive risks to the human male. *Fundam. Appl. Toxicol.,* **18,** 479–490.

Menard, R. H., Guenthner, T. M., Kon, H., and Gillette, J. R. (1979). Studies on the destruction of adrenal and testicular cytochrome P450 by spironolactone. *J. Biol. Chem.,* **254,** 1726–1733.

Moghissi, K. S., Segal, S., Meinhold, D., and Agronow, S. J. (1982). *In vitro* sperm cervical mucus penetration: studies in human and bovine cervical mucus. *Fertil. Steril.,* **37,** 823–827.

Moore, C. R. and Price, D. (1932). Gonad hormone functions, and the reciprocal influence between gonads and hypophysis with its bearing on the problem of sex antagonism. *Am. J. Anat.,* **50,** 13–67.

Morrissey, R. E., Lamb, J. C., Schwetz, B. A., Teague, J. L., and Morris, R. W. (1988). Association of sperm, vaginal cytology, and reproductive organ weight data with results of continuous breeding reproduction studies in Swiss CD-1 mice. *Fundam. Appl. Toxicol.,* **11,** 359–371.

Orgebin-Crist, M. C., Danzo, B. J., and Davies, J. (1975). Endocrine control of the development and maintenance of sperm fertilising ability in the epididymis. *In*: 'Handbook of Physiology', Section 7, Volume V. *Eds* D. W. Hamilton and R. O. Greep. American Physiological Society, Washington DC., pp. 319–335.

Overstreet, J. W. (1983). Evaluation and control of the fertilizing power of sperm. *In*: 'The Sperm Cell'. *Ed.* J. Andre. Martinus Nijhoff Publishers, Boston, p. 1.

Overstreet, J. W. (1984). Laboratory tests for human male reproductive risk assessment. *Teratogen. Carcinogen. Mutagen.,* **4,** 67–82.

Penhoat, A., Darbeida, H., Bernier, M., Saez, J. M., and Durand, Ph. (1988). Inhibition of hormonally induced cAMP and steroid production by inhibitors of pregnenolone metabolism in adrenal and Leydig cells. *Mol. Cell. Endocrinol.,* **60,** 55–60.

Reader, S. C. J. and Foster, P. M. D. (1990). The *in vitro* effects of four isomers of dinitrotoluene on rat Sertoli and Sertoli-germ cell co-cultures: Germ cell detachment and lactate and pyruvate production. *Toxicol. Appl. Pharmacol.*, **106**, 287–294.

Ritzen, E. M., Boitani, C., Parvinen, M., French, F. C., and Feldman, M. (1982). Age-dependent secretion of ABP by rat seminiferous tubules. *Mol. Cell Endocrinol.*, **25**, 25–33.

Rommerts, F. F. G., Bakker, G. H., and van der Molen, H. J. (1983). The role of phosphoproteins and newly synthesised proteins in the normal regulation of steroidogenesis in Leydig cells. *J. Steroid Biochem.*, **19**, 367–373.

Schrag, S. D. and Dixon, R. L. (1985). Occupational exposures associated with male reproductive dysfunction. *Annu. Rev. Pharmacol. Toxicol.*, **25**, 567–592.

Setchell, B. P. (1978). 'The Mammalian Testis'. New York, Cornell University.

Shaffer, C. B., Carpenter, C. P., and Smyth, H. R. (1945). Acute and sub-acute toxicity of di-(2-ethylhexyl)phthalate with notes upon its metabolism. *J. Ind. Hyg. Toxicol.*, **27**, 130–135.

Sharpe, R. M. (1982a). The homonal regulation of the Leydig cell. *In*: 'Oxford Reviews of Reproductive Biology'. Vol. 4. *Ed.* C. A. Finn. pp. 241–317.

Sharpe, R. M. (1982b). Cellular aspects of the inhibitory actions of LHRH on the ovary and testis. *J. Reprod. Fertil.*, **64**, 517–527.

Sharpe, R. M., Frazer, H. M., Cooper, I., and Rommerts, F. F. G. (1981). Sertoli-Leydig cell communication *via* an LHRH-like factor. *Nature (London)*, **290**, 785–787.

Sharpe, R. M. (1983). Regulation of the Leydig cell by luteinising hormone-releasing hormone. *In*: 'Recent Advances in Male Reproduction: Molecular Basis and Clinical Implications' *Eds* R. D'Agata, M. B. Lipsett, P. Polosa, and H. J. van der Molen. Raven Press, NY. pp. 197–204.

Shen, R. S. and Lee, I. P. (1977). Developmental patterns of enzymes in mouse testis. *J. Reprod. Fertil.*, **48**, 301–305.

Shivanandappa, T. and Krishnakumari, M. K. (1983). Hexachlorocyclohexane-induced testicular dysfunction in rats. *Acta Pharmacol. Toxicol.*, **52**, 12–17.

Sjöberg, P., Bondesson, U., Gray, T. J. B., and Plöen, L. (1986). Effects of di-(2-ethylhexyl)phthalate and five of its metabolites on rat testes *in vivo* and *in vitro. Acta Pharmacol. Toxicol.*, **58**, 225–233.

Steinberger, E. and Steinberger, A. (1974). Hormonal control of testicular function in mammals. *In*: 'Handbook of Physiology', **4(2)**, *Eds* K. Knobil and W. H. Sawyer. American Physiological Society, Washington D.C. pp. 325–345.

Steinberger, A. and Steinberger, E. (1976). Secretion of an FSH-inhibiting factor cultured Sertoli cells. *Endocrinology*, **99**, 918–921.

Steinberger, E., Browning, J. Y., and Grotjan, H. E. (1983). Another look at steroidogenesis in testicular cells. *In*: 'Recent Advances in Male Reproduction: Molecular Basis and Clinical Implications'. *Eds* R. D'Agata, M. B. Lipsett, P. Polosa, and H. J. van der Molen. Raven Press, NY. pp. 113–120.

Steinberger, E. (1976). Biological action of gonadotropins in the male. *Pharmacol. Ther., B*, **2**, 771–786.

Takizawa, K. and Mattison, D. A. (1983). Female reproduction. *Am. J. Ind. Med.*, **4**, 17–30.

Tash, J. S., Welsh, M. J., and Means, A. R. (1980). Protein kinase inhibitor as an intracellular marker of FSH action in the Sertoli cell. *In*: 'Testicular

Development, Structure and Function'. *Eds* A. Steinberger and E. Steinberger. Raven Press. NY, pp. 159–167.

Tcholakian, R. K. and Steinberger, A. (1978). Progesterone metabolism by cultured Sertoli cells. *Endocrinology, 103*, 1335–1343.

Thomas, J. A., Curto, K. A., and Thomas, M. J. (1982). MEHP/DEHP: Gondal toxicity and effects on rodent accessory sex organs. *Environ. Health Perspec., 45*, 85–88.

Van Thiel, D. H., Gavaler, J. F., Cobb, C. F., Sherins, R. J., and Lester, R. (1979). Alcohol-induced testicular atrophy in the adult male rat. *Endocrinology, 105*, 888–895.

Van Thiel, D. H., Sherins, J. R., Myers, G. H., and DeVita, V. T. (1972). Evidence for a specific seminiferous tubule factor affecting follicle-stimulating hormone secretion in man. *J. Clin. Invest., 51*, 1009–1019.

Waller, D. P., Killinger, J. M., and Zaneveld, L. J. D. (1985). Physiology and toxicology of the male reproductive tract. *In*: 'Endrocrine Toxicology' *Eds* J. A. Thomas *et al*. New York, Raven Press.

Waller, D. P. and Nikurs, A. R. (1986). Review of the physiology and biochemistry of the male reproductive tract. *J. Am. College Toxicol., 5*, 209–223.

Weiss, L. and Greep, R. O. (1977). 'Histology' 4th Edn. McGraw Hill Book Co., New York.

Welsh, M. J. and Wiebe, J. P. (1978). Sertoli cell capacity to metabolise C_{19}-steroids. Variation with age and the effect of follicle-stimulating hormone. *Endocrinology, 103*, 838–844.

WHO (1987). Laboratory manual for the examination of human semen and semen-cervical mucous interactions. Cambridge University Press, Cambridge, UK.

Willis, B. R., Anderson, R. A., Oswald, C., and Zaneveld, W. D. (1983). Ethanol-induced male reproductive tract pathology as a function of ethanol dose and duration of exposure. *J. Pharm Exp. Therap., 225*, 470–478.

Wyrobek, A. J. and Bruce, W. R. (1975). Chemical induction of sperm abnormalities in mice. *Proc. Natl. Acad. Sci. USA, 72*, 4425–4429.

Yanagimachi, R. (1981). Mechanisms of fertilization in mammals. *In*: 'Fertilization and Embryonic Development *In vitro*'. *Eds* L. Mastroianni and J. D. Biggers. Plenum Publishing Corp. New York, p. 81.

Yanahara, T. and Troen, P. (1972). Studies of the human testis I. Biosynthetic pathways for androgen formation in human testicular tissue *in vitro*. *J. Clin. Endocrinol. Metab., 34*, 783–792.

Zenick, H. and Goeden, H. (1988). Evaluation of copulatory behaviour and sperm in rats: role in reproductive risk assessment. *In* 'Physiology and Toxicology of Male Reproduction'. Chapter 8. *Eds* J. C. Lamb and P. M. D. Foster. Academic Press Inc., San Diego, USA.

CHAPTER 18

Statistics

PETER N. LEE

1 Introduction

This chapter is aimed at providing some insight into statistical aspects of
the design, conduct, analyses, and interpretation of toxicological data and
is intended mainly for the reader not qualified in statistics. In general,
mathematical details are kept to a minimum with emphasis being given to
the principles involved. While the chapter should give the reader
sufficient information to choose and carry out appropriate analyses in a
number of situations, the need to have the advice of an expert statistician
available when dealing with toxicological data cannot be over-
emphasised. Scientific journals are full of papers describing studies where
the authors' conclusions cannot be supported, owing to deficiencies in the
statistical methodology which would not have occurred had a qualified
statistician been available to the researcher or indeed had one refereed
the paper.

2 General Principles and Terminology

A Sources of Treatment/Control Difference

Essentially, the objective of a toxicological study is to determine whether
a treatment elicits a response, but the observation of a difference in
response between a treated and a control group does not necessarily
mean that the difference is a result of treatment. There may be two other
causes of difference, *bias,* that is systematic differences other than
treatment between the groups, and *chance,* or random differences. Good
experimental design and analysis should seek to avoid bias where possible
by ensuring that like is compared with like.

Chance factors cannot be wholly excluded since identically treated
animals will not all respond identically, however carefully the experiment
is conducted. It is impossible to be absolutely certain that even the most

extreme treatment/control differences are not due to chance, but the appropriate statistical analysis will allow the experimenter to assess the probability of a 'false positive', that is of the observed difference having occurred had there been no effect of treatment at all. The smaller the probability is, the more the experimenter will be confident of having found a real effect. In order to improve the likelihood of detecting a true effect with confidence, it is necessary to try to minimise the role of chance. Experimental design should aim also at trying to ensure that the 'signal' can be recognised above the 'noise'.

B   Hypothesis Testing and Probability Values

Toxicological reports often include statements such as 'the relationship between treatment and blood glucose levels was statistically significant $(p = 0.02)$'. What does this actually mean? There are three considerations.

Firstly there is a difference in meaning between biological and statistical significance. It is quite possible to have a relationship that is unlikely to have happened by chance and therefore is statistically significant, but of no biological consequence at all, the animal's well-being not affected. Similarly, an observation may be biologically but not statistically significant, as when one or two of an extremely rare tumour type are seen in treated animals. Overall judgment of the evidence must take into account both biological and statistical significance.

Secondly, '$p = 0.02$' does not mean that the probability that there is no real treatment effect is 0.02. The true meaning is that, given the treatment actually had no effect whatsoever (or, to phrase it more technically, 'under the null hypothesis'), the probability of observing a difference as great or greater than actually seen is 0.02.

Thirdly, there are two types of probability value (p-value). A 'one-tailed (or one-sided) p-value' is the probability of getting, by chance alone, a treatment effect in a specified direction as great as or greater than that observed. A 'two-tailed p-value' is the probability of getting, by chance alone, a treatment difference in either direction — positive or negative — as great as or greater than that observed. Normally, two-tailed p-values are appropriate, and by convention p-values are assumed to be two-tailed unless the contrary is stated. However, when there is prior reason to expect only a treatment effect in one direction, a one-tailed p-value is normally used. If a one-tailed p-value is used, differences in the opposite direction to that assumed are ignored.

While, in an unbiased study, a p-value of 0.001 or less can on its own provide very convincing evidence of a true treatment effect, less extreme p-values, such as $p = 0.05$, should be viewed as providing indicative evidence of a possible treatment effect, to be reinforced or supported by other evidence. If the difference is similar to one found in a previous study, or if the response, based on biochemical considerations, is expected, a larger *p*-value would be needed than if the response was unexpected or not found at other dose levels.

Some laboratories, when presenting results of statistical analyses, assign a magical significance to the 95% confidence level ($p < 0.05$) and simply mark results as significant or not significant at this level. This is poor practice as it gives insufficient information and does not enable the distinction to be made between a true effect and one which requires other confirmatory evidence. While it is not necessary to give *p*-values exactly, it is essential to give some idea of the degree of confidence. A useful method is to use plus signs to indicate positive differences (and minus signs to indicate negative differences) with $+++$ meaning $p < 0.001$, $++$ $p < 0.01$ and $\geqslant 0.001$, $+ p < 0.05$ and $\geqslant 0.01$, and (optionally) $(+)$ meaning $p < 0.1$ and $\geqslant 0.05$. This makes it easier to assimilate findings when results for many variables are presented.

C Multiple Comparisons

Toxicology studies frequently involve making treatment/control comparisons for large numbers of variables. If no treatment effect exists at all, it is possible that, purely by chance, one or more variables will show differences significant at the 95% confidence level. For example, with 100 independent variables at least one variable would show significance 99.4% of the time. Because of this, some have suggested making the critical values required to achieve significance more stringent with increasing numbers of variables studied, so that, in testing at the 95% confidence level, 19 times in 20 *all* the variables in the test should show non-significance if the treatment is without effect. This approach is not to be recommended because it is frequently the case in toxicological studies that a compound has only one or two real effects and has no effect on a large number of other variables studied. Such multiple-comparison tests would make it much more difficult to demonstrate statistical significance for the real effects. In any case there is something unsatisfactory about a situation where the relationship between a treatment and a particular response depends arbitrarily on which other response happens to be investigated at the same time. For this reason Section 4, which describes methods of analysis, does not give details of such procedures.

An alternative multiple-comparison problem relates to the situation where more than one test can be carried out in relation to the same response variable, because there is information either under different experimental situations (*e.g.* males, females) or from a number of different dose levels of treatment. This is discussed in some detail in Section 4A.

D Some Statistical Terms

Before going into details of experimental design and analysis it is helpful to give brief definitions of a number of the more commonly used statistical terms, some of which are used in the text that follows and some of which the reader may come across in other statistical documents. The terms are considered in groups rather than alphabetically. For a much more extensive alphabetical list see Marriott (1990).

Frequency Distribution

The frequency distribution is the relationship between the values of the data that occur and how often they occur. For observed data the distribution is usually given as a table with some grouping for continuous variables. Data are often assumed to satisfy some theoretical underlying frequency distribution in which the relationship between the data values and their relative frequency is defined by an equation depending on one or more parameters. For such distributions the formulae are usually defined so that the total frequency is equal to 1, the relative frequency for discrete data then being equivalent to the probability of the value occurring. An example of a theoretical distribution for discrete data is the *binomial distribution,* where the probability of an observed value x occurring is given by:

$$\text{prob}(x) = \frac{n!}{x!\,(n-x)!} p^x (1-p)^{n-x}$$

This distribution occurs when an event has a probability p of occurring at any one trial and n independent trials are carried out, x being the number of times the event occurs.

For continuous data the most commonly found distribution is the *normal distribution,* known otherwise as the Gaussian or Laplacian distribution. It has a bell-shaped curve with the equation:

$$dF(x) = \frac{1}{\sigma \sqrt{(2\pi)}} e^{-1/2 \left(\frac{x-m}{\sigma}\right)^2} dx$$

where m is the mean and σ the standard deviation (see below). The probability of an observation occurring within the range (x_1, x_2) can be obtained by integrating the curve over this range.

Histogram

A histogram is a diagrammatic representation of a frequency distribution in which rectangles proportional in area to the class frequencies are erected on sections of the horizontal axis, the width of each section representing the corresponding class interval of the variate.

Statistic

A statistic is a figure calculated from a set of data intended to summarise some feature of it. Usually it is an estimate of some parameter of the underlying population from which the sample is drawn.

Average

Statistics are often used to indicate central tendency. Average is a familiar word, not usually used in technical statistical reports, to describe such statistics. In common usage, average is understood to be the mean, but the median and the mode are other types of average.

Mean

The mean, or the arithmetic mean, is calculated by summing the observations and dividing by their number. For a theoretical frequency distribution, the mean is calculated by the integral $\int_{-\infty}^{\infty} x \, dF(x)$.

Median

The median is that value which divides the total frequency into two halves so that as many observations are greater as are less.

Mode

The mode is the value which occurs most frequently.

Variance

For a theoretical frequency distribution, the variance is calculated by the integral $\int_{-\infty}^{\infty} (x - \mu)^2 dF(x)$ where μ is the mean. This is a measure of the extent to which the data are dispersed about the mean. For a normal distribution the variance can be estimated from a sample by the formula:

$$\sigma^2 = \left(\sum_{i=1}^{n} x_i^2 - n\mu^2 \right) \bigg/ (n - 1)$$

where n is the number of samples and μ is the sample mean.

Standard Deviation

The standard deviation is the positive square root of the variance.

Standard Error

The standard error is the positive square root of the variance of the sampling distribution of a statistic, commonly the sample mean.

Skewness

Skewness is a measure of asymmetry of a frequency distribution.

Kurtosis

Kurtosis is a measure of 'peakedness' of a frequency distribution.

Chi-squared Distribution

The chi-squared distribution is the distribution of the sum of squares of n independent normal variates each with mean 0 and standard deviation 1.

F-distribution

The F-distribution is the distribution of the ratio of two independent variates, each distributed like a variance in normally distributed samples.

Student's t-Distribution

The Student's t-distribution is the distribution of the ratio of a simple mean to its simple variance, in samples from a normal population.

Degrees of Freedom

Degrees of freedom are the number of independent comparisons which can be made between members of a sample. For example, if a sample of a constant size n is grouped into r intervals, there are $r - 1$ degrees of freedom as, given $r - 1$ of the frequencies, the other one is specified.

Contingency Table

A contingency table is a table giving the breakdown of a sample according to two or more characteristics, each with two or more possible values. Thus the number of animals in an experiment may be presented in a $2 \times 5 \times 2$ contingency table according to sex (male or female), group (control or four dose levels), and presence of a tumour (yes or no).

Confounding

A study design is said to be confounded if the effects of two or more possible determinants of a response cannot be separated. A study in which the slides from the test group are examined by pathologist A and the slides from the control group are examined by pathologist B is confounded since any difference seen in response cannot be attributed with certainty either to an effect of treatment or to a difference between pathologists.

Interaction

Two factors are said to interact if the difference in response between levels of one factor depends on the level of the other factor. Thus, if treatment increased body weight by 30 g in males but only by 10 g in females, there would be a sex–treatment interaction.

Latin Square

A latin square is an experimental design which allows comparison of k experimental treatments without confounding from two other sources of variation in response, which are identified with rows and columns of a $k \times k$ square in which each treatment occurs exactly once in each row or column. Thus 25 animals in 5 treatment groups of 5 animals each may be assigned to the following design to avoid possible effects of time (row) and day (column) of injection.

	Mon.	Tue.	Wed.	Thurs.	Fri.
10.30 am	A	B	C	D	E
11.30 am	D	E	B	A	C
12.30 am	E	C	D	B	A
2.30 pm	C	D	A	E	B
3.30 pm	B	A	E	C	D

Regression

A regression is a relationship between a response variable, y, and one or more predictor variables, x_i. In simple linear regression the relationship is of the form $y = \beta_0 + \beta_1 x$ where β_0 and β_1 are regression coefficients (the intercept and the slope) to be estimated. In multiple linear regression the regression equation can be written as $y = \beta_0 + \beta_1 x_1 + \beta_2 x_2 + \cdots + \beta_p x_p$.

Efficiency

One estimator is regarded as more efficient than another if it has smaller variance.

Power

The power of a statistical test is the probability that it rejects the hypothesis that there is no effect of treatment where a true effect of treatment exists.

Sensitivity and Specificity

Consider a 2×2 table relating the number of times an event actually occurs or not to the number of times the recording device claims it occurs.

		Observed situation	
		Yes	No
True	Yes	a	b
situation	No	c	d

The sensitivity of the recording device is measured by $a/(a + b)$, the specificity by $d/(c + d)$. High sensitivity implies a low false negative rate, that is the device (or test) is sensitive to what is being measured. High specificity implies a low false positive rate, that is the device is specific for the effect being measured.

3 Experimental Design

The question of experimental design of toxicological studies is considered in detail in Chapter 4, but in view of the importance of statistical consideration in experimental design some of the main issues involved will be discussed briefly here. For convenience, the principles are illustrated with reference to the design of a long term carcinogenicity study. Sub-sections A–F examine decisions mainly involved with minimising the role of chance while sub-sections G–J discuss decisions mainly related to avoidance of bias.

A   Choice of Species

While maximising the 'signal' means avoiding a species where the response of interest is very rare, the use of an over-responsive species can bring penalties also. Thus, to have a 50% chance of achieving statistical significance at the 95% confidence level in an experiment in which the treated group has a 5% incidence of the lesion and the control group a 0% incidence, two groups of about 80 animals are required. To achieve the same level of statistical significance where the incidences are 55% and 50%, respectively, group sizes some 10 times larger are required. Further, it is not certain that an increased incidence of a lesion that is common spontaneously in the animal species used (such as pituitary tumours in Wistar rats) provides biological evidence of an effect which can be extrapolated to other species.

B   Dose Levels

The problem of dose selection is one of the most controversial and important elements in the development of a protocol for a toxicology bioassay. On biological grounds it would be ideal to test only at dose levels comparable to those to which humans are exposed. On statistical and economic grounds this is not usually practicable because the effect will be too small to detect without very large numbers of animals. To avoid the possibility of an effect which would occur in a small proportion of millions of exposed humans being missed in a study of hundreds or even thousands of animals, the solution usually employed is to test in animals at a dose level many times greater than maximum human exposure. Then, assuming that any effect that exists is dose related, the demonstration of a non-significant increase in response at a high dose level, though not providing evidence of absolute safety (an impossible goal), can give reasonable grounds for believing that any effects that occur at a dose level very much lower will be at most very small.

A particular problem with this procedure is deciding the dose level. In long term carcinogenicity studies the dose should clearly be one which is not so great that the animals die from toxic effects before they have a chance to get cancer. Based on these principles, the International Agency for Research on Cancer (1980) has recommended that the high dose to be used should be one expected, on the basis of an adequate sub-chronic study, to produce some toxicity when administered for the duration of the study but which should not induce the following:

(a) overt toxicity, *i.e.* appreciable death of cells or organ dysfunction, as determined by appropriate methods,
(b) toxic manifestations which are predicted materially to reduce the lifespan of the animals except as a result of neoplastic development, or
(c) 10% or greater retardation of body weight gain compared with control animals.

If the substance seems completely non-toxic, the high dose may represent about 5% of the diet or even more for substances such as food ingredients.

For a number of reasons it is important to have more than one dose level. One is to cater for the possibility that a misjudgement has occurred and the highest dose proves toxic. A second purpose is to cater for the possibility that the metabolic pathways at the high dose may differ from those at lower doses. A third is that the whole point of the study may be to obtain dose–response information. Finally, it may be necessary to ensure that no large effect occurs at dose levels in the range to be used by man.

C Number of Animals

The number of animals to be used is clearly an important determinant of the precision of the findings. The calculation of the appropriate number depends on the following:

(a) the critical difference, *i.e.* the size of the effect to be detected,
(b) the false positive rate, *i.e.* the probability of an effect being detected when none exists (known also as the 'type I error' or the 'α-level'), and

(c) the false negative rate, *i.e.* the probability of no effect being detected when one of exactly the critical size exists (known as the 'type II error' or the 'β-level').

Reducing any of these increases the number of animals required.

The method of calculation of the number depends on the experimental design and the type of statistical analysis envisaged. Tables are available for a number of standard situations. To give an idea how the numbers depend on the critical difference and on the α- and β-levels, Tables 1 and 2 give examples for two common situations, both of which relate to an experiment where there is a control and a treated group. In the first there is a normally distributed variable and the critical difference is expressed in terms of the number of standard deviations (δ) by which the treated group differs from the control group. Thus, given a control response known from past experience to have a mean value of 50 units with a standard deviation of 20 units, one would need 2 groups of 36 animals to have a 90% chance ($\beta = 0.10$) of detecting a difference in response of 10 units ($\delta = 10/20 = 0.5$) at the 95% confidence level ($\alpha = 0.05$).

In the second situation two proportions are compared. Here the numbers of animals depend not only on the ratio of proportions but also on the assumed proportion in the controls. Thus, where the control response is expected to be 10%, the number of animals required in each group to detect an increase response by a factor of 1.5 (r) is 920, assuming again an α-level of 0.05 and a β-level of 0.1, whereas, if the control response is expected to be 50%, the number required would be 79 per group.

For more complex situations the advice of a professional statistician should be sought, though a general rule is that, to increase precision (*i.e.* decrease the size of the critical difference) by a factor n, the number of animals required will have to be increased by a factor of about n^2.

Where there are a number of treatments to be tested in an experiment, each to be compared with a single, untreated control group, it is advisable that more animals be included in the control group than in each of the treated groups, since the precision of the control results is relatively more important. A useful rule in many situations is to have approxi-

Table 1 *Number of animals required in each of a control and a treated group in order to have a probability $(1 - \beta)$ of picking up a difference of σ standard deviations as significant at the $100\,(1 - \alpha)\%$ confidence level for a normally distributed variable*

| Single-sided test | $\alpha = 0.005$ | | | $\alpha = 0.025$ | | | $\alpha = 0.05$ | | |
| Double-sided test | $\alpha = 0.01$ | | | $\alpha = 0.05$ | | | $\alpha = 0.01$ | | |
$\beta =$	0.01	0.1	0.5	0.01	0.1	0.5	0.01	0.1	0.5
$\sigma = 0.5$	100	63	30	76	44	18	65	36	13
0.75	47	30	16	35	21	9	30	17	7
1.0	28	19	10	21	13	6	18	11	5
1.5	15	11	7	11	7		9	6	
2.0	10	8	5	7	5		6		

Table 2 *Number of animals required in each of a control and a treated group in order to have a probability* $(1 - \beta)$ *of picking up a proportional increase by a factor r as significant at the* $100\,(1 - \alpha)\%$ *confidence level for a binomially distributed variable*

Single-sided test	$\alpha = 0.005$			$\alpha = 0.025$			$\alpha = 0.05$		
Double-sided test	$\alpha = 0.01$			$\alpha = 0.05$			$\alpha = 0.1$		
$\beta =$	0.01	0.1	0.5	0.01	0.1	0.5	0.01	0.1	0.5
Control level = 10%									
$r = 1.25$	7679	4754	2120	5871	3358	1228	5039	2737	865
1.5	2103	1302	581	1608	920	337	1380	750	237
2.0	613	380	170	469	268	98	403	219	69
Control level = 20%									
$r = 1.25$	3353	2076	926	2563	1466	536	2200	1195	378
1.5	902	558	249	689	395	145	592	322	102
2.0	253	157	70	193	111	41	166	90	29
Control level = 50%									
$r = 1.25$	757	469	209	579	331	122	497	270	86
1.5	181	112	50	138	79	29	119	65	21
2.0	37	23	10	28	16	6	24	13	5

mately $\sqrt{k}$ times as many animals in the control group as in each of the k treated groups.

One point that is frequently misunderstood by experimentalists relates to the number of animals required in studies where more than one treatment is investigated in a crossed design. If, for example, one is comparing compounds A and B, each dissolved in two different solvents, in a 2×2 design with 4 groups, calculations of sample size to gain an overall verdict on the difference between the two compounds should generally be based on the overall numbers of animals on each compound for both solvents combined unless there is reason to expect compound/solvent interaction, *i.e.* that the compound A/compound B difference depends on which solvent is used. Conversely, if it has been decided that 2 groups of 100 animals are sufficient for attaining a given level of precision concerning the difference in effects of a treatment, additional information can be obtained on another factor (or factors) of interest, without requiring any additional animals.

D Duration of the Experiment

Duration of the experiment can also affect the sensitivity of the tests markedly. This is particularly so in long term carcinogenicity studies where the great majority of cancers are seen in the latter half of an animal's lifetime. While studies should therefore not be terminated too early, it is also important not to go on too long. This is because the last few weeks or months may produce relatively little extra data at a disproportionate cost, and diseases of extreme old age may be of little interest in themselves but render it more difficult to detect tumours and other conditions that are of interest. Where the study is of the prevalence of an age-related, non-lethal condition that ultimately occurs in all or nearly all of the animals, early termination is required. In this situation the greatest sensitivity is obtained when the average prevalence is around 50%.

E Accuracy of Determinations

Accuracy of observation is clearly important in minimising error. The advent of Good Laboratory Practice and Quality Control Units has done much to improve the quality regarding observations, but the quality depends still on interested and diligent personnel.

F Stratification

To detect a treatment difference with accuracy, the groups being compared should be as homogeneous as possible with respect to other known causes of the response of interest. For example, a set of animals thought to be homogeneous, but which in fact consisted of two genetically different sub-strains, might provide the following measurements of body weight in groups of 10 treated and 10 control animals, the underlined readings relating to the first of the two sub-strains:

Control: 181 192 217 290 321 292 307 347 276 256
Treated: <u>222</u> <u>249</u> <u>232</u> 284 270 215 265 378 328 391

Ignoring sub-strain increases the variability of the data so that more controls are required to detect a treatment effect and, if unequal numbers of each sub-strain are present in each group, may bias the comparison. In the example given the means are as follows:

	Sub-strain 1	*Sub-strain 2*	*Total*
Control	196.7	298.4	267.9
Treated	248.1	365.7	283.4
Difference	51.5	67.2	15.5

Although in each sub-strain the treatment results in an increase in body weight by over 50 units, the greater number of the lower weight strain in the treated groups results in the difference observed in the totals being much less.

There are methods to take account of the sub-strain as a 'stratifying variable' at the analysis stage. As described further in Section 4A, these involve carrying out separate analyses at each level of the variable considered and combining the results for an overall conclusion about the treatment effect. This does not, however, preclude the possibility that the proportions of each sub-strain in each group are so different that the data provide substantially less comparative information than might otherwise be achieved. In the extreme case where all control animals are sub-strain 1 and all treated animals sub-strain 2, the experiment would be worthless to determine whether differences were due to treatment or sub-strain. To obviate this possibility, sub-strain can be used as a 'blocking factor' in the design. In this case animals in each sub-strain are allocated equally to control and treated groups. Although this removes bias, it is still necessary to treat strain as a stratifying variable in the analysis.

Where more than one known factor affects the response, all can be taken into account simultaneously. At the design stage, or retrospectively in the analysis, the results are treated at each combination of levels of the factors. Thus, to block for sub-strain, sex, and room where 3 experimental rooms were needed to house the animals, 12 mini-experiments, one for each of the 2 (sub-strains) $\times$ 2 (sexes) $\times$ 3 (rooms) combinations, would be set up.

G Randomisation

Random allocation, or randomisation, of animals to treatment groups is an essential of good experimental design. If not carried out, it cannot be ascertained whether a difference between groups is a result of differences in the treatment applied or is due to some other relevant factor. It is a fundamental on which the statistical methodology is based that the probability of a particular response occurring is equal for each animal, regardless of group. The ability to randomise easily is one great advantage that animal experiments have over epidemiology.

The process of randomisation eliminates bias, so that the statistical analysis is concerned only with assessing the probability of an observed difference happening by chance. The smaller the probability is, the more it suggests a true treatment effect. The procedure used for randomisation should genuinely ensure that all possible assignments of animals to treatment groups are equally probable. Such equal probabilities are best achieved with pseudo-random numbers as found in tables or produced by computer, it being difficult to ensure that apparently random devices such as dice or playing cards really are random. Randomisation should never be based on a system of testing animals haphazardly as they come and assigning them to successive treatment groups. Not only do humans find it virtually impossible to generate random sequences unaided, but it is well known that the first animals selected may differ markedly from the last, who are more active and avoid being caught.

In many experimental situations it is adequate to allocate randomly all the animals to treatment groups, but in some the technique of stratified random sampling is to be preferred. In this technique the animals are divided into sub-groups ('strata') according to known factors believed to be strongly related to the response, and random allocation to treatment groups is carried out within each stratum. Sex is normally treated as a stratifying variable. In a large experiment where animals are delivered in batches, a batch could also be treated in this way, each batch forming a smaller experiment, the results of which can be combined in the analysis.

This discussion on randomisation and stratification has been concerned primarily with the allocation of animals to treatment groups. The same principles apply to anything that can affect the recorded response. Thus, in a two-group experiment, measurements of some biochemical parameter should not be made for the first group in the morning and for the second group in the afternoon. While the major part of such potential bias can be averted fairly easily by various simple procedures, such as doing alternate measurements on treated and control animals, randomisation is preferable. Although many different procedures throughout an experiment (feeding, weighing, observation, clinical chemistry, pathology) require consideration in this way, the same random number can usually be applied to all the procedures. Thus, if the cage position of the animals is randomly allocated and does not depend on treatment, the animals can always be handled in the same cage sequence.

H Adequacy of Control Group

The principle of comparing like with like implies that the control groups should be randomly allocated from the same source as the treatment groups. While historical control data can occasionally be of value in interpretation of findings in treated animals, there is so much evidence of quite large systematic differences in response between apparently identical untreated control groups tested at different times that it is often impossible to be sure whether a difference seen between a treated group and a historic group is really due to treatment at all.

It is also essential to be sure that the treated group differs from the control group only in respect of the treatment of interest. Thus, having a treated group where animals are exposed to cigarette smoke in a smoking machine and an untreated control group does not allow evaluation of the effect of cigarette smoke itself. The stress of being placed in a smoking machine regardless of whether smoke or air is passed through to the animal may have a quite marked effect on response. The appropriate control group to use in this case would be one in which animals are placed in a machine for the same length of time as the treated animals but are not exposed to smoke.

I Animal Placement

The general underlying requirement to avoid systematic differences between groups other than their treatment also demands that attention be given to the question of animal placement. If all treated animals are placed on the highest racks or at one end of the room, differences in heating, lighting, or ventilation may produce effects which are erroneously attributed to treatment. Such systematic differences should be avoided, and in many cases randomisation of cage positions is desirable. It may not be possible in some circumstances, for example in studies involving volatile agents where cross-contamination can occur.

An example from my personal files highlights the reality of the dangers of ignoring such problems. In a carcinogenicity study on albino rats the incidence of severe retinal atrophy among animals housed in cages located on the top shelf was 100% whilst among rats located on the bottom shelf it was only about 5%, with intermediate incidences seen in the central shelves. It was fortunate in this study that all groups were represented on each shelf. A design and analysis that ignored animal placement could easily have produced a false-positive effect of treatment.

J Data Recording

The application of the principle of comparing like with like means the avoidance of systematic bias in data-recording practices. Two distinctly different types of bias can occur. The first is a systematic shift in the standard of measurement with time, coupled with a tendency for the time of measurements to vary from treatment to treatment. The second is that awareness of the treatment may affect the values recorded by the measurer, consciously or subconsciously. The second bias is circumvented by the animal's treatment not being known to the measurer, *i.e.* the readings being carried out 'blind'. Although not always practical (that an animal is treated may be obvious from its appearance), it is recommended that laboratories organise their data-recording practices so that, at least for subjective measurements, the observations are made 'blind'.

The problem of avoidance of bias due to differences in time of observation is a particularly important one in histopathological assessment, especially for the recording of lesions of a graded severity, and in large experiments where the slides may take the pathologist more than a year to read. Where more than one pathologist reads the slides, there should be discussion between them as to standardisation of terminology and data to be recorded, and each should read a random or a stratified sub-set of the slides to avoid bias.

4 Statistical Analysis

A Introduction

In the simplest situation animals are randomly assigned to a treated or a control group and one observation is made for each animal, the objective of the statistical analysis being to determine whether the distribution of responses in the treated group differs from that in the control group. There are five characteristics of this situation which require discussion before detailed methods are described, as all should be clearly understood before an appropriate technique of analysis can be selected.

Univariate or Multivariate Data?

The analyses described in Sections B to D are appropriate for univariate data, *i.e.* where one analysis concerns a single response measure per animal. Methods for multivariate data, involving more than one response measure per animal, are discussed more briefly in Section E, which also covers some other methods not considered elsewhere.

Experimental and Observational Units

In the simple example cited the animal is both the 'experimental unit' and the 'observational unit'. This is not always so. In the case of feeding studies, the cage, rather than the animal, is usually the experimental unit in that it is the cage, rather than the animal, that is assigned to the treatment. In the case of histopathology, observations are often made from multiple sections per animal and here the section rather than the animal is the observational unit. For the purposes of determining treatment effects by the methods described below it is important that each experimental unit provides only *one* item of data for analysis, as the methods are all based on the assumption that individual data items are statistically independent. If there are multiple observations per experimental unit, these observations should be combined in some suitable way into an overall observation for that experimental unit before analysis. Thus, in an experiment in which 20 animals were assigned into 2 treatment groups of 10 animals and in which measurements of the weight of both kidneys were individually made, it would be wrong to carry out an analysis in which the 20 weights in group 1 were compared with the 20 weights in group 2 since the individual kidney weights are not independent observations. A valid method would be to carry out an analysis comparing the 10 average kidney weights in group 1 with the 10 average kidney weights in group 2.

Type of Response Variable

Responses measured in toxicological studies can normally be classified as being one of the following three types:
   (a) *Presence/absence,* in which a condition either occurs or it does not.
   (b) *Ranked,* in which a condition may be present in various degrees. For example, severity may be classed as minimal, slight, moderate, severe or very severe.
   (c) *Continuous,* in which a condition may take any value, at least within a given range.
Each type of response demands a different sort of statistical technique, described in Sections B to D respectively. It should be noted that by using the methods of Section B it is possible to analyse ranked or continuous data by defining values above a given cut-off point as present and those below as absent. In general, this is not to be recommended; it is wasteful of information and is less precise. Continuous data may also be analysed by the methods described in section C for ranked data. While again this may be somewhat wasteful, it may have advantages in avoiding

possibly unjustified assumptions about the form of statistical distribution underlying the response variable. This is explained more fully in Section D.

Number of Groups Being Compared

Sections B to D each start with methods appropriate for the comparison of two groups. They then go on to consider methods appropriate for the comparison of k (more than two) groups. While only one comparison is possible between two groups, more than one comparison is possible between k groups. Two particularly important methods which are described for each type of response are the test for *heterogeneity* and the test for *trend*. The test for heterogeneity determines whether, taken as a whole, there is significant evidence of departure from the (null) hypothesis that the groups do not differ in their effect. The test for trend is only applicable in experiments where the groups receive different doses of the same substance (or have some other natural ordering). It determines whether there is a tendency for a response to rise in relation to the dose of the test substance applied. It can be a much more sensitive method of picking up a true effect of treatment than an overall test for heterogeneity or the comparison of individual treated groups with the control. Graphically, the test for trend can be seen as determining whether a sloped straight line through the dose–response (x-and y-axis, respectively) relationship fits the data significantly better than a horizontal straight line. That the trend statistic is significantly positive does not necessarily imply that the treatment increases response at all dose levels, though it is a particularly good test if the true situation is a linear non-threshold model. In a true situation in which treatment increases response at higher dose levels but has no effect at lower levels, it is easy to construct data (see Section B) where only the trend test is able to detect any significant effect.

Stratification

In the simplest situation all the animals in the experiment differ systematically only in respect of the treatment applied. It is often the case, however, that there are a number of sets of animals, each of which differs systematically only in respect of treatment, but where the characteristics of the sets differ. The commonest situation relates to male and female animals, but there are many other possibilities, such as different treatment rooms, different secondary treatments or even different conditions under which the response variable was measured. While it is often useful to look within each set of animals, or 'stratum', as it is often called, to determine the effects of treatment in each situation, it is also often useful to determine whether, based on the data from all the strata, an effect of treatment can be seen overall. In some situations a relatively small number of animals in each stratum can make it difficult to pick up an effect of treatment as significant within individual strata. The essence of stratification lies in making comparisons within strata and then accumulating treatment differences over strata. As shown in Section B, the incorrect alternative of combining the data over strata and then making a single analysis can lead to erroneous conclusions. Methods for analysing stratified data are described in Sections B to D for the different types of response data.

In the sections that follow, three levels of statistical detail are given. Where the method is likely to be of frequent application, both the mathematics and an example are given. Where the method is of less frequent application or is more lengthy to perform, only the mathematics are given. Where the method is rarer still, or likely to need the assistance of a statistical expert, only a brief verbal description with references is given.

B   Presence/Absence Data

Unstratified analysis for 2 groups

The data can be expressed in the form of a 2×2 table as follows:

	Treated	Control	Total
Number with condition present	a	b	n_1
Number with condition absent	c	d	n_0
Total number at risk	m_1	m_0	N

A test of whether the proportion with the condition in the treated group (a/m_1) differs from that in the control group (b/m_0) is given by the *corrected chi-squared statistic*: (Breslow and Day, 1980, p. 131).

$$\frac{(|ad - bc| - \frac{1}{2}N)^2(N - 1)}{n_0 n_1 m_0 m_1}$$

Referring this statistic to tables of percentiles of the chi-squared distribution (see Beyer, 1966) with one degree of freedom gives an approximate two-tailed significance level, which may be halved to give the corresponding one-tailed test (Example 1).

EXAMPLE 1

In a feeding study the incidence of liver tumours discovered among rats killed at the end of the experiment was as follows:

	Treated	Control	Total
Tumour	22	15	37
No tumour	62	59	121
Total	84	74	158

$$\chi^2 = \frac{[22(59) - 15(62) - \frac{1}{2}(158)]^2 157}{121(37)74(84)} = 0.47$$

The chi-squared table value for one degree of freedom is 3.84 at the 0.05 probability level. There is therefore no significant association between liver tumour incidence and treatment (at $p = 0.05$).

The corrected form of the chi-squared statistic yields a closer approximation to the proper *p*-value than that given by the uncorrected chi-squared statistic in which the $-\frac{1}{2}N$ in the above formula is not included (Mantel and Greenhouse, 1968). Where the sample is small the corrected chi-squared statistic yields probabilities which are poor approximations to reality, and Fisher's exact test should be used instead. Armitage (1971) notes that approximations to significance levels of about 0.05 are reasonably good, provided that the 'expected' numbers for all four cells in the 2×2 tables are at least 5. [Here and subsequently the expected numbers are calculated by multiplying the proportion present (or absent) in the total sample by the total number examined in the group in question. Thus, the expected number of controls with the condition present is given by $m_0 n_1 / N$.]

Given that there is no relationsip between treatment and the condition, and given the marginal totals m_1, m_0, n_1, and n_0, the exact probability of an observed 2×2 table occurring is given by the formula:

$$P(a, b, c, d) = \frac{m_1!\, m_0!\, n_1!\, n_0!}{a!\, b!\, c!\, d!\, N!}$$

This formula can then be used to calculate the exact probability of a result the same as or more extreme than that actually observed (see Example 2) In the past exact calculations were only feasible with small 2×2 tables but modern high speed computers now allow them to be conducted in all cases.

EXAMPLE 2

In the same study the incidence of brain tumours was as follows:

	Treated	Control	Total
Tumour	5	1	6
No tumour	79	73	152
Total	84	74	158

The exact probability for this table is given by:

$$P(5, 1, 79, 73) = \frac{84!\, 74!\, 6!\, 152!}{5!\, 1!\, 79!\, 73!\, 158!}$$

$$\text{which by cancellation} = \frac{84(83)82(81)80(74)6}{158(157)156(155)154(153)} = 0.1164$$

Given the marginal totals, the only table showing a more extreme difference is the one where all 6 tumours occurred in the treated group. The exact probability for this table is given by:

$$P(6, 0, 78, 74) = 0.0207$$

Thus the overall (one-tailed) probability of observing a difference as large as or larger than that seen, given no true effect of treatment, is $0.1164 + 0.0207 = 0.1371$.

Note that the calculation of the second probability, 0.0207, would not normally be necessary in this situation as the first probability, 0.1164 was already larger than needed for significance.

Had a corrected chi-squared statistic been calculated, a value of 1.19 would have been obtained, with a corresponding probability of 0.2758, equivalent to a very similar one-tailed probability of 0.1379.

Unstratified Analysis for k Groups

The data can be expressed in the form of a $2 \times k$ table as follows:

<table>
<tr><th></th><th colspan="6">Group</th></tr>
<tr><th></th><th>1</th><th>2</th><th>j</th><th>k</th><th>Total</th></tr>
<tr><td>Present</td><td>a_1</td><td>a_2</td><td>... a_j ...</td><td>a_k</td><td>n_1</td></tr>
<tr><td>Absent</td><td>c_1</td><td>c_2</td><td>... c_j ...</td><td>c_k</td><td>n_0</td></tr>
<tr><td>At risk</td><td>m_1</td><td>m_2</td><td>... m_j ...</td><td>m_k</td><td>N</td></tr>
<tr><td>Dose applied</td><td>d_1</td><td>d_2</td><td>... d_j ...</td><td>d_k</td><td></td></tr>
</table>

An overall test of *heterogeneity* is given by

$$\chi^2 = \frac{N(N-1)}{n_0 n_1} \sum_{j=1}^{k} [(a_j - e_j)^2 / m_j]$$

where e_j is the expected number with the condition in group j ($= m_j n_1 / N$), the associated probability being given by reference to tables of the chi-squared distribution on $k - 1$ degrees of freedom (Armitage, 1971).

A test for *trend* is given by:

$$\chi^2 = \frac{N^2(N-1)[\sum_{j=1}^{k} d_j(a_j - e_j)]^2}{n_0 n_1 [N \sum_{j=1}^{k} d_j^2 m_j - (\sum_{j=1}^{k} d_j m_j)^2]}$$

which should be referred to tables of chi-squared on one degree of freedom (Armitage, 1955; Mantel, 1963) (see Example 3).

EXAMPLE 3

In a short term toxicity study the numbers of rats showing evidence of pigment deposition in the kidney were as follows:

	0 p.p.m.	10 p.p.m.	30 p.p.m.	100 p.p.m.	Total
Deposition	5	8	10	11	34
No deposition	15	12	9	5	41
Total	20	20	19	16	75

Given no relationship between treatment and pigment deposition, the expected numbers with deposition and the difference between observed and expected numbers are as follows:

Expected	9.067	9.067	8.613	7.253	34
Observed − expected	−4.067	−1.067	+1.387	+3.747	0

An overall test of heterogeneity is given by

$$\chi^2 = \frac{75(74)}{41(34)} \left(\frac{(-4.067)^2}{20} + \frac{(-1.067)^2}{20} + \frac{(1.387)^2}{19} + \frac{(3.747)^2}{16} \right)$$

which equals 7.42 on three degrees of freedom. From tables the associated probability is between 0.05 and 0.10 so that there is some weak evidence of between-treatment variation.

To test for trend we note that:

$$\sum_{j=1}^{k} d_j(a_j - e_j) = 10(-1.067) + 30(1.387) + 100(3.747) = 405.64$$

$$\sum_{j=1}^{k} d_j^2 m_j = 10^2(20) + 30^2(19) + 100^2(16) = 179,100$$

$$\sum_{j=1}^{k} d_j m_j = 10(20) + 30(19) + 100(16) = 2370$$

so that

$$\chi^2 = \frac{75^2(74)\,405.64^2}{41(34)[75(179,100) - (2370)^2]} = 6.29$$

The two-tailed probability for this single degree of freedom chi squared is 0.01, so the trend statistic gives much stronger evidence of an effect of treatment than does the test for heterogeneity.

Where expected values are low this χ^2 may be poorly approximated by a chi-squared statistic. In this case it may be useful to carry out an exact trend test noting that, given there is no relationship between treatment and condition, and given the marginal totals $m_1, m_2, \ldots, m_k, n_1$, and n_0, the exact probability of an observed $2 \times k$ table occurring is given by the formula:

$$P(a_1, c_1, \ldots, a_k, c_k) = \frac{m_1! \ldots m_k! \, n_1! \, n_0!}{a_1! \, c_1! \ldots a_k! \, c_k! \, N!}$$

The formula can then be used to calculate the exact probability of a result the same as or more extreme than that actually observed. The problem with this test is not only that the computing effort can become large, even with tables with small numbers, but also that the definition of what is 'the same or more extreme' may be difficult. One method is to combine the $2 \times k$ tables with the given marginal totals which yield values of $\sum_{j=1}^{k} d_j(a_j - e_j)$ equal to or greater than the observed table. The method is illustrated in Example 4. It is interesting to note that with these data a significant trend is seen, although none of the individual differences between any treatment group and control are significant.

EXAMPLE 4

In a small short term study the numbers of dogs showing evidence of congestion in the lung were as follows:

	0 mg	1 mg	3 mg	5 mg	10 mg	*Total*
Congestion	0	0	0	1	3	4
No congestion	4	4	4	3	1	16
Total	4	4	4	4	4	20

Given the number of cases of congestion that were seen, the exact probability of observing the table actually seen is given by:

$$P = \frac{4!\,4!\,4!\,4!\,4!\,16!\,4!}{4!\,4!\,4!\,3!\,3!\,20!}$$

(omitting terms of 0! and 1! which equal 1). Cancellation yields:

$$P = \frac{4(4)4(3)2}{20(19)18(17)} = 0.0033$$

The only other table which could have demonstrated a trend that was at least as strong is the one in which all 4 cases occurred in the top dose group. This has a probability

$$P = \frac{16! \, 4!}{20!} = 0.0002$$

The overall one-tailed exact probability for trend is therefore 0.0035, which gives quite strong support to the belief that a true treatment effect occurred. It is noteworthy that the simple comparison of top dose with control only gives a one-tailed exact probability of 0.07, which would not normally be considered significant. The trend test is a much more sensitive test as it takes into account all the data and knowledge about the rank ordering of the doses applied.

Stratified Analysis

The basis of stratification methods lies in accumulating over the strata both the number observed to have the condition and that expected. Summed over all treatment groups, the total observed and total expected numbers are mathematically constrained to be the same, so that an excess of observed over expected numbers in a particular treatment group indicates a positive effect in that group, whilst a deficiency indicates the effect is below average.

In the *two-group situation* the test of whether the treated group differs from the control group is given by referring:

$$\chi^2 = \frac{(|O - E| - \frac{1}{2})^2}{\Sigma V}$$

to tables of chi squared with one degree of freedom. This is the analogue of the *corrected chi-squared statistic* for unstratified data. In this formula O (the total observed) $= \Sigma a$ and E (the total expected) $= \Sigma m_1 n_1 / N$, (summation without subscripts being over strata). V is the variance of the expected number present in the treated group in a stratum and is given by

$$V = \frac{n_0 n_1 m_0 m_1}{N^2 (N - 1)}$$

Example 5 illustrates the method and also shows how totally misleading conclusions can be gained by ignoring the stratifying variable.

EXAMPLE 5

In a long-term carcinogenicity study the data on numbers of rats showing evidence of testicular atrophy (an age-related change which does not cause death) can be laid out as follows:

Period of death/weeks	Treated a c m_1			Control b d m_0			Expected	Variance of expected
1–52	0	4	4	0	10	10	0.0000	0.0000
53–78	1	8	9	2	36	38	0.5745	0.4443
79–104	1	11	12	5	29	34	1.5652	1.0284
Terminal kill	23	52	75	5	13	18	22.5806	3.0878
Total	25	75	100	12	88	100	24.7203	4.5604

For period 53–78 weeks the expected (number of treated rats with atrophy) is given by $9(1 + 2)/(9 + 38) = 0.5745$ while its variance is given by $9(38)3(44)/(47)^2 46 = 0.4443$. Similar calculations for each period allow completion of the right-hand columns of the table:

$$\chi^2 = \frac{(|25 - 24.7203| - 0.5)^2}{4.5604} = 0.01$$

There is thus no evidence of an association between testicular atrophy and treatment, after adjusting for period of death. Had an unstratified analysis been carried out on the totals, a corrected chi-squared statistic value of 4.75 which is significant at the 95% confidence level would have been obtained. This would have reflected the fact that survival was vastly better in the treated group so that more rats reached termination where the frequency of atrophy (about 30% in each group) was much higher than in rats dying earlier.

In the *k-group situation* an overall test for *trend* is given by referring:

$$\chi^2 = \frac{(\Sigma\, T)^2}{\Sigma\, \text{var}\, T}$$

to tables of chi squared with one degree of freedom. In this formula, for each stratum:

$$T = \sum_{j=1}^{k} d_j(a_j - e_j)$$

and its variance

$$\text{var}\, T = \frac{n_0 n_1}{N^2(N - 1)} \left[N \left(\sum_{j=1}^{k} d_j^2 m_j \right) - \left(\sum_{j=1}^{k} d_j m_j \right)^2 \right]$$

An overall test of *homogeneity* involves matrix manipulations and will be beyond the less mathematical reader. It is described by Breslow and Day (1980).

Age Adjustment

One important application of stratified analysis for presence/absence data relates to the use of age as a stratifier. For many conditions, such as tumour incidence, frequency increases markedly with age (and concomitantly with length of exposure to the agent), and the overall frequency in a treatment group can depend as much on the proportion of animals surviving a long time as on the actual ability of the treatment to cause the condition. A simple, unstratified analysis ignoring differences in survival between treatments can therefore give a false impression as to the relative effect of treatment on the condition. Stratifying for time at death in the context of long term carcinogenicity studies is discussed at length by Peto *et al.* (1980). They point out that the way time of death should be used as a stratifying variable depends on the circumstances in which the tumour is seen. There are three different situations:

(a) *Visible tumours.* When the time of onset (ideally to some defined size) can be observed exactly, the experimental time period is divided into small time periods (usually of a week). Then within each period the 'number at risk' (referring back to the terminology used above) is defined by the number of animals alive and tumour-free at the beginning of the period while the 'number with condition present' is defined by the number for which a tumour was first seen during the period.

(b) *Fatal tumours.* Other tumours are only observable at necropsy, and in order to be able to analyse them properly one has to be able to distinguish between those which (directly or indirectly) caused the death of the animal and those which were seen only because the animal died of an unrelated cause. In the case of the former group, 'fatal tumours', the analysis is similar to that for visible tumours, with the 'number at risk' being the number alive at the beginning of the period and the 'number with condition present' being the number dying because of the tumour in the period.

(c) *Incidental tumours.* For tumours only observable because the animal died of an unrelated cause, the methodology is somewhat different because one does not know whether living animals have or do not have a tumour at any given point. Here the 'total number at risk' is defined as the number dying in the period, while the 'number with condition present' is the number dying with the tumour in the period. In order to maintain adequate numbers examined in each time period, longer intervals must be used, perhaps 20 weeks early in a 2-year rat study, reducing to 8 weeks later. Interim or terminal kills are treated as special time periods in this analysis.

While there can be problems in defining individual tumours as fatal or incidental, it is important to realise that the conclusions can, on occasion, depend critically on the definitions used, as is clearly demonstrated by Peto *et al.* (1980).

C   Ranked Data

Unstratified analysis

An appropriate method to use when comparing observations in two or more groups suitable for analysis by ranking methods is the *Kruskal–*

Wallis one-way analysis of variance. This test is equivalent in the two-group situation to the *Mann–Whitney U-test* or the *Wilcoxon rank sum test.* In this method the N observations in all the groups combined are ranked in a single series from 1 to N. Observations which are tied are given the average of the ranks, so that if there are four equal lowest observations each scores $(1 + 2 + 3 + 4)/4 = 2.5$. For each group the sum of the ranks R_j is then computed. From this a statistic H is calculated as follows:

$$H = \frac{12}{N(N+1)} \sum_{j=1}^{k} (R_j^2/N_j) - 3(N+1)$$

where N_j is the number of observations in group j.

Where, as is usually the case, there are tied observations a correction factor C is computed as follows:

$$C = 1 - \frac{\Sigma (t^3 - t)}{N^3 - N}$$

where t is the total number of observations in all groups combined with the same score and summation is over the number of scores where such ties occur.

The statistic H/C, or simply H if there are no (or just a few) ties, can be referred to tables of chi squared on $k - 1$ degrees of freedom, provided that the sample sizes are sufficiently large. Where sample sizes are small, reference may be made to tables giving the exact probabilities which are given in Iman *et al.* (1975).

As noted by Conover (1980), there should no longer be any hesitation to apply this rank test to situations that have many ties, the Kruskal–Wallis test being an excellent test to use in a contingency table, where the data consists of numbers of animals in each group with one of a limited range of values, *e.g.* when observations are scored on a five-point scale.

A measure of response in a group can be calculated by the difference, D_j, between the scaled rank sum in that group, $2R_j/N$, and its expected value, $N_j(N+1)/N$. Using this, a one degree of freedom chi-squared statistic for *trend* can be computed from the formula:

$$\chi^2 = T^2/\text{var } T$$

where the trend T is given by:

$$T = \sum_{j=1}^{k} d_j D_j$$

and the variance is given by:

$$\text{var } T = C(N+1)\left[N\left(\sum_{j=1}^{k} d_j^2 N_j\right) - \left(\sum_{j=1}^{k} d_j N_j\right)^2 \right] \Big/ 3N^2$$

Stratified Analysis

For some reason little attention has been given to the use of rank tests in the analysis of stratified data. However, there seems little reason why methods analogous to those for presence/absence data cannot be carried out. Thus a stratified test for *trend* can be calculated by accumulating T and var T over strata and referring:

$$\chi^2 = (\Sigma\ T)^2/\Sigma\ \text{var}\ T$$

to tables of chi squared with one degree of freedom. In the two-group situation also, noting that within a stratum:

$$\text{var}\ D_j = C(N+1)(N-N_j)N_j/3N^2$$

one can refer:

$$\chi^2 = (\Sigma\ D_j)^2/\Sigma\ \text{var}\ D_j$$

to tables of chi squared with one degree of freedom as an overall test of difference between treated and control groups.

As for presence/absence data, a stratified test for homogeneity involves matrix manipulation and is beyond the scope of this chapter.

EXAMPLE 6

Times to death (in hours) in a rat study of three dose levels of an antidote to a given dose of toxin might be recorded as follows:

3 mg kg^{-1}	2 mg kg^{-1}	1 mg kg^{-1}
18, 48, 22, 5, 14, 48	10, 2, 1, 14, 14, 48	12, 1, 4, 4, 3, 8

To analyse these data by the Kruskal–Wallis one-way analysis of variance the times are first replaced by their ranks yielding:

3 mg kg^{-1}	2 mg kg^{-1}	1 mg kg^{-1}
14, 17, 15, 7, 12, 17	9, 3, $1\frac{1}{2}$, 12, 12, 17	10, $1\frac{1}{2}$, $5\frac{1}{2}$, $5\frac{1}{2}$, 4, 8

Totalling these gives rank totals of 82, $54\frac{1}{2}$, and $34\frac{1}{2}$. As a check, the overall total of ranks should be computed to see that it equals $N(N+1)/2 = 18(19)/2 = 171$, which it does. H is then calculated as follows:

$$H = \frac{12}{18(19)} \left[\frac{(82)^2}{6} + \frac{(54\frac{1}{2})^2}{6} + \frac{(34\frac{1}{2})^2}{6} \right] - 3(19) = 6.65$$

Noting that there are 2 rats with a tied time of 1 hour, 2 with a tied time of 4 hours, 3 with a tied time of 14 hours, and 3 with a tied time of 48 hours, the correction factor C is calculated:

$$C = 1 - \frac{(2^3 - 2) + (2^3 - 2) + (3^3 - 3) + (3^3 - 3)}{18^3 - 18} = 0.9897$$

An overall test for *heterogeneity* is therefore given by referring $H/C = 6.72$ to tables of chi squared on 2 degrees of freedom, which is significant at the 95% confidence level.

Testing for *trend*:

$$D_1 = \frac{2 \times 82}{18} - \frac{6 \times 14}{18} = 2.7778, \; D_2 = -0.2778, \; D_3 = -2.50$$

$$T = (3 \times 2.7778) + (2 \times -0.2778) + (1 \times -2.50) = 5.2778$$

$$\text{var } T = \frac{0.9897 \times 19}{3 \times 18 \times 18} \times [18[(3^2 \times 6) + (2^2 \times 6) + (1^2 \times 6)] - (18 + 12 + 6)^2]$$

$$= 4.1787$$

$$\chi^2 = (5.2778)^2 / 4.1787 = 6.666$$

which, on 1 degree of freedom, is significant at the 99% confidence level.

D Continuous Data

Assumptions

Whereas analysis of presence/absence or ranked data makes no assumption concerning the frequency distribution of the underlying data, analysis of continuous data does of necessity involve some assumptions. It is not the intention here to discuss the various sorts of distribution mentioned in the literature. Suffice it to say much biological data can be analysed by methods which assume that the observations in each group are normally distributed with a common variance. The methods outlined below are all based on these assumptions but are 'robust' in the sense that they are not highly contingent on minor departures from them.

Outliers

One common situation where the assumptions are not met is where there are 'outliers', *i.e.* one or two values in the data which are very different from the rest. While these have relatively little effect on the results for presence/absence or ranked data analysis, a single outlier can dramatically distort the conclusions when continuous data are analysed, not only by affecting the mean value of the group in which the outlier occurs but also by increasing the overall estimate of the variability of the data. This makes it important to screen the data with the possibility of outliers in mind. Choice of an appropriate formal test for outliers is a difficult subject to which Barnett and Lewis (1978) have devoted a whole book. Of the tests they discuss, that based on the sample kurtosis is one of the most useful, and the interested reader should refer to the book for the methodology.

Homogeneity of Variance

The second common situation where the assumptions are not met relates to the situation where the larger the value, the more inherently variable it tends to be. Thus animals in one group, with a mean response of 100, may have values varying in the range 80–120, while animals in another group, with a mean response of 300, may have values varying over the wider range of 240–360. Where the standard deviation of the data is approximately directly proportional to the mean, it is appropriate to take logarithms of the individual data items before carrying out the analysis. This not only makes the variability independent of the mean but also tends to ensure that the distribution of $\log x$ more closely resembles the symmetrical bell-shaped normal distribution than does the distribution of x. Where the variability increases with the mean but not so markedly, a square-root rather than a logarithmic transformation may tend to be better in making the variability more independent of the mean. A formal test as to whether results in different groups are equally variable, often referred to as a test of homogeneity of variance or of homoscedasticity, is Bartlett's test. Where, even after logarithmic or square-root transformation, there is still evidence of marked heterogeneity of variance, it will often be preferable to use methods for ranked data as described in Section 4C. However, it should be noted that it is in general preferable to use the same method of analysis when analysing similar data at different times, *e.g.* routine blood chemistry measurements. This is partly to make comparisons of different data sets easy and partly because whether a variable, or its logarithm or square root, is distributed normally is to some extent a property of an underlying biological process and therefore likely to be inherent. If experience suggests a variable is generally normally distributed, it is not justified to take logarithms just because on one specific occasion it happens to 'cure' significant heterogeneity of variance shown up in Bartlett's test. Remember that some significant results are expected by chance and that the analysis of variance procedures to be described below are robust against slight departures from the assumptions.

Bartlett's Test

Suppose we have k groups of animals and within each group of animals there are N_i observations x_{ij}, the subscript i referring to group and j to observations within group.

In order to test for homogeneity of variance, the procedure is as follows. For each group calculate the sum of the observations:

$$T_i = \sum_j x_{ij}$$

the sum of squares of the observations:

$$S_i = \sum_j x_{ij}^2$$

and hence the within-group variance:

$$\sigma^2 = \frac{N_i S_i - (T_i)^2}{N_i(N_i - 1)}$$

Defining the degrees of freedom in each group by $df_i = N_i - 1$, the statistic:

$$\chi^2 = \frac{\left(\sum_i df_i\right) \ln\left(\frac{\sum_i df_i \sigma_i^2}{\sum_i df_i}\right) - \sum_i (df_i \ln \sigma_i^2)}{1 + \frac{1}{3(k-1)}\left[\sum_i\left(\frac{1}{df_i}\right) - \frac{1}{\sum_i df_i}\right]}$$

can then be looked up in tables of the chi-squared distribution on $k - 1$ degrees of freedom. Instead of using logarithms to base e (natural logarithms), logarithms to base 10 may also be used in the above formula, the resultant statistic then being multiplied by 2.306 (see Example 7).

EXAMPLE 7

In a study of 18 rats, body weights at one point in time were recorded as follows:

Group 1 ($N = 5$) 124, 161, 143, 182, 202
Group 2 ($N = 6$) 182, 304, 194, 262, 135, 188
Group 3 ($N = 7$) 221, 204, 197, 283, 230, 250, 270

From these data calculate:

	df_i	T_i	S_i	σ_i^2	$df_i\sigma_i^2$
Group 1	4	812	135,674	951.30	3805.20
Group 2	5	1265	285,389	3736.97	18684.83
Group 3	6	1655	397,655	1060.95	6365.71
Total	15				28855.74

$$\sum_i (df_i \ln \sigma_i^2) = 110.363$$

$$\sum_i\left(\frac{1}{df_i}\right) = \tfrac{1}{4} + \tfrac{1}{5} + \tfrac{1}{6} = 0.61667$$

substituting into the formula:

$$\chi^2 = \frac{15 \ln\left(\frac{28855.74}{15}\right) - 110.363}{1 + \frac{1}{3 \times 2}(0.61667 - 1/15)} = \frac{3.0672}{1 + 0.09167} = 2.81$$

Referring this to tables of chi squared on 2 degrees of freedom, it can be seen that this is not significant even at the 90% confidence level, and the variances can therefore be accepted as homogeneous.

One-way Analysis of Variance

Given that there is no significant heterogeneity of variance, one-way analysis of variance can then be used to test whether there is evidence of any difference between the group means. Using the same notation as before, T_i, S_i, and σ_i^2 are calculated as for Bartlett's test and the group means are calculated by:

$$\mu_i = T_i/N_i$$

Then, summing over groups calculate the total number of observations:

$$N = \sum_i N_i$$

the sum of the observations:

$$T = \sum_i T_i$$

the sum of squares of the observations:

$$S = \sum_i S_i$$

and the sum of squares of the group total divided by the number of observations:

$$G = \sum_i (T_i^2/N_i)$$

Then calculate the 'correction for the mean':

$$C = T^2/N$$

and form the analysis of variance table as follows:

Source of variation	Degrees of freedom	Sum of squares	Mean square
Between groups	$k - 1$	$G - C$	$(G - C)/(k - 1)$
Within groups	$N - k$	$S - G$	$(S - G)/(N - k)$

The ratio (between-groups mean square)/(within-groups mean square) is then distributed on the null hypothesis as an F-statistic on $k - 1$ and $N - k$ degrees of freedom and the associated probability can be looked up using standard statistical tables. When there are only two groups ($k - 1 = 1$) the square root of this statistic can also be looked up in tables of Student's t-statistic on $N - k$ degrees of freedom—this is equivalent to the well known 't-test'.

When there are data from more than two groups, but results from two particular groups (i_1 and i_2) are to be compared, the t-test can be used, involving data from these groups only. In general, however, it is better to use the following formula:

$$t = MD/\sqrt{VD}$$

where MD is the difference between the two means and VD, the variance of the difference, is given by:

$$VD = \left(\frac{1}{N_{i_1}} + \frac{1}{N_{i_2}}\right)\frac{S - G}{N - k}$$

as this uses information about variability of the data from other groups.

Given that dose levels, d_j, are applied in each group, and given an observed difference, D_j, between the mean in each group and the overall mean, an approximate one degree of freedom Student's t-statistic testing for trend can be calculated by the formula:

$$T = \sum_{j=1}^{k} d_j N_j D_j$$

and its variance is given by:

$$\text{var } T = \frac{S - G}{N(N - k)} \left[N \sum_j (d_j^2 N_i) - \left(\sum_j d_j N_j \right)^2 \right]$$

Alternatively, linear regression analysis (see, for example, Johnson and Leone, 1964) may be used to test for a linear relationship between dose and response. See Example 8 for an example of a one-way analysis of variance with test for trend.

EXAMPLE 8

To compare percentage body weight changes over a defined period in a study in 4 groups of dogs, the following calculations would be carried out. First, based on the individual data items, calculate:

	Control	50 p.p.m.	100 p.p.m.	150 p.p.m.
	21.2	19.5	17.4	13.4
	22.8	23.1	19.0	16.5
	23.5	22.0	20.2	18.1
	20.7	18.4	16.5	12.3
			18.9	13.4
N_i	4	4	5	5
T_i	88.2	83.0	92.0	73.7
μ_i	22.05	20.75	18.40	14.74

and summing over groups: $N = 18$, $T = 336.9$
and hence the grand mean: $\mu = 18.717$
and the correction factor: $C = (336.9)^2/18 = 6305.645$

$$G = \frac{(88.2)^2}{4} + \frac{(83.0)^2}{4} + \frac{(92.0)^2}{5} + \frac{(73.7)^2}{5} = 6446.198$$

Next, based on the individual data items squared, calculate

	Control	50 p.p.m.	100 p.p.m.	150 p.p.m.
	449.44	380.25	302.76	179.56
	519.84	533.61	361.00	272.25
	552.25	484.00	408.04	327.61
	428.49	338.56	272.25	151.29
			357.21	179.56
S_i	1950.02	1736.42	1701.26	1110.27

and hence, summing over groups, $S = 6497.97$

Source of variation	Degrees of freedom	Sum of squares	Mean square	F
Between groups	3	140.553	46.851	12.67
Within groups	14	51.772	3.698	

Using tables of the F-distribution on 3 and 14 degrees of freedom the value of 12.67 obtained can be seen to be significant at the 99.9% confidence level. There is thus evidence of variation between the group means.

An idea of where the variation comes from can be seen by comparing each group mean with that for the controls, using the overall estimate of within group variance (mean square). The difference between high dose mean and the control is $(14.74 - 22.05) = -7.31$ and its standard error is given by $[(\frac{1}{4} + \frac{1}{5})3.698]^{\frac{1}{2}} = 1 \cdot 290$ yielding a t-value of -5.67 significant at the 99.9% confidence level. Thus a table can be set up:

Dose level	Difference from control Mean	Standard error	t
50 p.p.m.	-1.30	1.360	-0.96
100 p.p.m.	-3.65	1.290	-2.83
150 p.p.m.	-7.31	1.290	-5.67

Alternatively a test for trend can be carried out. Taking d_j as dose/50 for convenience, calculate:

Group	N_j	d_j	D_j	d_jN_j	$d_j^2N_j$	$d_jN_jD_j$
Control	4	0	$+3.333$	0	0	0
30 p.p.m.	4	1	$+2.033$	4	4	$+8.133$
100 p.p.m.	5	2	-0.317	10	20	-3.167
150 p.p.m.	5	3	-3.977	15	45	-59.650
Total				29	69	-54.683

$$T = -54.683$$

$$\text{var } T = \frac{3.698}{18}[(18 \times 69) - (29 \times 29)] = 82.383$$

$$\chi^2 = T^2/\text{var } T = 36.30 \text{ on 1 degree of freedom}$$

which is again significant at the 99.9% confidence level.

Stratified Analysis

Though, when analysing continuous data that is stratified, it is possible to devise global tests for homogeneity and trend in a manner similar to that used for presence/absence and ranked data, *i.e.* combining results of independent analyses carried out in each stratum, this technique is not normally employed in this situation. Where, as is often the case, variability of response is similar in different strata, it is more efficient, especially where numbers of observations may be small in particular strata, to use a method of analysis in which an overall estimate is made of variability rather than one in which separate estimates are made with strata. This brings us into the area of more complex analysis of variance, or more generally into the area of multiple regression and linear models, details of which are beyond the scope of this chapter but are discussed in numerous statistical textbooks. These methods normally require extensive calculations and the user will need the availability of appropriate software such as that provided by the BMD statistical package or the

General Linear Interactive Modelling (GLIM) program of Baker and Nelder (1978).

E   Other Methods

Relationships Between Two Variables

In the methods described in Sections 4B–4D interest has centred on whether there is evidence of variation between groups in respect of a single variable. When other variables have appeared, these have only been nuisance variables to be taken account of by stratification. In many situations, however, interest is centred on relationships between two variables. For a single group, relationships between two variables can be investigated by a 2×2 *chi-squared test* or *Fisher's exact test* in the case of presence/absence data. In the case of ranked data *Spearman's rank correlation coefficient* can be used. This is calculated by:

$$r = 1 - \frac{6\sum_{i=1}^{n} D_i^2}{n^3 - n}$$

where n animals are ranked separately in respect of two variables and D_i is the difference between the ranks for the ith animal. In the case of continuous data the *product–moment correlation coefficient* can be used. This is calculated by:

$$r = \frac{\sum_{i=1}^{n} x_i y_i}{(\sum_{i=1}^{n} x_i^2 \sum_{i=1}^{n} y_i^2)^{\frac{1}{2}}}$$

where x_i and y_i are the values of the two variables for the ith animal. For both the Spearman rank and the product–moment correlation coefficient observed values must lie between $+1$ (perfect positive relationship) and -1 (perfect negative relationship) with a value of 0 implying no association between the variables. Associated probability values for given values of r and n can be looked up in standard tables (*e.g.* Beyer, 1966).

Techniques are also available for determining whether these relationships differ for different treatment groups. Breslow and Day (1980) described methods appropriate for testing constancy of association over multiple 2×2 tables, whilst analysis of covariance (Bennett and Franklin, 1954) can be used for continuous data.

Multivariate Methods

Methods in which the inter-relationship of a large number of variables are studied will normally require the advice of a professional statistician and are not considered here in detail. Such techniques are most commonly applied to continuous data and include the following:

(a) *Multiple regression analysis.* In this, variation in a dependent variable, y, is related simultaneously to the values of a number of known predictor variables, x_k, and the significance of each is assessed.
(b) *Discriminant analysis.* In this, an individual is classified into one of g groups, depending on the values recorded of each of a number of known variables, x_k. The relative importance of the variables in discriminating individuals can be assessed.
(c) *Component analysis.* In this, the major part of the variation in a dependent variable, y, can be related to a reduced number of variables, formed from the original set of known variables, x_k. This technique is closely related to *factor analysis.* Where a number of variables are closely related, perhaps by having a common underlying cause, these techniques can simplify complex data, though in some cases the results can be difficult to interpret.

Multiple Observations on the Same Animal

For body weight and clinical-chemistry data it is common to take measurements on the same animals at regular intervals throughout a long term study. Clearly, data at any individual time point can be considered and a between-group comparison undertaken, but this only gives a limited amount of information in any one analysis. Often the interest is in how the measurements are changing rather than in their absolute level, as in the case of weight gain. An intuitively obvious approach is to use differences in measurements at two defined time points, but this is in fact not the best method. The problem is that, for any variable measured with some error, low values measured at one time point will tend to increase and high values decrease simply because of the 'regression to the mean' effect, even when no true change occurs. It follows that if there is group-to-group variation between the values measured at the first time point, estimation of between group differences in the change that has occurred owing to treatment will be contaminated by this effect. The correct technique to use is analysis of covariance where differences in mean values at the second time point after adjusting for the values observed at the first time point are sought.

In some circumstances a single analysis involving the data measured at all the time points may be worth attempting. If, for example, the growth curve for an animal can be expressed in terms of an equation depending on a limited number of parameters, analyses comparing the groups in respect of these parameters can be informative. However, such techniques tend to be complex.

Paired Data

It is not uncommon for an experiment to consist of a number of pairs of animals, one animal of each pair being a control and one a treated animal. Thus, to minimise the effects of between-litter variation, it can be advantageous to treat pairs of animals from the same litter in this way. Where the criterion on which pairing is carried out can affect the response, it is important to use appropriate statistical methods that take the pairing into account.

For presence/absence data the appropriate method to use is *McNemar's test,* in which the number of pairs in which the event occurs in the test animal and not in the control animal (N_1) is compared with the number of pairs in which the event occurs in the control animal and not in the test animal (N_2), the statistic:

$$\chi^2 = \frac{(|N_1 - N_2| - 1)^2}{N_1 + N_2}$$

being approximately distributed as chi squared on 1 degree of freedom. The same formula can also be used for continuous or ranked data where N_1 is the number of pairs in which the observation was greater in the test than in the control animal and N_2 the reverse. The exact probability associated with the observation can also be calculated, when the test is known as the *sign test.* For numbers of pairs up to 24, this is available in tabular form (*e.g.* Siegel, 1956).

This test uses information only about the direction of the differences within pairs. A technique which uses information about the relative magnitude as well as the direction of the difference is the *Wilcoxon matched-pair signed-ranks test.* In this method the difference in value between each pair is calculated (d_i) and these differences are then ranked regardless of sign (r_i). Then to each rank is affixed the sign of the difference to form r_i. Finally, the sums of the positive and the negative ranks are formed, which under the null hypothesis are equal to $N(N+1)/4$, the average of the ranks (N being the total number of *non-zero* differences). The smaller of these, T, is then either tested by taking:

$$\chi^2 = \frac{24[T - N(N+1)/4]^2}{N(N+1)(2N+1)}$$

as an approximate chi-squared distribution or, in the case of less than 25 pairs, by looking up the exact probability in tables (*e.g.* Siegel, 1956).

Alternatively, when the data are normally distributed, the matched-pair *t*-test can be used. Here the differences are used as raw data and their mean (μ) and variance (σ^2) calculated as in Section 4D. μ/σ is then looked up in tables of the *t*-distribution.

Probit Analysis

The objective of some toxicological experiments is to calculate the dose of a substance at which 50% of the animals will respond. This is known as the LD_{50} when the response is death (LD = lethal dose) and, more generally, as the median-effective dose. Typical data consist of results at a range of dose levels in which the number of animals exposed and the number responding are given. A single plot of dose against percentage response typically gives a relationship in the form of an elongated S. By plotting the logarithm of dose on the x-axis and the 'probit' of response on the y-axis, as can be easily done with special graph paper, this relationship usually becomes approximately linear and it is reasonably easy to get a rough estimate of the median-effective dose. The probit is defined by $Y + 5$, where Y is the normal equivalent deviate which is related to the proportion responding (P) by the formula:

$$P = \int_{-\infty}^{Y} \frac{1}{\sqrt{2\pi}} e^{-\frac{1}{2}x^2} \, dx$$

Techniques for formal estimation of the median-effective dose and its 95% confidence limits are laborious by hand. The reader is referred to Finney (1977) or for a more rapid, reasonably accurate procedure to Gad and Weil (1982).

Some Further Reading

This chapter has not given a detailed discussion of all the types of observation that the toxicologist may have to deal with and the appropriate methods for each. The interested reader may refer to Gad and Weil (1982) for some useful material, covering *inter alia* body and organ weights, clinical chemistry, haematology, reproduction, teratology, dominant lethal assay, and mutagenesis. For further details on methods for analysis of presence/absence data, Breslow and Day (1980) is useful, while both Siegel (1956) and Conover (1980) are valuable for the analysis of non-parametric data including rank methods. Many books are available on analysis of continuous data; two of these, Johnson and Leone (1964) and Bennett and Franklin (1954), though both rather old, are very clear. All these references contain numerous worked examples.

References

Armitage, P. (1955). Test for linear trend in proportions and frequencies. *Biostatistics*, **11**, 375–386.

Armitage, P. (1971). 'Statistical Methods in Medical Research'. Blackwell Scientific Publications, Oxford.

Baker, R. J. and Nelder, J. A. (1978). 'Generalised Linear Interactive Modelling'. Release 3. Numerical Algorithms Group, Oxford.

Barnett, V. and Lewis, T. (1978). 'Outliers in Statistical Data'. John Wiley and Sons, New York.

Bennett, C. A. and Franklin, N. L. (1954). 'Statistical Analysis in Chemistry and the Chemical Industry'. John Wiley and Sons, New York.

Beyer, W. H. (*Ed.*) (1966). 'CRC Handbook of Tables for Probability and Statistics'. The Chemical Rubber Company, Cleveland, Ohio.

Breslow, N. E. and Day, N. E. (1980). 'Statistical Methods in Cancer Research. Volume 1—The Analysis of Case-Control Studies'. IARC Scient. Publ. No. 32. International Agency for Research on Cancer, Lyon.

Conover, W. J. (1980). 'Practical Nonparametric Statistics'. 2nd Edn. John Wiley and Sons, New York.

Feron, V. J. *et al.* (1980). Basic requirements for long-term assays for carcinogenicity. *In*: 'Long-term and Short-term Screening Assays for Carcinogens: A Critical Appraisal'. IARC Monographs on the Evaluation of the Carcinogenic Risk of Chemicals to Humans. Suppl. 2. International Agency for Research on Cancer, Lyon.

Finney, D. J. (1977). 'Probit Analysis'. 3rd Edn. Cambridge University Press, Cambridge, England.

Gad, S. C. and Weil, C. S. (1982). Statistics for toxicologists. *In*: 'Principles and Methods of Toxicology'. *Ed.* A. W. Hayes. Raven Press, New York.

Iman, R. L., Quade, D., and Alexander, A. (1975). Exact probability levels for the Kruskal–Wallis test statistic. *Selected Tables in Mathematical Statistics,* **3,** 329–384 (5.2 Appendix).

Johnson, N. L. and Leone, F. C. (1964). 'Statistics and Experimental Design in Engineering and the Physical Sciences'. Vol. 1. John Wiley and Sons, New York.

Mantel, N. (1963). Chi-squared tests with one degree of freedom: extensions of the Mantel-Haenszel procedure. *J. Am. Stat. Assoc.,* **58,** 690–700.

Mantel, N. and Greenhouse, S. W. (1968). What is the continuity correction? *Am. Stat.,* **22,** 27–30.

Marriott, F. H. C. (1990). 'A Dictionary of Statistical Terms'. 5th Edn. Longman Scientific and Technical, Harlow.

Peto, R. *et al.* (1980). Guidelines for simple, sensitive significance tests for carcinogenic effects in long term animal experiments. *In*: 'Long-term and Short-term Screening Assays for Carcinogens: A Critical Appraisal'. IARC Monographs on the Evaluation of the Carcinogenic Risks of Chemicals to Humans. Suppl. 2. International Agency for Research on Cancer, Lyon.

Siegel, S. (1956). 'Nonparametric Statistics for the Behavioural Sciences'. McGraw-Hill Book Co., New York.

CHAPTER 19

Risk Assessment of Chemicals

DAVID P. LOVELL

1 Introduction

The basic objective of experimental toxicology is the identification and understanding of the properties of chemicals which are detrimental to the human population. These effects may be seen directly as adverse consequences on health, and indirectly through damage to economically important crops, livestock, the wider environment, and ecosystems. Such an objective is, of course, ambitious. Chemicals have a spectrum of effects, some considered beneficial to the human population, others deleterious. For instance, a pesticide may kill mosquitoes and reduce the burden of malaria but may leave residues which some consider unwelcome contaminants of food; or a drug may protect against heart disease but may cause damage to the patient's kidneys or eyes.

The procedures for determining the potential benefits and side effects of exposure to chemicals in the environment form the basis of *risk assessment*.

A What is Risk Assessment?

Unfortunately for a young discipline like risk assessment, the word 'risk' has a number of unspecific meanings. A more precise definition is necessary otherwise concepts such as risk estimation become nebulous.

A clear distinction is made between the words 'risk' and 'hazard'. This is not so in everyday speech where they are synonymous with chance, danger, peril, and related to concepts like harm.

The definitions below derive from those developed by a Royal Society Study Group (1983) and have subsequently been used by Lovell (1986) and Richardson (1986) in reviews of the Toxic Hazards of Chemicals.

B Definitions of Hazard, Risk, and Risk Assessment

A *hazard* is the circumstance where a particular incident could lead to harm, where harm is death, injury, or loss to an individual or society.

A more specific definition relevant to chemicals is the set of inherent properties of a chemical substance or mixture which makes it capable of causing adverse effects in man or the environment when a particular degree of exposure occurs.

Risk is the chance that a particular adverse effect actually occurs in a particular time period or after a specific challenge.

Relating this to chemicals: risk is the predicted or actual frequency of occurrence of an adverse effect of a chemical substance or mixture from a given exposure to humans or the environment.

Some examples should clarify the distinction. Ionising radiation is a known hazard to the human population causing harm in the form of cell death, genetic damage, and cancer in exposed individuals. The risk of such injuries actually occurring depends upon factors such as the level of exposure and individual susceptibility. Similarly, from epidemiological investigations, medical case reports, and experiments using volunteers, many chemicals are known to be toxic [for instance, TCDD (dioxin), methylisocyanate, and vinyl chloride] in various ways and degrees.

The risk that individuals will suffer toxic effects from exposure to such chemicals depends upon a range of variables. Hazard is therefore the potential for harm, risk the chance that harm will actually occur. Determining the chance, or probability, that harm will occur and the subsequent consequences form the basis of Risk Assessment.

Risk assessment is defined as the process of decision making applied to problems where there are a variety of possible outcomes and it is uncertain which event will happen. Risk assessment can be broken down into stages (Table 1). The problem has to be defined and the different risks estimated. These risks are evaluated in relation to other factors or interests involved and choices are made between the possible course of action. The total process of hazard identification, risk estimation, risk evaluation, decision making, and implementation makes up what is called risk assessment.

This definition is not the only one that has been developed. Confusion can arise because not all definitions use the same names or categories. However, the general ideas and principles are similar. For instance, the

Table 1 *The stages of risk assessment*

Hazard Identification
Risk Estimation
Risk Evaluation
Deciding on Acceptable Risk
Managing the Risk

US Environmental Protection Agency (EPA) have produced guidelines for risk assessment for carcinogenicity and mutagenicity. They make a clear demarcation between what they call risk assessment and risk management. The former consists of hazard identification, dose–response assessment, exposure assessment and risk characterisation; it defines the adverse health consequence using scientific methods. Risk management combines technical, socioeconomic, political, and other considerations in a decision making process (EPA, 1986).

This chapter will mainly concentrate on the methods available to the toxicologist to assist in risk estimation. The determination of toxic effects and identification of hazards is dealt with in other chapters in this book, while the more general aspects of evaluation, decision making and risk management are reviewed elsewhere (Lovell, 1986).

2   Estimating the Risks

There are four broad areas from which toxicological data can be used for estimating risk: human studies, theoretical considerations, short term testing, and animal testing.

A   Human Studies

Measures of the consequences of exposure of the human population to chemicals provide the most direct information available for risk estimation. Such exposures may be unintentional, such as a result of an industrial accident; deliberate, as in the case of a volunteer study; or a consequence of incidental exposure during the course of normal activity.

Risk estimates are based upon a calibration of the observed ill-health associated with an estimate of the exposure to the chemical to provide a dose–response relationship. Such studies are limited for chemicals but a major 'model' for the type of investigation is the prediction of the effects of radiation from various accidental or deliberate exposures. A major problem in studies of chemicals is obtaining reliable estimates of the exposures received under non-experimental conditons.

Estimates of the risks associated with exposure to a chemical can be drawn up based upon the results of these studies (sometimes in conjunction with the results of animal studies). They are then used to help manage the risks by the setting of occupational exposure levels such as 'Threshold Limit Values' (TLV).

These levels are often based upon the assumption that the effects of long term, low level exposure can be estimated from the results of short term, high concentration exposure. The main limitation with human studies are that they are, in general, retrospective as exposure has already occurred. Epidemiological studies of possible carcinogenic effects are particularly limited by the low statistical power of the designs, which make it difficult to detect any effects other than increases in rare diseases, the long term nature of the studies, and the consequences of the sub-group being exposed to a chemical before the risks associated with it can be evaluated.

B Theoretical Approaches and Short Term Tests

There is an increasing interest in the use of surrogates of either human exposure or animal studies to try to identify toxicological hazards. The results of such methods could then be used in the risk assessment process. Theoretical approaches include molecular modelling and quantitative structure–activity relationships (QSAR). Compounds with specific chemical structures which may pose a hazard can either be considered to need specific further testing or rejected as possible therapeutic or industrial substances.

Similarly, short term tests for toxicological activity such as potential genotoxicity or teratogenicity are used to try to predict potential hazards to the human population. The results of these tests also determine whether future testing is necessary, or whether the tested product will be developed further.

Theoretical approaches and short term tests are mainly of value in hazard assessment because of the difficulties of obtaining quantitative measures of risk applicable to man. This limits the replacement of animal studies by short term *in vitro* testing, which may be desirable on ethical grounds. Consequently, short tests are mainly used to supplement the results of animal and human studies.

C Animal Studies

Most new chemicals likely to be introduced into the environment have no relevant data on human exposure for assessing risks. Risk estimation is therefore based upon animal studies. These are also used to augment the results from studies in humans of chemicals already present in the environment.

Animal studies are characterised by an attempt to identify compound-related toxic effects following the administration of high doses of the compound to small numbers of animals. Such studies have two major complications when used for risk assessment. Firstly, the difficulty of estimating an effect at a low dose from the effects observed at high doses (a complication also of studies based on accidental exposures of groups of humans) and secondly, the transfer of conclusions from one species, often a rodent, to another, man.

The choice of high doses and low numbers of animals is forced on toxicologists because the obvious experiment of testing for effects at doses comparable with likely human exposures would require very large numbers of animals. Such experiments would be very expensive (the standard US National Toxicology Program assay using 800 animals has been costed at $1,000,000 at 1986 prices, Lave and Omenn, 1986). There are also increasing ethical objections to using large numbers of animals and such studies would probably be impractical to conduct. The largest experiment of this type carried out was, the ED_{01} (colloquially called the 'megamouse' experiment) study. This was a considerable strain on the resources and facilities of the US National Center for Toxicological Research.

Even with a design which has relatively large numbers of animals and high doses, the statistical power, or ability to detect successfully a true effect, of the experiment is limited. Figure 1 illustrates the probability of finding a condition induced by the test chemical at frequencies of between 1 and 5% for various different sample sizes. Over a third of samples of 100 animals would not include an animal with a particular condition even when the induced incidence was 1%. This incidence is considerably higher than many adverse drug reactions noted in patients who had been administered pharmaceutical products.

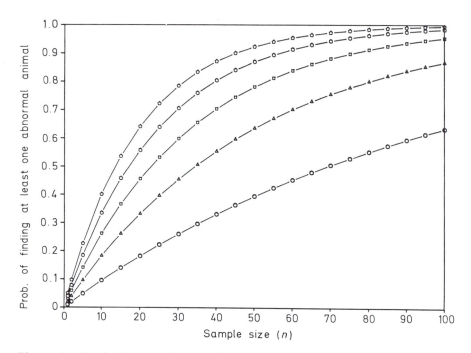

Figure 1 *Graph showing the probability of one or more animals carrying an abnormality (such as a tumour) being found in a group of animals for various samples sizes* (n) *and background incidences. The lines show the probabilities associated with background incidences of* 1% (O), 2% (△), 3% (□), 4% (◯) *and* 5% (□). *Note that with a sample size of 100 over a third (probability of* >0.33) *of samples will not contain a tumour-carrying animal when the background incidence is* 1%, *while over a tenth will not contain a tumour-carrying animal when the background incidence is* 2%

The use of high doses also has complications. The methods of deactivation occurring at high doses may be different from those operating at lower, more realistic dose levels while tissue distribution and pharmacokinetics may also vary. Criticism is particularly strong of the high dose levels used in the long term rodent carcinogenicity bioassay where the highest dose is called the maximum tolerated dose (MTD). This dose is based upon previous smaller experiments and is chosen so that no lethality is produced and body weight in the treated group is not reduced by more than 10% of the weight of the control animals. Critics have argued that toxic effects, particularly tumours, may be artefacts based upon chronic administration of the compound in high doses. This results in metabolic overload and pathological changes resulting from chronic repair processes.

A further complication of animal studies is that the route of administration may be different from how the human population is exposed. Dosing animals in a way that mimics human exposure is not always practicable and much of the safety evaluation of chemicals is done using either intraperitoneal or oral administration of the compound.

3 Approaches to Risk Estimation from Animal Studies

Two general approaches have been developed to use animal data to estimate the risk to human populations of a chemical.

A Safety Factor Approach

This derives from methods developed in the 1950s when an attempt was made to provide a legislative framework for the control of food additives. The Delaney amendment passed by Congress in the the USA, which states that no substance carcinogenic to animals should be deliberately added to human food, resulted in the application of safety factor methods only for non-carcinogenic findings. The most widely used method is the determination of an acceptable (sometimes referred to as allowable) daily intake (ADI) of a compound. This is obtained by dividing a value called the No Observable Effect Level (NOEL) by a safety factor (SF), *i.e.*

$$\text{ADI} = \frac{\text{NOEL}}{\text{SF}}$$

The NOEL is obtained from a study where groups of animals are exposed to different doses of the compound. The NOEL is the highest dose at which no observable effects are detected. The safety factor is traditionally, but not necessarily, 100. This is supposed to reflect the possibility that man may be 10-fold more sensitive than the test species and that there may be a 10-fold range in susceptibility within the human population.

The NOEL is not as objective a value as it might first appear as the following two descriptions quoted by Crump (1986) show. The first is a circular argument with NOEL defined 'as the level at which no effects were observed', while the second, used by the US Environmental Protection Agency (EPA), is 'the exposure level at which there are no statistically significant increases in frequency or severity of effects between the exposed population and its appropriate control'. Defining whether a statistically significant increase is a biologically important finding is a subjective decision partly determined by the choice of statistical test used. The NOEL could, for instance, be affected by the sample sizes at each dose. Table 2 illustrates how smaller sample sizes can produce apparently higher doses which are the NOEL. One objection, therefore, has been that the investigator who uses insensitive designs with too few animals is 'rewarded' with a higher safe dose.

In practice, the results of the statistical analysis of such tests is only one input into the assessment of the results. Other biological and toxicological considerations are considered in the evaluation. In this way the decision on the NOEL is based upon expert but nevertheless subjective decisions.

Use of a NOEL might seem to give support to the concept that there is a threshold dose, below which a chemical will not produce toxic effects. However, increasing the number of animals sampled at each dose and improving the sensitivity of the methods for detecting toxicity may find effects at lower doses. The choice of the NOEL is thus a reflection of relative rather than absolute safety. Others have suggested it should be called the no adverse effect level or that the lowest dose at which effects occur should be used in conjunction with a larger safety factor.

The size of the safety factor is also debated. Proposals have been made

Table 2 *Effect of varying sample size on the determination of a NOEL using a statistical test*

Data Set 1

	Control	Low	Medium	High	NOEL
A	$\frac{0}{10}$	$\frac{3}{10}$	$\frac{6}{10}$	$\frac{9}{10}$	Low
		NS	**	***	
B	$\frac{0}{20}$	$\frac{6}{20}$	$\frac{12}{20}$	$\frac{18}{20}$	None
		*	***	***	

Data Set 2

A	$\frac{0}{10}$	$\frac{1}{10}$	$\frac{2}{10}$	$\frac{3}{10}$	High
		NS	NS	NS	
B	$\frac{0}{20}$	$\frac{2}{20}$	$\frac{4}{20}$	$\frac{6}{20}$	Medium
		NS	NS	*	

Comparison between dose levels and control by one-sided Fisher exact test (NS Not significant; $*P < 0.05$; $**P < 0.01$; $***P < 0.001$).

Note that in both data sets results A are based upon sample sizes of 10 and results B on sample sizes of 20. The proportion of affected animals is the same in both A and B at each dose level with each data set.

for a factor of 10 to be used when chronic human exposure data are available, or a factor of a 1000 when there is no good human data (Hogan and Hoel, 1982). Weil (1972) has proposed that a safety factor of 5000 could be used when carcinogenic effects were found in the experiment.

Proponents of the safety factor approach argue that the method is satisfactory despite the lack of a real biological justification for a particular safety factor. They contend that the method has an added safety factor because the true dose level at which toxic effects would occur in practice may be considerably greater than the NOEL.

The approach incorporates both the estimation of low dose effects and the inter-species comparison in a single stage. However, the failure to take dose–response relationships into account, the arbitrary nature of the safety factor, and its unsuitability for the risk assessment of carcinogens have stimulated other approaches.

B Mathematical Modelling

The second approach has been to attempt to model mathematically the dose–response relationship in toxicological data. The aim is to try to provide quantitative risk estimates (this is confusingly called quantitative risk assessment by some US researchers).

The methods used attempt to extrapolate from the results of high doses to predict effects at low doses. (This should strictly be called interpolation as there is often an estimate of the incidence of the condition in the control group.)

The use of low-dose extrapolation is controversial. There are some statisticians who believe that 'mathematical models should be used to project results only within the range of observed data' (Salsburg, 1986). Others have developed a series of models for use in low-dose extrapolation. The problem has interested many leading biostatisticians and resulted in a large number of scientific publications spread through a range of journals. Such a large and disparate literature is impossible to summarise, reference, or review adequately in a chapter such as this. Much of the work depends upon a series of complex assumptions and is argued using statistical concepts and nomenclature. This makes the subject difficult for many toxicologists to follow. Worked examples of the various methods are rare and some of the models require access to specialised computer programs before they can be fitted to the data.

In this chapter an attempt will be made to determine the underlying concepts behind some of the methods, briefly discuss in non-statistical terms the more commonly encountered models, and then review the performance of the various approaches.

Virtually Safe Dose (VSD)

Mathematical modelling of carcinogenicity data arose out of the concerns described previously about the safety factor approach, together with an attempt to provide a rational alternative to the 'no safe dose of a carcinogen' concept behind the Delaney amendment. Mantel and Bryan (1961) proposed the use of a method of linear extrapolation based upon carcinogenicity data such that the dose of carcinogen was identified which caused a 10^{-8} increase in the risk of developing cancer. Such an increased risk was considered to be effectively zero and the dose causing the increased risk was called the virtual safe dose (VSD). Mantel and Bryan's method involved fitting the log-probit model developed for bioassay data to the data on the dose–response results from the chemical. The upper 99% confidence limits were calculated for the different doses and lines with a probit response of one drawn downwards (Figure 2). The most

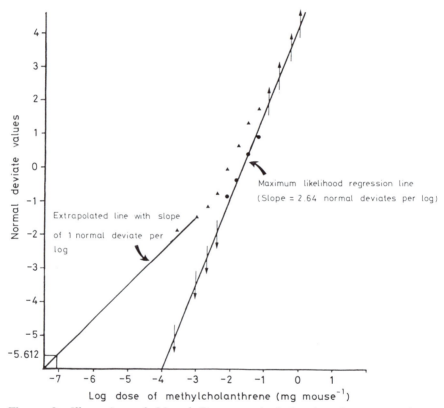

Figure 2 *Illustration of Mantel–Bryan method for low-dose extrapolation. Observed data are fitted using maximum likelihood regression line. The most conservative (i.e. gives lowest safe dose) extrapolated line with a slope of 1 normal deviate per unit log dose is drawn from the 99% upper confidence levels (▲) of the observed proportion of tumour-carrying animals (●). The virtual safe dose is calculated by reading off the dose which produces an increased tumour risk of (10^{-8}) or -5.162 normal deviates. This is $10^{-7.05}$ ($\approx 9 \times 10^{-8}$) mg methylcholanthrene per mouse (as shown in bottom left hand corner of graph).*

The graph and calculations are modified from those presented by Mantel and Bryan in their original 1961 paper. The arrows ⇅ indicate 100% and 0% respectively of animals were carrying tumours

conservative of these lines was then used to estimate the dose leading to a 10^{-8} extra risk.

Although the concepts of different individual tolerances underlying the probit model had some biological appeal, this method was shown not to be as conservative as the authors had initially expected. The probit method also proved to be a poor fit to many experimental data sets and the method had to be modified to include cases when there was a background incidence of the condition under study.

The level of acceptable risk chosen by Mantel and Bryan of 10^{-8} was also subsequently raised to 10^{-6} by the Food and Drug Administration, based upon consideration of the types of risk readily accepted by the population in general. The VSD which leads to a lifetime risk of death of 10^{-6} is equivalent to an effect which would cause about 3 extra deaths per year in USA based upon an average life span of 73 years and a population of 220 million or about one extra death per year in the United Kingdom.

Linear Extrapolation Method

A somewhat similar approach for low dose extrapolation is to find an upper confidence level for the lowest dose effect and then draw a straight line back to the spontaneous or control incidence (Figure 3). Such an approach has been popular because it is both simple and conservative. There has been some criticism that it may be too conservative and also does not reflect dose–response relationships found in multiple dose studies.

The Range of Mathematical Models

The failure of the Mantel–Bryan log-probit method to fit some of the observed sets of data stimulated other biomathematicians to try to find functions which were better fits to the observed dose–response relationships. The mathematical modelling has two main purposes. Firstly, to describe the observed data by some equation relating the incidence to the dose administered. Secondly, to try to estimate effects, using the equation rather than the linear extrapolation method, at doses not tested, particularly low doses. The types of models used fall into two categories: those based upon an assumed distribution of tolerances such as the probit used by Mantel and Bryan or the somewhat similar logit models, and the mechanistic models which are supposedly simulating biological events which lead to carcinogenicity. These include the one-hit, multi-hit, and multistage models.

One-hit models. This is one of the easier models to use. In its simplest form it assumes that cancer can result from a 'single hit', such as irreversible

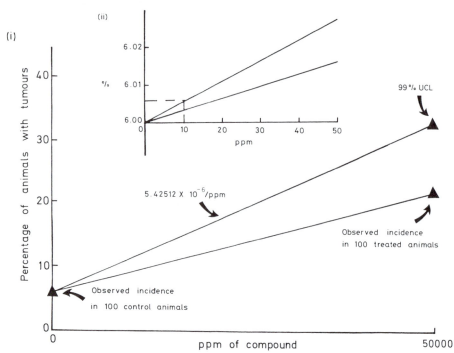

Figure 3 *Example of low dose extrapolation using 99% upper confidence level (UCL) of excess tumour rate in exposed animals. Control incidence in tumours is 6 in 100 animals and incidence in animals treated with 5000 p.p.m. is 22 in 100 animals. The 99% UCL is calculated as 33.13% (the increase over the control proportion = 0.271256). Figure 3(i) shows linear extrapolated line with a slope of 5.42512 × 10⁻⁶ excess tumours p.p.m.⁻¹ (0.271256 per 50000). Figure 3(ii) shows the estimated increased tumour risk associated with 10 p.p.m., i.e. 0.0054%. The dose associated with a risk of 10^{-6} is $\dfrac{10^{-6} \times 50000}{0.271256}$ or 0.1843 p.p.m. (The calculations upon which this graph is based can be found in Gad and Weil, 1986, p. 209)*

genetic damage or interaction with a receptor. Fitting the model does not require sophisticated computing facilities.

A consequence of the model is that in the low dose region it is approximately a simple linear model. This assumption of low dose linearity results in extremely low dose or conservative estimates of the virtual safe dose (VSD). In practical terms, using the one-hit model may be the same as applying the Delaney amendment as the VSD will be extremely small.

Although appealing in its relative simplicity, the one-hit model has only a single parameter to model what may be a complex shape of dose–response; the model, therefore, is often a poor fit to the data compared with the other models. The 'one-hit' concept of the mechanism of carcinogenicity may also be oversimple in view of developments in the theory of carcinogenicity.

Multi-hit models. This is considered to be a more realistic mechanistic model than the one-hit. It assumes that there are multiple hits (it is sometimes called the Gamma multi-hit model) required at the cellular levels to initiate carcinogenesis. Although this model has been used by a number of regulatory bodies and recommended by the Scientific Committee of the Food Safety Council (1980), there are critics who have questioned its use because of statistical problems and the production of inconsistent results. They have suggested that its use may lead to misleading conclusions in some circumstances.

Multistage models. This is another mechanistic model assuming that the development of cancer is a multistage process. This makes it biologically appealing but difficulties arise because the estimated parameters do not have any direct biological interpretation. It, too, has been criticised because assumptions implicit in its formulation may not actually apply in general. Various variants and modifications of the model have been developed but Park and Snee (1983) point to the potential of the model to overestimate risk and give misleading results under certain circumstances.

Nevertheless, the EPA (1986) suggest that in the absence of adequate information to the contrary, the linearised multistage procedure should be used. This is a variant of the multistage model which uses linear extrapolation. Extrapolation is from an upper 95% confidence bound calculated for a low dose down to the control levels (Crump, 1981). The EPA will allow results using other models to be presented but a rationale must be included to justify use of any model chosen.

Pharmacokinetic models. There is some concern that the mechanistic models developed for carcinogenicity oversimplify the pharmacological processing and metabolism of toxic compounds. Models have, therefore, been developed to account for such factors as: the dose reaching the target organ compared with that administered, thresholds, and inter-species differences in response. Their practical use is limited by the increased numbers of parameters needed to specify the models, which mean they can only be applied to studies with many doses. Development of models incorporating pharmacokinetic information is likely to continue.

The Weibull Model. The Weibull model is a tolerance distribution increasingly fitted to toxicological data. It is widely used in industry to study product reliability, such as the life span of electronic components. The Weibull model can be made more complex by the addition of extra parameters or simplified by specifying fewer (it eventually becomes a special case of the one-hit model). Models can be developed including parameters to take into account the length of time before a tumour develops or the length of administration of the treatment. Its use is limited by needing sufficient numbers of treated groups to allow parameters to be estimated. The use of the Weibull has increased in part because it can incorporate information on the time up to when a tumour develops. The inclusion of these data into the hypothesis testing procedure for carcinogenic effects (Peto *et al.*, 1980) was an important advance in the analysis of rodent carcinogenicity bioassay studies.

The performance of different mathematical models. Fitting of models to toxicological data is a complex and controversial subject with a voluminous literature. Most toxicologists do not need to know the underlying theoretical assumptions for the various methods. However, an appreciation of the practical results of applying various models is useful.

In general, many of the models fit the observed data satisfactorily, but they differ in their estimates of lower dose effects such as the VSD or the

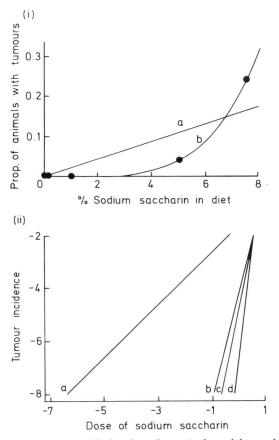

Figure 4 *Graphs to illustrate:* (i) *fit of mathematical models to observed data on the incidence of tumours in animals fed diet containing various percentages of sodium saccharin* (a) *is one-hit model;* (b) *is Weibull;* (ii) *low dose extrapolation of fitted lines for different models* (a) *one hit;* (b) *multi-stage;* (c) *Weibull;* (d) *multi-hit. Both axes on logarithmic scale.* (*Graphs reproduced by permission from Food Cosmet. Toxicol.,* 1980, **18,** 725).

10^{-6} extra risk. The Food Safety Council (1980) compared the results of 4 models on 14 different data sets. Three of the models, the multistage, multi-hit, and Weibull, provided reasonable fits to the data while the other, the one-hit model, has a poorer fit. Estimates of the VSDs differ quite appreciably between models, with the one-hit, in general, giving the smallest safe doses, the multistage, next lowest, then the Weibull, followed by the multi-hit. Figure 4 illustrates both the failure of the one-hit model to fit data for saccharin compared with the Weibull, and the differences in low dose extrapolations for the four models. Other studies have suggested that the probit and logit models give higher VSDs than the multi-hit, they are, in other words, less conservative.

Variability in the estimates of the results produced by the different models has resulted in the models having proponents. Considerable debate has ensued over their respective merits. In practice, the actual choice of a model must depend on the professional judgement of the investigator. The choice of method for low dose extrapolation must always be made with an appreciation of the statistical limitations of any such methods, as well as the limited biological applicability of any of the models.

4 Interspecies Comparisons

The problem of extrapolating from the test species to the human population remains, even if a satisfactory low dose extrapolation is achieved. This species scale-up has to take into account possible differences in distribution, metabolism, and excretion of a compound, together with factors such as differences in size and life span.

The three most frequently used approaches to the interspecies scale-up are: the conversion of results on the basis of the fraction of the diet, a body weight basis, and differences in surface area.

A Fraction of Diet Scaling Factor

This converts intake in mg kg^{-1} (diet) per day or parts per million (p.p.m.) or billion (p.p.b.) in the rat or mouse diet to the human intake on a similar basis. The human diet is estimated to weigh between 600 and 700 g day^{-1} while that of the rat is 21 g day^{-1} and the mouse 4 g day^{-1}. The effect of using mg kg^{-1} per lifetime is to produce a risk which is 35 times greater as it is based upon an average lifespan of 70 years for man and 2 years for rodents.

B Scaling Based Upon Body Weight

The administered dose is expressed in terms of mg kg^{-1} body weight of the test animal. Human exposure is then considered on the same basis of body weight. A conversion factor can be calculated which is a constant based on the ratio of average body weight of a rat or mouse to the average body weight of a human. The average body weights of a mouse, rate, and human being are assumed to be 20 g, 400 g, and 70 kg, respectively, giving a ratio of 3500 for the mouse and 175 for the rat. Clearly not everybody (or every mouse) conforms to these average weights.

C Scaling Based upon Surface Area

Dosage in terms of mg kg^{-1} body weight in the test animal is converted to mg per square metre of body surface area. Human exposure is then also expressed in terms of mg per square metre surface area.

Two alternative methods are used. The first calculates a term called the surface factor as the cubed root of the mass of human (M_h) divided by the mass of the test animal (M_t), *e.g.* $(M_h/M_t)^{1/3}$. In the case of the rat this factor is $\left(\dfrac{70000}{400}\right)^{1/3} \simeq 5.59$ and for mice it is $\left(\dfrac{70000}{20}\right)^{1/3} \simeq 15.18$.

The second method is based upon calculating the surface area by using the formula of the weight of the animal (W) to the power 2/3 multiplied by a species constant (K) provided by Spector (1956). [Algebraically the surface area is (cm)$^2 = K(W)^{2/3}$]. This results in a man:rat surface area ratio of 55.4 and for man:mouse, a ratio of 387.9.

D Choice of Method of Scaling

The choice of which of these methods should be used is a matter of debate. Hogan and Hoel (1982) have concluded that estimates based upon either mg kg^{-1} body weight day^{-1} or mg m^{-2} body surface area day^{-1} seem to lead to better projections of risk in man from carcinogenicity data than those based upon lifetime estimates. However, such methods have to be considered somewhat crude and experience of their use is limited. Considerable care is needed, therefore, in using such species scale-ups and consideration needs to be taken of species differences which could complicate such attempts.

The EPA (1986) suggest that extrapolation on the basis of surface area is considered appropriate mainly because some pharmacological effects appear to be scaled according to surface area.

Ames, Magaw, and Gold (1987) have taken an alternative approach for interspecies comparison. They have used rodent carcinogenicity data to try to rank possible carcinogenic hazards to the human population, rather than to predict absolute human risks. They used an index called HERP (the Human Exposure/Rodent Potency dose) to indicate the possible hazard. Human exposures were calculated for populations at risk and the Rodent Potency dose calculated using a method developed by Sawyer *et al.* (1984). They concluded that the carcinogenic hazards resulting from pesticides or water pollution are likely to be minimal compared with those from background levels of natural substances.

5 An Example of Risk Estimation in Practice

An example of the application of mathematical modelling, low dose extrapolation and species scale-up is illustrated by the approach used by Carlborg (1985) for estimating the risk associated with saccharin from the results of a long term carcinogenicity bioassay in rats.

Carlborg (1985) estimated the excess risk of a bladder tumour resulting from exposure to a dose of 0.01% saccharin fed in the solid food of rats for over two years. He converted this dose in rats to its human equivalent using a scaling based on body weights. The average rat body weight was taken as 742 g (see Table 3), and average daily intake of diet was 27.2 g. This resulted in an average daily intake per kg body weight for each rat of 3.666 mg of saccharin.

Table 3 *Steps used by Carlborg (1985) in estimating the risk of human consumption of saccharin from a long term rat feeding study*

(1) *Estimated from the rat study*
Average body weight of rat 742 g
Average daily intake of diet day^{-1} rat^{-1} 27.2 g

Estimated excess risk of bladder cancer in rats from addition of 0.01% saccharin to diet based upon Weilbull model is 4.1×10^{-8} to 2.5×10^{-10}

(2) *Estimated human exposure based upon body weight*
Average human body weight 70 kg
Average intake day^{-1} rat^{-1} of saccharin
   on 0.01% of diet $= 2.72$ mg day^{-1} rat^{-1}
Average intake day^{-1} kg^{-1} of rat $= \dfrac{2.72 \times 1000}{742}$
                      $= 3.666$ mg day^{-1} kg rat^{-1}
Converting intake for a 70 kg man $= 70 \times 3.666$ mg day^{-1} man^{-1}
                                $= 256.6$ mg day^{-1} man^{-1}

(3) *Estimating human exposure*
   1 can (12 fl. ozs) of 'diet-drink' contains
      110.4 mg saccharin
   256.6 mg saccharin is equivalent to 2.3 cans
   The 90 percentile for consumption of diet drinks in the USA is 2.3 cans per day. Therefore only 10% of the US population drink more than 2.3 cans per day or consume more than 0.01% saccharin in their diet.

Converting this to a daily intake for a 70 kg man gives a total dose of 256.6 mg per day for each person. A 12 fl. oz can of 'diet' soft drink contains 110.4 mg of saccharin, so a daily intake of 256.6 mg would be equivalent to 2.3 such cans. This is about the 90 percentile for the intake of saccharin within the USA. Carlborg used a series of mathematical models to try to fit the observed bladder cancer incidence in a long term rat carcinogenicity bioassay. The estimates of the risk obtained by extrapolation using the Weibull model for the 0.01% dose of saccharin were from 4.1×10^{-8} to 2.5×10^{-10} suggesting a very low and acceptable risk from saccharin if directly applied to comparable consumption in the human population.

Clearly such risk estimation involves a considerable number of assumptions which may be wrong to some degree. One check on the approach is to alter some of the estimates to see if changing the variables in the equations could alter the conclusions. This is called a sensitivity analysis.

It is important to realise that such estimates of risk are dependent upon the quality of data included in the equations. This is illustrated by another example.

6 An Example of Risk Estimation using Mouse Dominant Skeletal Mutations to Estimate Human Risk from Radiation

Dominant mutations in the offspring of individuals exposed to radiation or chemical exposure are of special concern because their effects are seen in this subsequent generation. Data from experiments on mice has been used to try to estimate the genetic risk to the human population from radiation. The model used was the production of dominant skeletal mutations in the mouse (Selby, 1982).

Estimates of the mutation frequencies of serious disorders per million livebirths for each 0.01 Gy (1 Rad) of radiation in the human population have been produced. The central estimate is about 20 per million with a range from about 5 to 66 per million, depending upon various assumptions made (Table 4). This compares with a current or spontaneous incidence of about 10,000 per million livebirths. The only direct experimental data used in the estimate are Selby's finding of 37 dominant skeletal mutations out of 2646 offspring of exposed male mice. The remaining 6 calculations are based upon corrections of the doses administered to make them applicable for low dose estimates, assumptions about the relative frequency and severity of skeletal mutations and the relative sensitivities of male and females to radiation. These represent the expert considerations of the UNSCEAR and BEIR III committees (UNSCEAR, 1977; BEIR III, 1980).

A sensitivity analysis was carried out in effect, because ranges of

Table 4 *Steps in the estimation of genetic risk in humans from induced skeletal mutations in mice*

	mutations/10^6 live births
(1) Dominant skeletal mutations after 1 Gy + 5 Gy at 0.60 Gy min^{-1} gamma radiation (37/2646)	13983.4
(2) Rate per 0.01 Gy (division by 600)	23.3056
(3) Dose-rate effect (division by 3)	7.7685
(4) Dose-fractionation effect (division by 1.9)	4.08871
(5) Total dominant mutations assuming skeletal mutations are a tenth (10%) of all dominant mutations (multiplication by 10)	40.8871
[range 5–15%]	[20.4435 to 61.3306]
(6) Proportion of mutations that are serious handicaps 50% (divide by 2)	20.4435
[range 25–75%]	[5.1109 to 45.9979]
(7) Relative sensitivity of maternal to paternal exposure. Assume same.	20.4435
[range: equal* risk to 44% greater than risk in male]	[5.1109 to 66.2370]

* The BEIR Committee (1977) gave a negligible risk as their lower band to the risk of protracted maternal irradiation

estimates were given for the proportion of dominant mutations that are skeletal, the proportion that produce severe disabilities, and the differential sensitivities of the two sexes. These give estimates of between 5 and 66 mutations per million livebirths.

These estimates are, of course, dependent upon the quality of the experimental data used in the equations. This may be open to interpretation. In this case, of the 37 presumed mutations, 6 were sterile and 3 others only identified by their small size. Subsequent confirmation and characterisation of the mutants was hindered by the loss of 13 of the mutants when the stocks were transferred from West Germany to the USA, while studies have shown that one of the strains used in these experiments have at some time become genetically contaminated (West, Peters, and Lyon, 1984).

These difficulties in the interpretation of the biological data do not necessarily alter the conclusions of the risk estimation process appreciably. However, they do highlight the need to evaluate critically the assumptions made in such methods and the potential dangers from over-interpreting data. Risk estimation is an important and useful procedure but must remain closely linked to data derived from good scientific experimentation.

7 Risk Evaluation and Management

The identification and estimation of the risks associated with particular hazards is followed by the stage of risk evaluation. This involves deciding on what levels of risk are acceptable and whether there are any offsetting benefits. Risk management is the series of procedures for controlling risk to this acceptable level. The steps of the risk estimation, evaluation, and management overlap. A more detailed discussion of the issues surrounding these concepts can be found in Lovell (1986).

One method which tries to evaluate the relative benefits and risks of actions is risk–benefit analysis. This is a special case of cost–benefit analysis in which the relative costs of the advantageous and disadvantageous outcomes are compared. Lave (1987) has identified two different types of such analyses. One, he termed 'risk-risk', is when both alternative actions have risks and benefits which have to be compared; the other, called 'how safe is safe enough', results in increasing cost for each successive reduction in risk. In the latter case a decision has to be made as to whether the costs of extra safety are worth the sacrifice of other desired actions. There are, of course, considerable difficulties in such analyses in providing monetary values to injury, illness, or death.

Lave and Omenn (1986) used a cost–benefit approach in an assessment

Table 5 *Outcomes of predicting carcinogenicity of chemicals using short term tests when incidence of carcinogens in a sample of 50,000 chemicals is (a) 2% and (b) 10%*

(Short term tests are assumed to have an accuracy of 90% with sensitivity and specificity both 90%)

(a) 2% carcinogens

		Actual carcinogenicity of chemical	
		Carcinogen	Non-carcinogen
Prediction from *in vitro* tests	Carcinogen	True positive 900	False positive 4900
	Non-carcinogen	False negative 100	True negative 44100
		1000 (2%)	49000 (98%)

(b) 10% carcinogens

		Actual carcinogenicity of chemical	
		Carcinogen	Non-carcinogen
Prediction from *in vitro* tests	Carcinogen	True positive 4500	False positive 4500
	Non-carcinogen	False negative 500	True negative 40500
		5000 (10%)	45000 (90%)

David P. Lovell

of the effectiveness of short term testing of chemicals for the prediction of carcinogenicity.

They illustrated the case (Table 5) of a series of 50,000 new chemicals of which either 2% or 10% were carcinogenic. They estimated the relative costs of incorrectly predicting the carcinogenicity of the chemical using *in vitro* short term tests. They assumed the cost of testing each chemical in the battery of tests was $8,000 and the cost of accidentally allowing a carcinogen (false negative) into the environment was $10 m, while the cost of unnecessary controls on a non-carcinogen incorrectly called a carcinogen (false positive) was $1 m. The balance sheets comparing the costs of testing against no testing, shown in Table 6, give net benefits of $3.7 bn if 2% of chemicals are carcinogens and $40.1 bn if 10% are carcinogens for short term testing.

Obviously such conclusions can be dramatically altered by changing the assumptions (such as that the short term tests have 90% accuracy) or the costs ascribed to different outcomes. However, although such approaches can be criticised, particularly when dealing with non-monetary outcomes, they do provide a framework for exploring the implications of various strategies.

Table 6 *Balance sheet of social costs of short term testing for carcinogenicity compared with no testing*

| | Percentage chemicals that are carcinogens | |
	2%	10%
	$ bn	$ bn
Cost of testing 50,000 chemicals at $8,000 per chemical	0.4	0.4
Cost to society of false positives ($1 m per chemical)	4.9	4.5
Cost to society of false negatives ($10 m per chemical)	1.0	5.0
Total cost to society of test strategy	6.3	9.9
Cost to society of not testing ($10 m per carcinogen)	10.0	50.0
Net saving to society by test strategy	3.7	40.1

(Tables 5 and 6 based upon Lave and Omenn, 1986)

8 Conclusions

Risk assessment is an imprecise procedure despite the statistical and mathematical aspects incorporated in it. The use of mathematical models and derivation of various estimates of risks from the data must not be used to provide a spurious precision or to create an appearance of scientific sophistication. Instead, such approaches can provide a framework to help and illustrate the implications of various evaluations and decision processes.

The multitude of conflicting interests of the producers, regulators, and consumers of chemicals is too complex a pattern of behaviour to be reduced to simple mathematical models. Individuals may not always act in a manner which, based upon an assessment, appears to be in their best interests. However, actions and ideas which may appear irrational have to be accommodated within the risk assessment processes. Risk assessment must always reflect the changing pattern of society's perception of its ideas and values.

References

Ames, B. N., Magaw, R., and Gold, L. S. (1987). Ranking possible carcinogenic hazards. *Science,* **236,** 271–280.

BEIR III Committee. Advisory Committee on the Biological Effects of Ionizing Radiation of the United States National Academy of Sciences, (1980). The effects on populations of exposure to low levels of ionising radiation, National Academic Press, Washington, DC.

Carlborg, F. W. (1985). A cancer risk assessment for saccharin. *Food Chem. Toxicol.,* **23,** 499–506.

Crump, K. S. (1981). An improved procedure for low-dose carcinogenic risk assessment from animal data. *J. Environ. Pathol. Toxicol.,* **5,** 339–348.

Crump, M. S. (1986). Letter to editor. *Fundam. Appl. Toxicol.,* **6,** 183–184.

Environmental Protection Agency (1986). Guidelines for carcinogen risk assessment. *Federal Register,* **51,** 33992–34003.

Food Safety Council (1980). Quantitative Risk Assessment. *Food Cosmet. Toxicol.,* **18,** 711–734.

Gad, S. C. and Weil, C. S. (1986). 'Statistics and Experimental Design for Toxicologists'. Telford Press, Caldwell, New Jersey.

Hogan, M. D. and Hoel, D. G. (1982). Extrapolation to Man. *In*: 'Principles and Methods of Toxicology'. *Ed.* A. Wallace Hayes. Raven Press, New York, pp. 711–731.

Lave, L. B. (1987). Health and Safety Risk Analyses: Information for better decisions. *Science,* **226,** 291–295.

Lave, L. B. and Omenn, G. S. (1986). Cost-effectiveness of short term tests for carcinogenicity. *Nature (London),* **324,** 29–34.

Lovell, D. P. (1986). Risk assessment—general principles. *In*: 'Toxic Hazard Assessment of Chemicals', *Ed.* M. L. Richardson. Royal Society of Chemistry, London pp. 207–222.

Mantel, N. and Bryan, W. R. (1961). 'Safety' testing of carcinogenic agents. *J. Natl. Cancer Inst.*, **27**, 455–470.

Park, C. N. and Snee, R. D. (1983). Quantitative risk assessment: state of the art for carcinogenesis. *Fundam. Appl. Toxicol.*, **3**, 320–333.

Peto, R., Pike, M. C., Day, N. E., Gray, R. G., Lee, P. N., Parish, S., Peto, J., Richards, S., and Wahrendorf, J. (1980). Guidelines for simple, sensitive significance tests for carcinogenic effects in long term animal experiments. *In*: 'Long Term and Short Term Screening Assays for Carcinogens'. A Critical Appraisal (IARC Monograph No. 2), Lyon, International Agency for Research on Cancer, pp. 311–426.

Richardson, M. L. (1986). Preface to 'Toxic Hazard Assessment of Chemicals'. *Ed*. M. L. Richardson. The Royal Society of Chemistry, London.

The Royal Society (1983). 'Risk Assessment—A Study Group Report'. The Royal Society.

Salsburg, D. S. (1986). 'Statistics for Toxicologists'. Marcel Dekker Inc., New York.

Sawyer, C., Peto, R., Bernstein, L., and Pike, M. C. (1984). Calculation of carcinogenic potency from long term animal carcinogenesis experiments. *Biometrics*, **40**, 27–40.

Selby, P. B. (1982). Induced mutations in mice and genetic risk assessment in humans. 'Progress in Mutation Research'. Vol. 3. *Ed*. K. C. Bora *et al.* Elsevier Biomedical Press, pp. 275–288.

Spector, W. S. (1956). 'Handbook of Biological Data'. W. B. Saunders, Philadelphia.

UNSCEAR Committee (United Nations Scientific Committee on the Effects of Atomic Radiation) (1977). Sources and effects of ionising radiation, Report to the General Assembly, with annexes, United Nations, Sales No. E77.lx.1 pp. 425–564.

Weil, C. S. (1972). Statistics *vs.* Safety Factors and Scientific Judgement in the Evaluation of safety for man. *Toxicol. Appl. Pharmacol.*, **21**, 454–463.

West, J. D., Peters, J., and Lyon, M. F. (1984). Genetic differences between two substrains of the inbred 101 mouse strain. *Genet. Res.*, **44**, 343–346.

CHAPTER 20

Epidemiology

G. M. PADDLE

1 Introduction

For present purposes, epidemiology will be defined as the study of patterns of health in groups of people. Besides brevity, this definition has the virtue that the key words all have colloquial meanings, which convey a picture of simplicity, and more esoteric meanings, which convey an air of profundity. Epidemiology is an observational, rather than experimental, pursuit, and the contrast between the pragmatism and opportunism of the barefoot epidemiologist and the purity of approach of the academics is one of its many fascinations. Many other definitions of epidemiology have been proposed (Lillienfield, 1978) but none is universally accepted. Indeed, as the range of epidemiological applications expands, there is a temptation to argue that, just as toxicology is what toxicologists do (Casarett and Doull, 1980), so epidemiology is what epidemiologists do. The aim of this chapter is to describe what epidemiologists do, and how they do it. The aim of this introduction is to link 'the study of patterns of health in groups of people' to the activities of other relevant disciplines.

Epidemiologists have to work closely with clinicians, who could be said to study the patterns of health of individuals. However, an epidemiologist's concern with the health of groups of people requires a different approach from that of a clinician concerned with the ill-health of an individual, and the goal of restoring it to normality. The information collected by epidemiologists and clinicians may often be similar. Each may record that John Smith was born in 1937, is 175 centimetres tall, weighs 86 kilograms, has worked for 20 years in the chemical industry, smokes 40 cigarettes a day, drinks 15 litres of beer a week, and is suffering from bronchitis. The clinician will use this information to assess John Smith's state of ill-health, to recommend a cure, and to counsel him about his future welfare. These consultations between the clinician and his patient usually occur only when the patient is unwell. The epidemiologist will use the same information to classify John Smith alongside other men of similar age, with similar occupational histories, and similar smoking and drinking habits, in order, perhaps, to gain a better understanding of bronchitis patterns, which for some reason John Smith is suffering from, and hypertension patterns which for some reason he is not suffering from. This accumulation of information by the epidemiologist about the subject in his study may occur when the subject is fit or unfit, in his presence or in his absence, or even after his death.

Epidemiologists and toxicologists, particularly those toxicologists who could be said to study patterns of health in groups of laboratory animals, often have common objectives. An epidemiologist and a toxicologist may have data about the effects of the same compound on the same target organ. The toxicologist may have conducted a long term rodent study with the aim of predicting the health effects that people will experience at the exposure levels prevailing in the workplace or in the environment. The epidemiologist may have studied the health patterns actually experienced by people exposed to the compound occupationally or at home. The total evidence about the risk to humans of this compound will consist of the toxicologist's precise, experimental data about the wrong species at the wrong exposure, and the epidemiologist's imprecise, observational data about the right species at the right exposure. The moulding of these data into a formal risk assessment for regulatory purposes is a problem receiving much attention currently (Shore *et al.*, 1992). At the development stage of a compound, and during the early days of its use, it is logical for the laboratory data to dominate the assessment, but, after several decades of use, it seems more logical for the human data to hold sway (Doll, 1987).

The definition of the exposure levels in an epidemiological study can be improved by the involvement of an industrial hygienist, whose role could be described as the study of patterns of exposure to hazards in groups of people. The ultimate aim of epidemiology is to establish the dose–response curve for human health effects *versus* exposure level of compound. Success in achieving this objective depends upon the collection of appropriate exposure data (Landrigan, 1982). Whereas it would be inconceivable to conduct a laboratory experiment without painstakingly specifying and controlling the conditions of exposure, it is the exception for an industrial hygienist to collect exposure data for epidemiological purposes, rather than to ensure compliance with regulations.

An alternative to the exposure data collected by a hygienist is the measurement of delivered dose by the analysis of body fluids, such as blood and urine. Collaboration with a biochemical laboratory is increasingly an important element in an epidemiological study. If the same analyses have been done on experimental animals, perhaps by the same laboratory using the same techniques, it becomes possible to compare the metabolic pathways and pharmacokinetics of different species, and hence arrive at a more rational and scientific risk assessment (Anderson *et al.*, 1987).

The aim of this introduction has been to show the extent to which epidemiology is reliant upon, and complementary to, other areas of activity. Epidemiology can make a substantial contribution to the improvement of health patterns in groups of people on its own. Greater improvements will be achieved with greater efficiency if epidemiology is appreciated and supported by other health science specialists.

2   Sources of Problems

Several of the chapters in this book have been devoted to the techniques available to the toxicologist, and to the systems which he studies when examining the toxicity of chemicals. The studies an epidemiologist can mount are similar in range, but they are conducted in a totally different fashion. The reasons for the differences are obvious. Whereas the toxicologist can contain his animals within a specified environment, and ensure that the food eaten and the air breathed by them is of a predetermined and standardised form; the epidemiologist has no such control over the people in his studies. Therefore, although their studies may be similar in basic design, the inability to impose an exposure regime, the absence of an ideal control group, and the need to analyse the data for factors additional to the one being studied, mean that the epidemiologist can never claim that his study approaches the rigour of that of the toxicologist.

The technique an epidemiologist applies will depend upon the source of the problem and the urgency attached to it. Some of the ways in which problems arise are as follows:

(a) An animal study points to a hazard to a human population
(b) A clinician has noticed a pattern of ill-health
(c) An epidemiological study prompts further work
(d) A body of health data has been accumulated
(e) Records for a large population are available
(f) People have expressed concern.

An example from each of these sources will help to introduce many of the principles and methods to be discussed in later sections.

A   An Animal Study Points to a Hazard in a Human Population

Whenever a laboratory study of a compound to which people are already exposed demonstrates an excess of disease in animals, it is a natural consequence to study the exposed people to protect them against the same or related health problems.

In 1981 it was reported that formaldehyde had been found to cause nasal squamous cell carcinomas in a chronic inhalation study in both rats and mice (C.I.I.T., 1981). Formaldehyde is a ubiquitous substance in the environment and it has a long history of widespread use in industry. There was obviously an urgent requirement to study the patterns of health in groups of people exposed to formaldehyde. The first epidemiological reports to appear were of case-control studies of nasal cancer. These are the easiest studies to complete quickly but their interpretation is always open to debate. Negative outcomes from these nasal cancer studies could not have been considered completely reassuring, because they left open the possibility that the target organ was different in people. Subsequently, two retrospective cohort studies have been published involving occupationally exposed people in the UK (Acheson *et al.*, 1984), and occupationally exposed people in the USA (Blair *et al.*, 1986). These studies examined mortality rates for all causes of death, not just nasal cancer, and have provided a sound basis on which to assess the risk attributable to formaldehyde exposure.

B A Clinician has Noticed a Pattern of Ill-health

Clinicians, through experience, become aware of the normal pattern of health of their patients, and are able, without sophisticated calculations, to detect departures from this pattern.

A cluster of cases of adenocarcinoma of the nose and sinuses was observed by two specialists in Buckinghamshire, and they were able to associate it with employment in the local furniture industry (Macbeth, 1965). The initial discovery that smoking is causally related to lung cancer was due to several clinicians independently reporting that a large proportion of the lung cancer cases referred to them were heavy smokers (*e.g.* Doll and Hill, 1950). Subsequent studies of workers have confirmed an excess risk of nasal cancer in woodworkers and progress has been made towards identifying the causative agent (MRC, 1981). Further studies of the patterns of health in smokers have confirmed the link between smoking and lung cancer, and helped to estimate the dose–response curve. One of the smoking studies was a prospective study which became an intervention study, inasmuch as many British doctors were persuaded to give up smoking, and their health patterns were compared with those of continuing smokers and non-smokers, (Doll and Peto, 1976). This study showed that cessation of smoking gave an immediate beneficial effect.

The observation by an astute clinician of an unusual pattern of health in his patients has been the single source of almost all the major epidemiological discoveries to date. When the excess risk is high, and particularly when it is concentrated in a small community, there is no reason to think that modern epidemiological techniques will prove to be a more successful method of detection.

C An Epidemiological Study Prompts Further Work

This is a major source of epidemiological projects, despite the fact that very few major health risks have been discovered by formal epidemiological studies.

An excess of a syndrome of such non-specific nature that it is often referred to as 'Danish painters disease' was reported, (Axelson *et al.,* 1976) and it was postulated that the syndrome was associated with chronic exposure to solvents. This led to the conduct of cross-sectional studies in Germany to test the hypothesis in a different environment and conferences in Copenhagen (WHO, 1985) and North Carolina (Cranmer and Golberg, 1986) at which the various aspects of the syndrome were discussed, so that a protocol could be established for further studies.

It is in the nature of observational studies to be inconclusive, and to pose more questions than they answer. The quest for indisputable conclusions, and the desire to resolve all the related issues necessitates a sequence of studies on the same topic. The time consumed by such activities, without any major decisions being reached, is justifiable because the excess risk is often small, or the health effect is not debilitating or mortal.

D A Body of Health Data has been Accumulated

Records collected on a routine basis by national and regional authorities permit patterns of health to be displayed, and variations to be identified and explored. The death records for England and Wales have been a source for epidemiological studies for several centuries (Graunt, 1662) and statistical methods for their analysis were first propounded over a century ago (Farr, 1839).

The Medical Research Council Environmental Epidemiology Unit has published maps of cancer mortality based on the England and Wales mortality data for 1968–1978 (Gardner *et al.,* 1983). These maps, an example of descriptive epidemiology, illustrate regional variations in cancer mortality and serve to target epidemiological effort into finding causes for the highest rates. As the maps clearly depict the excess cancer rates of known origin, it is reasonable to assume that there are underlying reasons for the so far unexplained peak figures. The same Unit has conducted a case-control study of cancers in young and middle-aged men in the North of England (Coggon *et al.,* 1986), because the British chemical industry is located there, and cancer rates are higher there than in the South of England.

E Records for a Large Population have Accumulated

Large chemical Companies that have provided a pension fund for several decades collect mortality records for deaths in service and deaths of pensioners in order to manage the fund. For many compounds, these records will relate to the most highly exposed and longest exposed members of the population.

It is, therefore, reasonable to suppose that analysis of these records will provide evidence about the health patterns generated by these compounds, and it would be uncaring of the Companies not to analyse them routinely.

Several Companies have published papers about their accumulated death records (*e.g.* O'Berg *et al.,* 1987). A standard analysis for a mortality study is used, and statistically significant clusters are detected and followed up, either by retrospective cohort studies, or nested case-control studies.

F   People have Expressed Concern

Just as individuals have access to a clinician to express concern about their health, so do groups of people have avenues they can explore when concerned about their mutual health pattern. All such expressions of concern have to be evaluated and resolved, even though a report from the responsible American authority (Schulte *et al.,* 1986) has shown that in their experience the concern is usually scientifically unfounded.

As heart disease is very common, and is increasing in incidence in men of working age in the UK, it is frequently a health concern among employed groups (Paddle and Bennett, 1983). In this example the employees were those of a power station run as part of a large chemical company, but separated from the neighbouring plants by a boundary fence. Through the normal consultative procedure, the men had expressed anxiety about a sequence of coronary illnesses, some of them fatal, and had inquired whether there was an occupational explanation. The Medical Officer felt that the frequency of the incidents may have been out of the ordinary, and initiated a detailed investigation. Careful scrutiny of the medical records for the workforce disclosed eleven coronary events in a period of eight years. Further research into this small cluster involved a variety of techniques. The frequency was found to be out of the ordinary, but it did not appear to be of occupational origin.

3   Guidelines

One consequence of epidemiological studies having a variety of backgrounds, being observational in character, and inconclusive in outcome, is that they are difficult to report clearly and concisely. Any two readers of a typical epidemiological report will perceive different real or imaginary flaws, will experience different levels of concern, and will recommend different courses of action. If a decision maker in industry, government, or academia is to do his job by co-ordinating the opinions of such readers, it is necessary for standards of reporting to be established.

Guidelines written by the Inter-agency Regulatory Liaison Group have been published (Epid. Work Group of the IRLG, 1981) that can be used to ensure that all important aspects of a study are reported. The guidelines cover the headings: Background and Objectives, Study Design, Study Subjects, Comparison Subjects, Data Collection Procedures, Analytical Methods and Statistical Procedures, Interpretation, Limitations and Inferences, and Supportive Documentation. Most of these headings are self-explanatory and clearly essential. Although these guidelines were written with reporting in mind, they are equally of value when planning studies. Perhaps the most frequently met failing in an epidemiological report is a lack of clear objectives and a consequential lack of justification for the interpretation and conclusions. If a study is mounted to determine whether smoking habits are related to baldness, it is inappropriate to conclude from it that alcohol intake is related to flat feet. The latter is at best a chance finding worthy of further study, and is at worst a predictable outcome of a study, that was designed for a different purpose.

4   Cohort Studies

Influential epidemiological output is dominated by two types of study; the cohort study and the case-control study. To a toxicologist, the former may look like an attempt to mimic a laboratory experiment, while the latter is more like an attempt to recreate the history of a laboratory from a study of the pathology reports. Let us look first at the most prestigious and most transparently valid of epidemiological techniques. It is known as the cohort study, where 'cohort' means a number of people grouped together by a common feature such as exposure to a chemical, occupation, domicile, or social habit. A classical cohort study is conducted prospectively, that is to say, the cohort is formed at the outset of the study and is followed for a period of time during which its health pattern will be recorded until such time as the entire cohort has died, or until sufficient data has been accumulated. This raises one obvious disadvantage which the epidemiologist faces compared with the toxicologist, namely the longer lifespan of humans compared to rodents. The epidemiologist runs the risk of not achieving results in his own lifetime. To overcome this difficulty, and to provide the scientific community with the data it needs to resolve today's problems, the epidemiologist can resort to a retrospective or historical cohort study. Using this technique, the epidemiologist recreates a cohort from the past and follows it towards the present day.

An advantage of the cohort study is that the cohort is usually known to be free of disease at the time of formation, or recruitment. As disease occurs, it can be assumed that it is either the result of everyday risks or the ageing process, and therefore equivalent to the incidence in a control group, or that it is due to the factor that exemplifies the cohort. Examples of retrospective cohort studies were referenced in the discussion of formaldehyde. A further example is that of employees in the United Kingdom exposed to vinyl chloride monomer (VCM) during the course of their work (Fox and Collier, 1977). In that study, an excess mortality from liver cancer was demonstrated, but, unlike studies of employees in other countries exposed to VCM, no excess mortality from cancer at other sites, or excess deaths due to other forms of disease, were found.

In that study, as in all such studies, it is possible to find ways in which the epidemiological work is of less validity than the equivalent rodent toxicology (Maltoni *et al.*, 1981). The humans were indeed exposed at work to VCM, but that was not the only potentially harmful substance to which they were exposed, either at work, or elsewhere. Many of them will have worked with other chemicals at the same time as they were exposed to VCM, many will have worked in different industries before and after employment in the plastics industry, and many will have hobbies or a social style which brought them into contact with other potentially harmful substances. Even the extent of their exposure to VCM had been sparsely recorded. Prior to the mid-1970s, when VCM was first shown to be harmful to people at high levels, the exposure had been measured very infrequently.

These drawbacks have tended to make the regulatory authorities dismiss epidemiological results as invalid evidence, despite the fact that epidemiology reflects the real risk at real exposures to real people.

The data from a cohort study is summarised by two tables. Table 1 sets out the deaths by age at death and year of death. Age at death is important because death rates increase rapidly with increasing age, and because the major causes of death of young people differ from the major causes of death of old people. Year of death is important because death rates have changed from one decade to the next. Some causes of death, such as lung cancer, have increased in frequency while others have

Table 1 *Deaths from all causes by age and calendar year* (Segment from a hypothetical study)

Age at death	Calendar year		
	1951–5	1956–60	1961–5
50–54	26	31	32
55–59	32	41	40
60–64	50	47	62
65–69	68	69	73

Table 2 *Person years at risk by age and calendar year* (Segment from a hypothetical study)

Age	Calendar Year		
	1951–5	1956–60	1961–5
50–54	2940	3410	4143
55–59	2407	2895	3290
60–64	2310	2380	2810
65–69	1976	2198	2304

decreased. Changes in diagnostic fashions and classification systems also have to be allowed for. Table 2 sets out the person-years at risk by age and calendar year. Each person, whether dead or alive at the end of the study, contributes to this table, according to his year of birth, his date of entry to the study, and his date of exit from the study. A person can leave the study because he dies, because the study ends or, unlike a laboratory study, because it becomes impossible to trace him. In the simplest report of a cohort study, age and calendar year, as displayed in Table 2, are sufficient to complete the analysis. If, however, there is an excess risk of cancer in the cohort due to exposure, any reasonable model of cancer in man will specify that there is no excess risk until a latent period has elapsed since first exposure, and that there is a higher risk at higher exposures. Latent periods for cancer are typically 15–30 years or more, which means that the first 15 or so years of exposure should be treated as if the people were unexposed. Several computer packages are available for partitioning the person-years at risk table into latency periods and exposure periods, so that the estimate of the excess risk due to exposure is compatible with a simple model of tumour development.

5 Case-control Studies

The other ubiquitous type of epidemiological study is the case-control or case-referent study. This is always a retrospective study, and will, therefore, produce results more quickly than a prospective approach. It is also a very efficient type of study because it only uses people who are known to be able to contribute valuable information. It can therefore produce results more rapidly and with less resources than a retrospective cohort study. It has a reputation of being a 'quick and dirty' technique, but reassessments and improvements to its design and analysis have led to a growing acceptance that it can be the most informative as well as the cheapest approach to a problem (Breslow and Day, 1980).

Typically, a case-control study will test the hypothesis that the presence of factor A in a person's life is related to his risk of contracting disease X.

Table 3 *Typical outcome of case-control study*

	Disease X cases	No disease X controls	Total
Factor A present	34	40	74
Factor A not present	62	152	214
Total	96	192	288

For example, factor A could be exposure to asbestos at work, chewing tobacco, or taking part in amateur dramatics, and disease X could be pleural mesothelioma, cancer of the larynx, or suicide. The first requirement to be met is a list of cases of X about whom the requisite information about factor A can be collected. The second requirement is a set of controls, or referents, for each case, who differ from the case only inasmuch as they have not contracted disease X. The outcome of the study is a table like Table 3.

In Table 3 the matching of controls to cases has been ignored. If a second factor is also important, for instance age, then a matched analysis can be much more powerful. The overwhelming advantage of this design is that the incidence of the disease is almost irrelevant. For a very rare disease, a prospective cohort study may require a thirty year follow-up of several million persons to produce worthwhile results. An equivalent case-control study can be productive even if there have only been forty cases, say, in England and Wales in the past twenty years.

An example of a case-control study is the survey of cancer and occupation in young and middle-aged men referred to in Section 2. In this project, the investigation of several diseases and many occupations was encompassed in a single study. Occupational and smoking histories were collected for all the cancer patients aged 18–54 in the area of the survey. When respiratory cancer was the disease specifying the cases, all the other patients were used as controls. A disadvantage of this approach is that so many hypothetical dose–response relationships are being tested simultaneously that some of them are bound to be statistically significant, even when no real relationships exist. This portmanteau approach to epidemiology can be very cost-effective if practised judiciously, but if applied without thought it can justifiably be described as a fishing expedition.

The case-control study is an approach that toxicologists can find counter-intuitive. This is because each animal they handle is taking part in a single organised experiment, and its demise is attributed to the presence or absence of the substance under test. There is little point in analysing all the liver tumours to determine what caused them, because the answer is already known. This is far from true of humans, each of whom is taking part, unwittingly, in a multitude of epidemiological tests by being, for example, a smoker, a VCM worker, a painter and decorator, and a heavy drinker.

The checkered history of case-control studies can be attributed to two features which prompt criticism. Firstly, the cases of disease X are often to be found in hospital or in the graveyard. This can make it difficult to obtain information about their exposure to factor A, and it can prevent them or their surrogate from being dispassionate about factor A. If factor A is to any extent suggested as a possible cause of disease X, the possibility that they admit to exposure can be increased. Secondly, if the cases are hospitalised or deceased, it is not obvious who to use as controls, differing only to the extent that they have not contracted disease X. Should the controls be correspondingly hospitalised or dead, and, if so, hospitalised with what or dead from what? If the controls are chosen from healthy persons, however, it is obvious that they differ in at least one important respect from the controls.

The rigour of a cohort study and the efficiency of a case-control study can both be obtained by using a case-control study nested within a cohort study (Kupper *et al.*, 1975). The cohort study is conducted in standard fashion but without reference to any exposure data. If it detects any categories of disease worthy of further investigation, a case-control study is mounted for each of these diseases. This approach is very economical if the exposure data for each individual is complicated by mixtures of chemicals, changes of job and changes in processes, as is usually the case.

6   Other Types of Study

This section describes the other types of study alluded to in Section 2.

A mortality study can be conducted when the results of a cohort study are required but either only details of deaths are available or there is not time or resource for a full study. Its disadvantage is that the results are not as valid as those of a cohort study and the extent of invalidity cannot be deduced.

The epidemiologist collects details of all the deaths that he can muster in the population of interest. The files of decedents collected by large chemical firms are examples. He extracts from the death certificates the underlying cause of death for each one, classifying them according to the International Classification of Diseases, (WHO, 1977) and tabulates them according to age at death. A hypothetical data set is shown in Table 4.

A difficulty in conducting this type of study is that some of the death records required will be easy to acquire, some less easy, and some will prove to be very elusive. The temptation will be to analyse the more accessible records, but this can mean that the study is biased towards some outcomes rather than others. As with cohort studies, the data in

Table 4 *Mortality data grouped by age and cause*

Cause of death	ICD codes	25–34	35–44	45–54	55–64	65–74	74–	All
Cancer	140–239	4	2	8	45	114	51	224
Coronary	393–429	2	5	28	67	124	106	332
Cerebral	430–439	0	2	2	7	42	60	113
Respiratory	460–519	0	0	1	13	62	61	137
Accidents	800–999	1	3	3	2	7	5	21
Other	Other	2	1	6	9	36	36	90
All	001–999	9	13	48	143	385	319	917

Table 4 can be partitioned in various ways to increase the validity of the analysis.

A cross-sectional study can be used when past records do not exist or cannot be accessed and a group of people available at the time have to be used.

A typical cross-sectional study is a morbidity prevalence study. This may be appropriate if a report is received of a possible link between a substance and a non-fatal disease. The epidemiologist will identify all those in the group in which he is interested, such as the painters in the Danish and German studies, and will have each individual assessed to decide whether or not he is suffering from the disease in question. He will categorise the sufferers and the non-sufferers according to their age at the time of examination. A hypothetical set of data is shown in Table 5.

A disadvantage of cross-sectional studies is that the health effect being studied must not be such as to affect who will and who will not remain in the exposed group, otherwise only fit people will be available for study. This may be true of most blood and urine parameters, and minor respiratory or coronary ailments, but it will not be true for debilitating or lethal diseases. The technical problem is that cross-sectional studies measure the prevalence of a disease, that is, the accumulation of cases that have not been cured, rather than the incidence, which is the frequency with which new cases occur.

Table 5 *Prevalence of disease grouped by age*			
Age of survey	Disease sufferer	No disease	Total
15–24	2	137	139
25–34	7	209	216
35–44	17	172	189
45–54	30	169	199
55–64	15	65	80

In the power station example, a cross-sectional study of coronary risk factors was conducted on the present employees at the power station and employees drawn from the neighbouring factor. Controls were chosen at random from those employees at the factory who matched each case for age, length of service, and staff category. The disease in question was 'suspect ischaemia' measured by the answers to a questionnaire or the results of an ECG. The study demonstrated no difference in prevalence of the condition in the two groups.

Descriptive studies can be either an end in themselves or they can form part of the background data for another form of study. The essence of a descriptive study is to describe the data as simply as possible but without introducing bias. Descriptive data can be presented as tables with summary statistics, as tables in which the results have been sequenced in descending order for easy reference, or as maps, in which shading or colours can be used to highlight the abnormal values.

An intervention study, such as the smoking study on British doctors, involves changing the exposure of some of the people in a study and observing whether their health patterns change as a result. In the industrial setting this could mean changing a regulatory exposure limit or ceasing using a compound altogether. This is likely to demonstrate a beneficial effect, because a chance finding will go away of its own accord, or a real excess risk will be reduced or eliminated. It can, however, take a long time for the benefit to become clear because the latency effect will continue to produce illness from past exposure. Dietary epidemiology is a particularly fertile area for intervention studies.

7   Control Groups

Having taken his study as far as a tabulation of the results for his exposed group, the epidemiologist will be faced by a problem which is met with equanimity by the toxicologist. He will need data for a control group with which to compare his figures. The toxicologist, in accordance with the principles laid out in previous chapters, takes another group of animals as his control group. These will be identical to those in the exposed group and, indeed, will have been allocated to the exposed and unexposed groups at random. This is far from being the way in which people conduct their lives. Where is the epidemiologist to find a control group identical to his exposed group except only that they do not live in the same village, or have not been employed in the same industry? The intricacies of this problem, and the measures that have been taken to alleviate it were discussed at a conference solely devoted to this topic (MRC, 1984).

Suppose that he chooses the population of the country in which the village or industry is situated. One major advantage of this apparently inappropriate choice may be that equivalent death, morbidity, or coronary statistics are available in published form. The disease pattern in his

Table 6 *Expected deaths from all causes by age and calendar year* (Segment from a hypothetical calculation)

Age of death	Calendar year		
	1951–5	1956–60	1961–5
50–54	29.1	32.6	34.3
55–59	36.7	43.8	45.9
60–64	49.6	53.4	67.8
65–69	62.4	73.7	80.2

study can now be compared with the pattern in the control population. In the common jargon, he will be able to calculate an 'expected' pattern of ill-health which he can compare statistically with the pattern 'observed'.

In a cohort study, the expected figures are calculated by multiplying the figures in the man-years at risk table (Table 2) by the age and calendar year death rates for the control group to arrive at the figures in Table 6. Although expected 'deaths', these figures will not be whole numbers.

The figures in Table 1 are summed to give a total 'observed' figure and those in Table 6 are summed to give a total 'expected' figure, and the ratio (total observed)/(total expected) is multiplied by 100 to give a standardised mortality ratio (or SMR) for all causes of death. The calculation can be repeated for any group of diagnoses, such as all cancers, or any specific diagnosis, such as chronic lymphatic leukaemia.

A similar calculation, employing proportions of deaths rather than man-years at risk, converts the figures in a mortality study into proportional mortality ratios (or PMRs).

From the observed and expected figures, it is simple to calculate whether the health of the exposed population is statistically significantly different from that of the control group.

Any 'observed' figures that are significantly in excess of the 'expected' figures are 'evidence' of an 'exposure' effect. Unfortunately, as there is no single control group which is obviously 'best', any choice of control group can be criticised. In the heart disease example, the population of England and Wales, the population of North West England, the employees of power stations, the employees of the company, and the employees of the neighbouring factory were all plausible control groups. If alternatives generate similar 'expected' values then the choice among them is largely academic, but if, as can happen, the expected values have a wide range, then the choice of control group will determine the outcome of the study. In the example, no sensible control group was found which gave an expected figure of more than about four, so that the observed figure of eleven clearly warranted further investigation. Had the 'expected' values spanned the range from two to twelve, however, it would first have been necessary to agree upon the most appropriate control group.

Whilst criticisms are always possible of any control group, it is unwise to dismiss all epidemiological studies on the grounds that the choice of control population is suspect. The study may still contain an important message about people, health, and the toxicity of chemicals.

8 Confounding Factors

In a laboratory study of formaldehyde, for example, the only difference between the exposed and control animals is their exposure to formaldehyde. In an epidemiological study of formaldehyde, the health pattern of the exposure group will be affected by their age, by their smoking habits, by their exposure to other hazards at work, such as wood dust, and by their state of health at recruitment to the cohort, and a host of other factors besides exposure to formaldehyde. It would be absurd to assume that these factors will not affect the study, and wildly optimistic to assume that they will have the same effects in whatever control group is chosen. Techniques are needed that will allow for the effects of these other factors and thus permit the effect of formaldehyde to be estimated. Unfortunately a single technique is not sufficient. Age is a factor that everyone has and knows. It can be allowed for by matching the control group to the exposed group, or by an age standardisation calculation. Smoking habits can be built into the analysis for both exposed and control groups, but as there may be a synergistic or antagonistic relationship between formaldehyde exposure and smoking, a simple linear correction may not be sufficient.

Wood dust exposure may be so closely linked to formaldehyde exposure that there separate effects cannot be disentangled. Wood dust would then be a confounding factor and the reliability of any separate estimates of the effects of formaldehyde and wood dust exposure would remain in some doubt.

The most demanding aspect of the statistical design and analysis of epidemiological data is the effort that has to be made to allow for these confounding, or nuisance, variables. Even in the best planned studies, there will be some need for sophisticated statistical analysis rather than the straightforward analysis that was originally planned.

9 Data Collection

At the data collection stage of a project, the epidemiologist has a great advantage over the laboratory experimenter; he is able to communicate with his subjects. The options of asking questions, recording answers, probing for additional information, and noting opinions are elements of most epidemiological studies. Data can be collected by telephone, correspondence, and direct computer interaction as well as by face to face interview. The questionnaire is a very versatile tool. It is also such a simple tool that it can easily be misused, and its complexities can easily be overlooked.

When collecting data the epidemiologist will be largely reliant upon three techniques—incident reporting, questionnaires, and quantitative measurement. Whichever of these techniques is used, it is important to avoid biased results by defining and following a rigorous protocol. Typically, incident reporting will cover birth and death, for which the data are usually adequate, but also ill-health, as measured perhaps by sickness absence. Before attaching too much importance to routine absence data, it should be realised that a period of sickness absence might be due to a football match taking place in the neighbourhood rather than to ill-health; and that a period of presence at work, even when sickness exists, may be due to a bonus being paid prior to a holiday period. It may be necessary to audit the routine sickness data before using it to assess patterns of health.

When a questionnaire has been completed, it is possible that the responder has not understood the questions, or is not providing truthful answers. When quantitative measurement is involved, it is possible that the patient has biased the result by not fasting before the sample is taken. The data then are subject to some doubt. It is possible to include validation techniques but these will usually involve further costs of one sort or another.

Recall for re-examination is hard to justify if management are requesting that the men are not withdrawn from work too frequently, or the budget for a study may be insufficient to cover repeat visits to check questionnaire responses. Nonetheless, the study should have been conducted according to a strict protocol and each item of information collected, (whether a response to a question, or an incident report, or a quantitative measurement) should have been checked for accuracy. If this has been the case, then the statistical analysis of the data should be a straightforward exercise. In the power station example all three methods of data collection were used at various stages. Initially, incident reports of coronary disease had to be counted. This may have been no easy task if the information had needed to be extracted from the medical records. Each record would have had to be examined, and there would have been no reason to expect the clinicians involved to have used consistent terminology. Fortunately the company had been operating a computerised medical records system for the period involved and the data for deaths and long term absences had been coded in a standardised manner. When a questionnaire was required for the latter part of the study, it was sensible to use the one devised and validated by the London School of Hygiene and Tropical Medicine. When ECGs were collected, it was preferable to rely on an internationally accepted method of classification rather than use the more detailed, yet less formalised, descriptions of the local clinicians.

10 Data Banks

One difference between epidemiology and toxicology is highlighted by a discussion of data banks.

It seems rather inconsistent that, with the benefit of huge data banks, epidemiology has managed to demonstrate few definite relationships between factors and diseases in humans, whereas toxicology with relatively small data banks has established many more relationships between factors and diseases in laboratory animals. A defence of epidemiology is that its data banks have only recently been made sufficiently detailed and comprehensive to contain the necessary evidence for relationships to be harvested in twenty years time. Previously the records had been kept in manual, unstandardised form. Occasional attempts to analyse toxicological data accumulated over a series of studies have revealed that animal data can be just as inconsistent and subject to just as many temporal variations as human data. The computerised files of health and occupational data kept in the Scandinavian countries, and linked by personal identification numbers are enabling epidemiological studies to be done almost wholly by modern technology. Whether the increase in sensitivity of toxicological studies and the more rapid access to epidemiological results will lead to a safer environment for mankind remains to be seen.

11 Interpretation

It is at the interpretation stage that arguments about the validity of an epidemiological study will arise. It is a straightforward statement in an animal study to say that exposure to a certain chemical has caused a certain incidence of cancer in a particular species of rodent; that is only bad news for the chemical species and for the rodents.

Indeed the experiment has been planned and conducted in such a way that very few interpretations are plausible. An excess of disease in the exposed group can only be attributed to exposure or to chance.

If the same sort of statement is made about the same chemical in humans, then one needs to be extremely sure that the statement is correct. Otherwise the unhappiness caused to those people exposed to that chemical but still alive, and to the best of their knowledge in good health, is considerable.

Thus the interpretation of the power station study was that the employees had been the subjects of a long run of bad luck, and the cross-sectional study gave no reason to suppose that the sequence would continue. The long history of healthy employees in power stations in all parts of the world makes this a more plausible explanation than the existence of some malevolent factor in this particular building. Of course, there could be a complicated multi-factorial explanation of the excess incidence, but all plausible simple explanations have been exhausted.

One difficulty in the interpretation of an epidemiological study is that very few studies stand on their own two feet. Usually, similar studies of similar groups of people exposed to similar chemicals in similar circumstances will already be in the literature.

It is extremely difficult in this situation to piece together the various bits of epidemiological information to reach a conclusion about the harmfulness of the chemical in question (Beaumont and Breslow, 1981).

A technique called meta-analysis can be used to combine the results of clinical trials to obtain a more precise estimate of the efficacy of a new drug. Whether meta-analysis can be used to combine the results of epidemiological studies, when the protocols are usually quite varied, remains a controversial issue (Spitzer, 1991). The extent to which the interpretation of a single study can be generalised to other situations, in which the epidemiology has yet to be done, is always uncertain.

This issue has been addressed in detail in Bradford Hill's fundamental book (Bradford Hill, 1971) but we may note that the requirements laid down by Bradford Hill are very rarely found in a single study. One of the remarkable features of the relationship in people between exposure to vinyl chloride monomer and the occurrence of angiosarcoma of the liver is that all of the requirements are met. In most cases only one or two of Bradford Hill's requirements obtain.

In the reporting of many epidemiological studies, no greater claim is made than that the study has managed to generate a hypothesis about a chemical and a disease that is worthy of further study. This manner of reporting may be justified by the extent of the data, or it may be that the author does not wish to risk subsequent discredit. Nonetheless, the difference between hypothesis generating studies and hypothesis testing studies is becoming part of the folklore. If the extent of hypothesis generating studies is of some embarrassment to the manufacturer of the chemicals, then the plethora of negative studies is becoming an equal embarrassment to the environmentalist in his quest for a world free of toxic chemicals. Many epidemiological studies are almost certainly negative from the day they commence.

In many cases, the population available for study, the degree of exposure within that population, and the background incidence of the disease being studied are all so small that the relative risk due to exposure would have to be astronomical for a statistically significant result to be achieved.

It may be asked whether these negative studies, of very low statistical power, can be justified on any grounds whatsoever. One justification for them is the extent to which they set at rest the minds of those who have to work with the chemicals and who have been subjected to disquieting reports based on studies of laboratory animals. Nonetheless, it has to be admitted that a great deal of epidemiological work doomed to be negative is reported as if there is an equal chance of it being positive or negative. It is of doubtful sense to rely simply on the statistical analysis of the data and the *p* value which results from that analysis. The other considerations listed by Bradford Hill should always be taken into account and in particular the biological plausibility of the findings must be considered, in the light of the available toxicological information.

12   Essential Skills

From a sample of the techniques that an epidemiologist will use, it is possible to draw up a list of skills that he will need to be able to draw upon. Ideally an epidemiologist will have been trained in both medicine and statistics, and many leading epidemiologists do have qualifications in both subjects. More commonly, a medically qualified person will acquire the specialised statistical knowledge necessary for the role of epidemiologist, or a statistician will perform the role by using medically qualified colleagues as consultants on the clinical aspects. All epidemiologists will, however, need to consult a higher level of expertise than their own on several facets of their work.

Consider again a mortality study. Suppose that the aim of the study is to test the hypothesis that the age corrected proportions of deaths due to respiratory disease in two towns are equal after allowing for smoking habits, despite the presence of more heavy industry in one town than the other. Clinical skill will be required to derive a tight definition of 'respiratory disease', and statistical skill will be required for the analysis and the hypothesis testing. In addition, the classification of disease is itself a science, called nosology, which requires training and experience. The collection of data on smoking habits and domicile will involve a questionnaire, which, as it clearly cannot be answered by the decedents, must be entrusted to a relative or neighbour. The extent to which ambiguities, leading questions and misunderstandings can invalidate questionnaire data is itself a topic that has generated books and papers (Kalton and Schuman, 1982). If the study is large and concerned with many factors it will prove essential to use a computer for the storage and analysis of the data. It then requires systems analysis skills to prevent incorrect data, imprecise programming, and inappropriate statistical packages from destroying what had so far been an immaculate study.

The danger is that so many skills have to be harnessed for a successful epidemiological study that the task becomes too daunting to be undertaken. Fortunately, such defeatism is not warranted. Although a perfect epidemiological study may never take place, it is possible to set out procedures which, if followed, will guarantee an acceptable study that will not deserve to be torn to shreds when reported.

13   Conclusion

That then is what epidemiologists, or at least some epidemiologists, do, and how they do it. They face many obstacles and they never achieve perfection. Are their efforts always in vain, and always to be discounted, or do they produce the most valid evidence about human toxicology? What can be said about the relative weight that can be attached to epidemiological information? On the credit side, the species is certainly the most appropriate one to study, and the dose levels are certainly those which are appropriate to that species. On the negative side are all the deficiencies admitted to in the course of this chapter. In many cases, those deficiencies outweigh the advantages. It does seem, however, that where epidemiological evidence exists that can rightly be called substantial, in terms of population size, latency, and dose, then the two prime advantages of that evidence must to a great extent outweigh conflicting findings in animal or bacterial species.

References

Acheson, E. D., Barnes, H. R., Gardner, M. J., Osmond, C., Pannett, B., and Taylor, C. D. (1984). Formaldehyde in the British chemical industry. An occupational cohort study. *Lancet*, **1**, 611–616.

Anderson, M. E., Clewell III, H. J., Gargas, M. L., Smith, F. A., and Reitz, R. H. (1987). Physiologically based pharmacokinetics and the risk assessment process for MeCl. *Toxicol. Appl. Pharmacol.*, **87**, 185–205.

Axelson, O., Hane, M., and Hogstedt, C. (1976). A case-reference study of neuropsychiatric disorders among workers exposed to solvents. *Scan. J. Work Environ. Health*, **2**, 14–20.

Beaumont, J. J. and Breslow, N. E. (1981). Power considerations in epidemiological studies of vinyl chloride workers. *Am. J. Epidemiol.*, **114**, 725–734.

Blair, A., Stewart, P., O'Berg, M., Gaffey, W., Walrath, J., Ward, J., Bales, R., Kaplan, S., and Cubit, D. (1986). Mortality among industrial workers exposed to formaldehyde. *J. Natl. Cancer Inst.*, **76**(6), 1071–1084.

Bradford Hill, A. (1971). 'Principles of Medical Statistics'. 9th Edn. *Lancet*.

Breslow, N. E. and Day, N. E. (1980). Statistical methods in cancer research. Volume 1—The analysis of case-control studies. International Agency for Research on Cancer, Lyon.

Casarett, L. J. and Doull, J. (1980). 'Toxicology. The Basic Science of Poisons'. Macmillan. pp. 3–10. 2nd Edn.

C.I.I.T. (1981). Battelle Columbus Laboratories, Final Report, Docket No. 10922. A chronic inhalation study in rats and mice exposed to formaldehyde. C.I.I.T. Research Triangle Park N.C.

Coggon, D., Pannett, B., Osmond, C., and Acheson, E. D. (1985). A survey of cancer and occupation in young and middle-aged men. I. Cancers of the Respiratory Tract. *Br. J. Ind. Med.*, **43**, 332–338.

Cranmer, J. M. and Goldberg, L. (1986). Proceedings of the Workshop on Neurobehavioural Effects of Solvents. *Neurotox.*, **7**(4).

Doll, R. (1987). Environmental chemicals and cancer. *Chem. Br.*, (Sept) 847–850.

Doll, R. and Hill, A. B. (1950). Smoking and carcinoma of the lung—preliminary report. *Br. Med. J.*, **2**, 739–748.

Doll, R. and Peto, R. (1976). Mortality in relation to smoking: 20 years' observation on male British doctors. *Brit. Med. J.*, **4**, 1525–1536.

Epid. Work Group of the IRLG. (1981). Guidelines for documentation of epidemiological Studies. *Am. J. Epidemiol.*, **114**, 609–613.

Farr, W. (1839). Letter to Registrar General, in first annual report of the Register General of Births, Deaths and Marriages in England. HMSO. London.

Fox, A. J. and Collier, P. F. (1977). Mortality experience of workers exposed to vinyl chloride monomer in the manufacture of polyvinylchloride in Great Britain. *Br. J Ind. Med.*, **34**, 1–10.

Gardner, M. J., Winter, P. D., Taylor, C. P., and Acheson, E. D. (1983). 'Atlas of Cancer Mortality in England and Wales 1968–1978', John Wiley and Sons.

Graunt, J. (1662). Preface to 'Observations upon the bills of mortality, 1662' in the Economic Writings of Sir William Petty. Vol. 2. *Ed.* C. H. Hull. p. 333. C.U.P. London, 1899.

Kalton, G. and Schuman, H. (1982). The effect of the question on survey responses: a review. *J. R. Stat. Soc.*, (A) **145**, 42–75.

Kupper, L. L., McMichael, A. J., and Spirtas, R. (1975). A hybrid epidemiological study design useful in estimating relative risk. *J. Am. Stat. Assoc.*, **70**, No. 351, 524–528.

Landrigan, P. J. (1982) Recent advances in assessment of workplace exposure— epidemiologic linkage of medical and environmental data. *J. Environ. Sci. Health*, A17, **4**, 499–513.

Lillienfield, D. E. (1978). Definitions of epidemiology. *Am. J. Epidemiol.*, **107**, 87–90.

Macbeth, R. G. (1965). Malignant disease of the paransal sinuses. *J. Laryngol.*, **79**, 592.

Maltoni, C., Lefemine, G., Ciliberti, A., Cotti, G., and Carretti, D. (1981). Carcinogenicity bioassays of vinyl chloride monomer: a model of risk assessment on an experimental basis. *Envir. Health Perspect.*, **41**, 3–29.

MRC Environmental Epidemiology Unit (1981) The carcinogenicity and mutagenicity of wood dust. Scientific Report No. 1. MRC, Southampton.

MRC Environmental Epidemiology Unit (1984) Expected numbers in cohort studies. Scientific Report No. 6. MRC, Southampton.

O'Berg, M.T., Burke, C. A., Chen, J. L., Walrath, J., Pell, S., and Gallie, C. R. (1987). Cancer incidence and mortality in the Du Pont Company: an update. *J. Occup. Med.*, **29**(3), 245–252.

Paddle, G. M. and Bennett, B. (1983). An investigation of an excess incidence of heart disease. *Stat. Med.*, **2**, 477–484.

Schulte, P. A., Ehrenberg, R. L., and Singal, M. Investigation of occupational cancer clusters: theory and practice. *Am. J. Public Health*, **77**(1), 52–55.

Shore, R. E., Iyer, V., Altshuler, B., and Pasternack, B. S. (1992). Use of human data in quantitative risk assessment of carcinogens: Impact on epidemiologic practice and the regulatory process. *Reg. Toxicol. Pharmacol.*, **15**(2), 190–221.

Spitzer, W. O. (1991). Meta–meta analysis: Unanswered questions about aggregating data. *J. Clin. Epidemiol.*, **44**(2), 103–107.

World Health Organisation (1977). 'Manual of the International Statistical Classification of Diseases, Injuries, and Causes of Death'. WHO.

World Health Organisation (1985). Neurobehavioural methods in occupational and environmental health. WHO, Copenhagen.

CHAPTER 21

Information and Consultancy Services in Toxicology

D. M. CONNING

1 Introduction

Toxicology holds a unique position among the natural sciences. Not only does the solution to a toxicological problem usually depend on the assembly of data from several disciplines, the solution is itself often used to resolve a practical problem concerning human health or safety. Such problems may arise in relation to the general public with respect to community health or consumer safety; they may be confined to the industrial setting and relate to occupational health; or they may be concerned with a potential environment hazard. Toxicology supplies the data with which such problems are resolved and thus contributes directly to decisions affecting human health.

Information Services in toxicology, therefore, have to meet two important obligations—to ensure first that available data are assimilated, and second that the quality of the data is subject to expert judgement. These obligations apply, of course, to any issue where scientific judgement is involved but are particularly onerous for the information consultant in toxicology because of the wide spectrum of technical data to be utilised and the added importance of the potential human danger.

A third important aspect of a toxicological information service are the possible vested interests of those using the resultant data or opinion. In a society that places a high value on the quality of the lives of its citizens, three types of vested interest can be identified. The first is the manufacturer eager to sell his wares. A great many obstacles in relation to health and safety are imposed on anyone attempting to bring a new chemical product to the market. These, increasingly, exist on an international as well as a national basis and, although great strides have been made by international regulatory authorities to 'harmonise' their regulations, there are still many areas where conflict exists and possible barriers to trade occur. There is always the possibility that a manufac-

Table 1 *Examples of the wide range of databases available on toxicology and related subjects*

	Examples of databases		Hosts/systems providing access include:
Toxicology:	TOXLINE	includes Environmental Mutagen Information Centre file, Environmental Teratology Information Centre file, Pesticide Abstracts [11 subfiles]	
	RTECS	NIOSH Registry of Toxic Effects of Chemical Substances	
	TDB	Toxicology Data Bank	BLAISE-LINK [NLM computer, Bethesda]
	SPHERE	Scientific Parameters for Health and the Environment	
	CTCP	Clinical Toxicology of Commercial Products	BRS [BRS, New York]
	ECDIN	Environmental Chemicals Data and Information Network	
	OHM-TADS	Oil and Hazards Materials Technical Assistance Data System	
Carcinogenicity:	CANCERLIT	Carcinogenesis Abstracts and Cancer Therapy Abstracts	DATA STAR [Radio-Suisse, Berne]
	CANCERPROJ	Cancer research projects in progress	
	CCRIS	Chemical Carcinogenesis Research Information System	
Biomedicine:	MEDLINE	includes Index Medicus	DIALOG [Lockheed, Palo Alto]
	EMBASE	Excerpta Medica	
	BIOSIS PREVIEWS	Biological Abstracts and Bioresearch Index	DIALTECH [IRS, Frascati, Italy]
Chemical Information:	CHEMLINE, CHEMNAME, *etc.*	Dictionary files	
	ECOIN	inventories	
	TOSCA		
	CA SEARCH	Chemical Abstracts	DIMDI [Köln, W. Germany]

Category	Database	Description	Host/Provider
Pharmaceuticals:	IPA	International Pharmaceutical Abstracts	ECDIN [Datacentralen, Copenhagen]
	MERCK INDEX		
	MARTINDALE ONLINE		
Agriculture:	AGRICOLA	represents holdings of US National Agricultural Library	EPA/NIH-CIS[ISC, Washington]
	CAB Abstracts	Commonwealth Agriculture Bureau	
	AGRIS	agriculture, forestry, fisheries, food	
Food:	FSTA	Food Science and Technology Abstracts	INFOLINE [Pergamon — Infoline, London]
	FOODS ADLIBRA	developments in food technology, packaging, etc.	
Environment:	ENVIROLINE	pollution, contamination, etc.	ORBIT [SDC, Santa Monica]
	POLLUTION ABSTRACTS		
Packaging:	RAPRA ABSTRACTS	commercial, technological and research aspects of rubber and plastics	QUESTEL [Telesystemes, Paris]
	PIRA ABSTRACTS	Paper and Board Abstracts, Printing Abstracts, Packaging Abstracts	
	PSTA	Packaging Sciences and Technology Abstracts	
Legislation/Guidelines:	CRGS	Chemical Regulations and Guidelines System; USA	and others
	FEDREG	Federal Register Abstracts	
	Paklegis	national, international and EEC regulations affecting packaging	
	HSE-line	legislation, industrial health and safety; UK	

and many more

turer will be impatient of what he sees as unnecessary bureaucracy and be tempted to diminish the value or importance of having his product properly evaluated for potential hazards. He must be protected against this very understandable tendency and one line of protection is the information service.

The second interest arises in that government departments charged with the protection of public health cannot afford to ignore any data that might have a bearing on the problem. The personnel involved must acquaint themselves with all aspects of the issues—no easy task when few have any formal training in the subject. In these circumstances there has been a tendency to establish the 'checklist' approach in which a series of items of data are devised and applied to any particular issue. Although this is very satisfactory administratively, it undermines the requirement for expert judgement of difficult biological issues and there has been a tendency in some departments for the 'checklists' to be sustained when the 'science' has left them far behind, because of the urge to protect the system of regulatory control.

The third vested interest is that of the research scientist in toxicology who sees a number of exciting biological problems which have to be resolved before definitive decisions on product-safety can be made. Such a man, dismissive or impatient of the need for pragmatism in resolving the issues in a reasonable time, may, if unrestrained, impose many difficulties in terms of unavailable or unexplained data.

An information service has a duty to serve all of these groups and has the added difficulty of remaining independent of each. The independent but accurate assessment of toxicological information is thus the main objective of any information service and this chapter outlines the basic considerations needed in the attempt to achieve it.

2 Data Acquisition

The data to be used by the information section should relate not only to available data but to data that is likely to accrue from on-going experiments—and that may become available by the time a given product is in use. Therefore, it is of great value if the information personnel are in touch with toxicology research workers who operate in the various disciplines concerned, and who are aware of the activities in their field.

With the advent of computerised systems, the acquisition of data has been revolutionised in the last 10 years. There are now a large number of data bases which can be tapped by anyone willing to purchase the access (Table 1). There are, in addition, sources of information on on-going experiments in toxicology being run by or on behalf of regulatory authorities. These sources are not yet widespread and there is still a problem of duplication of toxicity studies, but the position is likely to improve as regulatory requirements become more international in their scope.

In addition to computerised data, there is still a large amount of toxicological information in conventional form in the older literature and this cannot be ignored in any serious appraisal. Here too, it is now easier to acquire copies of older documents from centralised systems (such as National libraries), but the costs can be substantial.

3 Data Interpretation

Ideally, information personnel researching the data on a particular compound should be able to obtain papers reporting high-quality experimental work on biological effects pertinent to the intended use and thus to the expected human exposures, and to define safe dosage and conditions. In many respects this ideal was nearly achieved before the advent of computerised data bases, because information personnel then had to search personally to acquire relevant papers and consequently the objectives were more narrowly defined and the information was more restricted. Now it is easy to obtain a list of all the papers that might conceivably be connected and the problem is to sift them for their true pertinence. This task can assume massive proportions because of the considerable expansion of the data now considered relevant to toxicological assessments and the tendency for papers merely to record technical data while presenting a diminishing amount of truly scientific information. It remains of paramount importance to interpret results in the light of human need and, for this, good science is essential.

4 Data Dissemination

Most assessments of toxicological data are in response to specific queries on certain compounds or groups of compounds, or on certain functional effects. Apart from the accurate resolution of potential hazards, there is a need to ensure that the recipient of the report understands its content and implications.

Two factors must be considered in the latter connection. First, the recipient may not be a trained toxicologist and the data must be expressed in terms that he understands, *i.e.* in 'lay' language if necessary. This often requires considerable skill and is helped by a thorough understanding of the second factor—that the information provided must relate to the problem giving rise to the request. The information consultant must seek to relate the data he acquires to the intended application. Thus, a problem concerning the hazards of exposure by inhalation in the work setting will usually not benefit from a treatise on teratogenicity or on whether or why the compound shows greater carcinogenicity towards one tissue than to another.

Such considerations give added impetus to the assertion that a proper understanding of toxicology data in relation to the actual problem is imperative; hence the need for interpretive expertise to be available.

Apart from the response to specific enquiries, an information service is much enhanced in value by the provision of a service in current awareness. Regular bulletins covering contemporary developments in specific fields in respect of both new science and new regulatory developments can be very useful. It is helpful if these, too, are written in 'lay' language but, again, the emphasis should be on quality and the identification of data that allow real conclusions to be drawn on health and safety problems for man and do not merely stimulate academic debate about interesting biological problems.

Finally, the very nature of information work leads to the compilation of surveys of available data about particular compounds. Such surveys, particularly if conducted in critical fashion, represent invaluable compendia of available data and are too often allowed to languish in archives. This results in duplication of effort and the loss of opportunity for each survey to build on the last. Wide dissemination has the added value that surveys undertaken for a particular reason (and thus from a restricted viewpoint) can often stimulate a cross-fertilisation of ideas when the topic is approached from another direction. Thus, the evaluation of a compound for its genotoxic potential may present data of value to the toxicologists, to the expert in carcinogenicity, or to the enzymologist. Wherever possible, such surveys should be made available by publication or other means.

5   Summary and Conclusions

This chapter has attempted to enumerate the principles on which an information service should be based. The most important principle is that the scientific work generating the information should be of high quality and should be interpreted in relation to the human hazards envisaged. This requires a knowledge of the circumstances in which the opinion is to be used, and access to toxicological expertise which may be of a specialist type. It is essential that papers in the literature are read carefully and that opinions are not based mainly on available abstracts.

Every effort should be made, within the bounds of confidentiality, to disseminate the results of surveys as widely as possible, and advice and consultancy should be couched in non-specialist language where possible.

Current awareness surveys, especially involving data from different sub-specialists, are to be encouraged if resources allow. The acquisition of many references now possible through computerised data banks is useless if the results are not vetted scientifically.

CHAPTER 22

Regulations and Advisory Requirements in Relation to Food

D. M. CONNING

1 Introduction

The development of toxicology as a separate biological science has been dependent largely on its role in underpinning the laws and regulations designed to safeguard human health and safety. Such regulations are now commonplace throughout the world and cover most aspects of human wealth-generating activity. Thus, there are regulations designed to safeguard the health of the individual in industry, in agriculture, and in nearly all aspects of consumer affairs.

That branch of toxicology dealing with this aspect—namely Regulatory Toxicology—is beyond the scope of this book but it is important that any student of toxicology is aware of the ramifications of the regulatory system. What follows is an account of the principles and development of regulations as applied to the safety of foodstuffs, the most developed part of the regulatory system, which will serve as an illustration of the complexities involved.

A The Origin of Food Laws

The laws on food safety differ considerably from one country to another, reflecting their own particular needs and experiences. The basic principles of food law, however, are the same in all countries, and are designed to protect public health and to prevent fraud.

The countries of Western Europe and North America have the most comprehensive and complicated food laws, many of which have been widely adopted by other countries. Developing countries are showing an increasing tendency to develop specific food laws that are more in keeping with their ethnic requirements.

This chapter concentrates on the laws in developed countries with particular emphasis on the laws applicable in the USA and United Kingdom. The underlying principles of the laws are explained. Although food laws are subject to frequent change, it is unlikely that the broad principles will differ appreciably.

The present food laws evolved from attempts to legislate against the adulteration of foodstuffs, a practice that was fairly common in the nineteenth century. The slow progress between the first food acts of the mid-18th century and those of the modern era was largely due to the lack of analytical expertise with which to enforce the various acts. As such methods have been developed, so has the extent of the regulations been increased.

2   Evolution of Food Law Exemplified by the United Kingdom

The evolution of food law in the United Kingdom resembles that seen in most other developed countries. Early in the nineteenth century, no laws existed to protect a consumer from toxic substances that were present in their food. Fraudulent trade practices, such as the dilution of cocoa with farina, coffee with chicory, or milk with water, were not unusual. Formaldehyde and boric acid were permitted as food preservatives and lead compounds were used as food colours. The increasing demand for laws to prohibit the adulteration of food carried with it the need for government inspectors and analysts, and was dependent on the availability of analytical techniques.

As reliable methods became available medical practitioners, scientists, and social reformers began to press for governmental legislation, but there was little activity until the so-called 'Bradford Incident' of 1858 when as many as 200 people suffered an outbreak of poisoning.

The first General Act to Prevent the Adulteration of Food and Drink was passed in 1860 and several more parliamentary bills and acts culminated in the 'Sale of Food and Drugs Act, 1875'.

The first laws relating to food colouring materials were of limited value. A list of those substances which were not allowed in food included lead compounds, arsenic, picric acid, and Manchester yellow, which from experience were known to be hazardous. A list of substances permitted for use in food was not available so that compounds such as artificial dyes, which had been developed for the textile industry, could be used in food. In due course, a list of permitted food colours was produced but the list of prohibited additives was retained.

Significant factors in the decline of food adulteration were the improved means for enforcing government legislation, more stringent analytical methods, and better qualified personnel. The increase in scientific knowledge enabled specific laws on food additives to be made, relating the level of use of an additive to that which is unlikely to be associated with human risk.

By the time the Food and Drugs Act was passed in 1955, the attitudes of food manufacturers, wholesalers, and retailers had changed appreciably. As a consequence of improved means of enforcing food laws, and continuing surveillance by consumer protection groups, detailed and comprehensive food legislation was included in the Food Act, 1984. Since then, however, a number of problems relating to animal health and food hygiene, together with the increasing requirement to ensure compatibility throughout the European Community, resulted in the Food Safety Act, 1990. This is an enabling Act which allows the government to make regulations governing any part of the food chain.

3 Structure of Food Laws

A Food Additives

In general, two approaches have been used for the control of additives. Most common is the 'positive list', that is, a list of those substances which may be used in food for the purpose of processing, preserving, or colouring. In principle, this system is adopted by all developed countries for most additive classes.

To a lesser extent, the 'negative list' system, which specifies only those substances which may not be added to food, has been used and is increasingly being replaced by the positive list.

B Standard of Composition

In most countries where food quality is controlled by law, the staple foods and certain luxury items are frequently subject to a standard of composition. This normally specifies minimum levels of important constituents (for example, the meat or fruit content) and the additives that may be used. Standards of composition are intended usually to prevent fraud rather than to protect health. In recent times, with the advent of comprehensive food and nutrition labelling, composition standards are being phased out.

C Codes of Practice

To cover all aspects of food quality by specific laws would be a formidable task. In those areas where some common standards are required but specific legislation is not justified, Codes of Practice have often been adopted which, while not legally binding, are widely accepted by the food industry. They specify an agreed standard and if food of a lower standard is marketed it may be construed as an attempt to defraud the customer. Codes of Practice are popular with industry because they are flexible, less expensive to enforce, and are voluntary.

D   Enforcement

To be effective, any law should not only be readily understood by those to whom it applies, but should be enforcible.

Specific methods relating to food additives are frequently dependent on the technical ability of the industry (for example, the accuracy by which a curer can control the level of nitrite present in a finished product, or how evenly can the nitrite be distributed throughout the meat). The accuracy and reliability of the analytical methods available are also important, since the maximum permitted levels of a substance may be directly related to the limits of detection by the analytical methods available at the time. Purity criteria for food additives are usually decided in consultation with manufacturers, since they are in the best position to know what level of purity is achievable. Maximum levels of permitted food additives depend on the levels necessary to achieve the desired technological effect, within the constraints of safety.

Usually, the government department responsible for enforcing food law (which may be the local authority) will appoint analysts and inspectors who take samples at all places where the food is prepared, manufactured, or offered for sale. Failure to comply with food law usually constitutes a criminal offence; this is the case in the United Kingdom.

E   Amending Legislation

Food laws throughout the world are subject to frequent changes due to the addition of new additives, the restriction in the use list of substances which, on the basis of new information, are found to be toxic, and the revision of regulations on package labelling. Food manufacturers themselves may initiate changes by the process of petition of the government department concerned.

F   Analytical Methods

These may be specified but often this is not the case, as laws cannot be changed with sufficient speed to keep pace with technicological advances. If an analytical method is not specified by law, an analyst may use the best method available at the time.

4   Naturally Occurring Hazards in Food

Although most developed countries agree on the need to have laws to control the use of additives in food, there are few laws relating to naturally occurring hazards. Such laws that do exist usually relate to heavy metal contamination which, though unavoidable, it may be possible to reduce. For example, in the United Kingdom there is a maximum level of lead permitted in foods.

The presence of other contaminants is sometimes controlled by specifying the 'level' above which action will be taken on the basis of hazard to health. The FDA, for example, have listed a number of food groups with action levels. Amongst the food groups covered are: fish and shellfish; flours and cornmeals; fruit, chocolate, and chocolate products; nuts and spices; vegetables. Some examples of the defect action levels are insect fragments in cornmeal and *Drosophila* fly eggs in canned tomatoes.

Countries rarely specify microbiological standards for use in the food industry. (Exceptions include France and the USA, where certain foods are controlled by individual standards.) The principal reason for this is that methods by which microbiological contaminations are determined tend to vary greatly and they are insufficiently precise for legislative purposes. In practice, government inspectors concerned with food-quality adopt their own standards for regulating food. As a consequence, advantage may be taken of new microbiological assays as they become available without restriction by legislative processes. With continually changing methodology it would be counter-productive if the regulations specified a method that might quickly become out-dated.

The absence of microbiological standards in food law does not diminish the protection of the consumer, given that the basic Food Acts in all countries state that food must not endanger human health. Most countries have regulations controlling food hygiene and officials are authorised to inspect food premises to ensure that good hygienic practices are maintained.

The number of countries introducing legislation setting maximum tolerances in food for the common mycotoxins, particularly aflatoxins, is growing steadily. Countries without any legislation in this area do not necessarily tolerate mycotoxins; their presence at toxicologically un-acceptable levels would be an offence under the general provisions of Food Acts which forbid the sale of any food that constitutes a danger to human health.

5 Food Additive Legislation

In the USA a food additive has been defined as any substance, the intended use of which results or may reasonably be expected to result, directly or indirectly, either in their becoming a component of food or otherwise affecting the characteristics of food. A material used in the production of containers and packages is subject to the definition if it may reasonably be expected to become a component, or to affect the characteristics, directly or indirectly, of food packed in the container.

A Legislation on Food Additives in the USA

All comments relating to food legislation in the USA refer only to the Code of Federal Regulations. The separate food laws which exist in the individual states are not considered here. If the food complies with the federal law then it should be accepted in the individual states.

The Federal Food, Drug and Cosmetic Act specifies that 'no additive shall be deemed to be safe it is found to induce cancer when ingested by man or animal, . . .'.

This is the so-called 'Delaney clause,' which has been the cause of much debate since it was enacted in 1958. Also in 1958, the food additive amendment was passed and a regulation was made listing specified food additives with 'Generally Recognised as Safe' (GRAS) status. If an additive appeared on the list then it was considered suitable for food use.

The GRAS list consisted of about 500 food additives divided into categories, for example chemical preservatives, emulsifying agents, nutrient and/or dietary supplements, sequestrants, stabilizers, anticaking agents, spices and other natural seasonings and flavourings, essential oils, oleoresins, and natural extractives.

The food additives appearing in the GRAS list were those which had been in common use. Although most of the GRAS-listed substances were without maximum levels of use, other than those in accordance with good manufacturing practice, there were restrictions laid down for some, such as 0.1% maximum for benzoic acid or the prohibition of the use of sulphurous acid and its salts in meats or foods recognised as a source of vitamin B_1.

After the GRAS list regulation came into force, any food additives which were not prior sanctioned, that is in common use before 1958 and which were not on the GRAS list, could not be used until a specific regulation was passed permitting their use. Often such regulatory approval could only be achieved by providing detailed results from a comprehensive programme of toxicity testing.

A number of regulations were passed subsequently permitting a further large number of food additives, including emulsifiers (tweens and spans), stabilisers (carrageenan, furcelleran), chewing gum base, modified starches, and indirect additives including boiler water additives, chemicals used in the washing or lye peeling of fruits and vegetables, and chemicals for controlling micro-organisms in cane and beet-sugar mills.

A food ingredient of natural biological origin, which has been widely consumed for its nutrient properties in the United States prior to 1 January 1958, without known detrimental effcts, and which is processed by conventional means as practised prior to 1 January 1958 and for which no known safety hazard exists, is regarded as GRAS without it being specifically listed. (However, a food which has been widely consumed in another country may not necessarily be accepted as GRAS in the United States.)

In 1969, the FDA decided that all the substances on the 1958 GRAS list should be re-evaluated because they were originally added to the list without a detailed examination of all available information. The review was conducted by the Select Committee on GRAS Substances of the Life Sciences Research Office, Federation of American Societies for Experimental Biology (FASEB). Its reports on individual GRAS substances were made available for comment at public hearings.

It is inevitable that food ingredients which have been in use for some time, and hitherto considered safe, should sometimes have their safety questioned when additional information becomes available. New safety evidence may not be conclusive but once an additive has been banned, it is difficult to reinstate it. To cover this situation, the US authorities invoke the concept of an interim food additive regulation. Additives for which new information indicates a possible hazard, but where there is still reasonable certainty that the substance is not harmful, are given interim status and may continue to be used pending further study. An additive in this situation will eventually be deleted from the permitted list unless toxicological work is initiated to prove its safety. When the work is finished the additive will either be reinstated or deleted.

B Purity Criteria for Food Additives

The introduction of statutory purity criteria for food additives is increasing. In the USA, all GRAS listed additives must comply with the purity criteria laid down in the Food Chemicals Codex. Other additives often have purity criteria specified in the regulation permitting their use.

In the UK there are purity criteria for all the permitted additives and often these are the Food Chemicals Codex criteria or those specified in EEC Directives on food additives. The EEC has introduced criteria for all of the permitted additives covered by Directives and these have been adopted by the individual Member States.

One of the most important contributions on purity criteria has come from the Joint FAO/WHO Expert Committee on Food Additives (JECFA), a committee that has evaluated the acceptability of a wide range of food additives. Its reports are published annually, and are accompanied by monographs and recommended specifications. These purity criteria are used by many countries as guidelines in the absence of national legislation.

C Regulations on Specific Additives

Antioxidants

The regulations on the addition of antioxidants to foods are similar to those on food preservatives, in that it is usual to limit the use of antioxidants to a short list of specified foods and to set maximum levels of addition of each antioxidant in each food. The restriction on the 'natural' antioxidants, such as ascorbic acid, are usually less than those for the synthetic antioxidants.

The most commonly permitted antioxidants are: butylated hydroxyanisole (BHA); butylated hydroxytoluene (BHT); propyl, octyl, and dodecyl gallates (gallates); ascorbic acid; ascorbyl palmitate; tocopherols.

Artificial Sweeteners

Legislation on artificial sweeteners has received more publicity than any other food additive. The prohibition of cyclamates in the USA in 1969 was swiftly followed by a world wide prohibition. Even countries which previously had no food additive legislation introduced laws banning cyclamates. The subsequent doubts about the validity of the USA ban led national governments to be more circumspect in their consideration of permitted lists. Consequently when, in 1971, the USA authorities threatened to introduce severe restrictions on the use of saccharin amounting to a virtual ban, other countries were reluctant to follow suit and most of those which did permit saccharin still continue to do so.

In most European countries, the use of saccharin as a food additive is limited to soft drinks and slimming foods. In the United Kingdom, there is no such restriction and saccharin is generally permitted in foods, except those covered by a standard of composition; its use in soft drinks is subject to maximum permitted levels. Other artificial sweeteners, such as acesulfame and aspartame, have recently been added to the UK permitted list, a reflection of the more liberal regulatory stance to this additive class being seen world-wide.

The legislative history of sweeteners indicates the dilemma governments experience trying to balance the risk and benefits of additives. In theory, as saccharin has been shown to cause cancer, the Delaney clause should have ensured a ban under USA law. In practice, the American Congress recognises that the dangers from overweight are real and the dangers from saccharin, whilst not easy to define, are probably very small; they therefore wish to keep saccharin as a permitted additive until the position is clarified.

Colouring Material

Most countries have lists of permitted colouring materials, even if they have very little else in the way of other food legislation. The addition of colourings to food has long been a source of concern to consumer pressure groups in several countries. Safety testing has been carried out to a greater extent on food colours than any other class of additive with the consequent steady reduction in the numbers of permitted substances.

The most commonly permitted synthetic colourings are Ponceau 4R, Carmorsine, Amaranth, Erythrosine, Tartrazine, Sunset Yellow, Quinoline Yellow, Greens, Indigo Carmine, Patent Blue V, and Black DN. The 'natural' colourings often subject to legislative control include annatto, caramel, vegetable carbon, carmine, carenlenoids, iron oxide, and titanium dioxide.

Caramel. The name caramel covers a wide range of different products, most of which cannot be chemically defined. Some countries have attempted to distinguish between the different types of caramels and to introduce specific maximum levels for them. The Danish and Swedish colouring regulations differentiate between caramels made by the ammonia process and non-ammonia caramels, setting higher maximum permitted levels for the latter. Both countries also specify maximum levels of 4-methyl imidazole in the caramels themselves.

In the United Kingdom, there have been suggestions over the past few years from the Ministry of Agriculture, Fisheries and Food that caramels should be divided into four types for the purpose of regulations: burnt sugar; caustic caramel (catalysed by sodium hydroxide); ammonia caramel; and ammonium sulphite caramel. Within these types, the 'catalysed' caramels should fall within six basic specifications. Once the numbers of caramels available have been reduced in this manner, it will be possible to carry out suitable safety testing.

Emulsifiers and Stabilisers

The number of permitted emulsifiers and stabilisers is greater than any other group of food additives. They include: sorbitan esters; polyoxyethylated sorbitan esters; cellulose ethers; alginic acid and its salts; and phosphates.

Flavourings and Flavour Enhancers

When considering the legislations on flavourings, it is necessary to recognise a few commonly accepted definitions:

Natural flavourings are obtained from plant or animal sources and are used as such or extracted and concentrated by purely physical methods.

Nature-identical flavourings are substances obtained by synthesis or isolated by chemical processes from raw materials but are chemically identical to substances present in natural products.

Artificial flavourings are substances which have not yet been identified in nature.

The problem legislators face in this area is the very large number of substances involved. Many natural flavourings are complex mixtures and the composition varies according to the source, time of harvest, and other factors. Artificial flavourings and nature-identical synthetics can usually be precisely defined chemically, and purity criteria established.

The International Organisation of the Flavour Industry (IOFI) has recommended that flavourings should be controlled by a 'mixed list' system of legislation. This would consist of permitting all natural flavourings except those shown on a list of prohibited substances or on a restricted list (*e.g.* maximum tolerance in food or restricted to certain foods only). Nature-identical flavourings would be treated in the same way as natural flavourings. Artificial flavourings would be controlled by a positive list and only those substances on the list would be permitted, with any necessary restrictions. It is suggested by IOFI that this would solve the problem arising from the enormous number of flavourings. It would offer some legislative control of flavourings but would still permit freedom to use a wide variety of substances in food to suit ethnic tastes.

JECFA has evaluated a number of flavourings for food safety. The Council of Europe (which is also part of the United Nations) published in 1970 a Partial Agreement on Flavourings, drawn up by a Working Party supported by the Member States of the European Community and Switzerland. This lengthy document, which was revised and reissued in 1973 and is still under review, contains a list of artificial flavourings considered suitable for use in foodstuffs without hazard to health; a list of natural flavourings admissible for use in foodstuffs, based on their sources; a list of flavourings considered harmful; and criteria for the toxicological evaluation of new flavourings. The lists include natural and artificial flavourings which may be temporarily added to foodstuffs without hazard to health, and a list of artificial and natural flavouring substances not fully evaluated.

The 1976 United Kingdom recommendations for flavourings legislation are based largely on the Council of Europe document. They inspired much opposition and no move has been made to adopt them, partly due to the fact that the European Community is also working on proposals for flavourings legislation.

In Belgium, Denmark, and West Germany there are lists of flavouring substances which may not be used in food. These include: coumarin, tonka bean, safrole, sassafras wood, birch tar oil, bitter almond oil containing hydrocyanic acid, thujone, cade oil, and calamus oil containing more than 10% asarone.

In the USA there are a number of natural and artificial flavouring substances which are considered 'GRAS'. The extension of the original GRAS list was due in no small way to the Flavouring Extract Manufacturers Association of the USA (FEMA), which evaluated the safety of flavourings required by industry and petitioned the FDA to have them included in the regulations. Not all the flavourings considered to be safe by FEMA feature in the appropriate section of the Code of Federal Regulations; FEMA publish their list.

The most widely known and widely permitted flavour enhancer is monosodium glutamate (MSG). Repeated investigations into the safety of MSG do not appear to have greatly restricted its use, although in many countries it is not permitted in infant foods. In some cases, manufacturers have voluntarily ceased to use MSG.

Hydrolysed vegetable protein and, to a lesser extent, hydrolysed animal protein are used widely but they are usually exempt from food additive legislation, although their use may be controlled by standards of composition of the food. The lesser used flavour enhancers are the sodium salts of 5'-inosinic acid and 5'-guanylic acid.

Miscellaneous Food Additives

Food acids. The most commonly permitted food acids are: acetic, adipic, citric, fumaric, lactic, malic, tartaric, succinic, hydrochloric, phosphoric, and sulphuric.

The laws on food acids do not usually specify maximum levels and the use of food acids is often not restricted to specific foods.

Bases. The most commonly permitted bases are: carbonates, hydroxides, phosphates, citrates, lactates, tartrates, acetates, and malates.

These are treated in the same way as food acids.

Anticaking Agents. The most commonly permitted are: silicon dioxide, magnesium silicate, sodium silicoaluminate, calcium silicate, ferric ammonium citrate, sodium and potassium ferrocyanide, tricalcium phosphate, silica gel, and magnesium oxide. They are restricted usually to such as powder mixes for vending machines, instant desserts, and salt.

Antifoaming Agents. The most commonly permitted are: dimethylpolysiloxane and vegetable oils.

Firming agents. Calcium chloride and calcium lactate are often pemitted as firming agents in pickles and canned vegetables, for example.

Glazing agents. The most common use of these is in sugar confectionery; the typical glazing agents permitted are: beeswax, spermaceti, carnauba wax, magnesium and calcium stearates, paraffin wax, and shellac.

Release agents. Some countries do not permit release agents other than vegetable oils, others permit mineral hydrocarbons (UK and USA), oxystearin, and sperm oil.

Humectants. Sorbitol and mannitol are widely permitted for this purpose.

Clarifying or filtering agents. Diatomaceous earth, bentonite, kaolin, and tannic acid are commonly permitted in a few countries which allow polyvinylpolypyrrolidone.

Flour improvers and bleaches. Of the few countries which do permit flour treatment (*e.g.* USA, UK, and Canada) the permitted substances include: acetone peroxide, benzoyl peroxide, potassium bromate, ammonium or potassium persulphate, chloride dioxide, azodicarbonamide, L-cysteine hydrochloride monohydrate, sodium and calcium stearoyl-2-lactylate, oxides of nitrogen, nitrosyl chloride, and sulphur dioxide.

Wetting agents. Dioctyl sodium sulphosuccinate is a permitted wetting agent in a few countries.

Clouding agents for soft drinks. Brominated vegetable oils have largely fallen from grace world-wide. Glycerol ester of wood rosin, glyceryl ester of colophony, and sucrose diacetate hexaisobutyrate are permitted in a few countries.

Propellants. Carbon dioxide, nitrous oxide, chloropentafluoroethane, and octafluorocyclobutane are the most widely permitted propellants for food aerosols (mostly whipped cream); butane, propane, and chlorofluorocarbon propellants are also occasionally permitted.

Packaging gases. Nitrogen and oxygen are the usual gases permitted.

Sequestrants. Ethylenediamine tetraacetic acid (EDTA) and its salts, and calcium phytate sequestrants are permitted under specified conditions in a few countries. Phosphates are very widely permitted in many foods in most countries.

Liquid freezants in contact with food. Dichlorodifluoromethane is permitted in a few countries only.

Preservatives.

The legislation on preservatives in food is similar in most countries. The most commonly permitted preservatives are: benzoic, sorbic, and propionic acids and their sodium, potassium, and calcium salts; ethyl, propyl, and methyl *para*-hydroxybenzoic acid esters and their sodium salts; sulphur dioxide and various sulphur dioxide-generating salts; sodium and potassium nitrate and nitrite; nisin; thiabendazole; hexamine; *ortho*phenylphenol; and pimaricin.

As with the antioxidants, the use of preservatives is usually restricted to a narrow range of specific foods.

Nitrite. Since the demonstration of the possible formation of carcinogenic nitrosamines in nitrite-cured meat, governments have been trying to find ways of reducing the maximum levels of nitrite that can be safely permitted in cured meat, without incurring the danger of the growth of *Clostridium botulinum* whose toxin is fatal to man. The search for a safe processing method using minimal levels of nitrite is continuing.

Ethylene oxide. Ethylene oxide is used for the fumigation of certain foods such as spices. It is not a permitted preservative but rather an insecticidal substance. Doubts over safety have resulted in a recent reduction in legislative approval.

Processing Aids

There is no clear distinction made between a processing aid and a food additive. Processing aids are an unpredictable area as regards the pressure of legislation; some countries do not legislate for them at all and others, such as the USA attempt to legislate in considerable detail.

Specific usage additives. The following list of 'specific usage additives' taken from the USA regulations provides a picture of the type of substance which would fit the concept of processing aid: boiler water additives used in the preparation of steam which will contact food;

chemicals used in washing or to assist in the lye peeling of fruit and vegetables; chemicals for controlling micro-organisms in cane sugar and beet-sugar mills; defoaming agents; sanitising solutions; flocculents for the clarification of beet sugar and cane sugar juice and liquor; ion-exchange resins; molecular sieve resins; extraction solvents; for the extraction of spice oleoresins; for decaffeination of coffee; and enzymes.

6 Pesticides

A large number of insecticides, fungicides, and herbicides are used for treating crops in many parts of the world. Because the residues of these substances persist in human food, it is necessary for countries to control their use. It is essential to keep the level of residues as low as possible, therefore maximum tolerances are laid down.

International organisations such as the FAO/WHO Codex Alimentarius Commission, JECFA, and the EEC have produced lists of tolerances for pesticide residues in or on food. The maximum permitted residue levels generally lie within the range $0.1 \, \text{mg kg}^{-1}$ to $20 \, \text{mg kg}^{-1}$, with the majority in the $1 \, \text{mg kg}^{-1}$ zone.

7 Packaging Materials

This term includes any material or equipment intended to come into contact with food. Many countries do not have detailed regulations on packaging materials; the laws in the USA and the Netherlands and the recommendation of the BGA in West Germany often act as guidelines for those countries without their own packaging laws.

The first EEC Directive on materials and articles intended to come into contact with food was brief and merely specified the basic concept that nothing which is harmful to human health or detrimental to the food shall migrate from the packaging material into the food. It is intended that subsequent directives will specify details of constituents of plastics and regenerated cellulose, glass, ceramics, and stainless steel. A limit for the tolerance of vinyl chloride monomer in materials and articles in contact with food to and in foods has been agreed. Global limits for the migration of harmful substances from materials into food will probably be set. Maximum tolerances for migration of certain heavy metals from glass and ceramics have been set. Discussions are underway to produce EEC lists of approved monomers and additives for food-contact plastics.

A The Netherlands

The Dutch have a sophisticated system of regulations broadly similar in form to the BGA lists; a series of positive lists has been drawn up for each commonly used plastic as well as other packaging surfaces. Many of the additives are controlled by specific migration criteria.

B   United Kingdom

The regulations reflect the existing EEC directives, and as such generally lack specificity. There is a Code of Practice drawn up by the British Plastics Federation in conjunction with the British Industrial Biological Research Association. This code, first published in 1969 and revised periodically, provides recommendations on the safe use of ingredients of plastics for food contact applications. In drawing up some of the recommendations, the legislation in other countries is taken into account. The Code includes specifications for a number of polymer classes.

C   USA

Several types of food-contact materials (*i.e.* indirect food additives) are specifically covered by US regulations, including plastics, rubbers, paper and paperboard, cotton, and cotton fabrics. The indirect food additive regulations comprise positive lists which are published in Title 21 of the US Code of Federal Regulations. These regulations cover the two main components of packaging materials (adjuvants and base materials), which are either listed separately or as part of composite regulations on end use. Additional qualification (for example regarding maximum permitted concentration, restricted end uses, and migrational requirements) are also often stipulated.

D   West Germany

Although the German Food Law gives Ministers the power to legislate for control of the migration of substances from packaging and utensils, little detailed legislation has yet been enacted. At present 'recommendations' from the Bundesgesundheitsamt (BGA), whilst not having the status of law, are strictly followed by industry. The various BGA Recommendations over the years have resulted in a series of positive lists of ingredients, subdivided according to the different types of polymers or other packaging materials, such as paper and paperboard.

Acknowledgement

We gratefully acknowledge the contribution of Mrs Carol Burke, who prepared the data on which this chapter is, in part, based.

References

This chapter has not been referenced but is merely a guide to the legislative approach adopted by a number of countries.

CHAPTER 23

The Influence of a Growing Environmental Awareness on Laboratory Design

D. ANDERSON AND B. COPELAND

1 Introduction

Within the space of little more than a generation, the manufacture of all kinds of chemical products has become surrounded by an array of legislative and advisory requirements. These have been instituted by Governments or advisory bodies in many countries and have stemmed from a growing awareness of the possible hazards which attend some of the chemical compounds we use today.

Since the Food and Drug Association in the USA published its guidelines on toxicity testing in 1959 (Association of Food and Drug Officials, 1959) followed in 1965 in the UK (Ministry of Agriculture, Fisheries and Food, 1965), many authorities have devised schemes for assessing the safety of chemicals based on studies of structure, acute toxic effects, *etc.* All of these schemes until the last decade or so depended for toxicological information on animal studies. More recently schemes such as that of the EEC 6th amendment (Sixth Amendment to Directive, 1979) and the Health and Safety Executive in the UK on the Notification of New Substances (Health and Safety Commission, 1981) have included non-animal studies. The EEC Directive 86/609/EEC requires a reduction wherever possible in the numbers of animals used and encourages research into the development and validation of alternative techniques. The EEC intends to ban the use of animals for testing cosmetics in the Community from 1 January, 1988, but only if alternative methods of testing have been scientifically validated (CEC, 1993).

Because of the expanding numbers of new chemicals and the anti-vivisection movement, it is becoming increasingly impractical to use animals for all the necessary test procedures. Two year carcinogenicity studies, for example, are very demanding of resources, both in time and money. There has thus been a need to develop more rapid and cost-effective screening procedures. In response, new methods have emerged which involve less work with animals. IARC (the International Agency for Research into Cancer) is currently using data from non-animal studies, *i.e.* short term tests, in its evaluation of carcinogens. Earlier requirements were more concerned with standards of safety and accuracy within the laboratory than with the compounds themselves. The general concern for safety of workers was given legal status in the UK with the wide ranging Act of 1974 (Health and Safety at Work *etc.* Act, 1974) and for their offspring with the Act of 1976 [Congenital Disabilities (Civil Liability) Act, 1976]. To augment the intentions of this legislation, various specialist advisory groups have established recommendations regarding standards and methods within particular types of laboratory. Thus, regulations or recommendations have been made to protect the workers when handling dangerous organisms (Howie Report, 1978), against possible hazards from genetic engineering experimentation [Reports of the Genetic Manipulation Advisory Group, 1978, 1979; Health and Safety (Genetic Manipulation) regulations, 1978; consultative review of proposals to amend regulations, 1987], when handling carcinogens (Precautions for laboratory workers, 1966, Factories—HMSO, 1967; Department of Labour, 1974), against over-exposure to chemicals in the workplace (American Conference Governmental Hygienist from 1938; Guidance notes from Health and Safety Executive—to date; The Control of Substances Hazardous to Health Regulations—pending COSHH regulations), and against radiation (Workplace Health and Safety Service—TUC Proposals, 1981). The two aspects of regulatory controls, namely chemical and worker safety, have continued to evolve in parallel (Figure 1) and have substantially influenced optimal laboratory design.

The development of laboratories carrying out the new type of toxicological experiments has been relatively slow. Whereas the importance of good environmental control for animal accommodation (mainly with a view to keeping the animals comfortable and disease free) has long been recognised, laboratories have been given little attention. Many existing buildings equipped with low grade fume cupboards and 'open window' ventilation have had to be adapted as well as possible to the greater scale of activity and the more critical standards of hazard containment. The development of Genetic Toxicology (*e.g.* Ames *et al.,* 1973; Department of Health and Social Security, 1981; Department of Health, 1989) has in particular precipitated onerous environmental requirements for a whole range of procedures, from the microbial level to the complex animal studies, such as long term carcinogenicity tests. In carrying out these studies to accepted codes of safety practice, the need for controlled environmental conditions, which has been highlighted in the genotoxic field, has been recognised. The main criteria for correct laboratory design are:

(a) The provision of the correct conditions for the experiment.

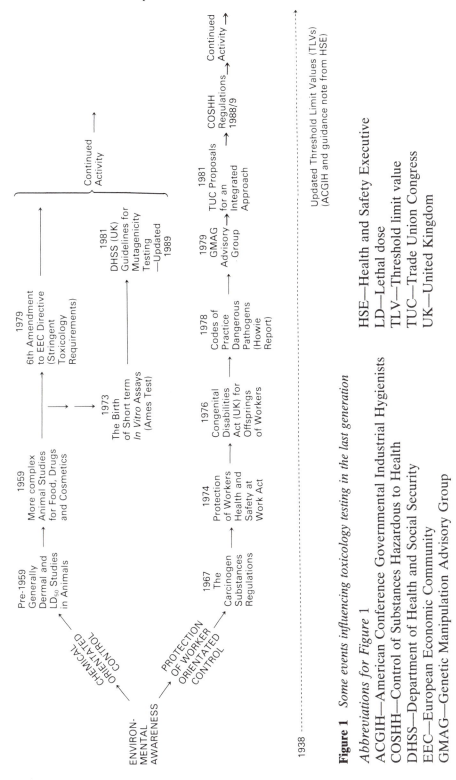

Figure 1 *Some events influencing toxicology testing in the last generation*

Abbreviations for Figure 1

ACGIH—American Conference Governmental Industrial Hygienists
COSHH—Control of Substances Hazardous to Health
DHSS—Department of Health and Social Security
EEC—European Economic Community
GMAG—Genetic Manipulation Advisory Group

HSE—Health and Safety Executive
LD—Lethal dose
TLV—Threshold limit value
TUC—Trade Union Congress
UK—United Kingdom

(b) The protection of the laboratory worker against hazard from the materials or organisms used.

(c) The protection of the external environment against noxious laboratory waste products.

Of these three aspects, (c) has least effect on the internal laboratory environment itself and will not be considered here.

Within the toxicology laboratory today a wide range of scientific disciplines are practised. These include;

Acute Toxicity/Long Term Toxicity   Immunology
Bacteriology   Pharmacology
Biochemistry   Pathology
Genetic Toxicology   Reproductive Toxicity
Haematology   Teratology
Inhalation Toxicology

Their requisites in terms of laboratory conditions separate basically into facilities for *in vivo* and for *in vitro* work.

2 Animal Laboratories

It has long been recognised that artificial, highly engineered environments, maintained at constant humidity and temperature, are desirable for animal accommodation. This is because each animal species has preferred environments and where long term tests are concerned, there is a need to keep animals free from disease. High efficiency air filtration and non-recirculatory ventilation systems are essential.

The subject of staff well-being has been of secondary concern and this has had consequences. For example, the incidence of allergies in humans caused by contact with animals or animal wastes, such as faeces or dander, is substantial. Proper attention to detailed design will eliminate many of the problems. Thus, (1) Careful integration of the airflow pattern with the room design, utilising a linear room layout, ensures that the operator (as well as each animal cage) has a continual supply of clean air (Figure 2) and eliminates 'Dead Spots' and low level extract grilles (where harmful particulates may collect in the room). (2) Particulates are removed by introducing a filter at the air extract point if required. (3) Surfaces should be constructed of materials and finishes which avoid ledges and joints, to facilitate cleaning. (4) Where protective clothing is worn within the animal room (gown, mask, boots, gloves) a strict operating protocol for personnel washing/changing and for storage/disposal of protective wear is necessary, and facilities for this procedure should be planned *en suite* with the laboratory accommodation. (5) To give adequate protection against the test substance, procedures such as dispensing or dosing of compounds should be carried out in a separate enclosure within the room. A simple negative pressure isolator design which can be used in such circumstances (Figure 3) has the advantage that, unlike a fume cupboard, it passes relatively small quantities of air and thus does not disrupt the air flow pattern (Figure 2).

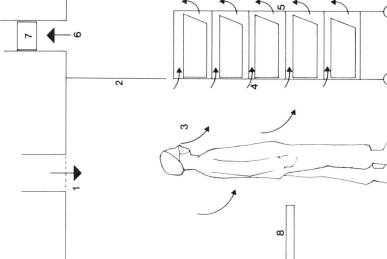

(1) ROOM SUPPLY AIR
(2) SUSPENDED VALANCE TO PREVENT SHORT-CIRCUITING OF AIR FROM SUPPLY TO EXTRACT
(3) FRESH AIR FLOW DOWN FACE OF OPERATIVE
(4) AIR PASSES BETWEEN EACH LEVEL OF CAGES APPROXIMATING TO LAMINA FLOW CONDITIONS
(5) MOBILE RACK CONTAINING CAGES SPECIALLY DESIGNED TO MEET AIR FLOW REQUIREMENTS
(6) AIR EXTRACT
(7) SAFE CHANGE FILTER ACCESSIBLE FROM OUTSIDE ROOM
(8) BENCH

Figure 2

(1) SIDE PORT FOR INTRODUCTION OF SPECIMENS, *ETC.*

(2) FLEXIBLE EXTRACT DUCT (CONNECTED TO CEILING OUTLET)

(3) PORT IN WORK TOP FITTED WITH LID

(4) DRAWER IN WHICH SPECIMENS ARE DEPOSITED FOR REMOVAL AFTER PROCEDURE

(5) WIDE DIAMETER FLEXIBLE PLASTIC GLOVE PORTS

(6) SEMI-FLEXIBLE CLEAR PLASTIC FRONT

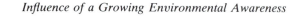

Figure 3

3 General Laboratories

The most common and familiar item of furniture within the laboratory is the bench and at one time—with great convenience but perhaps less accuracy and safety—virtually all laboratory work was carried out on the bench top. In order to achieve open bench top working to satisfy today's requirements, it would be necessary to guarantee a stable environment for the experiment and that dangerous byproducts are removed from the occupants of the room. This would require the installation of a comprehensive and costly mechanical ventilation system to constantly purge the whole room in a controlled way.

Another alternative, used in hospital operating theatres and manufacturing clean rooms where very specific activities may be undertaken in permanent locations, has been to install sophisticated whole-room air systems (using, for instance, vertical laminar flow techniques). The complexity and variety of activities within the toxicology laboratory generally preclude this solution, however, and in most instances a more flexible and cost-effective approach must be found. One method is to provide each worker with a 'space suit' complete with its own life support system, and leave the laboratory as a simple untreated room. This idea works in terms of building economics, but involves a great deal of inconvenience and manipulative difficulty for the worker and has the added risk of cross contamination between experiments.

The strategy generally adopted is to use an additional enclosure within the laboratory, either on the bench or instead of the bench, to provide a localised environment for each individual experiment.

A Fume Cupboards

The common general purpose enclosure is the fume cupboard. In its most basic form this is a box fitted on the bench in which an experiment is carried out, and through which air is drawn from the room and exhausted *via* a duct to the outside. The basic design has been greatly developed in recent years. The chief problem has been in the eddy currents set up by items of equipment inside the cupboard and by the operator's hand and body movements. In modern cupboards the surrounds have been streamlined to enable a smoother air entry, thus reducing the risk of turbulence. To ensure efficient purging, air speeds have been increased and, for the most demanding work, face velocities (of the air entering across the sash) normally fall within the range 0.4–0.8 m sec^{-1} (Figure 4).

The quantities of air to be removed from the room to satisfy this requirement are considerable. For instance, in a general purpose toxicology laboratory a ventilation rate of 25–30 air changes per hour may be necessary in order to provide a 1.5 m wide fume cupboard for each pair of workers.

Despite the usefulness of fume cupboards in the laboratory, there are a number of short-comings associated with them:

(1) EXTRACT DUCT (NOTE THAT FAN IS LOCATED REMOTELY AS PART OF THE BUILDING VENTILATION SYSTEM)

(2) VERTICALLY SLIDING LAMINATED GLASS SASH

(3) FLARED SIDES ACCOMMODATING SERVICES CONTROLS

(4) AEROFOIL SECTION AT SILL ENSURES GOOD SWEEP OF AIR ACROSS WORK SURFACE AND FORMS AN AUTOMATIC BYPASS TO PREVENT FACE VELOCITIES AT LOWER SASH OPENINGS FROM BECOMING TOO HIGH

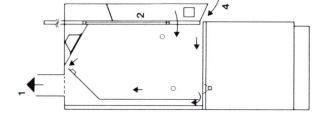

Figure 4

(a) The air movement in a fume cupboard can upset the accuracy of some experiments, for instance in the microbiological area.

(b) Despite modern fume cupboard design, personnel movement may cause eddy currents which allow fumes to spill from the front opening.

(c) The high extraction rate of non-recirculated room air causes high energy demands for heating and ventilation, with consequent effect on running costs.

B Safety Cabinets

A great many activities demanding special precautions within the laboratory may be better undertaken in the safety cabinet. Laminar flow cabinets have been used in tissue culture for some time. Their use, primarily, was to prevent the work from being contaminated by the environment, using horizontal laminar flow systems which did not provide the worker with any particular protection. Designs of cabinet are now available incorporating more sophisticated air flow arrangements, based on a vertical laminar flow system that reduces turbulence caused by the operator's hand movements to a minimum, whilst at the same time keeping the working environment at bench top level constant (Figure 5).

Some safety cabinets are designed primarily to protect the worker and make no provision for constant experimental conditions (Figure 6). The impetus for the development of these units in Britain, for example, has been created by the Howie Report (1978) which suggested basic safety standards for work with micro-organisms. This report has been followed by a new British Standard (BS 5726) which lays down specific requirements for safety cabinets. One general advantage of these cabinets is that they use a great deal less air than fume cupboards, either recycling all air back into the room *via* powerful (HEPA) filters, or exhausting only 10–20% of the air passing through the cabinet with consequent saving on energy consumption and heat loss.

C Isolators

A third type of enclosure is the isolator, which completely separates the operator from the workplace by a continuous membrane. The use of isolators is usually associated with work with highly dangerous pathogens in the medical field, and they range in size from a small glove box for the bench work, to a complete tent for the enclosure of a patient. Whilst offering extremely good protection to the operator, there is the additional advantage that the air inside is relatively still and for this reason, isolators are particularly useful for the weighing and dispensing of dangerous chemicals. One toxicological application for an isolator has been illustrated previously for use in an animal room (Figure 4). Another (Figure 7) is a proposed design for a combined isolator and fume hood for use in a laboratory involved in the handling of hazardous chemicals.

(1) FAN
(2) HEPA FILTERS
(3) DOWNWARD LAMINAR AIRFLOW
(4) AIRFLOW FROM ROOM
(5) HEPA FILTERED EXHAUST AIR RELEASED BACK INTO ROOM (AN EXHAUST SYSTEM COULD BE FITTED FOR DISCHARGE TO OUTSIDE IF REQUIRED)
(6) PERFORATED WORK SURFACE
(7) FLUORESCENT LIGHTS

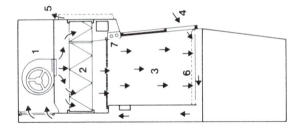

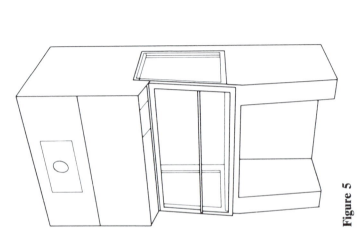

Figure 5

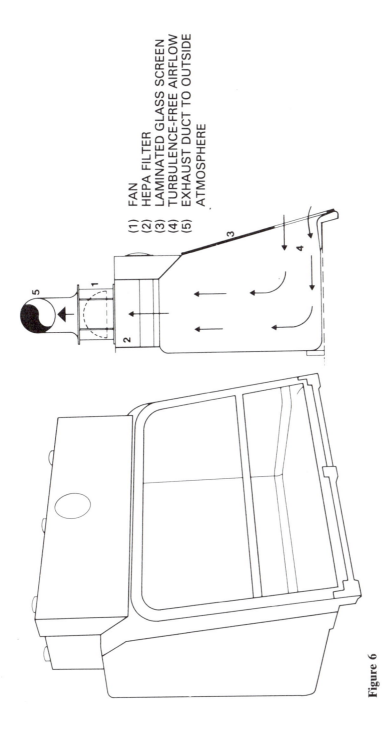

(1) FAN
(2) HEPA FILTER
(3) LAMINATED GLASS SCREEN
(4) TURBULENCE-FREE AIRFLOW
(5) EXHAUST DUCT TO OUTSIDE
     ATMOSPHERE

Figure 6

(1) EXHAUST DUCTS
(2) LARGE WALK-IN FUME CUPBOARD WITH HORIZONTAL SLIDING GLAZED DOORS
(3) SMALL GLASS SLIDING PANELS FITTED TO EACH DOOR AT BENCH HEIGHT
(4) SAFETY CABINET FOR WEIGHING
(5) VERTICALLY SLIDING SASH FITTED WITH GLOVES
(6) VERTICALLY SLIDING SASH BETWEEN FUME CUPBOARD AND SAFETY CABINET

SAFETY CABINET IS USED FOR STORAGE, WEIGHING, AND DISPENSING OF DANGEROUS CHEMICALS. NORMAL OPERATION IS WITH SASH UP, OPERATING LIKE A CLASS 1 CABINET. FOR WEIGHING, IN ORDER TO ENSURE MINIMUM AIR MOVEMENT SASH COMES DOWN AND WORK IS CARRIED OUT THROUGH GLOVES. CHEMICALS CAN BE PASSED THROUGH TO FUME CUPBOARD ADJACENT, AVOIDING CARRYING DANGEROUS COMPOUNDS THROUGH THE LABORATORY.

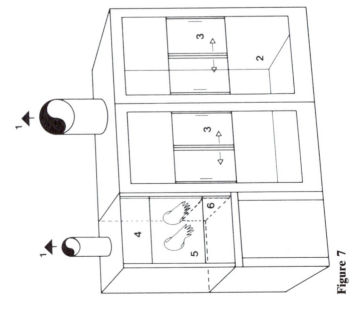

Figure 7

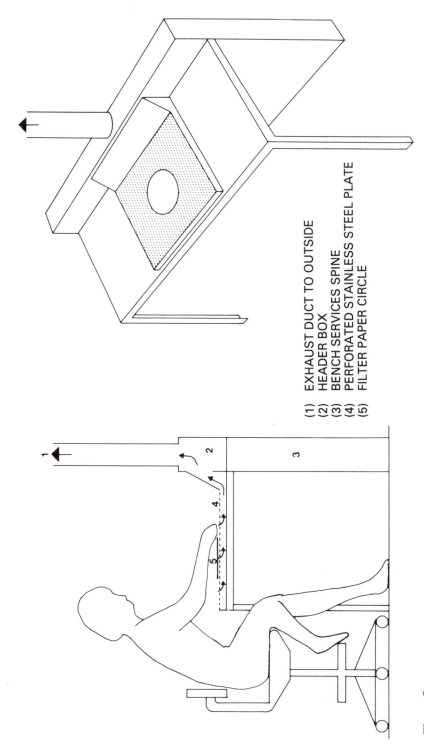

(1) EXHAUST DUCT TO OUTSIDE
(2) HEADER BOX
(3) BENCH SERVICES SPINE
(4) PERFORATED STAINLESS STEEL PLATE
(5) FILTER PAPER CIRCLE

Figure 8

Notwithstanding the various enclosures now available, there are still some tasks which, although hazardous, can only be carried out on the bench because of the dexterity required. In these instances there may be no alternative but to provide adequate personal protection, though it is still sometimes possible to devise apparatus to satisfy particular problems. An example is the simple bench equipment developed for use in the preparation of histology slides, where formaldehyde fumes are kept from affecting the worker by extracting them from below through a perforated metal plate (Figure 8).

D   Ventilation

These special environments for individual experiments depend entirely on the precise control of air flow and it is essential that the laboratory itself has a predictable environment in terms of temperature, humidity, filtration and air movement. Ventilating the laboratory by natural means (*i.e.* opening windows) cannot give an adequate degree of control; rather it is necessary to seal the room and install air conditioning to provide comfort conditions for the workers and equilibration of air for fume cupboards and other enclosures.

Not all the spaces in a laboratory building need to be air conditoned, and much cruder, and therefore cheaper, environmental systems can be

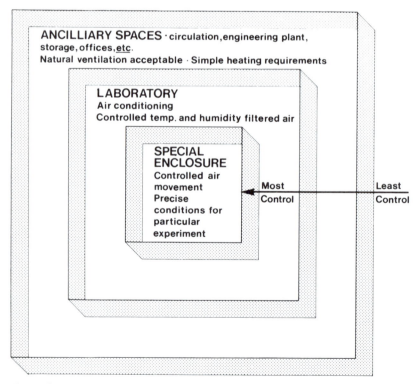

Figure 9

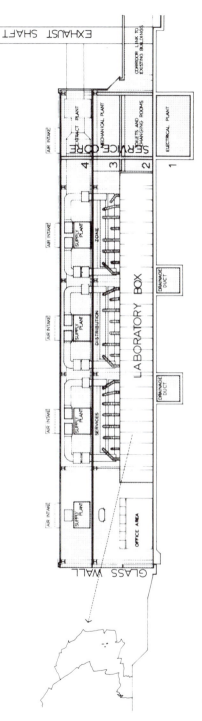

Figure 10

used for such areas as corridors, engineering workshops, stores, and offices. (Figure 9). The hierarchical principle illustrated has resulted in a design (Figure 10) in which comparatively expensive air conditioned laboratory space has been accommodated in a box inside a larger, comparatively inexpensive 'shed' building. The ancilliary accommodation is arranged within a buffer space between the inner laboratory box and the outside walls of the building. This arrangement significantly reduces the load on the laboratory air conditioning due to variations in outside weather conditions.

In such a building the costs of running the air conditioning can be reduced by installing a heat recovery system which utilises the cooling plant as a heat pump during the winter period. The design requires a large amount of space for the engineering installation. The quantities of plant and distribution systems installed are considerable and additional space is needed for accessibility for maintenance and replacement. To avoid disruption to laboratory work and to reduce potential hazards to non-scientific staff, it is desirable that maintenance access be from outside the laboratory space. It is possible to accommodate all these engineering requirements successfully but it would, in most cases, be extremely difficult to achieve the same standards by attempting to refurbish an existing laboratory building of a multi-storey layout with inadequate space between floors to accommodate the ventilation ductwork.

It is clear that the complexities of providing the correct scientific environments, together with adequate safety, has a radical effect on the building design required. Substantial investment in more controllable environments is necessary, together with a responsibility, placed on the scientist, to ensure that each laboratory task is carried out safely. The scientist will inevitably need to be aware of the standards achievable within his particular building, and what they cost.

In many cases where standards have to be improved, the cheapest and most effective solution in the long run may well be to build anew rather than refurbish. Keeping in mind the disruption caused to ongoing work by construction activities in an occupied building, the apparent additional costs of a new structure may be theoretical rather than real. Its advantages in terms of a well-laid out and efficient engineering system, as well as an internal environment designed flexibly to accommodate the many types of enclosures and other laboratory equipment now necessary, are self evident.

References

American Conference Governmental Industrial Hygienists (1983–to date). Threshold Limit Values for Chemical Substances and Physical Agents in the Workroom Environment with Intended Changes for a Particular Year.

Ames, B. N., Durston, W. E., Yamasaki, E., and Lee, F. D. (1973). Carcinogens and mutagens: a simple test system combining liver homogenates for activation and bacteria for detection. *Proc. Natl. Acad. Sci. (USA)*, **70,** 2281.

Association of Food and Drug Officials of the United States (1959). Appraisal of the Safety of Chemicals in Foods, Drugs and Cosmetics. Topeka, Kansas.

CEC (1993). Council Directive 93/35/EEC of 14 June, 1993 amending for the sixth time Directive 76/768/EEC on the approximation of the laws of the member states relating to cosmetic products. *Off. J. European Communities,* **36,** (L151), 32–37.

Congenital Disabilities (Civil Liability) Act, HMSO, London, (1976).

Department of Labour (1974). Occupational Safety and Health Standards, Subpart G.—Regulations on the Handling of Carcinogens. Part 1910 of Title 29. *Fed. Regist.,* **39,** 23502.

Department of Health and Social Security (1981). No. 24. Report on Health and Social Subjects. Guidelines for the Testing of Chemicals for Mutagenicity. Committee on Mutagenicity of Chemicals in Food, Consumer Products and the Environment. HMSO, London.

Department of Health (1989). No. 35. Report on Health and Social Subjects. Guidelines for the testing of Chemicals for Mutagenicity. Committee on Mutagenicity of Chemicals in Food Consumer Products and the Environment. HMSO. London.

European Economic Community (1986). Council Directive 86/609/EEC on the approximation of laws, regulations and administrative provisions regarding the protection of animals used for experimental and other scientific purposes. O.J. No. L358, 18.12.1986.

Factories, The Carcinogenic Substances Regulations. HMSO, London, (1967).

First Report of the Genetic Manipulation Advisory Group. (7215). HMSO, London, 1978 and 1979.

Guidance Note from the Health and Safety Executive. Threshold Limit Values for a Particular Year (to date) HMSO, London.

Health and Safety at Work *Etc.* Act. (1974) Chapter 37. Health Safety and Welfare in Connection with Work and Control of Dangerous Substances and Certain Emissions into the Atmsophere. HMSO, London.

Health and Safety Commission (1981). Consultative Document. Notification of New Substances. Draft Regulations and Approved Codes of Practice. HMSO, London.

Health and Safety (Genetic Manipulation) Regulations 1978 and Review of Health and Safety (Genetic Manipulation) Regulations 1978, consultative paper (1987).

Health and Safety Annex 1, 21 July, 1987. The Control of Substances Hazardous to Health Regulations 198–(Draft).

Health and Safety at Work (1987). What the COSHH Regs will mean in practice. **9**(11), 21–22.

Howie, J. (1978). Department of Health and Social Security—Working Party to Formulate a Code of Practice for the Prevention of Infection in Clinical Laboratories. HMSO, London.

Ministry of Agriculture Fisheries and Food. (1965). Memorandum on Procedure for Submissions of Food Additives and on Methods of Toxicity Testing. HMSO, London.

Precautions for Laboratory Workers who Handle Carcinogenic Aromatic Amines. Chester Beatty Research Institute, 1966.

Sixth amendment to Directive 67/548/EEC on the Approximation of the Laws, Regulations and Administrative Provisions Relating to the Classification, Packing and Labelling of Dangerous Substances. (1979). Annex VII. *Off. J. European Communities* (Sept.).

Second Report of the Genetic Manipulation Advisory Group. (1979). (7785). HMSO London.

Workplace Health and Safety Service (1981). TUC Proposals for an Integrated Approach, TUC Publications.

CHAPTER 24

Good Laboratory Practice

ROSEMARY I. HAWKINS

1 Introduction and Background

Good Laboratory Practice (GLP) Compliance is a legislative requirement which has completely altered in concept since its initiation in 1979, from being guidelines applicable only to American Food and Drug Administration (FDA)-related non-clinical studies, to a universal requirement for the International acceptance of toxicological data.

It was introduced by the FDA because this agency was dissatisfied with the quality of data submitted for safety-in-use evaluation. The legislation proposed in 1976 for GLPs was finalised in the Federal Register in December 1978 (Federal Register, 1978) with an enforcement date of 20 June 1979. While not applicable in retrospect, the ruling made life initially difficult because of its applicability to that part of any study underway at the time. Although aimed primarily at toxicological laboratories, the GLPs regulate any laboratory contributing data to non-clinical safety tests of FDA-related products (*e.g.* food and colour additives, drugs, medical devices, *etc.*), the exemption being method development and efficacy tests.

The legislation, which will be considered in detail later in this chapter, embodies those routinely expected standards of good scientific practice, together with certain additional requirements hitherto unpractised in toxicological laboratories. Any such legislation requires both implementation, and the FDA opted for a mechanism of both laboratory and study data inspection, including foreign as well as US facilities, and enforcement operating by eventual disqualification.

Following the FDA initiative, however, other legislative authorities in the US felt the need for drawing up similar guidelines, the most notable being the draft proposal from the Environmental Protection Agency (EPA). This agency proposed not one, but several sets of GLPs regulating for the various legislation under its jurisdiction, and thus the proposed Guidelines for the Toxic Substances Control Act (TSCA) (Federal Register, 1979), Federal Insecticide, Fungicide and Rodenticide

Act (FIFRA) (Federal Register, 1980) and the Physical, Chemical, Persistence and Ecological Effects Testing (Federal Register, 1980), all purported to include GLP standards. These multiple GLP requirements posed a potential threat to the toxicology laboratory. Firstly, the draft proposals were all sufficiently different to cause compliance problems for the testing facility, more especially for the laboratories conducting contract safety evaluation studies on clients' behalf, as the study submission agency had to be first clarified from the sponsor and it was inevitable that compliance of these laboratories must be to those most stringent requirements. More importantly, there was the confusion of the generic principles of laboratory practice with those specific requirements pertaining to a particular type of study. Furthermore, the draft GLPs were completely inflexible (in contrast to those of the FDA), being drafted in terms of absolute requirement rather than guidelines, and they were not internationally applicable, demanding for example adherence to American international legislation and personnel qualification. Such documents certainly did not fulfil the uniformity of interagency legislation as promised by the Interagency Regulatory Liaison Group (IRLG) established at that time to standardise American legislation.

Together with the multiplicity of GLPs there was, of course, the added burden of the multiple inspections by agency teams for their enforcement, and European laboratories were faced with a prospect of the disruption and subsequent cost burden of multinational and multiagency inspections.

In Stockholm in 1978, therefore, the Organisations for Economic Co-operation and Development (OECD) initiated the formation of an expert group to consider the whole question of GLP guidelines for its member nations and to formulate an international document. The requests of such a document are that it is mutually acceptable to all member states and that it is internationally applicable.

Any requirements for GLP must embody three aspects, namely to establish the guidelines themselves, a mechanism for the assurance of compliance, and a mechanism for enforcement. Initiated after the legislation from America, it seemed logical that the OECD requirements should be formulated along similar lines; industry had spent considerable resources both in time and capital expenditure to comply with the FDA and the acceptance by this country of any European system was paramount to the exercise.

2 Current Situation

The OECD Expert Committee has now finalised three documents (OECD, 1982):
(1) OECD Principles of Good Laboratory Practice.
(2) Implementation of OECD Principles of Good Laboratory Practice.
(3) Guidelines of national GLP inspection and study audits.

It has been agreed that study data generated under conditions of the OECD Principles of GLP will be internationally acceptable.

The OECD has therefore fulfilled its remit in this area and produced generic, internationally acceptable guidelines, from which it was intended that member nations would prepare their own national GLPs, so formulated to take into account internal legislation and needs. However, it was agreed that it was the National government's own responsibility to set up a suitable and mutually acceptable mechanism for the implementation and for the enforcement of compliance.

The World Health Organisation (WHO) considered GLP as part of the International Programme on Chemical Safety, and agreed to adopt the OECD approach, thus embracing a still wider range of countries.

OECD guidelines carry no member enforcement requirement and the GLP Principles were mostly used directly translated as recommendations by European governments to their national industries. However, an EEC Directive must be implemented by Member States and following the 6th Amendment to the Dangerous Substance Directive, 1981, the UK Health and Safety Executive (HSE) published a set of GLPs (HMSO, 1983) to regulate for those tests carried out under the Notification of New Substances Regulation, 1982. The authority also set up an implementation mechanism of faulty inspection and study audit similar to that of the FDA, and a GLP Monitoring Unit was established.

It must be remembered, however, that this was only operative in the UK for a limited area of work and not for other toxicological studies. A step in this direction was precipitated when the Japanese Pharmaceutical Affairs Bureau, Ministry of Health and Welfare (MHW), established 'standards for Conducting Safety Evaluation Tests on Drugs' (Ministry of Health and Welfare, 1981) which became fully applicable as from 1 April 1983, for those non-clinical studies conducted for new drug applications, for manufacturing and importing, and for application for re-examination. Studies submitted from foreign laboratories had to be accompanied by government assurance of compliance to national GLPs of similar or higher standards compared with the Japanese GLP requirements. Eventually there was agreement to accept FDA or OECD regulations as this equivalent, but the need for a government Statement of Compliance acted as a form of trade barrier to those countries where no such government's inspection mechanism had been established.

As the initial inspection mechanism operative in the UK was that for the HSE for monitoring compliance of organisations generating data in respect of the Notification of New Substances Regulations, the Department of Health and Social Security (DHSS) established, as an interim measure, a GLP Unit located in, but distinct from, the Medical Division concerned with Toxicology and Environmental Protection to ensure compliance with OECD Principles for the Pharmaceutical and Cosmetics Industry. Distinct from the HSE Monitoring Unit, the authority assured industry that it would avoid duplicate inspection, but this was clearly an unsatisfactory situation. However, as a result of discussions with the DHSS, an arrangement has now been made with the Japanese MHW as being the first step to producing a full agreement on GLP, by exchange of *notes verbal* in March 1984.

For UK industry there followed another problematical Japanese requirement for the GLP compliance of agrochemicals, issued by the Ministry of Agriculture, Forestry and Fisheries and made known to UK industry with little or no forewarning, with an effective date of 1 October 1984. As compliance to OECD GLPs was accepted by the Japanese Government as the equivalent to its national requirements, there was no problem with study conduct, but the preliminary registration of the testing facility with the Director of the Agricultural Production Bureau was a requirement fraught with difficulties. Details of *retrospective* GLP compliance over a period of 3 years, plus facility description and floor plans, instrument information and location, personnel *curriculum vitae,* and even financial aspects of the organisation were required—clearly a topic for discussion between governments, and one which is currently under careful review.

In October 1984, the UK Government finally announced the decision to base the long term comprehensive GLP monitoring scheme in the Toxicological and Environmental Protection Division of the DHSS. The Department arrangements for pharmaceuticals and cosmetics were then extended to include all laboratories involved in the health and environmental safety testing of agrochemicals and food additives, together with the HSE responsibility for GLP monitoring for the notification of new chemicals. In 1986 the DHSS published 'Good Laboratory Practice, The United Kingdom Compliance Programme', which was the explanatory document in the principles and operational procedures adopted by the UK GLP Monitoring Unit. To ensure maximum uniformity at an international level, the GLP principles are those requirements as developed by the OECD and reproducing and extending the provisions published initially by the HSE. The Monitoring Unit is structured to carry out laboratory inspection and study audits and is currently negotiating with the regulatory authorities, notably Japan and USA, for agreement of mutual acceptance of inspection programmes. The inspections are of two categories, regular monitoring every 2 to 3 years, and monitoring at the specific request of a regulatory authority, either national or foreign. The former is initially conducted at the request of the facility itself, but once inspected it is included in the Unit's on-going monitoring programme in order to maintain the knowledge of the laboratories compliance status. The operating costs are recovered from the participating laboratories which are allocated to one of three categories of scale of charge.

European countries are similarly in the position of requiring to formulate GLPs from the OECD guidelines and, more important, to establish a mechanism of government assurance that these have been adhered to and which will be acceptable to all authorities. Most national governments are in the position of publishing a set of GLPs, although it is not always clear as to the status. The situation is obviously one of continual change and the following resumé must, when at the time of going to press, be out of date. In France for example, the Ministere des Affairs Sociales et de la Solidarite Nationale Sante published an official text 'Bonnes Pratiques De Laboratoire' (BPL), instructions regarding principles of BPL (May, 1983) and Laboratory Inspection (September, 1984), but this applies at the moment only to pharmaceuticals and medicines. A memorandum of understanding is being negotiated between the FDA and the French Directorate for Pharmacies and Medicaments (*Food Chemical News*, 1984). Currently the separate inspectorate for studies regulated under the EEC 6th Amendment is not yet in place.

In Germany it is necessary to differentiate between chemicals and pharmaceuticals. While no GLP regulations have yet been issued, the Chemicals Act following the EEC Directive on Dangerous Substances allows for governmental action and the OECD principles have been published reflecting that non-compliance under the German legal system could be regarded as culpable negligence and failure to exercise reasonable care in study conduct. There is no requirement for GLP-compliance in the generation of data on pharmaceutical products but the local health authorities are prepared, on request and a fee-paying basis, to inspect and issue compliance certificates for the Japanese requirement.

The Italian Ministero della Sainta has started a GLP programme by sending questionnaires to industry, but meanwhile recommending compliance to either FDA or OECD GLPs.

In Scandinavia, the Swedish National Board of Health and Welfare published some guidelines in the summer of 1983 which apply to pharmaceuticals. In Denmark, public laboratories concerned with control of industrial chemicals, pesticides, and food additives are working towards preparation of determined guidelines and their implementation; private laboratories comply with OECD requirements by request, but so far there is no requirement by the National Board of Health for Pharmaceuticals.

When considering the European situation, its complexity must be viewed with the realisation that each country not only has established mechanisms of government agency authorities with responsibility for distinct areas, but also in some cases regional authorities. Thus, the OECD principles needed to be generic in order to allow flexibility of existing national regulations and mechanisms.

Further change in the GLP status of various countries has been initiated, led by new legislation, namely the EEC Pharmaceuticals Directive, December 1984, covering the requirement for GLPs for pharmaceuticals, and the proposed Directive on Good Laboratory Practice designed to ensure that all facilities in the Community comply with GLP by 1 July 1988. The draft Directive requires member states to ensure compliance by inspection and also outlines the need for authorisation of GLP requirements within the community.

The American scenario is also changing. While there is yet no update of the FDA GLPs as promised (for example, with the onerous requirements of sample retention), the final EPA GLPs have now been published (Federal Register, 1983), both from TSCA (applicable to both Section 4, the 6th Amendment of the Dangerous Substances Directive, and Section 5, the testing of existing chemicals if so requested), and from FIFRA. These requirements apply to any study conducted, initiated, or supported on or after 29 December 1983. Thankfully there is harmonisation with both the FDA requirements and with the OECD, and within the EPA itself the only differences in the 2 sets of laws relate to different statutory requirements. The most significant of these differences is in the scope; the TSCA regulations contain a special section to cover environmental toxicology as well as health effects.

The concept regarding GLP requirements has thus changed radically since 1978, when only those laboratories involved with US FDA submitted safety-in-use studies were concerned. Now these requirements are universal for the generation of toxicological data in non-clinical studies and for this data to be internationally acceptable, and this, of course, applies to all facilities conducting such studies, academic, industrial, or contract laboratories.

More and more countries are recognising the need for establishing National GLP principles and a mechanism of compliance assurance outside of OECD, EEC, and EFTA groups, an example being South Korea, thus allowing for the confidence in, and exchange of, reliable data on an ever-increasing scale.

Following these fragmented beginnings progress since the first edition of this book has happily unified the issue for all Member States of the European Economic Community (EEC). Requirements mutually agreed by all Member States finally became embodied in a Council Directive to be implemented into the National legislation of each State. Negotiations in Brussels, where finalised when the EEC accepted the OECD principles of GLP on behalf of all Member States and requirements harmonised in Directives 87/118/EEC of 17 January 1987 and 88/320/EEC of 11 June 1988 on the application of GLP principles to chemical testing and the inspection and verification, respectively. The latter was amended in 90/18/EEC to replace the reference and OECD guidance documents in order to facilitate interpretation throughout the Community, Annex A and B contain the guides for monitoring procedures for GLP and the conduct of laboratory inspection and study audits.

Now established in Community requirements, GLP has become an integral part of numerous other Directives on toxicological and safety issues (*e.g.* cosmetics, industrial chemicals, medicinal products, food additives, animal feed additives, pesticides, genetic engineering). All laboratories have a need to ensure all such relevant texts are part of its testing programme and be alert to all subsequent amendments which may follow.

3 Specific Requirements

The requirements for GLP compliance can be separated into those practices which are standard in any facility fulfilling good standards of study conduct, and those which are a new addition to the routine of the toxicological laboratory. However, as with most guidelines it is all a question of interpretation and of rational judgement as to the best way the requirements of each particular organisation can be served, so as to maintain compliance with the maximum cost-effectiveness. The FDA have given sufficient scope for company policy to be operative and to allow for managerial thought along these lines.

A Special Requirements

Standard Operating Procedures

Standard Operating Procedures (SOPs) are defined as 'those procedures in writing setting forth non-clinical laboratory study methods that management is satisfied are adequate to ensure the quality and integrity of the data generated in the course of a study'.

The FDA suggest areas which require SOPs to be established, but in practice every department needs to have SOPs for each and every procedure which is performed, from cleaning glassware and laboratory equipment, reagent and media preparation, to carrying out specific test methods. In initiating GLP in any laboratory it is the writing of SOPs that becomes the time-consuming tedium against which all scientists rebel. However, carefully constructed, they can become valuable tools as laboratory training documents and reference manuals.

Authorisation by management is required; the easiest technique is to have the signature of the scientist in charge to assure technical authenticity, and the signature of the manager responsible for facility compliance, both of which signatures must be dated. A sufficient number of copies of each SOP must be available to ensure that members of staff conducting each procedure have ready access to the SOP. With technological change, method development, acquisition of new equipment, progress with new techniques, *etc.,* continual updating of SOPs is required. A historical file of all SOPs must be maintained, thus providing a procedural document as reference for past studies.

In practice, in laboratories conducting studies on behalf of a sponsor, procedures are often developed which are specific to the nature of material under test. In these circumstances study-specific SOPs are written, but confidentiality must be respected so that they are not included in the departmental manuals but retained separately with other study data.

The format of SOPs depends upon the laboratory but it is best to keep them straightforward and easy to follow. Additions of technique can easily be made, and use of standard text book methods or maintained manuals are permissible either as a direct photocopied inclusion, or as a reference providing the location of the original is clearly indicated and adequate for the procedure to be carried out without detriment.

Quality Assurance Unit

The Quality Assurance Unit (QAU) is the mechanism whereby each study is monitored to assure management that it is in compliance with GLP requirements. Thus, it must monitor the facilities, equipment, personnel, methods, records, quality control, and performance. The QAU may be a separate department staffed by several individuals if the organisation is sufficiently large to demand this (and in practice most are) but in smaller laboratories it can be any member of staff, providing they are independent of the personnel engaged in the direction and conduct of the study in question.

Industry was familiar with quality control, but the QAU is distinguished from this by being the overall monitor of all those actions necessary to ensure the proper conduct of the study and the integrity of the data generated. Quality control is the responsibility of the head of the department; quality assurance ensures there is adequate quality control as part of its function, the other functions are specifically designated by the FDA as follows.

The QAU must maintain a master schedule identifying all the non-clinical studies being carried out, the nature of each study including documentation of the test material and test system, sponsor, study director, start date, current status, and final report status. The rationale behind such demands is to ensure that the facility is not overloaded and can give correct application to each individual study, but in reality maintenance of confidentiality only allows for an agency to be permitted access to information on studies relating to its own legislative remit, so that this requisite plus the retaining of past schedules loses some of its impetus in practice.

The QAU must maintain a copy of the protocol for each study for which it is responsible. The QAU must inspect each critical phase of each study, and report this inspection, identified with the date and phase indicated and any action required, recorded together with the date when it is reported to management. In effect, the definition of what constitutes a critical phase is a matter for decision for each organisation. For most safety evaluation studies, there will be certain standard phases but QAU inspections must be related to all individual studies. It is usually most helpful to designate the inspection phases on finalisation of the authorised protocol and at this time to determine any unusual or non-routine steps. The inspection programme can then be scheduled from the study plan. The actual QAU inspection should be conducted without prior knowledge of the staff carrying out the study whenever possible, and should be performed with the minimum of disturbance. In order to achieve this it is useful to design a check list for each inspection phase which can be completed simply and quietly by the QAU staff, and who then undertake a full inspection report on returning to the Unit. For studies lasting longer than 6 months, as inspection is required every 3 months, usually this can be designed to coincide with a phase inspection, for example, in a long term study involving animal parameters at the time of general observations or weight measurements, and thus reduce the time involved for QAU staff. Similarly, the requisite report on study status can be so timed.

Apart from the study inspection, the audit of the final report is one of the most important and time-consuming areas of the QAU requisite. The audit is to assure that the report correctly describes the methods used and that it accurately reflects the raw data. It must be remembered that in no way is the QAU involved with the scientific interpretation of the raw data, which is the responsibility of the study director, but it is responsible for reducing the number of errors reproduced in the final report from the raw data to the minimum possible level.

Final report audits can best be divided into two sections, raw data and the actual report.

Raw data. The raw data includes all the original observations, worksheets, records, memoranda, and notes generated during the conduct of the study, and may include photographs, microfilm or microfiche copies, computer printouts, magnetic media, dictated observations, and records from automated instruments. The raw data from any one study must be clearly and correctly annotated and authorised and should be in such a condition that the study report and conclusions could be reconstructed from the data alone in future years without the original report or personnel being available.

The raw data are typically included in the final report in the form of appendices. These must be checked for transcription and omission errors. In most toxicological studies this will include hundreds or even thousands of numerical information and descriptions, including the laboratory (including animal house environmental data), animal, and pathological data. A complete check on every item, while desirable, and I believe is the only sure way if the report size and staff permit, can become impractical and a sampling procedure must be utilised. The procedure used will depend upon the choice of each laboratory, but the two most frequently employed are the technique of sequential analysis (Moroney, 1956) and the British Standards Institution method (British Standards Institution, 1972).

Apart from the straightforward errors of transcription, one of the most frequent problems encountered with raw data audits is the error of omission. All the raw data must be included in the final report unless there is a specific reason for this exclusion. Often this will arise, especially with certain biochemical parameters or if not required by the sponsor, but the difficulty is to train the scientist responsible to annotate the raw data clearly and authoritatively as to what has been omitted and the reason why. Other common problems are the re-identification of groups of values which have not been clearly indicated, the non-inclusion of the derivation of manipulated data (especially where this has been generated using a hand-held calculator) and the different phraseology employed in pathology reporting. This latter is a matter of education of the pathologists to provide a code or key to those comparable diagnoses in the raw data which have been grouped together under a common term in the final report. While the FDA accept the final diagnosis as the study raw data, thus allowing the pathologist the necessary flexibility of thought, it is impossible for an auditor to understand or accept differences in terminology and this must be regarded as an error.

It is usual to check 10% of the calculations, but machine-derived data is of course not subject to re-calculation, although the data input must be checked.

The error level acceptable to the QAU auditor must depend upon the nature of the errors involved. Usually once an error level of greater than 10% is apparent the complete set of raw data is returned to the study director to carry out a 100% data check. If only a few errors are detected, and of a magnitude and nature not to affect the final outcome of the report, then these are identified and returned for correction. An audit report is always made, identifying fully the study, project number, study director, identity of raw data, the result of the audit, and any action taken or test required. This is authorised and dated by the auditor and by management.

The audit of computer-driven data is the most recent problem for the QAU with the increasing development of this instrumentation in toxicology. Direct on-line capture of data must be conceived with the advice of the unit staff so that some form of assurance will be built into the programme of the identification of the individual responsible (together with prevention of unauthorised data entry by personnel or study identification); date of entry; return capacity of incomplete or erroneous input; and any changes made, together with the data, reason, and personnel responsible for the change.

When observations are so entered, directly as input to a computer data base (either on-line, or as batch records) which is maintained on magnetic media (disc or tape), the computer data base is considered as raw data, or if preferred for economic reasons the resultant printout which contains the collected data (providing these are clearly authorised). When observations are captured in the form of the written record, it is this written record which forms the raw data and any subsequent steps, such as transfer of a computer data base, are not considered as raw data.

Report. The actual typed text of the report must be checked for procedural details, for completeness, and for accurate reflection of the data in a non-scientific context, such as dates, times, frequencies, values, and numerical conclusions. The tables and figures must also be checked for completeness and the accuracy of any annotation, and to ensure that this tallies with any references made to them in the text.

It is usually more expedient to check the final draft of the written report, as this is more cost-effective if changes have to be made. Re-audit of both text and appendices should take little time, the check for final editorial and typing errors is the responsibility of the study director.

Each final report must contain a compliance statement, prepared and signed by the QAU, which specifies the dates inspections were made and the findings reported to management. It should be noted that the inspection findings themselves are confidential to the laboratory and the FDA cannot ask to see them, but a laboratory may be requested to certify that inspections were implemented, performed, documented, and any action required fulfilled.

All the procedures and responsibilities of the QAU must be fully documented, SOPs are required for all aspects of the units remit, and all records maintained by the unit are required to be kept in one location.

Study management. This third requisite caused some laboratories to undergo an organisational re-think, while for others it was merely an adjustment of an already existing managerial system.

This requirement was for the establishment of a study director before the initiation of a study, to have overall responsibility for the technical conduct of the study, as well as for the interpretation, analysis, documentation, and reporting of results. In this way, a single point of study control is established and becomes the contact for external and internal liaison. It is the organisation's responsibility to ensure that the study director designated has the appropriate education, training and/or experience for such a position, and that replacement, if it beomes necessary to do so during the conduct of the study, is carried out promptly and with the required documentation. There are certain activities which have been specifically regulated for the study director: to assure that the protocol is approved and followed; that all experimental data are accurately recorded and verified; that unforeseen circumstances which may affect the study quality and integrity are noted and corrected if needed; that all applicable GLP regulations are followed and that all raw data, documentation, protocols, specimens, and final reports are eventually deposited in the archives.

B General Requirements

The remaining requirements can be considered as those which would be normal practice for the correct conduct of a non-clinical study in any properly constructed facility, but now with the necessity of the documentation of previously accepted conditions and the maintenance of records and observations in a specific way.

Personnel

Personnel conducting or supervising non-clinical studies should have the correct education/training/experience for the function performed, and the job description and training/experience records must be documented. Assurance is required that there are sufficient number of personnel, correctly clothed for the study conduct and taking the appropriate precautions against personal contamination. Any ill-health that may adversely affect the quality or integrity of the study should be documented and the necessary action taken.

Facilities

Organisations have now to be able to demonstrate that they are of suitable size, construction, and location to undertake non-clinical studies, and that there is an adequate degree of separation of various facilities to prevent any inter-contamination. This includes adequate space for administration and supervision and sufficient locker and toilet facilities for all of the staff, as well as the necessary general and specialised laboratory areas and space for cleaning and sterilisation. Specific requirements for animal care and facilities are designated, and it must be remembered that these requirements were stuctured in conditions existing in the USA and that in the UK, for example, such specifications appear superfluous in light of the existing inspections carried out by the Home Office. Thus, the facilities are required to be constructed and located to minimise any disturbances to the studies and there must be adequate space for separation of different species and studies. Disposal of animal waste and refuse must be such as to prevent any environmental contamination, and disease hazard. Separate and adequately furnished areas are required for storage of bedding, supplies, equipment, and animal feeds, which are protected from infestation from wild animals and contamination. Similarly, in order to prevent contamination and possible mix-up of materials for different studies, separate areas are required for receipt and storage of the test article (material under test) and any control article, for the mixing of the test and control article with a vehicle if required, and the storage of these mixtures.

Equipment

Equipment employed in the conduct of the studies needs to be demonstrated to be of appropriate design and adequate in capacity. The location needs to be that for correct operation, inspection, cleaning and maintenance. While the necessary SOPs for these functions are located by the instrument, written records of all inspection, maintenance, testing, calibration, and/or standardisation operations, together with the date and staff involved are needed, and any non-routine repairs and details of default are recorded.

In order to be cost-effective, compliance to any requirements should be carried out with the specifics of the individual organisation in mind. Thus, the combination of these first three requirements to form a departmental manual can be met with some success. Each department has specific facility and operational requirements, and these are detailed together with a ground-plan to demonstrate adequacy of that department.

Personnel details are confidential and as such should be restricted to location in the Personnel Department. However, each manual contains a list of the staff with their general background, current education details, date of organisation membership, signature, and initials—most important as they play a vital role on raw data sheets and identification is essential. Staff changes are indicated by a single line drawn through the resumé, together with the reason and date of change, thus maintaining a historical record for past studies. Equipment is listed with description, serial number, and date of purchase, together with the details of manufacturer and/or in-house maintenance of relevant SOPs detailing its cases. The in-house training is the responsibility of the head of that department and, as such, designation of a member of staff to carry out each department's and other department's SOPs is the indication that training is judged fulfilled and the training record for that member of staff is complete. Thus a check-board listing of SOPs against staff proceeds a list and copy of each SOP and completes the departmental manual.

Such a document, held by the QAU whose senioral staff as required fulfils requirements, and assures compliance, ensures confidentiality, and allows for rapid managerial analysis and decision making.

Reagents and Solutions

All reagents and solutions are required to be labelled for identity, titre or concentration, storage requirements, and expiration date. The rationale for this is pefectly clear, it ensures deteriorated or outdated material will not be inadvertently employed during the study, but it has not been fully thought out. It is impossible, for example, to place an expiry date for a stock bottle of analar grade sodium chloride. It is preferable that, for such reagents ordered after the GLP compliance date, the date of receipt from the suppliers is stamped onto the label as an indication of age, and any storage requirements, other than those of normal conditions, clearly indicated. Shelves in laboratories can easily be marked as being for those reagents requiring normal storage conditions only.

Animal Husbandry

The requirements set out in this section again must be considered in view of conditions and existing legislation in the USA, and the need for the

FDA to incorporate certain requirements to make good the deficits which do not necessarily exist in other countries.

Separate areas for quarantine and the isolation and treatment of sick animals are not usually provided in such organisations in the UK. Animals designated for a study are immediately housed in the room identified for that study alone, and any animal so ill as to need treatment would be killed, preventing possible spread of infection and/or removal from the test for medication.

However, other details of animal care are applicable: the appropriate identification of each animal with full details on the cage, and the adequate separation of different studies and animals species.

Cleaning of cages and accessory equipment, the use of adequate non-contaminating bedding, and recording of pest control are all routine. However, initially in the UK a quite problematical requirement was that of the analysis of animal feed and water to ensure contaminants capable of interfering with the study are not present. Animal feed manufacturers are now well informed of this requirement and provide adequate analytical data with each batch of foodstuffs. The FDA have agreed that the manufacturer batch analysis is acceptable, as most laboratories would not have the necessary equipment or expertise, but do require analysis if the test article is such that a possible food contaminant might interfere with the study. It should be remembered, however, that each batch of animal food requires an analysis document (and for each study supplied from this batch) and that these analyses must be kept together with the rest of the raw data for each study. A more difficult requirement for the UK was that of drinking water analysis; the water supply is more standardised than in America where private well, bore hole, or river supply is common. By this time most water authorities are used to the approach from laboratories and will offer 3 monthly analysis of random site samples.

Test and Control Articles

Many laboratories, which conduct safety evaluation studies on behalf of sponsors, are not able to specifically fulfil the FDA requirements for complete documentation on the test—and sometimes the control—article(s). The requirements for definition and documentation of identity, strength, purity, and composition (or characteristics to allow for definition) may well be confidential information and even if it can be made available, the laboratory may not have the equipment or expertise to ensure the necessary characterisation. One method of complying to this request with the limited means available is to define the sponsor's responsibility for having this information and for assurance that the material given to the facility to test fulfils this analysis. On receipt of the material from the sponsor, the laboratory should check the container and ensure that it is visibly undamaged in any way; the receipt form once signed should then be returned to the sponsor for his signature that the undamaged material so received fulfils the specification analysis and/or identity requirements. It is the laboratory's responsibility to obtain from the sponsor all knowledge of stability, storage requirements, and possible hazard handling precautions and whether these will be altered in any way following mixture with the vehicle, where applicable. Each sample must be correctly identified—and this must include the lid as well as the body of any large container, and for a study lasting longer than 4 week's duration, reserve samples should be kept for as long as they afford analysis—if mixed with the diet as vehicle this will be for the duration of the viability of that diet batch as specified by the manufacturer. When mixed with such a vehicle, the uniformity and concentration of the article/vehicle mix is determined.

Study Conduct

The requirements specified for the conduct of the non-clinical studies are not unusual in themselves but did require, for many laboratories, the redesign of worksheets for data capture.

Thus, the study is required to be in accordance with the protocol, specimens adequately identified, and all experimental procedures carried out as requested. All data generated is recorded directly, legibly, and in ink, or at least in an unerasible medium. All data entries are to be dated and authorised by the member of staff responsible for the entry, hence the importance of the recognition of initials. Changes in the entry must not obliterate the original entry, and must be dated and signed and the reason for the change given. In practice, no data sheets can be that large, so it is normal for a code of errors to be keyed for each set of data and this to be used to keep laboratory worksheets clear, simple, and legible. Computer data needs the data input to be identified and the same requirements for data change.

Protocol

The requirements set out by the FDA for the information to be contained in the study protocol are those which it considers minimal. These are as follows:
  (1) Descriptive title and study purpose.
  (2) Test and control article identification.
  (3) Name of the sponsor and the name and address of the testing facility.
  (4) Proposed starting and completion dates.
  (5) Justification of choice of test system.
  (6) Details of test system where applicable, such as number, body weight, sex, supply source, species, strain.
  (7) Method of identification of test system.
  (8) Description of experimental design (including randomisation).

(9) Description and/or identification of diet and any materials employed with the test and control article.

(10) Route of administration and its justification.

(11) Dosage level, method, and frequency of administration of test and control articles.

(12) If necessary, the method used to determine the degree of absorption of the test and control article by the test system.

(13) The type of frequency of the tests, analyses, and measurements to be made.

(14) The records to be maintained.

(15) The date of protocol approval by the sponsor, and the signature of the study director.

(16) The statistical methods to be used.

Any changes and revisions to the protocol are issued as a protocol amendment and signed and dated and maintained with the protocol.

It is often useful to include with the protocol the signature of the head of the QAU which assures both the management of the testing facility and the sponsor that the protocol is in compliance with GLP requirements. While some organisations list the SOPs involved in the study protocol this seems unnecessary and time-consuming, and a general statement that all procedures will be carried out according to the organisation's SOPs would seem to be sufficient.

Reports

Similarly to the protocol contents, the FDA also sets out the minimum contents that should be included in the final report of each study:

(1) Name and address of the facility conducting the study and the start and completion dates.

(2) The objectives and procedures as stated in the protocol (together with any amendments).

The archives are limited by requirement to authorised personnel only and require indexing for expedient retrieval by test article, date of study, test system, and nature of study. Most facilities will produce the system best suited for their organisation, but colour-coding is often applicable. In small organisations, the employment of a specialist archivist is not cost-effective and the librarian or member of the QAU staff can just as efficiently be designated for this responsibility. Material withdrawn from the archives at any time should be well documented and with the date and signature of the person responsible together with its temporary location, and countersigned on return.

4   Conclusions and Summary

Since its inception in 1978, the concept of GLP has altered considerably. No longer an American regulatory-agency-orientated exercise, GLP has become an established code of conduct of non-clinical studies on an international level. As far as industry was concerned, the original exercise to ensure compliance was extremely costly, both in real terms and in terms of time and manpower; it is estimated that technician time, for example has been increased by at least 30%. Continual monitoring and updating is necessary and costly, and the auditing of reports has necessarily increased the length of time taken, especially for long term studies. Animal suppliers and diet manufacturers have had to improve their records and documentation and improve analytical techniques all of which have added to greatly increasing their costs, which in turn have been passed on to the customer. However, against this must be balanced the greatly improved quality of the generated data and study, integrity and, more recently, the importance of the international acceptance of data.

Any mechanism which improves the quality of a toxicology study can only be a good thing; laboratories have certainly had to look to their housekeeping, and even the burdensome chores of SOPs can be proved useful in the prevention of slipshod habits being passed on to newly recruited staff, and be valuable training manuals. It must be remembered that these guidelines are sufficiently flexible to allow for company policy to operate. It is not a question of blindly following a set of rules but of reasoned thought of the rationale underlying the requirements. Surely the critics of GLP should be rather bemoaning the reason why these requirements were necessary, and strive to improve the standards of technique and scientific integrity of their laboratory.

This ending to the first edition has been retained simply to illustrate the problems, that as the original instigators of the GLP principle, we encountered both with staff and management.

It seems especially pertinent that the second edition should be published in 1993, a year which heralds the completion of the internal market and the removal of trade barriers. In such an environment it is essential for the free movement of goods to have a harmonised system which allows for the mutual recognition of test data. This can only be achieved by ensuring data is generated using standard and recognised test methods and that there are recognised procedures for monitoring compliance to these throughout the Community. Thus test data generated by one Member State are recognised by all other Member States.

One issue of great concern to all testing laboratories was data confidentiality. Reassurance can be found in Directive 88/320/EEC, Article 4, that all commercially sensitive and confidential information which might become available during inspection is so respected.

Lastly, but by no means of the least importance, is the reduction in manpower and resources of the testing laboratories by negating the need for duplicating effort. Most important of these is the consequent reduction of experiments on animals which has always been an issue of great concern to all involved, and is embodied in a separate Council Directive 86/609/EEC.

Any system that is new meets with certain opposition. Much hard work and lengthy negotiations with governments, both National and International, has produced a reasonable framework acceptable to both scientists and legislators which should enhance and not hinder the progress of toxicology.

References

British Standards Institution (1972). Specification for Sampling Procedures and Tables for Inspection by Attributes (BS.6001). Published by the British Standards Institution.

British Standards Institution (1972). Guide to the Use of BS.6001 (BS.6000). Published by the British Standards Institution.

European Economic Community (1987). Council Directive 87/118/EEC on the harmonisation of laws, regulations and administrative provisions of relating to the application of the principles of good laboratory practice and the verification of their applications for tests on chemical substances. O.J. No. L15, 17.1.1987.

European Economic Community (1988). Council Directive 88/320/EEC on the inspection and verification of good laboratory practice. O.J. No. L145, 11.6.1988.

European Economic Community (1990). Council Directive 90/18/EEC adapting to technical progress the Annex to Council Directive 88/320/EEC. O.J. No. L11, 13.1.1990.

European Economic Community (1986). Council Directive 86/609/EEC on the approximation of laws, regulations and administrative provisions regarding the protection of animals used for experimental and other scientific purposes. O.J. No. L358, 18.12.1986.

Fed. Regist. (1978), **43,** (247) 22 December.

Fed. Regist. (1979), **44,** (91), 9 May.

Fed. Regist. (1980), **45,** (77), 18 April.

Fed. Regist. (1980), **45,** (117), 1 November.

Fed. Regist. (1983), **48,** (230), 29 November.

Food Chem. News (1982), **24,** (12), 31 May.

Food Chem. News (1982), 17 October.

Food Chem. News (1984), **26,** (22), 6 August.

Ministry of Health and Welfare (1981). Notification No. 313 of Pharmaceutical Affairs Bureau.

Moroney, H. J. (1956). 'Fact from Figures.' Penguin Books, 3rd edn; reprint with revisions, 1969.

OECD (1982). 'Good Laboratory Practice in the Testing of Chemicals', Final Report of the OECD Expert Group on Good Laboratory Practice, Paris. (ISBN 92-69-12367-9).

HMSO (1983). Health and Safety Commission Approved Code of Practice COP7, Principles of Good Laboratory Practice. HMSO (ISBN 011 8836587).

HSE (1983). Health and Safety Executive. Notification of New Substances Regulations, 1982. Establishment of a good laboratory practice compliance programme (ISBN 0717601250).

DHSS (1983). Department of Health and Social Security. Arrangements to monitor compliance with Good Laboratory Practice, NS/14 March 1983.

DHSS (1986). 'Good Laboratory Practice'. The United Kingdom Compliance Programme. Published by DHSS, Hannibal House, Elephant and Castle, London.

Ministere Des Affaires Sonales Et De La Solidante Nationale, Emnes Pratignes De Laboratoire (B.P.L.) ISSN 0758–1998.

CHAPTER 25

*Ethics in Experiments on Animals**

DOUGLAS BROWN MCGREGOR

1   Introduction

Ethics is moral philosophy. It is a subject which is constantly reviewed, analysed, and synthesised, reflecting the spirit of the time. It attempts to identify the standards by which one 'ought' to behave.

Ethics is fraught with difficulties, as was recognised by Hume when he stated that there is no deductive relation between an 'is' and an 'ought' (MacNabb, 1962). Thus, there is no way of deducing how one ought to behave from observation of the world. But, if this view is correct, then it puts in jeopardy all claims to moral knowledge, leaving ethics at the mercy of subjective whim, against which no arguments will stand (Scruton, 1984). The difficulties have been restated many times, for example in this century by the pragmatist, C. I. Lewis (1946), 'Valuation is always a matter of empirical knowledge. But what is right and what is just, can never be determined by empirical facts alone.'

Kant (Abbott, 1909) believed that the 'is–ought' problem could be answered by practical reason, based upon the exercise of the Categorical Imperative: 'Act only on the maxim which you can at the same time will as a universal law.' Practical reason guides action and aims at rightness, while theoretical reason guides belief and aims at truth. Practical reason is nonsense without freedom of action and only rational beings can be free in the sense required of morality. It is against this background that we can attempt to answer the question, 'is it morally right to involve non-rational creatures (animals) in our efforts to understand the Universe and to safeguard our position in it?'

Philosophy, unlike science, does not progress as new results are established, replacing existing ones. Much of philosophy is restatement in a contemporary context of ideas and arguments which may have existed

* Modified from Ethics of Animal Experimentation. *Drug Metab. Rev.*, 1986, **17**, 349–361.

for centuries. For example, Judaism is the background from which Christianity developed and Christianity is at the foundation of our society, no matter how eroded this may have become today. Perhaps our concern for animals has its basis in these religions. In the Old Testament, ethics means conformity of human activity to the will of God and it is stated in the Old Testament that man shall have dominion over all living things. However—and it cannot be stressed too strongly—to have dominion over something is not the same as having no consideration for its welfare.

Of all the many ethical theories, the Judaeo-Christian ethic of love is one of the very few that might have anything at all to say about animal welfare. The ethic of love holds that there is only one basic ethical imperative—love—and that all others are to be derived from it. Thus, "'Love the Lord your God with all your heart, with all your soul, with all your mind". It comes first. The second is like it: 'love your neighbour as yourself'. Everything in the Law and the prophets hangs on these two commandments." (N.E.B.).

The second commandment would seem to set ethics of love on a teleological basis, the ultimate standard of what is morally right or wrong being the non-moral value that is brought about. The final appeal, directly or indirectly, must be to the comparative balance of good over evil produced. Thus, an act should be done only if it produces more good than evil than would result from not acting or from acting in an alternative manner. Teleologists may hold very different views about what is good in a non-moral sense. For example, a teleologist who is a hedonist will identify good with pleasure, evil with pain. Other teleologists may identify good with power, knowledge, self-realisation, or perfection. In the current context, the moral question is whether an experiment is right or wrong; the non-moral standard against which such a question is set is whether the knowledge gained from that experiment serves a purpose. The second commandment suggests that the purpose is benefit to Man.

The first commandment, on the other hand, is not a teleological principle. One can hardly say that loving God means promoting His good. Only if one identifies loving God with loving His creatures, and loving them by promoting for them the greatest balance of good over evil, can one interpret the Judaeo-Christian ethic of love as a teleological theory. It has a bearing on the use of animals in experiments only if one includes non-human as well as human animals in one's description of God's creatures. If such a broad interpretation is allowed, then the ethical basis for objection to animal experiments requires absolute obedience to God. For the person who rejects obedience to God, an alternative basis for an ethical theory decrying animal experiments could be the ancient reverence for a so-called life-force.

In the ancient world, people were fascinated by the 'life force' that made the corn grow and the olives and grapes ripen. This awe-inspiring 'life force' was also responsible for the generation of Man, so it is readily understandable that such an important concept should be worshipped. An anti-vivisectionist view might be an appeal to a life-force, or some modern version of it. If such a thing or concept is to be placed in high authority, commanding reverence (and one might question why this should be so), then all life is to be held in reverence, be it human, non-human, vegetable, or microbial. With such a principle, actions may be judged on the basis of those which are to bring about the greatest expression of the life-force. But it leaves open to question whether or not, for example, that expression is size, number, or level of consciousness.

It very quickly becomes obvious that reverence for life-force cannot be a basis for determining our actions. Many different living things inflict harm on other living things simply so that they can survive. All animal life is dependent upon this activity and evolution is a description of changes where some individuals thrive better than others and result in population imbalances, leading to either success of some or, with time, disappearance of others. Life as such is not imbued with qualities which can serve as a basis for our relationships with each other and with the rest of the living world.

There may be some characteristic of certain forms of life which may be a basis for a morality, in which case this should be identified. One such characteristic is that we can see something of ourselves in animals. But this characteristic is a qualitative, subjective one based upon *human* perception which stirs feelings of guardianship and is far from the objectivity sought in ethics. In the same way, a fit and able man can adopt a position of guardianship over a sick man in whom he sees a much more accurate reflection of himself than he would in any other animal. Consequently, if a situation arises whereby the sick man can benefit if the fit and able man causes harm to an animal, then the animal should be harmed. Appeal to a life-force without qualification will result in the animal remaining unharmed and the sick man remaining sick.

2 Human Experiments

The issue of experiments with animals could be avoided if human subjects were used. It is argued that non-human animals make poor models for Man anyway, therefore experiments with rodents, dogs, or primates are a waste of time and money. This argument does have some currency but experiments are performed on people also and it is necessary to consider whether these are ethical if supporting information is not already available from another source. Most human experiments are ethical, in the judgment of those involved with their sanctioning. Household products are tested upon human volunteers prior to market-ing, but only after at least some animal testing has been completed.

Other, nonpharmaceutical, products such as pesticides are sometimes tested in human volunteers; not necessarily to demonstrate some adverse response, but to gain important information on metabolism to help identify better non-human animal models with which to study the toxicology of the substance. Pharmaceutical products entering Phase 1 clinical (tolerance) trials are subject to lay and expert review—Ethics Review Committees in the UK, Institutional Review Boards in the US—and the healthy volunteers are not expected to derive any direct benefit from participation, or experience any significant health risk. The Phase 1 experiments initially involve rising dose exposures to the compound until the appearance of side effects. They are conducted under careful clinical supervision, the volunteers are well informed about the compound and the results of earlier animal tests, and these people are free to withdraw at any time.

Any surgical advance is an experiment. Thus, the transplantation of adrenal medullary tissues into the brains of two very advanced Parkinson's disease patients in Sweden, in an attempt to promote regeneration of dopamine secreting cells, was certainly an experiment and the result was far from certain at the outset (Backlund *et al.,* 1985). There was a risk of very serious side effects because the adrenal tissue secretes a whole range of highly active pharmacological substances in addition to dopamine, *e.g.* adrenalin, noradrenalin, opiates, *etc.* The effects of such ectopically secreted substances were essentially unknown prior to the experiment and there was virtually no improvement in the patients' condition. More recent experiments using a different technique were more successful in alleviating the symptoms of Parkinsonism, although the results are being met with skepticism (Madrazo *et al.*, 1987). It is unlikely that either experiment would have been undertaken without extensive investigation upon an animal model: 6-hydroxydopamine-denervated rats (*e.g.* Freed *et al.,* 1981).

Any patient given a new chemotherapeutic cocktail is a human guinea-pig. Heart and heart-lung transplantations are called 'routine' these days, but this is not so; they are still experimental, involving combinations of surgical skill (probably gained on pigs), immunosuppressive regimes, tissue-typing expertise, and live organ storage capability.

Unfortunately, some experiments upon people have not been either of this benign nature or patient benefit orientated.

In Germany, under Nazi domination, atrocities were perpetrated against prisoners of race and conscience, sometimes in circumstances where wily criminal prisoners had ingratiated themselves with the guards and were in 'trusted' positions. Little work of medical importance resulted from these experiments. One example, chosen because it was by no means the worst, and because it could be argued that benefits did result, was the study on typhus immunisation by Dr. E. Ding (Mellanby, 1973). He prepared various anti-sera for the purpose of protecting people from louse-borne typhus, which is caused by a rickettsia. He then deliberately infected groups of people with the organisms causing the disease; some of these had been protected by his different anti-sera while some were unprotected controls. Scientifically, it was a well-conducted experiment, and although many of those taking part died, including most of the controls of course, Ding's work formed the basis not only of German, but also of British and Allied anti-typhus vaccine policy. Ding appeared to think that the deaths caused by his experiments—about 250—were justified; about 5 million people died of typhus during and after World War II. It is to be hoped that those days are behind us. Ding's utilitarian view is indefensible since people have inviolable rights; there are certain actions which should never to taken against them. All innocent human beings have a right to life; it is wrong to kill innocent human beings; and it is wrong to deny innocent human beings reasonable straightforward protection against life threatening conditions (Gillon, 1985).

In contrast to this de-ontological position, another utilitarian view states that pains of the same intensity and duration are equally bad whether felt by humans or animals (Singer, 1976). In his arguments against 'speciesism', Singer claims the difference between men and other animals is our greater capacity for self-awareness, abstraction, complex communication, and extended social relationships. This claim has been taken up by a portion of the animal rights lobby—and unambiguously stated by Singer himself—as a position which must grant equal or greater right to life to animals than to retarded or senile human beings. Since these words were uttered, the prospect of experiments upon human embryos has become a reality, as a consequence of excess *in vitro* fertilisations and of abortions. Embryos in research have again focused attention upon our definition of humanness. Is there any fundamental change which occurs during development to permit us to say the conceptus is not human at one stage, but is human at another? It is unlikely that any absolute change will be identified to support our convenient definitions. One definition suggested has been the possession of a rudimentary sense of awareness which might be equated to sensitivity to pain. However, this has not prevented legal abortions at ages where this 'rudimentary sense of awareness' is almost certainly present. Such a definition would permit experiments upon suitably anaesthetised people of any age. While permitting *these* acts, are we to prohibit work with protozoans which do have sensory perception and will react to stimuli?

3   Animal Experiments

A   Their Necessity in Principle

The machinery of life is complex. Any electron micrograph of a cell demonstrates something of the extent of that complexity. Nevertheless, the picture is only a frozen moment in the death of that cell. It is a fossil. Look at another cell and one can see that it is different. Each cell is an individual representation of life, unique, in spite of the readily recognisable sub-cellular organisation within it. Attempted definitions of Life are irrelevant in this context and only detract from the essence of the concept. Knowledge and truth are different.

Microanatomists have identified the structures, distorted though they may be, in which chemists have recognised an abundance of reactions working in a co-ordinated way, permitting survival and propagation. Even the simplest cells are complex. In multi-cellular organisms, size dictates that there should be specialisation since some cells are physically better placed for some functions than others. The advantages of size are not immediately obvious, but clearly they were sufficient to allow, for example, *Volvox* to survive along side *Chlorella*.

The organisational step from *Chlorella* to *Volvox* is small, but the step becomes greater as the phylogenetic tree is climbed until, in a very short time, there is no single cell which is 'typical' of an entire organism. This is all so obvious that it hardly needs to be mentioned and, when the effects of a chemical upon any animal need to be known, it is also obvious that there is ultimately no substitute for the exposure of that animal. Animals are needed for research because it is currently impossible to model the complex interactions of the biological machinery used by mammalian systems by other than experimental animals or man. Perhaps the more fundamental issue is do we need the research?

The form which the research may take varies greatly. Research using animals may be in nutrition, *e.g.*, the study of energy utilisation, amino acid balance, the requirements of man or domesticated animals for vitamins and minerals; in physiology, *e.g.*, the study of renal or hepatic function; in pharmacology, *e.g.*, the study of muscular activity, neural transmission, CNS function, cardiac activity of drugs used in medicine or components of our diet; virology; bacteriology; parasitology, *e.g.*, to study host–parasite relationships and the relative responses of the host and parasite to various treatment regimes; psychology, *e.g.*, enabling us to understand components of complex behaviour patterns; surgery, *e.g.*, for the development of new techniques; and toxicology for the investigation of adverse responses to chemicals and the establishment of safe levels of exposure.

For all of these purposes combined, enormous numbers of animals are used, mainly in the industrialised, developed countries. There are strengthening voices which say that these numbers are too high and that they should be cut drastically. Some of the people involved in animal liberation groups are more willing to countenance experiments upon people who have been defined as mentally sub-normal (by how much is not made clear) than experiments upon animals, and there is a much larger proportion who are willing for experiments to be conducted on legally aborted foetuses. These appalling suggestions are unethical on the basis of most moral philosophies and dependent upon a background of knowledge gained from non-human sources.

B The Cruelty to Animals Act, 1876 and The Animals (Scientific Procedures) Act, 1986

One of the first pieces of legislation relating to animal experiments was created more than 100 years ago, in the Cruelty to Animals Act, 1876. At that time, perhaps a few hundred animals were used each year for experiments. When legislation was first introduced, the main use of animals was to extend the scientific basis for medical and veterinary work.

Public opinion, however, had been aroused by description of cruel and unnecessary experiments conducted in certain countries. Legislation was passed through parliament without serious opposition and has remained on the statute books ever since, without any substantial amendments. It has, however, been repealed and replaced by new legislation, The Animals (Scientific Procedures) Act, 1986. Only time will tell whether moderate opinions are contented with the changes encompassed by the new legislation; activist opinions are unlikely to be satisfied, but appeasement is an ephemeral process and should not form the basis for a document which may remain operative for many years.

The 1876 Act controlled all experiments which were likely to cause pain in living animals. The purpose of an experiment had to be for the advancement, by new discovery, of physiological knowledge or the acquisition of knowledge useful for saving or prolonging life or alleviating suffering. Experiments were absolutely prohibited in which the purpose was to attain manual skills. Within these constraints, the Act authorised the Home Secretary to license 'suitable qualified' persons to perform experiments. But there were safeguards. No experimental animal was allowed to suffer severe pain which was likely to endure.

In any experiment performed under licence alone, the animal had to be under the influence of an anaesthetic to prevent it feeling pain. If pain was likely to continue once the effect of the anaesthetic had worn off, or if any serious injury had occurred, the animal had to be killed before it recovered from the influence of the anaesthetic. Where these restrictions frustrated the objectives of the experiment, then certificates could be granted at the discretion of the Home Secretary which allowed the requirement for anaesthetic to be relaxed. In effect the Act was exercised through Cruelty to Animals Inspectors, who advised on all applications for authority to conduct animal experiments and visited the laboratories to ensure that the law was observed by the licensees.

Problems arose because the Act contained no definitions of terms and because the conditions and scope of animal experimentation today are very different from what they were in 1876. The Act had the object of controlling experiments, but the term 'experiments' was not defined and no definition was made in a court of law. Also, the definition of pain was a difficulty because this is subjective and it cannot be measured even on this basis in animals, except from an anthropocentric view point. This problem persists in the new legislation.

The new Act requires licensing of people, places, projects, and procedures involving 'protected animals' which may cause pain, suffering, distress, or lasting harm for scientific purposes. The definition of a 'protected animal' is any living vertebrate, other than man, which has developed beyond the mid-point of their foetal, embryonic, or larval stages. The Act also controls the breeding and use of animals with harmful genetic defects in scientific procedures, the production of antisera and other blood products, the maintenance and passage of tumours and parasites and, for purposes of scientific investigation, the administration of any substance to dull perception. Breeding and supplying establishments are also regulated for certain commonly used mammalian species.

The change in conditions which has occurred since 1876 is the vast increase in the tests on products from the pharmaceutical, cosmetic, food, pesticide, and chemical industries to meet statutory obligations. In the United Kingdom these may be in compliance with the Agriculture (Poisonous Substances) Act, 1952, the Food and Drug Act, 1955, the Medicines Act, 1968, Consumer Protection Act, 1971, and the Health and Safety at Work Act, 1974. Equivalent legislation is to be found in other countries. International bodies, such as the European Economic Community (EEC) and the Organisation for Economic Cooperation and Development (OECD) have developed chemical safety testing requirements or guidelines which have become law or need to be ratified by legislation within the member countries.

This type of legislation is mentioned to remind us that manufacturers would be breaking various laws if they did not test their products for hazardous properties and that such testing does require animals.

C Number of Animals Used

Moderate and widely held opinion asks that toxicologists use fewer animals and restrict the type of chemical which is tested on them; they should not be subjected at all to cosmetics. There is also concern about the type of animal used. There is certainly an emotional attraction about such views, but they appear to embrace conflicting ideas.

If the argument for fewer animals is based on a belief that Man abuses his privileged position by using other living creatures to serve his own ends then the number of animals used is irrelevant. Each animal is an individual and it does not diminish the individual animal's plight if the total is reduced. Man may persuade himself, by virtue of his reasoning empathy, that the ethical status of an experiment is improved by smaller numbers but this form of rationalisation involves a negation of the perceived ethical principle and is forgetful of the relative isolation in an animal's experience, because of its inability to communicate at a distance, and its lack of a highly developed language.

D Species

People are generally much more concerned about experiments involving animals with which they can develop an emotional rapport, than those involving animals where such relationships seldom occur. Thus, experiments on a species which constitute domestic pets provoke disproportionate reactions, when compared with rodents or even certain non-human primates. Yet, empathic behaviour is important among primates—particularly apes—while it is probably of little or no significance in rodents and is at best poorly developed in common domestic pets. So, if one of these animals in an experiment experiences well-being, discomfort, or suffering, these feelings are not shared by others unless there is close proximity of the individuals in the experiment and empathy is a possible experience. Even Man generally shows little concern for the plight of others of his own kind. And yet, no matter what uncertainties exist in our understanding of behaviour, Man's ability to contemplate is far higher than in any other animal. Nevertheless, because of these uncertainties, we should probably refrain totally from experimental toxicology on higher apes. Rarely is there argument for toxicology or metabolism involving these creatures, in spite of there being certain aspects of their physiology with which there are substantial human similarities.

Judgement of whether experiments are required with these highly intelligent animals or with domesticated animals should be made on a case-by-case basis. This is already the usual situation with large or expensive animals, but these considerations are made on practical or economic grounds. There is a formidable price barrier to the indiscriminate use of dogs, primates, or horses, but this hurdle is much lower for most other domesticated animals used in laboratories. Nevertheless, scientists would be foolish to disregard the strong emotional bonds which can develop between Man and domesticated animals. It is arguable that the real issue is not the ethics of experiments on animals, but the psychology of relationships between animals and men. Because they accept obligations, men have rights; animals also would have rights if they could accept responsibilities, but they cannot, living as they do outside the sphere of rights and in a state of innocence. Perhaps it is this innocence which is so beguiling to men's emotions, or is it an underlying gratitude for some received benefit, which may be as simple, but as welcome, as unquestioning affection? A person having experienced this relationship may then feel obliged to care for all individuals of a species.

E   Chemicals

Safety testing of cosmetics and toiletries is the topic of much anti-vivisection rhetoric. The arguments are: (a) these are trivial uses of chemicals and (b) there are enough available anyway. There is less criticism of testing pharmaceutical preparations. Here again the anthropocentric view is expressed. The animal experiencing discomfort, pain, or well-being is unconcerned about the human objectives of the experiment. It is irrelevant for the animal whether the chemical is a cosmetic, a pesticide, or a life-saving drug.

F   Pain

Various states of suffering can be recognised, ranging from discomfort (poor condition, reduced appetite, inactivity, avoidance behaviour) through stress to pain (struggling, screaming, convulsions, severe palpitations). Physical pain results from an interaction of neurological response to injury with mind. The perception of pain, then, is subjective and without some physical response it is not possible for another person to know whether pain is being experienced. The special problem with animal pain is that our projections of experience are even less likely to be correct, when made into animals, than when they are made into other people's lives. However, this problem extends beyond pain to all animal perception, be this visual, auditory, tactile, *etc*. It has been said that there might be circumstances in which it is legitimate to restrict the freedom of scientific investigation (Huxley, 1984). The position stated is that, 'Provided a proposal is scientifically promising, it should normally be permitted if the pain and distress involved is only 'mild', but if 'substantial', then the proposal should be scrutinised with care before being approved. Severe, but short lived pain should be permitted only exceptionally, *e.g.,* perhaps occasionally for the investigation of pain itself; severe and enduring pain, or excruciating pain, never.' While this opinion will have many supporters, one has still to determine the basis, from the animal's point of view, why it is occasionally, but not always, permissible to inflict 'substantial' pain.

   If the argument for never inflicting pain prevails, then it is possible that the good of animals is the guiding principle; but to argue for inconsistency suggests the presence of some other, usually unstated, principle.

4   Conclusions

It is important that we afford the same rights to all members of our species. If rights are to be gained only by the acceptance of responsibility, then rational thought is a requisite. This attribute develops as one progresses from infanthood to adulthood and may be lost with encroaching senility or mental defectiveness, be this either congenital/hereditary or the result of accident. Are animal rights to be acknowledged on a similar scale? To do so would present a situation open to all kinds of abuse.

The fundamental question is whether experiments on non-human animals are or are not to be done. The basis for this choice has been inadequately explored here. But, if they are to be done, it is argued here that objective concern for the animals used is not based upon numbers, value of the experiment to Man, or the type of chemicals being tested.

Concern for the use of animals is widespread, the husbandry of our fellow earth-dwellers being a matter to be treated seriously and responsibly. Members of animal welfare groups have significant contributions to make to these discussions of ethics in biological science, but some fringe members of these groups are only helping to antagonise moderate support when their morality allows them to harrass, frighten, or threaten the families of people engaged in biological research. The values of these demonstrators are not clear; they should take another look at the principles upon which their own behaviour is based, then ask whether this is how their fellow human beings should be treated. Is it right or justifiable for a human being to cause pain, discomfort, or distress to another human in the name of protecting animals from pain, discomfort, or distress?

We cannot deny the historical framework in which we live. We cannot dismantle the results of the industrial revolution. In that revolution, which had its origins in Man's curiosity about the physical universe, there was a great deal of human suffering and it has been a slow process to obtain general acceptance of the idea as a guiding principle that neither a company nor an individual should present to the public a product which will cause harm. Gewirth (1986) concludes that, 'the rights of workers to health and safety are of paramount importance and must override all other considerations that may be adduced to remove or limit them.' This is a restatement of the de-ontological moral principle referred to earlier. The choices open to us are:

(1) Abandon our industrialised society, leaving our future generations without its bad aspects or its benefits, or

(2) ensure as best we can that we do not harm, knowingly or through negligence, ourselves or our unborn children. If one accepts the second of these choices, then we must take action to bring about that situation. Toxicologists are not wedded to the use of animals in their work. Their objectives are to identify hazard and to quantify risk. If these objectives cannot be met, or can be obtained in some other way which is better (*i.e.* more accurate, or more efficient), then animals would be abandoned, but it is less likely that animals, as experimental subjects, could becomes redundant in many other branches of biology.

Acknowledgement

The author is grateful for the helpful suggestions made by Dr Roger Scruton, Department of Philosophy, Birkbeck College, University of London.

References

Abbott, T. K. (1909). Translator of I. Kant: 'Groundwork of the Metaphysic of Morals', 6th edn. London.

Backlund, E.-O., Grandberg, P.-O., Hamberger, B. *et al.* (1985). Transplantation of adrenal medullary tissue to straitum in Parkinsonism: first clinical trials. *J. Neurosurg.*, **62,** 169–173.

Freed, W. J., Morihisa, J. M., Spoor, H. E. *et al.* (1981). Transplanted adrenal chromaffin cells in rat brain reduce lesion-induced rotational behaviour. *Nature (London)*, **292,** 351–352.

Gewirth, A. (1986). Human rights and the workplace. *Am. J. Ind. Med.*, **9,** 31–40.

Gillon, R. (1985). An introduction to philosophical medical ethics: the Arthur case. *Br. Med. J.*, **290,** 1117–1119.

Huxley, A. (1984). Anniversary address by the President. *Proc. R. Soc. A.*, **391,** 215–230.

Lewis, C. I. (1964) 'An Analysis of Knowledge and Valuation'. London.

MacNabb, D. G. C. (1962). *Ed.* of D. Hume: 'A Treatise of Human Nature', 1740. Book 3, Part 1, Section 1. Collins, Glasgow.

Madrazo, I., Drucker-Colín, R., Díaz, V., *et al.* (1987). Open microsurgical autograft of adrenal medulla to the right caudate nucleus in two patients with intractable Parkinson's disease. *New Engl. J. Med.*, **316,** 831–834.

Mellanby, K. (1973). 'Human Guinea Pigs'. Merlin Press, London.

New English Bible (1970). St. Matthew Chapter 22, verses 37–40. Oxford University and Cambridge University Press.

Scruton, R. (1983). 'A Short History of Modern Philosophy from Descartes to Wittgenstein'. Ark, London.

Index